Practical Handbook
of
Marine
Science
Third Edition

Marine Science Series

The CRC Marine Science Series is dedicated to providing state-of-the-art coverage of important topics in marine biology, marine chemistry, marine geology, and physical oceanography. The series includes volumes that focus on the synthesis of recent advances in marine science.

CRC MARINE SCIENCE SERIES

SERIES EDITOR

Michael J. Kennish, Ph.D.

PUBLISHED TITLES

Artificial Reef Evaluation with Application to Natural Marine Habitats, William Seaman, Jr.

Chemical Oceanography, Second Edition, Frank J. Millero

Coastal Ecosystem Processes, Daniel M. Alongi

Ecology of Estuaries: Anthropogenic Effects, Michael J. Kennish

Ecology of Marine Bivalves: An Ecosystem Approach, Richard F. Dame

Ecology of Marine Invertebrate Larvae, Larry McEdward

Environmental Oceanography, Second Edition, Tom Beer

Estuary Restoration and Maintenance: The National Estuary Program, Michael J. Kennish

Eutrophication Processes in Coastal Systems: Origin and Succession of Plankton Blooms and Effects on Secondary Production in Gulf Coast Estuaries, Robert J. Livingston

Handbook of Marine Mineral Deposits, David S. Cronan

Handbook for Restoring Tidal Wetlands, Joy B. Zedler

Intertidal Deposits: River Mouths, Tidal Flats, and Coastal Lagoons, Doeke Eisma

Morphodynamics of Inner Continental Shelves, L. Donelson Wright

Ocean Pollution: Effects on Living Resources and Humans, Carl J. Sindermann

Physical Oceanographic Processes of the Great Barrier Reef, Eric Wolanski

The Physiology of Fishes, Second Edition, David H. Evans

Pollution Impacts on Marine Biotic Communities, Michael J. Kennish

Practical Handbook of Estuarine and Marine Pollution, Michael J. Kennish

Seagrasses: Monitoring, Ecology, Physiology, and Management, Stephen A. Bortone

Practical Handbook
of
Marine
Science
Third Edition

Edited by

Michael J. Kennish, Ph.D.
Institute of Marine and Coastal Sciences
Rutgers University
New Brunswick, New Jersey

CRC Press
Boca Raton   London   New York   Washington, D.C.

Library of Congress Cataloging-in-Publication Data

Practical handbook of marine science / edited by Michael J. Kennish.--3rd ed.
    p. cm.-- (Marine science series)
  Includes bibliographical references (p.).
  ISBN 0-8493-2391-6
   1. Oceanography. 2. Marine biology. I. Title: Marine science. II. Kennish, Michael J.
III. Series.

GC11.2 .P73 2000
551.46--dc21

00-060995

Visit the CRC Press Web site at www.crcpress.com

© 2001 by CRC Press LLC

No claim to original U.S. Government works
International Standard Book Number 0-8493-2391-6
Library of Congress Card Number 00-060995
Printed in the United States of America 2 3 4 5 6 7 8 9 0
Printed on acid-free paper

Dedication

This book is dedicated to the Jacques Cousteau National Estuarine Research Reserve at Mullica River–Great Bay, New Jersey

Preface

This third edition of *Practical Handbook of Marine Science* provides the most comprehensive contemporary reference material on the physical, chemical, and biological aspects of the marine realm. Since the publication of the second edition of this book 5 years ago, there have been significant advances in nearly all areas of marine science. It is the focus of this volume to examine these developments and to amass significant new data that will have appeal and utility for practicing marine scientists and students engaged in investigations in oceanography and related disciplines. Because of its broad coverage of the field, this volume will be valuable as a supplemental text for undergraduate and graduate marine science courses. In addition, administrators and other professionals dealing in some way with the management of marine resources and various problems pertaining to the sea will find the book useful.

Much of this third edition consists of updated material as evidenced by the large number of recent references (1995 to 1999) cited in the text. This edition contains a systematic collection of selective physical, chemical, and biological reference data on estuarine and oceanic ecosystems. It is comprised of six chapters: Physiography, Marine Chemistry, Physical Oceanography, Marine Geology, Marine Biology, and Marine Pollution and Other Anthropogenic Impacts. Each chapter is arranged in a multisectional format, with information presented in expository, illustrative, and tabular formats.

The main purpose of this handbook is the same as the two previous editions: to serve the multidisciplinary research needs of contemporary marine biologists, marine chemists, marine geologists, and physical oceanographers. It is also hoped that the publication will serve the academic needs of a new generation of marine science students.

I wish to acknowledge my colleagues who have been instrumental in enabling me to gain new insights into the complex and fascinating world of marine science. In particular, in the Institute of Marine and Coastal Sciences at Rutgers University I am thankful to Kenneth W. Able, Michael P. DeLuca, Richard A. Lutz, J. Frederick Grassle, John N. Kraeuter, and Norbert P. Psuty. I also express my deep appreciation to the editorial staff of CRC Press, especially John B. Sulzycki, who supervised all editorial and production activities on the book, and Christine Andreasen, who provided technical editing support on the volume. Finally, I am most appreciative of my wife, Jo-Ann, and sons, Shawn and Michael, for recognizing the importance of my commitment to complete this volume and for providing love and support during its preparation.

Preface

The Editor

Michael J. Kennish, Ph.D., is a research professor in the Institute of Marine and Coastal Sciences at Rutgers University, New Brunswick, New Jersey.

He graduated in 1972 from Rutgers University, Camden, New Jersey, with a B.A. degree in Geology and obtained his M.S. and Ph.D. degrees in the same discipline from Rutgers University, New Brunswick, in 1974 and 1977, respectively.

Dr. Kennish's professional affiliations include the American Fisheries Society (Mid-Atlantic Chapter), American Geophysical Union, American Institute of Physics, Atlantic Estuarine Research Society, New Jersey Academy of Science, and Sigma Xi.

Dr. Kennish has conducted biological and geological research on coastal and deep-sea environments for more than 25 years. While maintaining a wide range of research interests in marine ecology and marine geology, Dr. Kennish has been most actively involved with studies of marine pollution and other anthropogenic impacts in estuarine and coastal marine ecosystems as well as with biological and geological investigations of deep-sea hydrothermal vents and seafloor spreading centers. He is the author or editor of ten books dealing with various aspects of estuarine and marine science. In addition to these books, Dr. Kennish has published more than 100 research articles and book chapters and has presented papers at numerous conferences. His biographical profile appears in *Who's Who in Frontiers of Science and Technology*, *Who's Who Among Rising Young Americans*, *Who's Who in Science and Engineering*, and *American Men and Women of Science*.

Introduction

Marine science constitutes a broad field of scientific inquiry that encompasses the primary disciplines of oceanography—marine biology, marine chemistry, marine geology, and physical oceanography—as well as related disciplines, such as the atmospheric sciences. Its scope is extensive, covering natural and anthropogenic phenomena in estuaries, harbors, lagoons, shallow seas, continental shelves, continental slopes, continental rises, abyssal regions, and mid-ocean ridges. Because of the breadth of its coverage, the study of marine science is necessarily a multidisciplinary endeavor, requiring the efforts of many scientists from disparate disciplines.

Chapter 1 describes the topography and hypsometry of the world oceans. Emphasis is placed on the major physiographic provinces (i.e., continental-margin, deep-ocean, and mid-ocean ridge provinces). The characteristics of the benthic and pelagic provinces in the sea are also discussed.

Chapter 2 examines marine chemistry, focusing on the major, minor, trace, and nutrient elements in seawater. Information is also presented on dissolved gases and organic compounds in the hydrosphere. In addition, vertical profiles of the various chemical constituents are detailed. The major nutrient elements (i.e., nitrogen, phosphorus, and silicon) are particularly noteworthy because of their importance to plant growth, with nitrogen being the principal limiting element to primary production in estuarine and marine waters. However, phosphorus may be the primary limiting element to autotrophic growth in some estuaries during certain seasons of the year. Low silicon availability, in turn, can suppress metabolic activity of the cell, can limit phytoplankton production, and can reduce skeletal growth of diatoms, radiolarians, and siliceous sponges. Hence, these three nutrient elements play a critically important role in regulating biological production in the sea.

Chapter 3 deals with physical oceanography. It investigates the physical properties of seawater and the circulation patterns observed in open-ocean, coastal-ocean, and estuarine waters. Waves, currents, and tidal flow, as well as the forcing mechanisms responsible for their movements, are assessed. Ocean circulation is divided into two components: (1) wind-driven (surface) currents and (2) thermohaline (deep) circulation. These components are described for all the major oceans. A detailed account is given on conspicuous wind-driven circulation patterns (e.g., gyres, meanders, eddies, and rings). In estuaries, water circulation depends greatly on the magnitude of river discharge relative to tidal flow. Turbulent mixing in these shallow systems is a function of river flow acting against tidal motion and interacting with wind stress, internal friction, and bottom friction. Surface wind stress and meteorological forcing also play a vital role in modulating circulation in coastal ocean waters.

Chapter 4 addresses marine geology. Major structural features of the seafloor (e.g., mid-ocean ridges, transform faults, and deep-sea trenches) are explained in light of the theory of plate tectonics, which represents the unifying paradigm in geology. The dynamic nature of the ocean crust and seafloor is coupled to the movement of lithospheric plates. The genesis of ocean crust occurs along a globally encircling mid-ocean ridge and rift system through an interplay of magmatic construction, hydrothermal convection, and tectonic extension. The destruction of ocean crust, in turn, takes place at deep-sea trenches. The relative motion of lithospheric plates is responsible for an array of tectonic and topographic features on the seafloor. Moving away from the mid-ocean ridges, the deep ocean floor exhibits the following prominent topographic features: abyssal hills and plains, seamounts, aseismic ridges, and trenches. The continental margins are typified by continental rises, slopes, and shelves. The seafloor is blanketed by sediments in most areas. Terrigenous sediments predominate on the continental margins. The deep ocean floor contains a variety of sediment types, with various admixtures of biogenous, terrigenous, authigenic, volcanogenic, and cosmogenic components. The relative concentration of these sediment components at any site depends on water depth, the proximity to landmasses, biological productivity of overlying waters, volcanic activity on the seafloor, as well as other factors.

Marine biology is treated in Chapter 5. Major taxonomic groups of plants and animals found in estuarine and oceanic environments are reviewed. Included here are various groups of phytoplankton, zooplankton, benthic flora and fauna, and nekton. Data are compiled on the abundance, biomass, density, distribution, and diversity of these organisms. Information is also chronicled on estuarine and marine habitats.

Chapter 6 conveys the seriousness of pollution and other anthropogenic impacts on biotic communities and habitats in estuarine and marine environments. Acute and insidious pollution problems encountered in these environments are commonly linked to nutrient and organic carbon loading, oil spills, and toxic chemical contaminant inputs (i.e., polycyclic aromatic hydrocarbons, halogenated hydrocarbons, heavy metals, and radioactive substances). Many human activities disrupt and degrade habitats, leading to significant decreases in abundance of organisms. For example, uncontrolled coastal development, altered natural flows, overexploitation of fisheries, and the introduction of exotic species can have devastating effects on estuarine and coastal marine systems. The impacts of human activities—especially in the coastal zone—will continue to be a major issue in marine science during the 21st century.

Contents

Chapter 1   Physiography
  I. Ocean Provinces...1
    A. Dimensions...1
    B. Physiographic Provinces 2
      1. Continental Margin Province...2
      2. Deep-Ocean Basin Province 2
      3. Mid-Ocean Ridge Province 3
    C. Benthic and Pelagic Provinces 3
      1. Benthic Province 3
      2. Pelagic Province..4
  1.1 Conversion Factors, Measures, and Units 5
  1.2 General Features of the Earth..10
  1.3 General Characteristics of the Oceans 14
  1.4 Topographic Data...17

Chapter 2   Marine Chemistry
  I. Seawater Composition 45
    A. Major Constituents...45
    B. Minor and Trace Elements...46
    C. Nutrient Elements 47
      1. Nitrogen 47
      2. Phosphorus 48
      3. Silicon 48
    D. Gases 48
    E. Organic Compounds..49
    F. Dissolved Constituent Behavior 49
    G. Vertical Profiles 49
      1. Conservative Profile...49
      2. Nutrient-Type Profile 49
      3. Surface Enrichment and Depletion at Depth 50
      4. Mid-Depth Minima 50
      5. Mid-Depth Maxima 50
      6. Mid-Depth Maxima or Minima in the Suboxic Layer 50
      7. Maxima and Minima in Anoxic Waters 50
    H. Salinity 51
  2.1 Periodic Table..54
  2.2 Properties of Seawater 59
  2.3 Atmospheric and Fluvial Fluxes...68
  2.4 Composition of Seawater...76
  2.5 Trace Elements...88
  2.6 Deep-Sea Hydrothermal Vent Chemistry 98
  2.7 Organic Matter 107
  2.8 Decomposition of Organic Matter...126
  2.9 Oxygen 128
  2.10 Nutrients 136
  2.11 Carbon 153

Chapter 3   Physical Oceanography

    I. Subject Areas..167
   II. Properties of Seawater 167
      A. Temperature...167
      B. Salinity 168
      C. Density 168
  III. Open Ocean Circulation..169
      A. Wind-Driven Circulation..169
        1. Ocean Gyres...169
        2. Meanders, Eddies, and Rings 169
        3. Equatorial Currents..170
        4. Antarctic Circumpolar Current 171
        5. Convergences and Divergences 171
        6. Ekman Transport, Upwelling, and Downwelling.............................171
        7. Langmuir Circulation 172
      B. Surface Water Circulation...172
        1. Atlantic Ocean 172
        2. Pacific Ocean 173
        3. Indian Ocean 173
        4. Southern Ocean 173
        5. Arctic Sea...173
      C. Thermohaline Circulation 174
        1. Atlantic Ocean 174
        2. Pacific Ocean 174
        3. Indian Ocean 175
        4. Arctic Sea...175
  IV. Estuarine and Coastal Ocean Circulation 176
      A. Estuaries 176
      B. Coastal Ocean 177
        1. Currents..177
        2. Fronts 178
        3. Waves 178
          a.   Kelvin and Rossby Waves 178
          b.   Edge Waves..179
          c.   Seiches 179
          d.   Internal Waves 179
          e.   Tides...179
          f.   Surface Waves..180
          g.   Tsunamis 182
3.1   Direct and Remote Sensing (Oceanographic Applications).....................................185
3.2   Light 193
3.3   Temperature...196
3.4   Salinity 201
3.5   Tides 206
3.6   Wind 211
3.7   Waves and Their Properties 215
3.8   Coastal Waves and Currents 220
3.9   Circulation in Estuaries...238
3.10 Ocean Circulation 258

Chapter 4   Marine Geology

I. Plate Tectonics Theory 279
II. Seafloor Topographic Features 280
  A. Mid-Ocean Ridges 280
  B. Deep Ocean Floor 282
    1. Abyssal Hills 282
    2. Abyssal Plains 283
    3. Seamounts 283
    4. Aseismic Ridges 283
    5. Deep-Sea Trenches 284
  C. Continental Margins 285
    1. Continental Shelf 285
    2. Continental Slope 285
    3. Continental Rise 286
III. Sediments 286
  A. Deep Ocean Floor 286
    1. Terrigenous Sediment 287
    2. Biogenous Sediment 288
      a. Calcareous Oozes 288
      b. Siliceous Oozes 289
    3. Pelagic Sediment Distribution 289
    4. Authigenic Sediment 290
    5. Volcanogenic Sediment 291
    6. Cosmogenic Sediment 291
    7. Deep-Sea Sediment Thickness 291
  B. Continental Margins 291
    1. Continental Shelves 292
    2. Continental Slopes and Rises 293
4.1 Composition and Structure of the Earth 297
4.2 Ocean Basins 303
4.3 Continental Margins 305
4.4 Submarine Canyons and Oceanic Trenches 311
4.5 Plate Tectonics, Mid-Ocean Ridges, and Oceanic Crust Formation 335
4.6 Heat Flow 353
4.7 Hydrothermal Vents 362
4.8 Lava Flows and Seamounts 384
4.9 Marine Mineral Deposits 397
4.10 Marine Sediments 404
4.11 Estuaries, Beaches, and Continental Shelves 410

Chapter 5   Marine Biology

I. Introduction 441
II. Bacteria 441
III. Phytoplankton 444
  A. Major Taxonomic Groups 445
    1. Diatoms 445
    2. Dinoflagellates 445
    3. Coccolithophores 445
    4. Silicoflagellates 446
  B. Primary Productivity 446

    IV. Zooplankton 447
       A. Zooplankton Classifications...447
         1. Classification by Size..447
         2. Classification by Length of Planktonic Life 448
           a. Holoplankton...448
           b. Meroplankton 448
           c. Tychoplankton 449
    V. Benthos...450
       A. Benthic Flora...450
         1. Salt Marshes...452
         2. Seagrasses 453
         3. Mangroves 454
       B. Benthic Fauna 454
         1. Spatial Distribution 455
         2. Reproduction and Larval Dispersal......................................456
         3. Feeding Strategies, Burrowing, and Bioturbation.................457
         4. Biomass and Species Diversity 458
           a. Biomass 458
           b. Diversity 459
       C. Coral Reefs..459
   VI. Nekton...460
       A. Fish..460
         1. Representative Fish Faunas 461
           a. Estuaries..461
           b. Pelagic Environment 461
             i. Neritic Zone...461
             ii. Epipelagic Zone..461
             iii. Mesopelagic Zone 461
             iv. Bathypelagic Zone...461
             v. Abyssopelagic Zone 461
           c. Benthic Environment 461
             i. Supratidal Zone...461
             ii. Intertidal Zone 461
             iii. Subtidal Zone..462
             iv. Bathyal Zone...462
             v. Abyssal Zone 462
             vi. Hadal Zone..462
       B. Crustaceans and Cephalopods 462
       C. Marine Reptiles..462
       D. Marine Mammals...463
       E. Seabirds...463
5.1   Marine Organisms: Major Groups and Composition...................470
5.2   Biological Production in the Ocean 479
5.3   Bacteria and Protozoa...491
5.4   Marine Plankton...497
5.5   Benthic Flora..504
5.6   Benthic Fauna 541
5.7   Nekton..561
5.8   Fisheries 571
5.9   Food Webs..572

5.10 Carbon Flow 580
5.11 Coastal Systems 587
5.12 Deep-Sea Systems 594

Chapter 6   Marine Pollution and Other Anthropogenic Impacts
  I. Introduction 621
  II. Types of Anthropogenic Impacts 622
    A. Marine Pollution 622
      1. Nutrient Loading 622
      2. Organic Carbon Loading 624
      3. Oil 625
      4. Toxic Chemicals 627
        a. Polycyclic Aromatic Hydrocarbons 627
        b. Halogenated Hydrocarbons 628
        c. Heavy Metals 630
        d. Radioactive Substances 631
    B. Other Anthropogenic Impacts 632
      1. Coastal Development 632
      2. Marine Debris 633
      3. Dredging and Dredged Material Disposal 634
      4. Oil Production and Marine Mining 635
      5. Exploitation of Fisheries 635
      6. Boats and Marinas 636
      7. Electric Generating Stations 637
      8. Altered Natural Flows 638
      9. Introduced Species 638
  III. Conclusions 639
6.1 Sources of Marine Pollution 647
6.2 Watershed Effects 654
6.3 Contamination Effects on Organisms 671
6.4 Nutrients 683
6.5 Organic Carbon 694
6.6 Sewage Waste 709
6.7 Pathogens 721
6.8 Oil 724
6.9 Polycyclic Aromatic Hydrocarbons 738
6.10 Halogenated Hydrocarbons 754
6.11 Heavy Metals 785
6.12 Radioactive Waste 807
6.13 Dredging and Dredged Material Disposal 826

Index 837

CHAPTER 1

Physiography

I. OCEAN PROVINCES

A. Dimensions

The world oceans including the adjacent seas cover ~71% of the earth's surface (~3.6 × 10^8 km²), and they have a total volume of ~1.35 × 10^9 km³. The mean depth of all the oceans amounts to ~3700 m, with the Pacific Ocean being deepest (4188 m), followed by the Indian Ocean (3872 m) and the Atlantic Ocean (3844 m). Nearly 75% of the ocean basins lie within the depth zone between 3000 and 6000 m. The seas are much shallower, being ~1200 m deep or less. The Pacific Ocean is by far the largest and deepest ocean, comprising 50.1% of the world ocean and occupying more than one third of the earth's surface. By comparison, the Atlantic Ocean and Indian Ocean constitute 29.4 and 20.5% of the world ocean, respectively. The oceans range from ~5000 km (Atlantic) to 17,000 km (Pacific) in width.

Ocean water is not evenly distributed around the globe. In the southern hemisphere, the percentage of water (80.9%) to land (19.1%) far exceeds that of the percentage of water (60.7%) to land (39.3%) in the northern hemisphere. This uneven distribution greatly affects world meteorological and ocean circulation patterns.

The mean temperature of the oceans is 3.51°C, and the mean salinity 34.7‰. Excluding the Southern Ocean as a separate entity, the Pacific Ocean exhibits the lowest temperatures and salinities with mean values of 3.14°C and 34.6‰, respectively. In contrast, highest mean temperatures (3.99°C) and salinities (34.92‰) exist in the Atlantic Ocean despite its large volume of riverine inflow. This is particularly true in the North Atlantic, where the mean temperature (5.08°C) and salinity (35.09‰) exceed those of all other major ocean basins. The Indian Ocean has intermediate mean temperature (3.88°C) and salinity (34.78‰) values.

The major oceans also include marginal seas. Some of these smaller systems are bounded by land or island chains (e.g., Caribbean Sea, Mediterranean Sea, and Sea of Japan). Others not bounded off by land are distinguished by local oceanographic characteristics (e.g., Labrador, Norwegian, and Tasman seas).[1] Marginal seas can strongly influence temperature and salinity conditions of the major ocean basins. For example, the warm, saline waters of the Mediterranean Sea can be detected over thousands of kilometers at mid-depths in the Atlantic Ocean.

Comparing oceanic depths and land elevations on earth, it is quite clear that relative to sea level, the landmasses are not as high as the oceans are deep. As demonstrated by a hypsographic curve, 84% of the ocean floor exceeds 2000 m depth, while only 11% of the land surface is greater than 2000 m above sea level. The maximum oceanic depth, recorded in the Mariana Trench in the western Pacific, amounts to 11,035 m. The highest elevation on land, Mt. Everest, is 8848 m.

B. Physiographic Provinces

1. Continental Margin Province

The ocean floor is divided into three major physiographic provinces—the continental margins, deep-ocean basins, and mid-ocean ridges—characterized by distinctive bathymetry and unique landforms.[2,3] The continental margins represent the submerged edges of the continents, and they consist of the continental shelf, slope, and rise, which extend seaward from the shoreline down to depths of ~2000 to 3000 m. The continental shelf is underlain by a thick wedge of sediment derived from continental sources. Here, the ocean bottom slopes gently seaward at an angle of ~0.5°. Although typified by broad expanses of nearly flat terrain, many shelf regions also exhibit irregularly distributed hills, valleys, and depressions of low to moderate relief. Continental shelves, which range from <5 km in width along the Pacific coasts of North and South America to as much as 1500 km along the Arctic Ocean, average ~7.5 km in width worldwide. They cover ~7% of the total area of the ocean floor. The outer margin of continental shelves lies at a depth of ~150 to 200 m, where the slope of the ocean bottom increases abruptly to ~1 to 4°, marking the shelf break.

The continental slope occurs seaward of the shelf break, being inclined at ~4°. It descends to the upper limit of the continental rise. Sediments underneath the continental slope are commonly incised by submarine canyons having steep-sided, V-shaped profiles. These erosional features, with a topographic relief of 1 to 2 km, are usually cut by turbidity currents. They serve as conduits for the transport of sediment from the continental shelf to the deep-ocean basins. Examples are the Hudson, Baltimore, LaJolla, and Redondo canyons.

The continental rise is a gently sloping apron of sediment accumulating at the base of the continental slope and spreading across the deep seafloor to adjacent abyssal plains. It is a topographically smooth feature, sloping at an angle of ~1° and covering hundreds of kilometers of the ocean floor down to depths of ~4000 m. Clay, silt, and sand turbidite deposits underlying the rise may accumulate to several kilometers in thickness, being transported by turbidity currents from the nearby continental shelf and slope.

2. Deep-Ocean Basin Province

The deep-ocean basins are found seaward of the continental rises at a depth of ~3000 to 5000 m. Significant topographic features within this province include abyssal plains, abyssal hills, deep-sea trenches, and seamounts. As such, the topography is more variable than in the continental margin physiographic province, ranging from nearly flat plains to steep-sided volcanic edifices and deep narrow trenches.

With slopes of less than 1 m/km, the abyssal plains are the flattest areas on earth. They consist primarily of fine-grained sediments ranging from ~100 m to more than 1000 m in thickness. These level ocean basin regions form by the slow deposition of clay, silt, and sand particles (turbidites) transported via turbidity currents off the outer continental shelf and slope, ultimately burying a significant amount of the volcanic terrane.[3] A large fraction of the sediments are transported to the deep sea through submarine canyons.

Abyssal hills and seamounts frequently dot the seafloor in the deep-ocean basin province. Typically of volcanic origin, the abyssal hills rise as much as 1000 m above the seafloor, but generally have an average relief of only ~200 m. Some appear as elongated hills, and others as domes ranging from ~5 to 100 km in width. In contrast, seamounts with circular, ovoid, or lobate shapes usually protrude more than a kilometer above the surrounding seafloor, and typically have slopes of 5 to 25°. Large intraplate volcanoes may approach 10 km in height. Occasionally, seamounts merge into a chain of aseismic ridges. Flat-topped forms, referred to as guyots, develop as the volcanic peaks are eroded during emergence. Seamounts are most numerous in the Pacific Ocean basin, where as many as 1 million edifices cover 13% of the seafloor.[4]

The greatest ocean depths (>11,000 m) have been recorded in deep-sea trenches. These long, narrow depressions are on average 3000 to 5000 m deeper than the ocean basins. They are bordered by volcanic island arcs or continental margin magmatic belts, and mark the sites of major lithospheric subduction zones. Deep-sea trenches represent tectonically active areas associated with strong earthquakes and volcanism.

3. Mid-Ocean Ridge Province

The largest and most volcanically active chain of mountains on earth occurs along a globally encircling mid-ocean ridge (MOR) and rift system that extends through all the major ocean basins as seafloor spreading centers. It is along the 75,000-km global length of MORs where new oceanic crust forms through an interplay of magmatic construction, hydrothermal convection, and tectonic extension.[5,6] These spreading centers are sites of active basaltic volcanism, shallow-focus earthquakes, and high rates of heat flow.

The seafloor at MORs consists of a narrow neovolcanic zone (1 to 4 km wide) flanked successively by a zone of crustal fissuring (0.5 to 3 km) and a zone of active faulting out to a distance of ~10 km from the spreading axis. The neovolcanic zone is the region of most recent volcanic activity along the ridge. The summit of the ridge is marked by a rift valley, axial summit caldera, or axial summit graben. The MOR system is segmented along-axis by transform faults, which commonly extend far into deep-ocean basins as inactive fracture zones. The irregular partitioning of the ridge axis by transform faults creates a hierarchy of discontinuities.[7]

The mid-ocean ridge physiographic province lies at a depth of 2000 to 3000 m. The volcanic ridges comprising this mountain system are comparable in physical dimensions to those on the continents. Most seamounts form on or near mid-ocean ridges. The inner valley floor of the northern Mid-Atlantic Ridge is composed of piled-up seamounts and hummocky pillow flows, representing the product of crustal accretion.[8-10] They account for highly variable and rugged volcanic landscapes. As this volcanic material and the remaining newly formed lithosphere cool and subside on either side of the MOR, the elevations of the submarine volcanic mountains decline, and the topography becomes less rugged. The original volcanic topography also is gradually buried under a thick apron of sediments as the lithosphere moves away from the MOR.

C. Benthic and Pelagic Provinces

1. Benthic Province

The oceans can also be subdivided on the basis of major habitats on the seafloor (benthic province) and in the water column (pelagic province). The benthic province consists of five discrete zones: littoral, sublittoral, bathyal, abyssal, and hadal. The littoral (or intertidal) zone encompasses the bottom habitat between the high and low tide marks. Immediately seaward, the sublittoral (or subtidal) zone defines the benthic region from mean low water to the shelf break at a depth of ~200 m. The seafloor extending from the shelf break to a depth of ~2000 m corresponds to the bathyal zone. The deepest benthic habitats include the abyssal zone from 2000 to 6000 m depth and the hadal zone below 6000 m. These zones roughly conform with the aforementioned physiographic provinces. For example, the sublittoral zone represents the benthic environment of the continental shelf. The bathyal zone corresponds to the continental slope and rise, and the abyssal zone to the deep-ocean basins exclusive of the trenches, which are represented by the hadal zone. The abyssal zone accounts for 75% of the benthic habitat area of the oceans, and the bathyal and sublittoral zones 16 and 8% of the area, respectively.

2. Pelagic Province

Pelagic environments are subdivided into neritic and oceanic zones. The neritic zone includes all waters overlying the continental shelf, and the oceanic zone, all waters seaward from the shelf break. Waters of the oceanic zone are further subdivided into the epipelagic, mesopelagic, bathypelagic, abyssalpelagic, and hadalpelagic regions. Epipelagic waters constitute the uppermost portion of the water column extending from the sea surface down to a depth of 200 m. The waters between 200 and 1000 m constitute the mesopelagic zone, and those between 1000 and 2000 m, the bathypelagic zone. Deepest ocean waters of the pelagic province occur in the abyssalpelagic zone, located between 2000 m and 6000 m depth, and in the underlying hadalpelagic zone, occupying the deep-sea trenches. The abyssalpelagic, mesopelagic, and bathypelagic zones contain the greatest volume of seawater, amounting to 54, 28, and 15% of all water present in the oceanic zone, respectively.

REFERENCES

1. Pickard, G. L. and Emery, W. J., *Descriptive Physical Oceanography: An Introduction,* 4th ed., Pergamon Press, Oxford, 1985.
2. Millero, F. J., *Marine Chemistry,* 2nd ed., CRC Press, Boca Raton, FL, 1997.
3. Pinet, P. R., *Invitation to Oceanography,* Jones and Bartlett Publishers, Sudbury, MA, 1998.
4. Smith, D. K., Seamount abundances and size distribution, and their geographic variations, *Rev. Aquat. Sci.,* 5, 197, 1991.
5. Macdonald, K. C. and Fox, P. J., The mid-ocean ridge, *Sci. Am.,* 262, 72, 1990.
6. Cann, J. R., Elderfield, H., and Laughton, A. S., Eds., *Mid-Ocean Ridges: Dynamics of Processes Associated with Creation of New Ocean Crust,* Cambridge University Press, New York, 1998.
7. Macdonald, K. C., Scheirer, D. S., and Carbotte, S. M., Mid-ocean ridges: discontinuities, segments, and giant cracks, *Science,* 253, 968, 1991.
8. Smith, D. K. and Cann, J. R., The role of seamount volcanism in crustal construction at the Mid-Atlantic (24°–30°), *J. Geophys. Res.,* 97, 1645, 1992.
9. Smith, D. K. and Cann, J. R., Building the crust at the Mid-Atlantic Ridge, *Nature,* 365, 707, 1993.
10. Smith, D. K., Mid-Atlantic Ridge volcanism from deep-towed side-scan sonar images, 25–29°N, *J. Volcanol. Geotherm. Res.,* 67, 233, 1995.

1.1 CONVERSION FACTORS, MEASURES, AND UNITS

Table 1.1–1   Recommended Decimal Multiples and Submultiples

Multiples and Submultiples	Prefixes	Symbols
10^{18}	exa	E
10^{15}	peca	P
10^{12}	tera	T
10^{9}	giga	G
10^{6}	mega	M
10^{3}	kilo	k
10^{2}	hecto	h
10	deca	da
10^{-1}	deci	d
10^{-2}	centi	c
10^{-3}	milli	m
10^{-4}	micro	μ
10^{-9}	nano	n
10^{-12}	pico	p
10^{-15}	femto	f
10^{-18}	atto	a

Source: Beyer, W. H., Ed., *CRC Standard Mathematical Tables,* 28th ed., CRC Press, Boca Raton, FL, 1987, 1. With permission.

Table 1.1–2   Conversion Factors

To Obtain	Metric to English Multiply	By
Inches	Centimeters	0.3937007874
Feet	Meters	3.280839895
Yards	Meters	1.093613298
Miles	Kilometers	0.6213711922
Ounces	Grams	$3.527396195 \times 10^{-2}$
Pounds	Kilograms	2.204622622
Gallons (U.S. liquid)	Liters	0.2641720524
Fluid ounces	Milliliters (cc)	$3.381402270 \times 10^{-2}$
Square inches	Square centimeters	0.1550003100
Square feet	Square meters	10.76391042
Square yards	Square meters	1.195990046
Cubic inches	Milliliters (cc)	$6.102374409 \times 10^{-2}$
Cubic feet	Cubic meters	35.31466672
Cubic yards	Cubic meters	1.307950619

Table 1.1–2   Conversion Factors (continued)

To Obtain	English to Metric[a] Multiply	By
Microns	Mils	**25.4**
Centimeters	Inches	**2.54**
Meters	Feet	**0.3048**
Meters	Yards	**0.9144**
Kilometers	Miles	**1.609344**
Grams	Ounces	28.34952313
Kilograms	Pounds	**0.45359237**
Liters	Gallons (U.S. liquid)	**3.785411784**
Milliliters (cc)	Fluid ounces	29.57352956
Square centimeters	Square inches	**6.4516**
Square meters	Square feet	**0.09290304**
Square meters	Square yards	**0.83612736**
Milliliters (cc)	Cubic inches	**16.387064**
Cubic meters	Cubic feet	$2.831684659 \times 10^{-2}$
Cubic meters	Cubic yards	0.764554858

To Obtain	Conversion Factors—General[a] Multiply	By
Atmospheres	Feet of water @ 4°C	2.950×10^{-2}
Atmospheres	Inches of mercury @ 0°C	3.342×10^{-2}
Atmospheres	Pounds per square inch	6.804×10^{-2}
BTU	Foot-pounds	1.285×10^{-3}
BTU	Joules	9.480×10^{-4}
Cubic feet	Cords	**128**
Degree (angle)	Radians	57.2958
Ergs	Foot-pounds	1.356×10^{7}
Feet	Miles	5280
Feet of water @ 4°C	Atmospheres	33.90
Foot-pounds	Horsepower-hours	1.98×10^{6}
Foot-pounds	Kilowatt-hours	2.655×10^{6}
Foot-pounds per min	Horsepower	3.3×10^{4}
Horsepower	Foot-pounds per second	1.818×10^{-3}
Inches of mercury @ 0°C	Pounds per square inch	2.036
Joules	BTU	1054.8
Joules	Foot-pounds	1.35582
Kilowatts	BTU per minute	1.758×10^{-2}
Kilowatts	Foot-pounds per minute	2.26×10^{-5}
Kilowatts	Horsepower	0.745712
Knots	Miles per hour	0.86897624
Miles	Feet	1.894×10^{-4}
Nautical miles	Miles	0.86897624
Radians	Degrees	1.745×10^{-2}
Square feet	Acres	43560
Watts	BTU per minute	17.5796

Temperature Factors

$$°F = 9/5 \,(°C) + 32$$
Fahrenheit temperature = 1.8 (temperature in kelvins) − 459.67
$$°C = 5/9 \,[(°F) − 32]$$
Fahrenheit temperature = 1.8 (Celsius temperature) + 32
Celsius temperature = temperature in kelvins − 273.15

[a] Boldface numbers are exact; others are given to ten significant figures where so indicated by the multiplier factor.

Source: Beyer, W. H., Ed., *CRC Standard Mathematical Tables*, 28th ed., CRC Press, Boca Raton, FL, 1987, 21. With permission.

Table 1.1–3 Metric and U.S. System, Measures, Units, and Conversions

Lengths

Metric System		U.S. System
10^{-8} cm	1Å	$3.937 \cdot 10^{-9}$ in.
10^{-4} cm	1 μ	$3.937 \cdot 10^{-5}$ in.
	1 cm	0.3937 in.
	2.540 cm	1 in.
	0.3048	1 ft
	0.9144 m	1 yd
10^{2} cm	1 m	1.09361 yd
	1.8288 m	1 fathom
10^{5} cm	1 km	0.62137 mi
	1.60935 km	1 mi
	1.852 km	1 int. nautical mi

Area

Metric System	U.S. System
1 mm^2	0.00155 $in.^2$ (sq. in.)
1 cm^2	0.155 $in.^2$
6.45163 cm^2	1 $in.^2$
0.0929 m^2	1 ft^2
0.83613 m^2	1 yd^2
1 m^2	10.7639 ft^2
1 km^2	0.3861 mi^2
2.58998 km^2	1 mi^2

Volume

Metric System	U.S. System
1 mm^3	$0.6102 \cdot 10^{-4}$ $in.^3$ (cu. in.)
1 cm^3	0.06102 $in.^3$
16.3872 cm^3	1 $in.^3$
0.02831 m^3	1 ft^3
0.76456 m^3	1 yd^3
1 m^3	1.30794 yd^3

Liquid Measures

Metric System	U.S. System	
1 ml	0.0610 $in.^3$	
0.473 l	28.875 $in.^3$	1 pt
0.946 l	57.749 $in.^3$	1 qt
1 l	61.0 $in.^3$	1.0567 qt
3.7853 l	231 $in.^3$	1 gal

Mass

Metric System	U.S. System
1 g	0.035 oz av (ounce av)
28.349 g	1 oz av
453.59 g	1 lb av (lb av)[a]
1 kg	2.20462 lb av
907.1848 kg	1 ton sh (short ton)
1 t	1.1023 ton sh
1016.047 kg	1 ton 1 (long ton)

Density

Metric System	U.S. System
1 g/cm^3	0.036127 $lb/in.^3$
27.68 g/cm^3	1 $lb/in.^3$
0.0160 g/cm^3	1 lb/ft^3

(continued)

Table 1.1–3 Metric and U.S. System, Measures, Units, and Conversions (continued)

Energy

	erg	Joule$_{mt}$	kW$_{int}$h	kcal$_t$	Liter-atmos.	BTU
erg	1	0.9997×10^{-7}	2.7769×10^{-14}	2.389×10^{-11}	9.8692×10^{-10}	9.4805×10^{-11}
Joule$_{int}$	1.0002×10^{7}	1	2.7778×10^{-7}	2.390×10^{-4}	9.8722×10^{-3}	9.480×10^{-4}
kW$_{int}$h	3.6011×10^{13}	3.6000×10^{6}	1	8.6041×10^{2}	3.5540×10^{4}	3.413×10^{3}
Kcal$_{15}$	4.1853×10^{10}	4.186×10^{3}	1.1622×10^{-3}	1	4.1306×10^{1}	3.9685
Liter-atmos.	1.0133×10^{9}	1.0133×10^{2}	2.8137×10^{-5}	2.421×10^{-2}	1	9.607×10^{-2}
BTU	1.0548×10^{10}	1.0548×10^{3}	2.930×10^{-4}	2.5198×10^{-1}	1.0409×10^{1}	1

Pressure

	bar	Torr	atm.	at	lb/in.^{2}
1 bar (10^6 dyn/cm^2)	1	750	0.98692	1.0197	14.504
1 torr	0.00133	1	0.00131	0.001359	0.01934
1 atm	1.0133	760	1	1.033	14.696
1 at (1 kg/cm^2)	0.98067	735.56	0.96784	1	14.223
1 lb/in.^{2}	0.06895	51.7144	0.068046	0.07031	1

Temperature

Absolute Centigrade or Kelvin (K) $\times$ °K = T°C + 273.18

Degrees Centigrade (°C) $\times$ °C = 5/9 (T°F − 32)

Degrees Fahrenheit (°F) $\times$ °C = 5/4 T°R

$\times$ °F = 9/4 T°R + 32

$\times$ °F = 9/5 T°C + 32

Degrees Réaumur (°R) $\times$ °R = 4/9 (T°F − 32)

$\times$ °R = 4/5 T°C

Centigrade to Fahrenheit

°C	°F	°C	°F
−200	−328	60	140
−150	−238	70	158
−100	−148	80	176
−50	−58	90	194
0	+32	100	212

°C	°F	°C	°F
		200	392
		250	482
		300	572
		400	752
		500	932

Centigrade to Fahrenheit

°C	°F	°C	°F
+10	50	110	230
20	68	120	248
30	86	130	266
40	104	140	284
50	122	150	302

°C	°F	°C	°F
		600	1112
		700	1292
		800	1472
		900	1652
		1000	1832

[a] 1 lb av = 1 pound avoirdupois is the mass of 27.662 in.^{3}.

Source: Heydemann, A., *Handbook of Geochemistry*, Vol. 1, Wedepohl, K. H., Ed., Springer-Verlag, Berlin, 1969. With permission.

1.2 GENERAL FEATURES OF THE EARTH

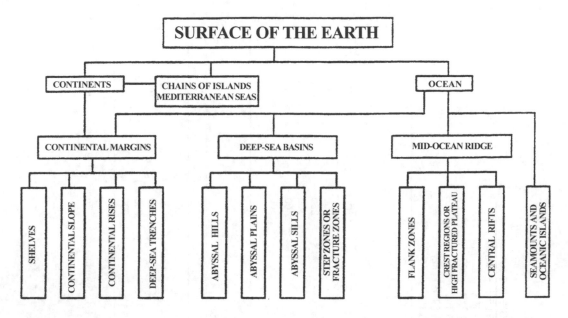

Figure 1.2–1 Divisions of the earth's surface. (From Millero, F. J., *Chemical Oceanography*, 2nd ed., CRC Press, Boca Raton, FL, 1996, 4. With permission.)

Table 1.2–1   Mass, Dimensions, and Other Parameters of the Earth

Quantity	Symbol	Value	Unit
Mass	M	$5.9742 \cdot 10^{27}$	g
Major orbital semi-axis	a_{orb}	1.000000	AU
		$1.4959787 \cdot 10^{8}$	km
Distance from sun at perihelion	r_x	0.9833	AU
Distance from sun at aphelion	r_α	1.0167	AU
Moment of perihelion passage	T_x	Jan. 2, 4 h 52 min	
Moment of aphelion passage	T_α	July 4, 5 h 05 min	
Siderial rotation period around sun	P_{orb}	$31.5581 \cdot 10^{6}$	s
		365.25636	d
Mean rotational velocity	U_{orb}	29.78	km/s
Mean equatorial radius	$\bar{a}$	6378.140	km
Mean polar compression (flattening factor)	α	1/298.257	
Difference in equatorial and polar semi-axes	$a - c$	21.385	km
Compression of meridian of major equatorial axis	α_a	1/295.2	
Compression of meridian of minor equatorial axis	α_b	1/298.0	
Equatorial compression	ε	1/30,000	
Difference in equatorial semi-axes	$a - b$	213	m
Difference in polar semi-axes	$c_N - c_S$	~70	m
Polar asymmetry	η	$\sim1.10^{-5}$	
Mean acceleration of gravity at equator	g_e	9.78036	m/s²
Mean acceleration of gravity at poles	g_p	9.83208	m/s²
Difference in acceleration of gravity at pole and at equator	$g_p - g_e$	5.172	cm/s²
Mean acceleration of gravity for entire surface of terrestrial ellipsoid	g	9.7978	m/s²

Table 1.2–1   Mass, Dimensions, and Other Parameters of the Earth (continued)

Quantity	Symbol	Value	Unit
Mean radius	R	6371.0	km
Area of surface	S	$5.10 \cdot 10^8$	km^2
Volume	V	$1.0832 \cdot 10^{12}$	km^3
Mean density	ρ	5.515	g/cm^3
Siderial rotational period	P	86,164.09	s
Rotational angular velocity	ω	$7.292116 \cdot 10^{-5}$	rad/s
Mean equatorial rotational velocity	v	0.46512	km/s
Rotational angular momentum	L	$5.861 \cdot 10^{33}$	J s
Rotational energy	E	$2.137 \cdot 10^{29}$	J
Ratio of centrifugal force to force of gravity at equator	q_c	$0.0034677 = 1/288$	
Moment of inertia	I	$8.070 \cdot 10^{37}$	kg m^2
Relative braking of earth's rotation due to tidal friction	$\Delta\omega_e/\omega$	$-4.2 \cdot 10^{-8}$	century^{-1}
Relative secular acceleration of earth's rotation	$\Delta\omega_i/\omega$	$-1.4 \cdot 10^{-8}$	century^{-1}
Not secular braking of earth's rotation	$\Delta\omega/\omega$	$-2.8 \cdot 10^{-8}$	century^{-1}
Probable value of total energy of tectonic deformation of earth	E_t	$\sim 1 \cdot 10^{23}$	J/century
Secular loss of heat of earth through radiation into space	$\Delta' E_k$	$1 \cdot 10^{23}$	J/century
Portion of earth's kinetic energy transformed into heat as a result of lunar and solar tides in the hydrosphere	$\Delta'' E_k$	$1.3 \cdot 10^{23}$	J/century
Differences in duration of days in March and August	ΔP	0.0025 (March–August)	s
Corresponding relative annual variation in earth's rotational velocity	$\Delta^*\omega/\omega$	$2.9 \cdot 10^{-8}$ (Aug.–March)	
Presumed variation in earth's radius between August and March	$\Delta^* R$	-9.2 (Aug.–March)	cm
Annual variation in level of world ocean	Δh_o	~ 10 (Sept.–March)	cm
Area of continents	S_C	$1.49 \cdot 10^8$	km^2
		29.2	% of surface
Area of world ocean	S_o	$3.61 \cdot 10^8$	km^2
		70.8	% of surface
Mean height of continents above sea level	h_C	875	m
Mean depth of world ocean	h_o	3794	m
Mean thickness of lithosphere within the limits of the continents	$h_{c.l.}$	35	km
Mean thickness of lithosphere within the limits of the ocean	$h_{o.l.}$	4.7	km
Mean rate of thickening of continental lithosphere	$\Delta h/\Delta t$	10–40	m/10^6 y
Mean rate of horizontal extension of continental lithosphere	$\Delta l/\Delta t$	0.75–20	km/10^6 y
Mass of crust	m_1	$2.36 \cdot 10^{22}$	kg
Mass of mantle		$4.05 \cdot 10^{24}$	kg
Amount of water released from the mantle and core in the course of geological time		$3.40 \cdot 10^{21}$	kg
Total reserve of water in the mantle		$2 \cdot 10^{23}$	kg
Present content of free and bound water in the earth's lithosphere		$2.4 \cdot 10^{21}$	kg
Mass of hydrosphere	m_h	$1.664 \cdot 10^{21}$	kg
Amount of oxygen bound in the earth's crust		$1.300 \cdot 10^{21}$	kg
Amount of free oxygen		$1.5 \cdot 10^{18}$	kg
Mass of atmosphere	m_a	$5.136 \cdot 10^{18}$	kg
Mass of biosphere	m_b	$1.148 \cdot 10^{16}$	kg
Mass of living matter in the biosphere		$3.6 \cdot 10^{14}$	kg
Density of living matter on dry land		0.1	g/cm^2
Density of living matter in ocean		$15 \cdot 10^{-8}$	g/cm^3

Table 1.2–1   Mass, Dimensions, and Other Parameters of the Earth (continued)

Age of the earth	$4.55 \cdot 10^9$	y
Age of oldest rocks	$4.0 \cdot 10^9$	y
Age of most ancient fossils	$3.4 \cdot 10^9$	y

Note: This table is a collection of data on various properties of the earth. Most of the values are given in SI units. Note that 1 AU (astronomical unit) = 149,597,870 km.

Source: Lide, D. R. and Frederikse, H. P. R., Eds., *CRC Handbook of Chemistry and Physics*, 79th ed., CRC Press, Boca Raton, FL, 1998, 14–6. With permission.

REFERENCES

1. Seidelmann, P. K., Ed., *Explanatory Supplement to the Astronomical Almanac,* University Science Books, Mill Valley, CA, 1992.
2. Lang, K. R., *Astrophysical Data: Planets and Stars*, Springer-Verlag, New York, 1992.

Table 1.2–2   Density, Pressure, and Gravity as a Function of Depth within the Earth

Depth km	ρ g/cm^3	p kbar	g cm/s^2	Depth km	ρ g/cm^3	p kbar	g cm/s^2
	Crust				**Mantle (solid)**		
				1771	4.96	752	994
				2071	5.12	903	1002
0	1.02	0	981	2371	5.31	1061	1017
3	1.02	3	982	2671	5.45	1227	1042
3	2.80	3	982	2886	5.53	1352	1069
21	2.80	5	983				
					Outer Core (liquid)		
	Mantle (solid)						
				2886	9.96	1352	1069
21	3.49	5	983	2971	10.09	1442	1050
41	3.51	12	983	3371	10.63	1858	953
61	3.52	19	984	3671	11.00	2154	874
81	3.48	26	984	4071	11.36	2520	760
101	3.44	33	984	4471	11.69	2844	641
121	3.40	39	985	4871	11.99	3116	517
171	3.37	56	987	5156	12.12	3281	427
221	3.34	73	989				
271	3.37	89	991		**Inner Core (solid)**		
321	3.47	106	993				
371	3.59	124	994	5156	12.30	3281	427
571	3.95	199	999	5371	12.48	3385	355
871	4.54	328	997	5771	12.52	3529	218
1171	4.67	466	992	6071	12.53	3592	122
1471	4.81	607	991	6371	12.58	3617	0

Note: This table gives the density ρ, pressure p, and acceleration due to gravity g as a function of depth below the earth's surface, as calculated from the model of the structure of the earth in Reference 1. The model assumes a radius of 6371 km for the earth. The boundary between the crust and mantle (the Mohorovicic discontinuity) is taken as 21 km, while in reality it varies considerably with location.

Source: Lide, D. R. and Frederikse, H. P. R., Eds., *CRC Handbook of Chemistry and Physics*, 79th ed., CRC Press, Boca Raton, FL, 1998, 14–10. With permission.

REFERENCES

1. Anderson, D. L. and Hart, R. S., *J. Geophys. Res.*, 81, 1461, 1976.
2. Carmichael, R. S., *CRC Practical Handbook of Physical Properties of Rocks and Minerals,* CRC Press, Boca Raton, FL, 1989, 467.

Table 1.2–3   Abundance of Elements in the Earth's Crust and Sea

Element	Abundance Crust (mg/kg)	Sea (mg/l)	Element	Abundance Crust (mg/kg)	Sea (mg/l)
Ac	5.5×10^{-10}		N	1.9×10^1	5×10^{-1}
Ag	7.5×10^{-2}	4×10^{-5}	Na	2.36×10^4	1.08×10^4
Al	8.23×10^4	2×10^{-3}	Nb	2.0×10^1	1×10^{-5}
Ar	3.5	4.5×10^{-1}	Nd	4.15×10^1	2.8×10^{-6}
As	1.8	3.7×10^{-3}	Ne	5×10^{-3}	1.2×10^{-4}
Au	4×10^{-3}	4×10^{-6}	Ni	8.4×10^1	5.6×10^{-4}
B	1.0×10^1	4.44	O	4.61×10^5	8.57×10^5
Ba	4.25×10^2	1.3×10^{-2}	Os	1.5×10^{-3}	
Be	2.8	5.6×10^{-6}	P	1.05×10^3	6×10^{-2}
Bi	8.5×10^{-3}	2×10^{-5}	Pa	1.4×10^{-6}	5×10^{-11}
Br	2.4	6.73×10^1	Pb	1.4×10^1	3×10^{-5}
C	2.00×10^2	2.8×10^1	Pd	1.5×10^{-2}	
Ca	4.15×10^4	4.12×10^2	Po	2×10^{-10}	1.5×10^{-14}
Cd	1.5×10^{-1}	1.1×10^{-4}	Pr	9.2	6.4×10^{-7}
Ce	6.65×10^1	1.2×10^{-6}	Pt	5×10^{-3}	
Cl	1.45×10^2	1.94×10^4	Ra	9×10^{-7}	8.9×10^{-11}
Co	2.5×10^1	2×10^{-5}	Rb	9.0×10^1	1.2×10^{-1}
Cr	1.02×10^2	3×10^{-4}	Re	7×10^{-4}	4×10^{-6}
Cs	3	3×10^{-4}	Rh	1×10^{-3}	
Cu	6.0×10^1	2.5×10^{-4}	Rn	4×10^{-13}	6×10^{-16}
Dy	5.2	9.1×10^{-7}	Ru	1×10^{-3}	7×10^{-7}
Er	3.5	8.7×10^{-7}	S	3.50×10^2	9.05×10^2
Eu	2.0	1.3×10^{-7}	Sb	2×10^{-1}	2.4×10^{-4}
F	5.85×10^2	1.3	Sc	2.2×10^1	6×10^{-7}
Fe	5.63×10^4	2×10^{-3}	Se	5×10^{-2}	2×10^{-4}
Ga	1.9×10^1	3×10^{-5}	Si	2.82×10^5	2.2
Gd	6.2	7×10^{-7}	Sm	7.05	4.5×10^{-7}
Ge	1.5	5×10^{-5}	Sn	2.3	4×10^{-6}
H	1.40×10^3	1.08×10^5	Sr	3.70×10^2	7.9
He	8×10^{-3}	7×10^{-6}	Ta	2.0	2×10^{-6}
Hf	3.0	7×10^{-6}	Tb	1.2	1.4×10^{-7}
Hg	8.5×10^{-2}	3×10^{-5}	Te	1×10^{-3}	
Ho	1.3	2.2×10^{-7}	Th	9.6	1×10^{-6}
I	4.5×10^{-1}	6×10^{-2}	Ti	5.65×10^3	1×10^{-3}
In	2.5×10^{-1}	2×10^{-2}	Tl	8.5×10^{-1}	1.9×10^{-5}
Ir	1×10^{-3}		Tm	5.2×10^{-1}	1.7×10^{-7}
K	2.09×10^4	3.99×10^2	U	2.7	3.2×10^{-3}
Kr	1×10^{-4}	2.1×10^{-4}	V	1.20×10^2	2.5×10^{-3}
La	3.9×10^1	3.4×10^{-6}	W	1.25	1×10^{-4}
Li	2.0×10^1	1.8×10^{-1}	Xe	3×10^{-5}	5×10^{-5}
Lu	8×10^{-1}	1.5×10^{-7}	Y	3.3×10^1	1.3×10^{-5}
Mg	2.33×10^4	1.29×10^3	Yb	3.2	8.2×10^{-7}
Mn	9.50×10^2	2×10^{-4}	Zn	7.0×10^1	4.9×10^{-3}
Mo	1.2	1×10^{-2}	Zr	1.65×10^2	3×10^{-5}

Note: This table gives the estimated abundance of the elements in the continental crust (in mg/kg, equivalent to parts per million by mass) and in seawater near the surface (in mg/l). Values represent the median of reported measurements. The concentrations of the less abundant elements may vary with location by several orders of magnitude.

Source: Lide, D. R. and Frederikse, H. P. R., Eds., *CRC Handbook of Chemistry and Physics*, 79th ed., CRC Press, Boca Raton, FL, 1998, 14-14. With permission.

REFERENCES

1. Carmichael, R. S., Ed., *CRC Practical Handbook of Physical Properties of Rocks and Minerals,* CRC Press, Boca Raton, FL, 1989.
2. Bodek, I. et al., *Environmental Inorganic Chemistry,* Pergamon Press, New York, 1988.
3. Ronov, A. B. and Yaroshevsky, A. A., Earth's crust geochemistry, in *Encyclopedia of Geochemistry and Environmental Sciences,* Fairbridge, R. W., Ed., Van Nostrand, New York, 1969.

1.3 GENERAL CHARACTERISTICS OF THE OCEANS

Table 1.3–1   Area, Volume, Mean, and Maximum Depths of the Oceans and Their Adjacent Seas

Sea	Area[a] (10⁶ km²)	Volume[a] (10⁶ km³)	Depth Mean[a] (m)	Depth Maximum[b] (m)
Oceans without adjacent seas				
Pacific Ocean	166.24	696.19	4,188	11,022[a]
Atlantic Ocean	84.11	322.98	3,844	9,219[b]
Indian Ocean	73.43	284.34	3,872	7,455[c]
Total	323.78	1,303.51	4,026	—
Mediterranean seas				
Arctic[d]	12.26	13.70	1,117	5,449
Austral-Asiatic[e]	9.08	11.37	1,252	7,440
American	4.36	9.43	2,164	7,680
European[f]	3.02	4.38	1,450	5,092
Total	28.72	38.88	1,354	—
Intracontinental Mediterranean seas				
Hudson Bay	1.23	0.16	128	218
Red Sea	0.45	0.24	538	2,604
Baltic Sea	0.39	0.02	55	459
Persian Gulf	0.24	0.01	25	170
Total	2.31	0.43	184	—
Marginal seas				
Bering Sea	2.26	3.37	1,491	4,096
Sea of Okhotsk	1.39	1.35	971	3,372
East China Sea	1.20	0.33	275	2,719
Sea of Japan	1.01	1.69	1,673	4,225
Gulf of California	0.15	0.11	733	3,127
North Sea	0.58	0.05	93	725[g]
Gulf of St. Lawrence	0.24	0.03	125	549
Irish Sea	0.10	0.01	60	272
Remaining seas	0.30	0.15	470	—
Total	7.23	7.09	979	—
Oceans, including adjacent seas				
Pacific Ocean	181.34	714.41	3,940	11,022[a]
Atlantic Ocean	106.57	350.91	3,293	9,219[b]
Indian Ocean	74.12	284.61	3,840	7,455[c]
World ocean	362.03	1,349.93	3,729	11,022[a]

[a] Vitiaz Depth in the Mariana Trench.
[b] Milwaukee Depth in the Puerto Rico Trench.
[c] Planet Depth in the Sunda Trench.
[d] Consisting of Arctic Ocean, Barents Sea, Canadian Archipelago, Baffin Bay, and Hudson Bay.
[e] Including Aegean Sea.
[f] Including Black Sea.
[g] In the Skagerrak area.

Source: Millero, F. J. and Sohn, M. L., Chemical Oceanography, CRC Press, Boca Raton, FL, 1992, 6. With permission.

Table 1.3–2 Depth Zones in the Oceans

Ocean Area[a]	Depth Zone (km)												World Ocean
	0–0.2	0.2–1	1–2	2–3	3–4	4–5	5–6	6–7	7–8	8–0	9–10	10–11	
Pacific Ocean[b]	1.631	2.583	3.250	6.856	21.796	34.987	26.884	1.742	0.188	0.063	0.019	0.001	45.919
Austral Asiatic	51.913	9.255	10.433	12.151	6.698	7.780	1.636	0.076	0.058	0	0	0	2.509
Mediterranean Seas[c]													
Bering Sea	46.443	5.975	7.623	10.330	29.629	0	0	0	0	0	0	0	0.625
Sea of Okhotsk	26.475	39.479	22.383	3.403	8.260	0	0	0	0	0	0	0	0.384
East China Sea[d]	81.305	11.427	5.974	1.239	0.055	0	0	0	0	0	0	0	0.332
Sea of Japan	23.498	15.176	19.646	20.096	21.551	0.033	0	0	0	0	0	0	0.280
Gulf of California	46.705	20.848	25.891	6.556	0	0	0	0	0	0	0	0	0.042
Atlantic Ocean[b]	7.025	5.169	4.295	8.590	19.327	32.452	22.326	0.738	0.067	0.012	0	0	23.909
Arctic Med.[e]	47.083	17.427	9.317	11.153	12.834	2.195	0	0	0	0	0	0	3.386
American Med.	23.443	10.674	13.518	15.313	20.796	13.440	2.572	0.193	0.051	0	0	0	1.203
European Med.[f]	22.868	20.814	18.362	30.326	7.426	20.204	0	0	0	0	0	0	0.834
Baltic Sea	99.832	0.168	0	0	0	0	0	0	0	0	0	0	0.105
Indian Ocean[b]	3.570	2.685	3.580	10.029	25.259	36.643	16.991	1.241	0.001	0	0	0	20.282
Red Sea	41.454	43.058	14.920	0.568	0	0	0	0	0	0	0	0	0.125
Persian Gulf	100.000	0	0	0	0	0	0	0	0	0	0	0	0.066
World Ocean	7.492	4.423	5.376	8.497	20.944	31.689	21.201	1.232	0.105	0.032	0.009	0.001	100.00

a As a percentage of the surface of each ocean.
b Without adjacent seas.
c Including Aegean Sea.
d Including Yellow Sea.
e Consisting of Arctic Ocean, Barents Sea, Canadian Archipelago, Baffin Bay, and Hudson Bay.
f Including Black Sea.

Source: Millero, F. J., *Chemical Oceanography,* 2nd ed., CRC Press, Boca Raton, FL, 1996, 4. With permission. Originally adapted from Dietrich, G. et al., *General Oceanography,* 2nd ed., John Wiley & Sons, New York, 1980.

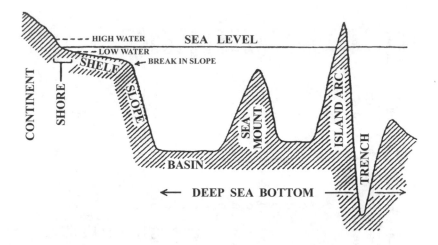

Figure 1.3–1   Generalized structure of the ocean bottom. (From Millero, F. J., *Chemical Oceanography*, 2nd ed., CRC Press, Boca Raton, FL, 1996, 5. With permission.)

Table 1.3–3   Benthic and Pelagic Marine Environments

Zone	Depth (m)	Volume (%)
Pelagic Environments[a]		
Epipelagic	0–200	3
Mesopelagic	200–1000	28
Bathypelagic	1000–2000	15
Abyssalpelagic	2000–6000	54
Hadalpelagic	>6000	<1

Zone	Depth (m)	Area (%)
Benthic Environments[a]		
Sublittoral	0–200	8
Bathyal	200–2000	16
Abyssal	2000–6000	75
Hadal	>6000	1

[a] Excludes the Arctic Ocean and includes all seas adjacent to other oceans.

Source: Pinet, P. R., *Invitation to Oceanography,* Jones and Bartlett Publishers, Boston, MA, 1998, 318. With permission.

Table 1.3–4   Average Temperature and Salinity of the Oceans[a]

	Temperature (°C)	Salinity (parts per thousand)
Pacific (total)	3.14	34.60
North Pacific	3.13	34.57
South Pacific	3.50	34.63
Indian (total)	3.88	34.78
Atlantic (total)	3.99	34.92
North Atlantic	5.08	35.09
South Atlantic	3.81	34.84
Southern Ocean	0.71	34.65
World Ocean (total)	3.51	34.72

[a] Excluding marginal seas.

Source: Worthington, L. V., in *Evolution of Physical Oceanography*, Warren, B. A. and Wunsch, C., Eds., MIT Press, Cambridge, MA, 1981, 42. With permission.

Table 1.3–5 Water Sources for the Major Ocean Basins[a]

Ocean	Precipitation	Runoff from Adjoining Land Areas	Evaporation	Water Exchange with Other Oceans
Atlantic	78	20	104	6
Arctic	24	23	12	35
Indian	101	7	138	30
Pacific	121	6	114	13

[a] Values in cm/year.

Source: Budyko, M. I., The Heat Budget of the Earth's Surface, Office of Technical Services, Department of Commerce, Washington, D.C., 1958.

Table 1.3–6 Major River Discharges to the World Ocean Basins

River	Discharge 10^9 m³/year	Discharge mi³/year	Ocean Basin
Amazon	5550	1330	Atlantic
Congo	1250	300	Atlantic
Yangtze	688	165	Pacific
Ganges	590	141	Indian
Yenisei	550	130	Arctic
Mississippi	550	180	Atlantic
Lena	490	118	Arctic
St. Lawrence	446	107	Atlantic
Mekong	350	84	Pacific
Columbia	178	42.7	Pacific
Yukon	165	39.6	Pacific
Nile	117	28	Atlantic
Colorado	5	1.2	Pacific
Total world	30,000	7200	

Source: Based on data from the U.S. Geological Survey, Washington, D.C.

1.4 TOPOGRAPHIC DATA

HYPSOMETRY OF OCEAN BASIN PROVINCES

Physiographic provinces—These "provinces" are regions or groups of features that have distinctive topography and usually characteristic structures and relations to other provinces. Province boundaries are based on detailed physiographic diagrams where available (Heezen and Tharp, 1961, 1964; Heezen et al., 1959; Menard, 1964) supplemented by more generalized physiographic and bathymetric charts. Provinces do not overlap nor are they superimposed in this study. Thus the area of a volcano rising from an ocean basin is included only in province VOLCANO and excluded from province OCEAN BASIN. The provinces identified in this study, the capitalized province names used in the text, the abbreviations used in data processing, and the corresponding numbers appearing in illustrations are

1. Continental SHELF AND SLOPE (CONS), the whole region from the shoreline to the base of the steep continental slope. Shelf and slope are grouped because they are merely the top and front of the margins of continental blocks.
2. CONTINENTAL RISE and partially filled sedimentary basins (CNRI). Gently sloping or almost flat, they appear to have characteristic features resulting from the accumulation of a thick fill of sediment eroded from an adjacent continent and overlying an otherwise relatively normal oceanic

crust. In this respect, the Gulf of Mexico and the western basin of the Mediterranean differ from the continental rise off the eastern U.S. only because they are relatively enclosed.

3. OCEAN BASIN (OCBN), the remainder after removing all other provinces. Abyssal plains and abyssal hills and archipelagic aprons are common features of low relief.

4. Oceanic RISE and RIDGE (RISE), commonly called "mid-ocean ridges" despite the fact that they continue across ocean margins. They form one worldwide system with many branches. Boundaries are taken in most places as outer limit of essentially continuous slopes from crest.

5. RIDGE NOT KNOWN TO BE VOLCANIC (RIDG), relatively long and narrow and with steep sides. Most have unknown structure and some or most may be volcanic.

6. Individual VOLCANO (VOLC), with a boundary defined as the base of steep side slopes.

7. Island ARC AND TRENCH (TNCH), includes whole system of low swells and swales subparallel to trenches. Continental equivalents or extensions of islands arcs, such as Japan, are excluded.

8. Composite VOLCANIC RIDGE (VRCM), formed by overlapping volcanoes and with a boundary at the base of steep side slopes.

9. POORLY DEFINED ELEVATION (BLOB), with nondescript side slopes and length no more than about twice width. Crustal structure unknown; may be thin continental type.

Tabulation of data and measuring procedure—Data were tabulated by 10° squares of latitude and longitude. Squares containing more than one ocean were split, and each ocean was treated separately. Within a square, the areas between the depth intervals 0–200 m, 200–1,000 m, and between 1-km contours down to 11 km were compiled for each physiographic province.

The polar planimeters (Keuffel and Esser models 4236 and 4242) used for measuring areas were read to the nearest unit on the vernier scale, and measurements were tabulated directly. These values were converted to square kilometers during computer processing by a scale factor derived from a measurement of the total number of units in the square. The area of a square was calculated assuming a spherical earth with a radius of 6,371.22 km.

Depth distribution in different oceans as a function of provinces—The depth distribution in provinces in different ocean basins, as seen in Figure 1.4–12, closely resembles the composite distribution in the world ocean (Figure 1.4–8). The sum of the depth distribution of all provinces in an ocean basin is double peaked, but for the individual provinces it is single peaked and relatively symmetrical. However, the depth distributions are sufficiently different to warrant some discussion. The mean depths of all provinces in the three major ocean basins, including marginal seas, range from 3575 m for the Atlantic Ocean to 3940 m for the Pacific (Table 1.4–4). The range of mean depths of the OCEAN BASIN province in each of these ocean basins is similar. The smallest mean OCEAN BASIN depth of 4530 m in the Indian Ocean may be the result of epirogenic movement of the oceanic crust, but it is also partially attributable to sedimentation. The eastern and southwestern parts of the Indian Ocean are deeper than 5000 m and are thus below the mean depth of the world ocean. The northwestern and southeastern parts, however, are exceptionally shallow. Seismic stations and topography show that the northwestern region has been shoaled by deposition of turbidities spreading from the mouths of the great Indian and east African rivers (Menard, 1961; Heezen and Tharp, 1964).

The mean depths of RISE and RIDGE have a limited range: from 3945 m in the Indian Ocean to 4008 m in the Atlantic Ocean (Table 1.4–4). This uniformity seems remarkable considering the widespread and diverse evidence that oceanic rises and ridges are tectonically among the more unstable features of the surface of the earth. It is all the more remarkable because the local, relief, or elevation above the adjacent OCEAN BASIN differs substantially in different oceans. The relief of RISE AND RIDGE in an ocean basin can be estimated by subtracting the mean depth from that of OCEAN BASIN or by determining the deepening required to give a best fit of individual hypsometric curves for each province. Comparing the means gives the relief of RISE AND RIDGE ABOVE OCEAN BASIN as 585 m in the Indian Ocean, 662 m in the Atlantic, and 928 m in the Pacific. The reliefs from matching curves are 800, 900, and 1200 m, respectively. The greater relief obtained by the curve-matching method results from ignoring the shallow tails of the depth distribution. Thus the range in relief is about six times as great as the range in mean depths of RISE AND RIDGE, which may be explained if the seafloor is not only elevated by epirogeny, but is also depressed. It seems reasonable to assume that the depth intervals in OCEAN BASIN with the largest areas (4 to 5 and 5 to 6 km) are those underlain by normal crust and mantle. A uniform oceanic process in the mantle acting on a uniform oceanic crust at a uniform depth may produce oceanic rises and ridges of uniform depth. Many current hypotheses for the origin of rises and ridges suggest just such an elevation. However, it is at least implicit that the mantle under the ocean basins cannot become denser and thus epirogenically depress the crust. Moreover, it is assumed by advocates of convection that if the crust is dragged down dynamically it forms a long narrow oceanic trench. The symmetrical distribution curves for OCEAN BASIN indicate that a considerable area is below the most common depth interval. Very extensive regions deeper than 6000 m exist in the northwestern Pacific and eastern Indian oceans, and there may be places where the normal oceanic crust is epirogenically depressed by more than a kilometer below the 4753-m mean depth of the OCEAN BASIN for the world ocean. Formation of broad depressions would alter the depth distribution in OCEAN BASIN and thereby vary the relief of RISE AND RIDGE in different oceans. If these broad epirogenic depressions exist, they may have a significant effect on the possible range of sea level changes relative to continents. This will be considered under "Discussion."

Depth distribution in island arc and trench provinces—These provinces have been defined to include not only trenches, but also the subparallel low swells and the island arcs which rise above some of them. The justification for this definition is that these features probably are caused by the same process; one question that can be answered by this type of study is whether the process elevates or depresses the seafloor. The volcanoes, some capped with limestone, which form most islands in this province, have a rather minor volume and have hardly any effect on the hypsometry.

The median depth for island ARC AND TRENCH is somewhat less than 4 km, which is less than the median depth for all ocean basins and considerably less than for the OCEAN BASIN province. The average depth would be much shallower if it were possible in some simple way to include the elevations above normal continents of the mountain ranges parallel to the Peru-Chile, Central America, Japan, and Java trenches. This would require some elaborate assumptions, but it is clear that the process that forms trenches and related features generally elevates the crust.

Volume of the ocean—Murray (1888) calculated the volume of the ocean at 323,722,150 cubic miles, which equals about 1.325×10^9 km³. Kossinna (1921) obtained 1.370×10^9 km³, and we obtain 1.350×10^9 km³. It appears unlikely that this value is in error by more than a few percent. Our method of calculation is essentially the same as that of Murray and Kossinna. The midpoint value of a depth interval is multiplied by the area of that interval, and the volumes of the intervals are summed.

DISCUSSION

Seafloor epirogeny and sea level changes—Seafloor epirogeny is only one of a multitude of causes of sea level change of which the wax and wane of glaciers is probably the most intense. Epirogeny is especially important because it may have occurred at any time in the history of the earth in contrast to relatively brief periods of glaciation. That eustatic changes in sea level have occurred during geological time is suggested by widespread epicontinental seas alternating with apparently high continents.

The hypothesis that oceanic rises are ephemeral (Menard, 1958) provides a basis for quantitative estimates of epirogenic effects on sea level. If the approximate volume of existing rises and ridges is compared with the area of the oceans, it appears that uplift of the existing rises has elevated the sea level 300 m. Likewise, subsidence of the ancient Darwin rise has lowered it by 100 m (Menard, 1964).

The present study suggests that the seafloor may be depressed epirogenically in places where this movement does not merely restore the equilibrium disturbed by a previous uplift. The argument derives from the fact that the mean depth in the OCEAN BASIN province is about 4700 m. Considering that the crust has about the same thickness everywhere in the province, variations from this depth generally are caused by differences in density in the upper mantle. (We assume that where the mean depth of the crust is "normal" it is underlain by a "normal" mantle.) Thus the deeper regions, which are roughly 70 million km^2 in area, have been depressed by a density increase in the mantle. If large areas of the seafloor can be depressed as well as elevated, the resulting changes in sea level would be highly complex.

Only the most general conclusions can be drawn from this analysis, but they may be significant. First, a plausible mechanism is available to explain the eustatic changes in sea level observed in the geological record. At present, the mechanism places no constraints on the sign of a change, but appears to limit the amount to a few hundred meters. Second, in large regions the upper mantle may possibly become denser than normal. Substantial evidence exists that it is less dense than normal under rises and ridges (Le Pichon et al., 1965). If it can also be more dense than normal in large regions, these facts can provide very useful clues regarding the composition of the upper mantle and processes acting below the crust. The implications of possible densification of normal mantle can be avoided by defining the "normal" depth as the deepest that is at all widespread. If this definition is accepted as reasonable (it does not appear so to us), small decreases in density of the upper mantle occur under most of the world ocean. The volume of ocean basin elevated above normal is consequently large, and the possible range of sea level changes is thus at least 1 km.

Seafloor spreading and continental drift—Several aspects of our data appear to have some bearing on modern hypotheses of global tectonics. The relationships are not definitive, however, and at this time we prefer merely to indicate some of the equations which have arisen.

1. The proportion of RISE AND RIDGE to OCEAN BASIN in a basin could range from zero to infinity, but it is 0.84 for the Pacific, 0.82 for the Atlantic, and 0.61 for the Indian Ocean. The sample is very small, and consequently the similarity of the proportions may be coincidental. However, it suggests that the area of RISE AND RIDGE is proportional to the whole area of an ocean basin. This in turn suggests that the size of the basin is related to the existence of rises and ridges.

2. The proportion of SHELF AND SLOPE to OCEAN BASIN plus RISE AND RIDGE is relatively constant for large ocean basins and quite dif021ferent from the proportion for small ocean basins. This relationship may require modification of at least many of the details of the hypothesis that the Atlantic Ocean basin was formed when an ancient continent split. When the supposed splitting began, the whole basin was SHELF AND SLOPE. Consequently, the proportion of SHELF AND SLOPE has since decreased. In the Pacific basin, on the other hand, the proportion of SHELF AND SLOPE to OCEAN BASIN plus RISE AND RIDGE was smaller than now and has since increased. If the Atlantic split apart at a constant rate and is still splitting as the Pacific contracts, the present equality of the proportions of SHELF AND SLOPE in the two ocean basins requires a striking coincidence. No coincidence is necessary if the splitting occurred relatively rapidly until it reached some dynamic equilibrium state, perhaps when the proportion of RISE AND RIDGE to OCEAN BASIN in each ocean basin reached about 0.8 to 0.9.

Source: Menard, H. W. and Smith, S. M., Hypsometry of ocean basin provinces. *J. Geophys. Res.,* 71, 4305, 1966. With permission of American Geophysical Union.

REFERENCES

Heezen, B. C. and Tharp, M., *Physiographic Diagram of the South Atlantic,* Geological Society of America, New York, 1961.

Heezen, B. C. and Tharp, M., *Physiographic Diagram of the Indian Ocean,* Geological Society of America, New York, 1964.

Heezen B. C., Tharp, M., and Ewing, M., The Floors of the Ocean. 1. The North Atlantic, Geological Society of America, New York, Special paper 65, 1959.

Kossinna, E., Die Tiefen des Weltmeeres, *Inst. Meereskunde, Veroff. Geogr. Naturwiss.,* 9, 70, 1921.

Le Pichon, X., Houtz, R. E., Drake, C. L., and Nafe, J. E., Crustal structure of the mid-ocean ridges, 1, Seismic refraction measurements, *J. Geophys. Res.,* 70(2), 319, 1965.

Menard, H. W., Development of median elevations in ocean basins, *Bull. Geol. Soc. Am.,* 69(9), 1179, 1958.

Menard, H. W., *Marine Geology of the Pacific,* McGraw-Hill, New York, 1964.

Murray, J., On the height of the land and the depth of the ocean, *Scot. Geogr. Mag.,* 4, S. 1, 1888.

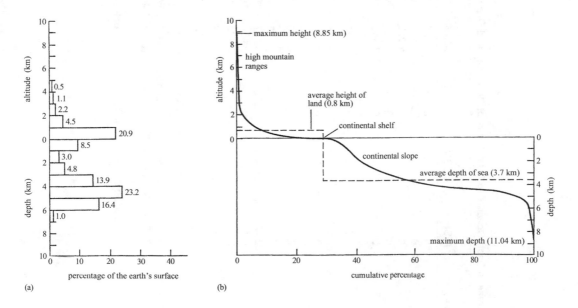

Figure 1.4–1   The distribution of levels on the earth's surface. (a) A histogram showing the actual frequency distribution. (b) A hypsographic curve for the earth. (From Brown, J., et al., *The Ocean Basins: Their Structure and Evolution*, Pergamon Press, Oxford, 1989, 27. With permission.)

Table 1.4–1 Province Areas in Each Ocean and Total Areas of Provinces and Oceans (10^6 km²)

Oceans and Adjacent Seas	RISE	OCBN	VOLC	CONS	TNCH	CNRI	VRCM	RIDG	BLOB	Total Area of Each Ocean
Pacific Ocean	65.109	77.951	2.127	11.299	4.757	2.690	1.589	0.494	0.227	166.241
Asiatic Mediterranean	0	0	0.003	7.824	0.023	1.233	0	0	0	9.082
Bering Sea	0	0	0	1.286	0.281	0.694	0	0	0	2.261
Sea of Okhotsk	0	0	0	1.254	0.023	0.115	0	0	0	1.392
Yellow and East China Seas	0	0	0	1.119	0.082	0	0	0	0	1.202
Sea of Japan	0	0	0.005	0.798	0	0.210	0	0	0	1.013
Gulf of California	0.042	0	0	0.111	0	0	0	0	0	0.153
Atlantic Ocean	30.519	35.728	0.882	12.658	0.447	5.381	0	0.412	0.530	86.557
American Mediterranean	0	1.346	0.060	1.889	0.201	0.861	0	0	0	4.357
Mediterranean	0	0	0	1.465	0	1.046	0	0	0	2.510
Black Sea	0	0	0	0.263	0	0.245	0	0	0	0.508
Baltic Sea	0	0	0	0.382	0	0	0	0	0	0.382
Indian Ocean	22.426	36.426	0.358	6.097	0.256	4.212	0.407	2.567	0.679	73.427
Red Sea	0	0.070	0	0	0.383	0	0	0	0	0.453
Persian Gulf	0	0	0	0.238	0	0	0	0	0	0.238
Arctic Ocean	0.513	0	0	5.874	0	2.267	0.302	0	0.528	9.485
Arctic Mediterranean	0	0	0	2.483	0	0.289	0	0	0	2.772
Total area each province	118.607	151.522	3.435	55.421	6.070	19.242	2.298	3.473	1.965	362.033

Note: CONS, Continental shelf and slope; CNRI, continental rise and partially filled sedimentary basins; OCBN, ocean basin; RISE, oceanic rise and ridge; RIDG, ridge not known to be volcanic; VOLC, individual volcano; TNCH, island arc and trench; VRCM, composite volcanic ridge; BLOB, poorly defined elevation.

Source: Menard, H. W. and Smith, S. M., *J. Geophys. Res.*, 71, 4305, 1966. With permission of American Geophysical Union.

Table 1.4–2 Percent of Provinces in Oceans and Adjacent Seas

Oceans and Adjacent Seas	RISE	OCBN	VOLC, VRCM, RIDG, BLOB	CONS	TNCH	CNRI	Percent of World Ocean in Each Ocean Group
Pacific and adjacent seas	35.9	43.0	2.5	13.1	2.9	2.7	50.1
Atlantic and adjacent seas	32.3	39.3	2.0	17.7	0.7	8.0	26.0
Indian and adjacent seas	30.2	49.2	5.4	9.1	0.3	5.7	20.5
Arctic and adjacent seas	4.2	0	6.8	68.2	0	20.8	3.4
Percent of world ocean in each province	32.7	41.8	3.1	15.3	1.7	5.3	

Note: CONS, Continental shelf and slope; CNRI, continental rise and slope; CNRI, continental rise and partially filled sedimentary basins; OCBN, ocean basin; RISE, oceanic rise and ridge; RIDG, ridge not known to be volcanic; VOLC, individual volcano; TNCH, island arc and trench; VRCM, composite volcanic ridge; BLOB, poorly defined elevation.

Source: Menard, H. W. and Smith, S. M., *J. Geophys. Res.*, 71, 4305, 1966. With permission of American Geophysical Union.

Figure 1.4–2   Pacific Ocean—physiographic provinces. Text contains key to province numbers. Individual volcanoes (VOLC) in black. (From Menard, H. W. and Smith, S. M., *J. Geophys. Res.*, 71, 4305, 1966. With permission of American Geophysical Union.)

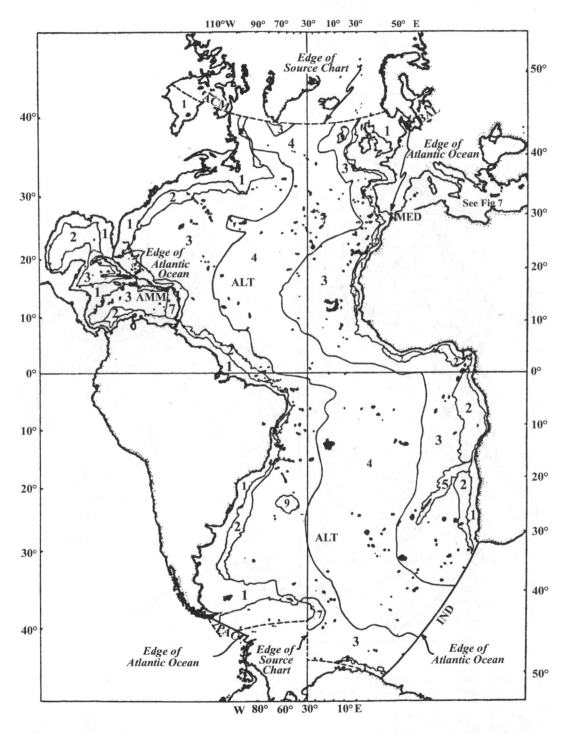

Figure 1.4–3 Atlantic Ocean—physiographic provinces. (From Menard, H. W. and Smith, S. M., *J. Geophys. Res.*, 71, 4305, 1966. With permission of American Geophysical Union.)

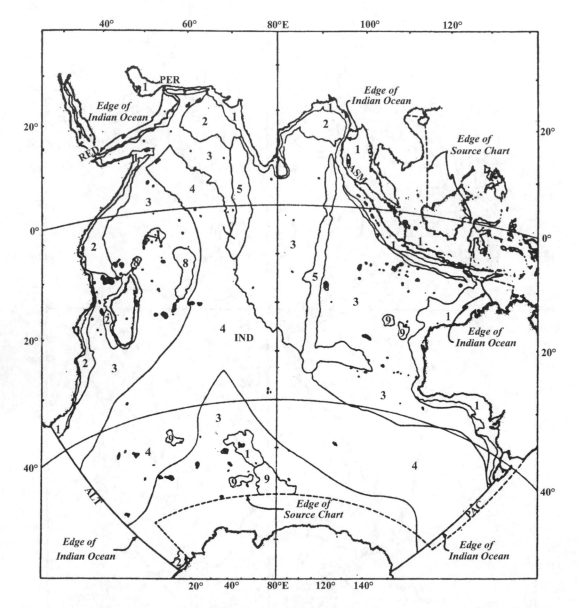

Figure 1.4–4   Indian Ocean—physiographic provinces. (From Menard, H. W. and Smith, S. M., *J. Geophys. Res.*, 71, 4305, 1966. With permission of American Geophysical Union.)

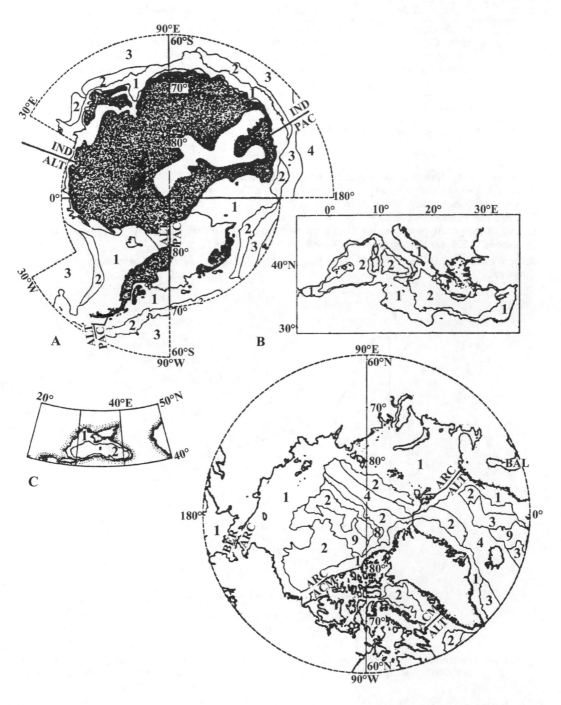

Figure 1.4–5 Smaller oceans and seas—physiographic provinces. Antarctic sub-ice in white is below sea level. (From Menard, H. W. and Smith, S. M., *J. Geophys. Res.*, 71, 4305, 1966. With permission of American Geophysical Union.)

Table 1.4–3   Bathymetric Charts Used for Hypsometric Calculations

Source No.	Title	Scale	Projection	Ref.
1	Pacific Ocean	1:7,270,000[a]	Lambert azimuthal equal-area	A
2	Indian Ocean	1:7,510,000[a]	Lambert azimuthal equal-area	B
3	Antarctica	1:9,667,000	Polar azimuthal equal-area	C
4	Atlantic Ocean	1:10,150,000[a]	Lateral projection with oval isoclines	B
5	Tectonic Chart of the Arctic	1:10,000,000	Polar azimuthal equal-area	D
6	Mediterranean Sea	1:2,259,000	Mercator	E
7	Northern Hemisphere	1:25,000,000	Polar azimuthal equal-area	B

[a] Scale of photographic enlargement used for measuring.

REFERENCES

A. Menard, H. W., *Marine Geology of the Pacific,* McGraw-Hill, New York, 1964.
B. Main Administration in Geodesy and Cartography of the Government Geological Committee, USSR.
C. American Geographical Society, New York.
D. Geological Institute, Academy of Science, Moscow.
E. Unpublished chart of the Mediterranean, modified from contours compiled by R. Nason from various sources. U.S. Navy Hydrographic Office chart 4300 used as base.

Source: Menard, H. W. and Smith, S. M., *J. Geophys. Res.*, 71, 4305, 1966. With permission of American Geophysical Union.

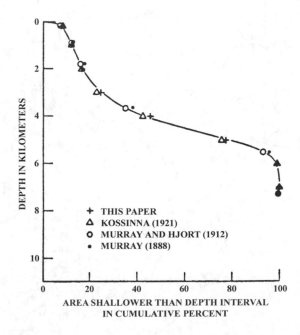

Figure 1.4–6   Hypsometry of all ocean basins according to various studies (see References). (From Menard, H. W. and Smith, S. M., *J. Geophys. Res.*, 71, 4305, 1966. With permission of American Geophysical Union.)

REFERENCES

1. Menard, H. W. and Smith, S. M., Hypsometry of ocean basin provinces, *J. Geophys. Res.*, 71, 4305, 1966.
2. Kossinna, E., Die Tiefen des Weltmeeres, *Inst. Meereskunde, Veroff. Geogr. Naturwiss.*, 9, 70, 1921.
3. Murray, J. and Hijort, J., *The Depths of the Ocean,* Macmillan, London, 1912.
4. Murray, J., On the height of the land and the depth of the ocean, *Scot. Geogr. Mag.*, 4, S. 1, 1888.

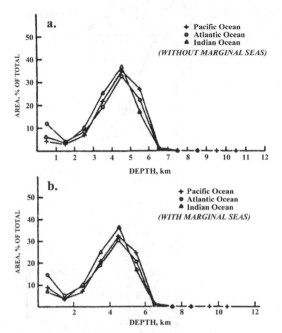

Figure 1.4–7 Hypsometry of individual major ocean basins. (From Menard, H. W. and Smith, S. M., *J. Geophys. Res.*, 71, 4305, 1966. With permission of American Geophysical Union.)

Figure 1.4–8 Hypsometry of all ocean basins. This diagram is for all provinces combined (ALLP) and for individual major provinces: CONS, continental shelf and slope; CNRI, continental rise; RISE, oceanic rise and ridge; OCBN, ocean basin. (From Menard, H. W. and Smith, S. M., *J. Geophys. Res.*, 71, 4305, 1966. With permission of American Geophysical Union.)

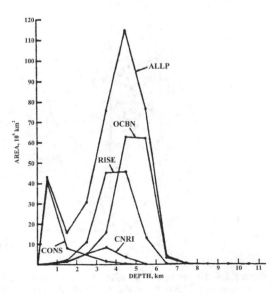

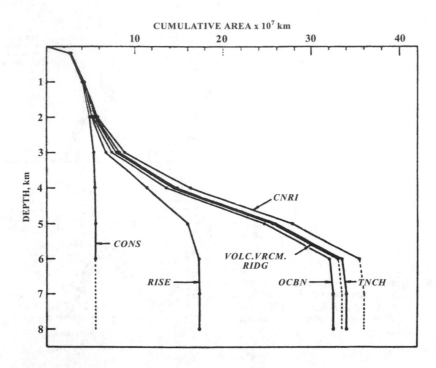

Figure 1.4–9   Hypsometry of ocean basins plotted cumulatively by provinces. CONS, continental shelf and slope; RISE, oceanic rise and ridge; CNRI, continental rise; VOLC, individual volcano; VRCM, composite volcanic ridge; RIDG, ridge not known to be volcanic, OCBN, ocean basin; TNCH, island arc and trench. (From Menard, H. W. and Smith, S. M., *J. Geophys. Res.*, 71, 4305, 1966. With permission of American Geophysical Union.)

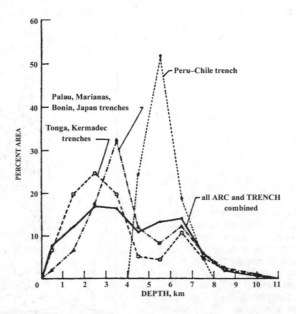

Figure 1.4–10   Hypsometry of all arc and trench provinces and of some groups of arcs and trenches. (From Menard, H. W. and Smith, S. M., *J. Geophys. Res.*, 71, 4305, 1966. With permission of American Geophysical Union.)

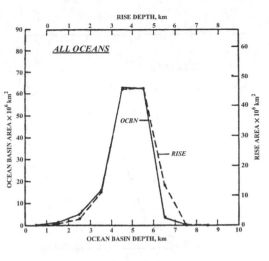

Figure 1.4–11 Hypsometric curve of ocean basins. Hypsometric curve of all ocean basins for RISE and RIDGE province normalized to curve for OCEAN BASIN province to show close similarity. (From Menard, H. W. and Smith, S. M., *J. Geophys. Res.*, 71, 4305, 1966. With permission of American Geophysical Union.)

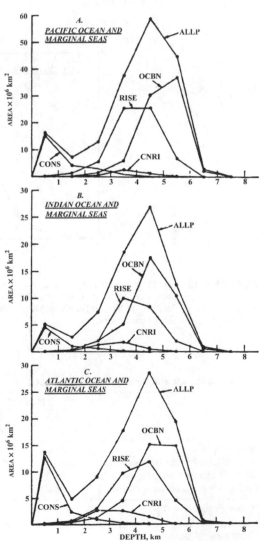

Figure 1.4–12 Hypsometry of all provinces (ALLP) and individual provinces in major basins. (From Menard, H. W. and Smith, S. M., *J. Geophys. Res.*, 71, 4305, 1966. With permission of American Geophysical Union.)

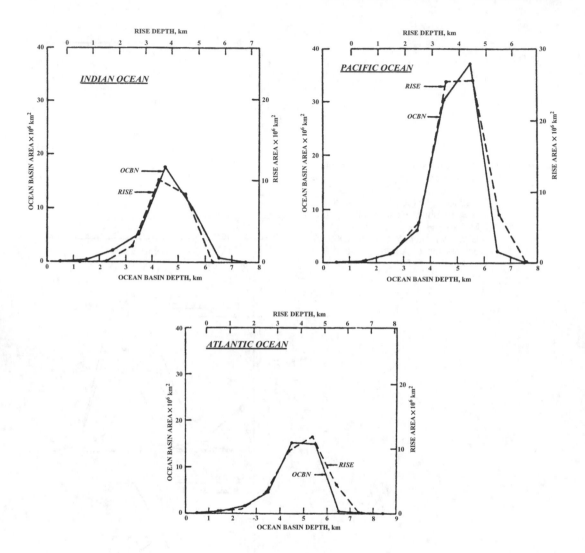

Figure 1.4–13 Hypsometric curves for rise and ridge provinces. Hypsometric curves for RISE and RIDGE provinces in major ocean basins normalized to curves for OCEAN BASIN provinces. (From Menard, H. W. and Smith, S. M., *J. Geophys. Res.*, 71, 4305, 1966. With permission of American Geophysical Union.)

Table 1.4–4   Characteristics of Ocean Rises

	Mean Depth, m			Relief, m	
	All Provinces	OCBN (1)	RISE (2)	Mean (1) – Mean (2)	Shift of Distribution Curves[a]
World ocean	3729	4753	3970	783	1000
Pacific Ocean and marginal seas	3940	4896	3968	928	1100
Atlantic Ocean and marginal seas	3575	4670	4008	662	900
Indian Ocean and marginal seas	3840	4530	3945	585	800

[a] See Figure 1.4–11.

Source: Menard, H. W. and Smith, S. M., *J. Geophys. Res.*, 71, 4305, 1966. With permission of American Geophysical Union.

Table 1.4–5   **Major Deeps and Their Location, Size, and Depths**

Trench	Max Depths, m	Ref.
Marianas Trench (specifically Challenger Deep)	11,034 ± 50	5
	10,915 ± 20[a]	
	10,915	8
	10,863 ± 35	1
	10,850 ± 20[a]	
Tonga	10,882 ± 50	5
	10,800 ± 100	4
Kuril-Kamchatka	10,542 ± 100	13
	9,750 ± 100[b]	2
Philippine (vicinity of Cape Johnson Deep)	10,497 ± 100	6
	10,265 ± 45	16
	10,030 ± 10[a]	
Kermadec	10,047	14
Idzu-Bonin		
(includes "Ramapo Deep" of the Japan Trench)	9,810	13
(vicinity of Ramapo Depth)	9,695	11
Puerto Rico	9,200 ± 20	9
New Hebrides (North)	9,165 ± 20[a]	
North Solomons (Bougainville)	9,103	14
	8,940 ± 20[a]	
Yap (West Caroline)	8,527	7
New Britain	8,320	14
	8,245 ± 20[a]	
South Solomons	8,310 ± 20[a]	
South Sandwich	8,264	10
Peru-Chile	8,055 ± 10	3
Palau	8,054	14
	8,050 ± 10[a]	
Aleutian (uncorrected, taken with nominal sounding		
velocity of 1500 m/s)	7,679	
Nansei Shoto (Ryuku)	7,507	
Java	7,450	15
New Hebrides (South)	7,070 ± 20[a]	
Middle America	6,662 ± 10	

[a] These soundings were taken during Proa Expedition, April–June, 1962, aboard R. V. *Spencer F. Baird.* A Precision Depth Recorder was employed, and the ship's track crossed over (within the limits of celestial navigation) points from which maximum depths had been reported.

[b] This is the maximum sounding obtained in the vicinity of the Vitiaz Depth (Udintsev, 1959) by French and Japanese vessels in connection with dives of the bathyscaph *Archimède,* July, 1962.

REFERENCES

1. Carruthers, J. N. and Lawford, A. L., The deepest oceanic sounding, *Nature,* 169, 601, 1952.
2. Delauze, personal communication, 1962.
3. Fisher, R. L., in *Preliminary Report on Expedition Downwind,* I.G.Y. General Report Ser., 2, I.G.Y. World Data Center A, Washington, D.C., 1958.
4. Fisher, R. L. and Revelle, R., A deep sounding from the southern hemisphere, *Nature,* 174, 469, 1954.
5. Hanson, P. P., Zenkevich, N. L., Sergeev, U. V., and Udintsev, G. B., Maximum depths of the Pacific Ocean, *Priroda (Mosk),* 6, 84, 1959 (in Russian).
6. Hess, H. H. and Buell, M. W., The greatest depth in the oceans, *Trans. Am. Geophys. Union,* 31, 401, 1950.

7. Kanaev, V. F., New data on the bottom relief of the western part of the Pacific Ocean, *Oceanol. Res.,* 2, 33, 1960 (in Russian).

8. Lyman, J., personal communication, 1960.

9. Lyman, J., The deepest sounding in the North Atlantic, *Proc. R. Soc. London,* A222, 334, 1954.

10. Maurer, H. and Stocks, T., Die Echolötungen des Meteor. *Wiss. Ergebn., Deut. Atlant. Exped. 'Meteor,' 1925–27,* 2, 1, 1933.

11. Nasu, N., Iijima, A., and Kagami, H., Geological results in the Japanese Deep Sea Expedition in 1959, *Oceanogr., Mag.,* 11, 201, 1960.

12. Udintsev, G. B., Discovery of a deep-sea trough in the western part of the Pacific Ocean, *Priroda (Mosk),* 7, 85, 1958 (in Russian).

13. Udintsev, G. B., Relief of abyssal trenches in the Pacific Ocean (abstr.), *Intern. Oceanogr. Cong. Preprints,* Am. Assoc. Adv. Sci., Washington, D.C., 1959.

14. Udintsev, G. B., Bottom relief of the western part of the Pacific Ocean, *Oceanol. Res.,* 2, 5, 1960 (in Russian).

15. van Riel, P. M., The bottom configuration in relation to the flow of bottom water, in *The 'Snellius' Exped.,* E. J. Brill, Leiden 2(2), chap. 2, 1933.

16. Wiseman, J. D. H. and Ovey, C. D., Proposed names of features on the deep-sea floor, *Deep-Sea Res.,* 2, 93, 1955.

Source: Hill, M. N., Ed., *The Sea,* Vol. III, Wiley-Interscience, New York, 1963. With permission. In part after the compilation of Wiseman and Ovey[16].

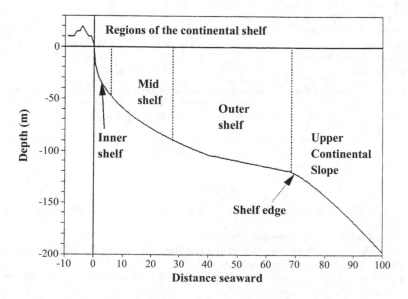

Figure 1.4–14   Subdivisions of the continental shelf. (From Wright, L. D., *Morphodynamics of the Inner Continental Shelf,* CRC Press, Boca Raton, FL, 1995, 2. With permission.)

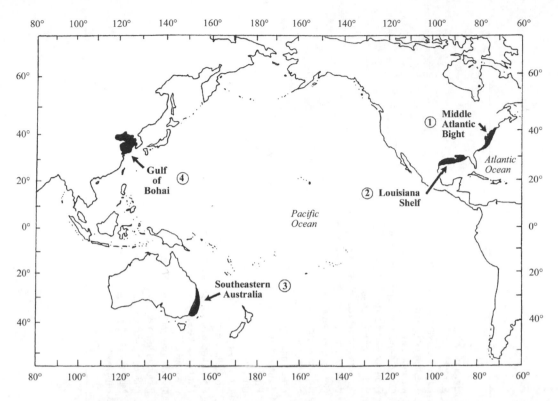

Figure 1.4–15 World map showing the locations of four contrasting shelves. (From Wright, L. D., *Morpho-dynamics of the Inner Continental Shelf*, CRC Press, Boca Raton, FL, 1995, 33. With permission.)

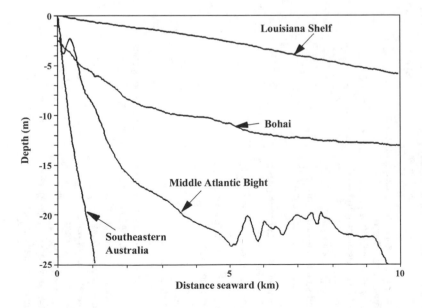

Figure 1.4–16 Four contrasting inner continental shelf profiles: (1) Middle Atlantic Bight; (2) Louisiana Shelf; (3) Southeastern Australia (Sydney region); and (4) Western Gulf of Bohai (China). (From Wright, L. D., *Morphodynamics of the Inner Continental Shelf*, CRC Press, Boca Raton, FL, 1995, 33. With permission.)

Table 1.4–6   Submarine Canyons: Location, Size, Depth

Symbol Key for Table

1. Length of canyon measured along axis (nautical miles)
2. Depth at canyon head (feet)
3. Depth at canyon terminus (feet)
4. Character of coast inside canyon head
   A. Heads in estuary
   B. Heads off embayment
   C. Heads off straight beach or barrier
   D. Heads off relatively straight cliff
   E. Uncertain
5. Relation of canyon head to points of land
   A. On upcurrent side of point
   B. Relatively near upcurrent side of point
   C. No relation to point
6. Relation of canyon head to river valleys
   A. Probable connection
   B. No connection
   C. Uncertain
7. Source of sediments to canyon head
   A. Receives good supply
   B. Supply restricted now, greater during lowered sea level stages
   C. Little known supply of sediment because of depth
8. Gradient of axis in meters per kilometer
9. Nature of longitudinal profile
   A. Generally concave upward
   B. Generally convex upward
   C. Relatively even slope
   D. Local steplike steepening along axis
10. Maximum height of walls in feet

11. Channel curvature
    A. Straight
    B. Slightly curving
    C. Twisting or winding
    D. Meandering
    E. One meandering bend
    F. Right-angled bends
12. Abundance of tributaries
    A. As common as typical land valleys
    B. Less common than typical land valleys
    C. Confined to canyon head
    D. No known tributaries
13. Character of transverse profile
    A. Predominantly V-shaped
    B. V-shaped inner canyon, trough-shaped outer canyon
    C. Predominantly trough-shaped
    D. Uncertain
14. Nature of canyon wall material
    A. Crystalline rock dredged
    B. Rock dredged, but all sedimentary
    C. Mud only dredged on wall
    D. Unknown
15. Nature of core sediment from axis
    A. Includes sand layers
    B. Includes sand and gravel layers
    C. Mud cores only
    D. Unknown

Canyon Name and Location	1 Canyon Length	2 Depth at Canyon Head	3 Depth at Canyon Terminus	4 Coast Character	5 Relation to Points	6 Relation to River Valleys	7 Sediment Sources at Head	8 Gradient in m/km
California								
Coronado	8.0	240	5,580	C	C	C	B	58
La Jolla	7.3	50	1,800	C	B	A	A	40
Scripps (tributary)	1.45	60	900	D	C	A	A	97
Redondo	8.0	30	1,920	C	B	B	A	39
Dume	3.0	120	1,860	C	A	A	A	97
Mugu	8.0	40	2,400	C	B	C	A	49
Sur Partington	49.0	300	10,200	D	C	C	A	34
Carmel (tributary)	15.0	30	6,600	A	A	A	A	73
Monterey	60.0	50?	9,600?	C	C	B	A	26.5
Delgada	55.0	90	8,400	D	C	B	A	25
Mattole	16.0	60	5,720	B	B	A	A	59
Eel	27.0	250	8,500	C	B	A	B	51
Total or Average (12)	21.5	110	5,290	1A, 1B, 7C, 3D	2A, 4B, 6C	6A, 3B, 3C	10A, 2B	54
Oregon-Washington								
Columbia	37.0?	360	6,130?	B	B	A	C	26
Willapa	60.0	500+	7,000	B	C	A	C	24
Gray	30.0	500	6,440	B	C	A	C	33
Quinault	25.0	500±	5,750	C	C	A	C	35
Juan de Fuca	31.0+	800±	4,520+	B	B	C	C	20
Total or Average (5)	36.6	532	5,968	4B, 1C	2B, 3C	4A, 1C	5C	27.6
Bering Sea								
Umnak	160.0	900	10,850	B	C	C	C	10.4
Bering	220.0	600±	11,160	B	C	C	C	8
Pribilof	86.0	500±	10,700	C?	C	C	C	20
Total or Average (3)	155.3	667	10,903	2B, 1C	3C	3C	3C	12.8
U.S. East Coast								
Corsair	14+	360	5,400	E	C	C	B	23
Lydonia	16+	370	4,400+	E	C	C	B	42
Gilbert	20+	480	7,680+	E	C	C	C	60
Oceanographer	17+	600+	7,230+	E	C	C	C	65
Welker	27+	400	6,450+	E	C	C	B	38

(continued)

Table 1.4–6   Submarine Canyons: Location, Size, Depth (continued)

Canyon Name and Location	1 Canyon Length	2 Depth at Canyon Head	3 Depth at Canyon Terminus	4 Coast Character	5 Relation to Points	6 Relation to River Valleys	7 Sediment Sources at Head	8 Gradient in m/km
U.S. East Coast (continued)								
Hydrographer	27+	450+	6,600+	E	C	C	B	37
Hudson	50	300	7,000	B	C	A	B	25
Wilmington	23+	320	6,940+	E	C	C	B	48
Baltimore	28+	400	6,110+	B?	C	C	B	34
Washington	28+	360	6,740+	E	C	C	B	38
Norfolk	38	320	8,300	E	C	C	B	35
Total or Average (11)	26.2	395	6,623	2B, 9E	11C	1A, 10C	9B, 2C	40
Hawaiian-Molokai								
Halawai	6.0+	300±	3,540+	B	C	A	B	90
Naiwa	7.5	380	4,880	D	C	A	B	100
Waikolu	9.0	<600	6,540	B	C	A	B	110
Pelekunu	10.0	<320	6,320	B	C	A	A	100
Hawaiian-Kauai								
Hanakapiai	6.0+	280	7,480	D	C	A	B	200
Hanakoa	3.7	600±	4,820	D	C	C	C	190
Hanopu	3.6	300	5,100	D	C	C	B	220
Total or Average (7)	6.5	397	5,526	3B, 4D	7C	5A, 2C	1A, 5B, 1C	144
Western Europe								
Shamrock	30+	1,200±	14,400	B?	C	C	C	28
Black Mud	30+	900±	12,200	E	C	C	B	57
Audierne	27	600±	10,500	B	C	C	C	60
Cap Ferret	50+	800	11,647	C	C	C	C	31
Cap Breton	135 or 70	400±	13,100	D	C	C	B	58
Aviles	65	60±	8,000	C	B	A (old river)	A	20 or 16
Llanes	38	450	13,300	D	B	C	B	45
Nazare	93	200±	14,764	C	A	A	A	36
Lisbon	21	400	6,450	B	B	A?	B	48
Setubal	33+	350	6,880	C	B	A	B	33
Total or Average (10)	52.2 or 45.7	536	11,224	3B, 4C, 2D	1A, 3B, 5C	4A, 5C	2A, 4B, 3C	41.6 or 41.2

Mediterranean Mainland							
Grand Rhone	15+	600±	5,550	C	C	A	C	55
Marseille	20+	600±	6,840±	B	B	C	C	52
Canon de la Cassidaigne	19+	360	6,630+	B	B	A	B	55
Toulon	12+	260	6,600	A borderline	C	C	B	110
Stoechades	17+	300	4,380+	A	C	C	A	40
St. Tropez	25+	60?	5,750	A borderline	C	A	A	38
Cannes	17+	100?	6,600?	C or A	C	B	A	65
Var	15	160	6,550	D delta	A	A	A	71
Nice	12	150	5,840	D	C	B	A	79
Cap d'Ail	14	320	6,870	B	C	C	B	78
Nervia	16	330	6,280	C	C	A	B	62
Taggia	12	300±	7,500	C	C	A	B	100
Mele	31	200	6,150	C curving	C	B	B	32
Noli	14	120	4,990	D curving	C	A?	A	58
Polcevera	49	300	8,830	C	C	A?	B	29
Genoa	20	260	6,260	B	C	A	B	50
Total or Average (16)	17.4	276.5	6,351	3A, 4B, 5C, 3D	1A, 1B, 14C	9A, 3B, 4C	7A, 7B, 2C	60.9
Mediterranean Islands							
Crete	4	<300	3,300	B?	C	A	A	200
West Corsica							
St. Florent	25	150±	7,850	A	A	A	A	51
Calvi	13	200±	7,800	B	C	A	B	97
Porto	20	150±	8,200	A	C	A	A	67
Sagone	29	150±	6,200	A	C	A	A	35
Ajaccio	34	150±	8,200	A	C	A	A	39
Valinco	35	150±	8,000	A	C	A	A	37
Total or Average (7)	22.8	178	7,078	5A, 2B	1A, 6C	7A	6A, 1B	75
Baja, California							
San Pablo	20+	<400	8,400+	D	A	B	A	67
Cardonal	16+	<450	7,500+	C	B	B	B	73
Vigia	10	?	7,200	C + D	C	A or B	A	115
San Lucas–	19	30	6,900?	A-bay	C	A	A	70
Santa Maria	(24)		8,000					56
San Jose	32	50	7,200?	C	C	A?	A	41
Vinorama–Salado	9	200	6,300	C	A	A	B	113
Los Frailes	9.5	10	5,200	A-bay	A (S. wind)	A	A	91

(continued)

Table 1.4–6 Submarine Canyons: Location, Size, Depth (continued)

Canyon Name and Location	1 Canyon Length	2 Depth at Canyon Head	3 Depth at Canyon Terminus	4 Coast Character	5 Relation to Points	6 Relation to River Valleys	7 Sediment Sources at Head	8 Gradient in m/km
Baja, California (continued)								
Saltito	6	1,200	5,100	B?	C	B	C	108
Palmas–	13	100?	5,300	C	A (S. wind)	C	A	65 or 91
Pescadero	9.3							
Total or Average (9)	15.2	305	6,710	2A, 1B, 4½ C, 1½ D	4A, 1B, 4C	4½ A, 3½ B, 1C	6A, 2B, 1C	81
East Honshu								
Ninomiya	4.8	400	2,600	C	C	C	B	77
Sagami	5.0	310	3,300	C	C	C	B	100
Enoshima	6.7	450	3,250	E	C	C	B	70
Hayama	13	300	4,600	E	C	C	B	54
Miura	15	173		E	C	C	B	48
Misaki	14.5	330	4,600	E	C	C	B	49
Jogashima	10	330	4,900	A	C	C	B	71
Tokyo	30	300	5,500	B(bay)	B	C	B	26
Mera	20	190	9,100	B	C	A	B	44
Kamogawa	25	200						59
Total or Average (10)	29.25	298	4,731	1A, 2B, 2C, 5E	1B, 9C	1A, 9C	10B	60
Miscellaneous								
Great Bahama	125	4,800	14,060	B	C	C	C	13
Congo	120	80	7,000	A	C	A	A	96
Ceylon Trincomalee	20+	30+	9,500+	A	A?	A	A	79
Manila	31+	300	7,800+	B	C	A	B	40
Bacarra NW Luzon	15	300	6,000+	C + D	C	A	B	63
San Antonio, Chile	20+	<150	2,700+	C?	C	A?	A	32
Total or Average (6)	55	943	7,843	2A, 2B, 1½ C, ½ D	1A, 5C	5A, 1C	3A, 2B, 1C	54

Canyon Name and Location	9 Nature of Long Profile	10 Max. Wall Heights to Nearest 1000 ft	11 Channel Curvature	12 Abundance of Tributaries	13 Transverse Profile Character	14 Name of Canyon Wall Material	15 Sediment Found in Axial Cores
California							
Coronado	A	1,000	C	B	A	B	A
La Jolla	A	1,000	C	A	A	B	B
Scripps (tributary)	A	<1,000	B	C	A	B	B
Redondo	A	1,000	C	B	C	B	A
Dume	A	1,000	C	B	A	A	D
Mugu	A	<1,000	D	A	A	B	A
Sur Partington	A	2,000	C	A	A	B	D
Carmel (tributary)	A	2,000	C	A	A	A	A
Monterey	A + D	6,000	C or D	A	B	A	B
Delgada	A	2,000	D	B	A	D	D
Mattole	A	3,000	C	B	A	B	D
Eel	B	4,000	E	B	A	D	B
Total or Average (12)	10A, 1B, 1D	2,083	½B, 8½C, 2½D, 1E	5A, 6B, 1C	10A, 1B, 1C	3A, 7B, 2D	4A, 3B, 5D
Oregon-Washington							
Columbia	A	2,000	C	B	D	D	D
Willapa	A	2,000	C	A	D	B	A
Gray	A	1,000	C	A	D	D	D
Quinault	A	3,000	C	A	D	D	D
Juan de Fuca	A	2,000	C	D	D	D	D
Total or Average (5)	5A	2,400	5C	3A, 1B, 1D	5D	4D, 1B	4D, 1A
Bering Sea							
Umnak	A	4,000	C	A	D	D	D
Bering	A + D	6,000	C	A	A	B?	D
Pribilof	A	7,000	C	A	D	D	D
Total or Average (3)	2½A, ½D	5,667	3C	3A	1A, 2D	1B, 2D	3D
U.S. East Coast							
Corsair	A	2,000	B	B	A	B	D
Lydonia	A	3,000	B	B	A	B	C
Gilbert	A + D	3,000	B	A	A	B	D
Oceanographer	A	2,000	B	B	A	D	D

(continued)

Table 1.4–6 Submarine Canyons: Location, Size, Depth (continued)

Canyon Name and Location	9 Nature of Long Profile	10 Max. Wall Heights to Nearest 1000 ft	11 Channel Curvature	12 Abundance of Tributaries	13 Transverse Profile Character	14 Name of Canyon Wall Material	15 Sediment Found in Axial Cores
U.S. East Coast (continued)							
Welker	A + D	4,000	B	B	A	B	A
Hydrographer	A + D	3,000	B	B	A	C	A
Hudson	A + D	4,000	B	A	A	B	A
Wilmington	A	3,000	C	A	A	C	C
Baltimore	A + D	3,000	B	B	A	C	C
Washington	A + D	2,000	B	A	A	C	C
Norfolk	A + D	3,000	B	B	A	B	A
Total or Average (11)	7½A, 3½D	2,900	10B, 1C	4A, 7B	11A	6B, 4C, 1D	4A, 4C, 3D
Hawaiian-Molokai							
Halawai	C	1,000	C	D	A	D	D
Naiwa	C	1,000	B	D	A	A	A
Waikolu	A	2,000	B	B	A	A	B
Pelekunu	A	1,000	C	B	A	D	D
Hawaiian-Kauai							
Hanakapiai	B + D	1,000	B	B?	A	A	D
Hanakoa	B	1,000	B	B?	A	A	A
Hanopu	B + D	2,000	B	B?	A	B	A
Total or Average (7)	2A, 2B, 2C, 1D	1,286	5B, 2C	5B, 2D	7A	4A, 1B, 2D	3A, 1B, 3D
Western Europe							
Shamrock	C + D	3,000	B	B?	D	D	D
Black Mud	D	3,000	B	B	B	B	A
Audierne	D	4,000	B	D	D	D	D
Cap Ferret	A	3,000	C	A	D	D	D
Cap Breton	A	6,000	B	B	D	D	D
Aviles	C + D	5,000 or 6,000	B	A?	B	B	D
Llanes	C	5,000	C	A	D	D	D
Nazare	D	5,000	F	B	D	A?	D
Lisbon	B + D	4,000	B	D	D	B	D
Setubal	C?	2,000	C?	B	D	B?	D
Total or Average (10)	2A, ½B, 3C, 3½D	4,000	5B, 3C, 1F	3A, 4B, 2D	1B, 8D	1A, 3B, 5D	9D

		Elevation					
Mediterranean Mainland							
Grand Rhone	C?	2,000	B	C	A	D	D
Marseille	A + D	2,000	C	A	A	D	D
Canon de la Cassidaigne	D	3,000	C	B	A	D	D
Toulon	D?	4,000	B	A	A	D	D
Stoechades	A	4,000	B	A	A	B?	D
St. Tropez	C?	3,000	C	A	A	D	D
Cannes	A	3,000	C	A	A	D	D
Var	A	3,000	C?	A	B	D	B
Nice	A + D	2,000	C?	B	B	B?	B
Cap d'Ail	D	1,000	B	B	B	D	D
Nervia	D	2,000	B	B	B	D	B
Taggia	A	2,000	C	B	A	D	D
Mele	A + D	1,000	C	B	B	D	D
Noli	A	2,000	B	B	B	D	D
Polcevera	A + D	3,000	B	B	B	D	D
Genoa	D	2,000	B	B	C?	D	D
Total or Average (16)	7A, 2C, 7D	2,400	7B, 9C	7A, 8B, 1C	8A, 7B, 1C	2B, 15D	3B, 13D
Mediterranean Islands							
Crete	D	1,000	C	B	A	D	B
West Corsica							
St. Florent	A	3,000	C	A	A	D	A
Calvi	A + D	3,000	C	A borderline	A	D	D
Porto	A	4,000	C	A	A	A?	D
Sagone	A	3,000	C	A	A	D	D
Ajaccio	A	4,000	B	A	A	D	D
Valinco	A + D	4,000	B	A	A	D	D
Total or Average (7)	6A, 1D	3,100	2B, 5C	5½A, 1½B	7A	1A, 6D	1A, 1B, 5D
Baja, California							
San Pablo	B + D	3,000	C	B	A	B	D?
Cardonal	A + D	3,000	C	A	A	B	D
Vigia	A	3,000	B	C	B	A	B
San Lucas–Santa Maria	A + D	3,000	C	A	A	A	B
San Jose	A + D	3,000	C	A	A	A (sed.)	B
Vinorama–Salado	A	1,000	C	A	A	A	B
Los Frailes	A + D	2,000	C	C	A	A	B

(continued)

Table 1.4–6 Submarine Canyons: Location, Size, Depth (continued)

Canyon Name and Location	9 Nature of Long Profile	10 Max. Wall Heights to Nearest 1000 ft	11 Channel Curvature	12 Abundance of Tributaries	13 Transverse Profile Character	14 Name of Canyon Wall Material	15 Sediment Found in Axial Cores
Baja, California (continued)							
Saltito	A	1,000	C?	A	A	A	D
Palmas–Pescadero	A	2,000	C	A	A	A	A
Total or Average (9)	6A, ½B, 2½D	2,333	1B, 8C	6A, 1B, 2C	8A, 1B	7A, 2B	1A, 5B, 3D
East Honshu							
Ninomiya	A	<1,000	B	C	A	B	D
Sagami	A	<1,000	C	A	B	B	D
Enoshima	A	<1,000	B	C	A	B	B
Hayama	A	2,000	C	B	A	A	D
Miura	A	2,000	C	B	A	B	B
Misaki	A	2,000	C	A	A	B	A
Jogashima	D	1,000	C	A	A	A	D
Tokyo	A + D	3,000	C	A	B	A	B
Mera	C + D	2,000	C	A	A	B	B
Kamogawa	A + D	5,000	B	B	B	B	A
Total or Average (10)	7A, ½C, 2½D	2,000	3B, 7C	5A, 3B, 2C	7A, 3B	3A, 7B	2A, 4B, 4D
Miscellaneous							
Great Bahama	A	14,000	C	A	A	C	B
Congo	A	4,000	B	C	A?	D	A sand
Ceylon Trincomalee	A	4,000	F	A	A	A?	D
Manila	A + D	6,000	C	A	D	D	D
Bacarra NW Luzon	C	3,000	C	A	B?	D	D
San Antonio, Chile	A	3,000	B + C	B	D	D	D
Total or Average (6)	4½A, 1C, ½D	5,666	1½B, 3½C, 1F	4A, 1B, 1C	3A, 1B, 2D	1A, 1C, 4D	1A, 1B, 4D

Source: Shepard, F. P. and Dill, R. F., *Submarine Canyons and Other Sea Valleys*, Rand McNally & Co., Chicago, 1966. With permission.

CHAPTER 2

Marine Chemistry

I. SEAWATER COMPOSITION

Seawater is more than simply dissolved ions in water. It consists of a complex mixture of inorganic salts, atmospheric gases, traces of organic matter, and small amounts of particulate matter. Seawater components can be subdivided into four phases:

1. Dissolved solutes
   a. Inorganic solutes
      1. Major (>1 ppm)
      2. Minor (<1 ppm)
   b. Organic solutes
2. Colloids (passes through a 0.45-μm filter, but is not dissolved)
   a. Organic
   b. Inorganic
3. Solids (does not pass through a 0.45-μm filter)
   a. Particulate organic material (plant detritus)
   b. Particulate inorganic material (minerals)
4. Gases
   a. Conservative (N_2, Ar, Xe)
   b. Nonconservative (O_2 and CO_2)[1]

Nearly all natural elements exist in seawater, most in the form of dissolved ions.[2] By weight, chloride (Cl^-), sodium (Na^+), sulfate (SO_4^{2-}), and magnesium (Mg^{2+}) are most abundant, accounting for 55.04, 30.61, 7.68, and 3.69% of all dissolved substances in seawater, respectively. These four major constituents plus calcium (Ca^{2+}) and potassium (K^+) constitute more than 99% of the dissolved ions in seawater. In terms of total mass, the ten most important elements comprising the dissolved material in seawater include chlorine (2.5×10^{16} t), sodium (1.4×10^{15} t), magnesium (1.7×10^{15} t), calcium (5.4×10^{14} t), potassium (5.0×10^{14} t), silicon (2.6×10^{12} t), zinc (6.4×10^9 t), copper (2.6×10^9 t), iron (2.6×10^9 t), manganese (2.6×10^8 t), and cobalt (6.6×10^7 t).

A. Major Constituents

The major dissolved ions in seawater generally behave conservatively; that is, their concentrations vary little except at ocean boundaries, where they may be changed by input or output processes (e.g., in estuaries due to dilution by river discharges, anthropogenic activity, and atmospheric deposition), as well as at anoxic and evaporated basins and hydrothermal vents, where evaporation, dissolution, and oxidation can alter the composition of the major constituents. The residence time of the major ions ranges from approximately 10^5 years for bicarbonate to more than 10^8 years for sodium, and approaches infinity for chloride. The residence time of many minor ions is significantly shorter. Because the residence time of most major ions is great relative to the mixing time of the ocean, the ratios of the concentrations of the major ions are effectively constant. Minor ions with residence times substantially shorter than the mixing time of the ocean have variable concentrations in seawater. They are more reactive than the major ions, being removed rather quickly. Thus, the minor ions often act nonconservatively in the oceans.

B. Minor and Trace Elements

Minor and trace elements in seawater exist at concentrations less than 2 mg/kg (2 ppm). Some trace elements have concentrations less than 1 part per trillion (ppt). Minor elements with concentrations between 0.05 and 50 μmol/kg include the alkali metal ions lithium (Li^+) and rubidium (Rb^+), the alkaline earth cation barium (Ba^{2+}), the transition metal molybdenum (as MoO_4^{2-}), and the halogen iodine (as IO_3^- and I^-). The nutrient elements nitrogen (as NO_3^-) and phosphorus (as HPO_4^{2-}) also are found at concentrations between 0.05 and 50 μmol/kg. The average concentrations of 22 trace elements range from 0.05 to 50 μmol/kg as well, whereas 35 others occur at concentrations <50 pmol/kg.

Riverine and atmospheric influx, anthropogenic input, and hydrothermal activity at seafloor spreading centers represent the principal sources of trace elements in the ocean.[3-7] Although present in dissolved, colloidal, and particulate forms, trace metals in solution exist in very low concentrations that are generally variable. Estuaries act as a filter for many trace elements, particularly metals. In these coastal ecotones, bottom sediments serve as a reservoir for trace metals, which may be released to overlying waters by dissolution, desorption, and autolytic biological processes. Trace metal mobilization is facilitated by changes in pH and redox potential in the sediment column. Physical factors (e.g., wave action, currents, dredging) also promote mobilization of the metals from bottom sediments.

Trace elements can be critically important to the life processes of marine organisms, despite their extremely low concentrations. Cobalt, copper, iron, manganese, molybdenum, vanadium, and zinc are essential elements for functioning of marine flora. Cobalt and iron are required for nitrogen metabolism; iron, manganese, and vanadium are needed for photosynthesis.[8] In addition, cobalt, copper, and manganese are necessary for other metabolic functions. Most of these trace metals are likewise essential for life processes of marine fauna.

Various trace metals, while essential to the growth of marine organisms at low concentrations, may be toxic at elevated concentrations.[9] The toxicity of trace metals to marine organisms increases in the following order: cobalt < aluminum < chromium < lead < nickel < zinc < copper < cadmium < mercury.[10] There is considerable variation in trace metal toxicity of marine organisms because the uptake, storage, detoxification, and removal of the metals vary greatly among different marine species. According to Rainbow,[11] intrinsic and extrinsic factors that affect trace metal uptake by these organisms include (1) intra- and interspecifically variable intrinsic factors (e.g., nutritional state, stage of molt cycle, throughput of water by osmotic flux, and surface impermeability); and (2) extrinsic physical–chemical factors (e.g., temperature, salinity, dissolved metal concentration, presence or absence of other metals, and presence or absence of chelating agents). These factors can also affect the bioavailability of the metals (i.e., their accessibility to the organisms).

Viarengo[12] divided trace metals into two classes: (1) those essential for effective function of biochemical processes (e.g., cobalt, copper, iron, magnesium, manganese, and zinc); and (2) those with no established biological function but which are significant as contaminants in the environment (e.g., cadmium, chromium, lead, and mercury). The transition metals (i.e., cobalt, copper, iron, and manganese), although essential for metabolic function of marine organisms at low concentrations, may be toxic at high concentrations. The metalloids (i.e., arsenic, cadmium, lead, mercury, selenium, and tin), in contrast, are not required for metabolic activity but may be toxic at low concentrations.[13]

Many estuarine and marine organisms have the capacity to internally regulate, sequester, and detoxify trace metals. For example, trace metals are commonly bound to metallothioneins—low-molecular-weight, sulfhydryl-rich, metal-binding proteins—which can significantly mollify toxic effects.[14,15] Aside from metallothioneins, lysosomes (i.e., cellular structures involved in intracellular digestion and transport) also play a role in trace metal regulation and homeostasis.[16] Engel and Brouwer[17] confirm the involvement of metallothioneins in sequestration of elevated levels of trace metals, but hypothesize that their primary function is in regulating normal metal metabolism of estuarine and marine organisms.

Trace metal accumulation in estuarine and marine organisms depends on several factors, most notably the chemical behavior of the element, composition of food ingested, and habitat occupied. The toxicity of trace metals is strongly coupled to the free metal ionic activity. The bioavailability of a metal, together with its absolute concentration in the environment, must also be considered in assessing trace metal impacts on the organisms.

Urbanized estuaries and coastal marine waters receiving industrial and municipal wastewaters have historically exhibited the highest concentrations of trace metals.[18] Boston Harbor, Newark Bay, Santa Monica Bay, San Francisco Bay, and Commencement Bay provide examples. However, natural fluid emissions from deep-sea hydrothermal vents at seafloor spreading centers also yield significant concentrations of these metals.[19] The levels of trace metals in open ocean waters by comparison are much lower.

C. Nutrient Elements

Phytoplankton and other marine plants require nutrient elements for growth, chiefly nitrogen, phosphorus, and silicon. Several trace metals are also essential for plant growth (e.g., cobalt, copper, iron, manganese, molybdenum, and zinc), and they generally occur in sufficiently high concentrations to satisfy growth demands. Nitrogen is usually the major limiting nutrient element to plant growth in marine waters.[1,5,20,21] Enhanced nitrogen loading in estuaries can cause eutrophication problems.[22] In some estuaries, plant growth may be phosphorus limited during certain seasons. A number of biotic groups (e.g., diatoms, siliceous sponges, and radiolarians) need silicon for skeletal growth. The concentrations of nitrogen, phosphorus, and silicon in near-surface waters of the ocean amount to 0.07, 0.5, and 3 ppm, respectively.[23]

1. Nitrogen

Approximately 10% of the total nitrogen in seawater consists of inorganic and organic compounds, the remainder being N_2. Nitrate (NO_3^-), nitrite (NO_2^-), and ammonia ($NH_3 + NH_4^+$) are the primary inorganic nitrogen forms. The concentrations are as follows:

1. NO_3^- (1 to 500 μM)
2. NO_2^- (0.1 to 50 μM)
3. $NH_3 + NH_4$ (1 to 50 μM)[1]

Other inorganic nitrogen forms existing in seawater, albeit in small concentrations, include hydroxylamine, hyponitrite, and nitrous oxide ions. Dissolved organic nitrogen (e.g., urea, amino acids, and peptides) also occurs in seawater and appears to be useful for autotrophic growth. Particulate nitrogen in the sea is largely in organic form.

Nitrogen enters seawater principally from the atmosphere (NO_2^- from nitrogen fixation), volcanic activity (NH_3), and riverine inflow (fertilizers).[1] Anthropogenic nitrogen inputs in estuaries and coastal waters are significant. Here dissolved inorganic nitrogen concentrations range from near 0 to greater than 100 μM.

2. Phosphorus

The concentration of phosphorus in seawater varies from near 0 to $\sim$70 μg/l.[24] Seawater contains both dissolved and particulate forms, although little information is available on the nature of particulate phosphorus in this medium.[1] The ionized products of H_3PO_4 (i.e., $H_2PO_4^-$, HPO_4^{2-}, and PO_4^{3-}) constitute the dissolved inorganic phosphorus forms. Of these fractions, HPO_4^{2-} attains highest concentrations, exceeding that of PO_4^{3-} by approximately one order of magnitude.[20] Maximum values of PO_4^{3-} in the Atlantic and Pacific Oceans are about 1.5 and 3.2 μM, respectively. Most of the organic phosphorus compounds consist of sugar phosphates, phospholipids, phosphonucle-otides, and their hydrolyzed products, as well as aminophosphoric acids and phosphate esters.[1]

The ratio of nitrogen to phosphorus in marine waters ($\sim$16:1) approximates that in phytoplank-ton. In the Atlantic Ocean, this ratio declines from 16:1 in surface waters to 15:1 in deep layers. In the Pacific Ocean, the ratio decreases to $\sim$14:1 in the oxygen-minimum layer.[1] Nitrogen-to-phosphorus ratios in estuarine waters exhibit a much greater range.[25]

3. Silicon

As in the case of nitrogen and phosphorus, silicon is present in both dissolved and particulate forms in seawater. The concentrations of SiO_2 range from near 0 to 200 μM in marine waters, being substantially greater in the North Pacific than the North Atlantic Ocean.[1] The dissolution of siliceous skeletal remains (e.g., diatoms, radiolarians, and siliceous sponges), as well as the release of silica from quartz, feldspar, and clay minerals, regenerate dissolved silicon.[26,27] Hydrothermal vents at seafloor spreading centers also release significant concentrations of SiO_2 to the oceans.[19]

D. Gases

Although all atmospheric gases are dissolved in seawater, only three (i.e., nitrogen, oxygen, and carbon dioxide) account for more than 98% of all gases dissolved in surface waters of the ocean. Hydrogen, the noble gases (i.e., helium, neon, argon, krypton, and xenon), and unstable minor gases (e.g., carbon monoxide and methane) comprise the remainder. Aside from the atmosphere, submarine volcanic emissions, as well as various biological and chemical processes in the sea (e.g., photosyn-thesis, respiration, organic matter decomposition, and radioactive decay) represent the principal sources of seawater gases.

Physical–chemical factors (e.g., salinity, temperature, and water circulation) greatly influence the concentration and distribution of gases in seawater. Dissolved gaseous solubility depends on temperature, salinity, and partial pressure. According to Henry's law, the dissolved gas concentration is related to the partial pressure by

$$P_i = k_i [i]$$

where k_i is Henry's law constant and $[i]$ is the concentration of the dissolved gas. The value of k_i is a function of temperature, salinity, total pressure, and the gas in question.[1] Gaseous solubility in seawater decreases as temperature and salinity increase.

The concentrations of oxygen and carbon dioxide vary in seawater not only because of changes in temperature, salinity, and partial pressure, as discussed above, but also because of biological activity. Primary sources of dissolved oxygen in marine waters are gaseous exchange with the atmosphere across the air–sea surface interface and photosynthesis by phytoplankton and other autotrophs. Submarine volcanic activity is a secondary source. Dissolved carbon dioxide originates mainly from gaseous exchange with the atmosphere, respiration of organisms, decomposition of organic matter, inorganic chemical reactions, and riverine inputs.

Life processes of marine organisms significantly affect the concentrations of dissolved oxygen and carbon dioxide in seawater. For example, due to seasonal and geographic differences in the rates of photosynthesis and respiration, both gases exhibit substantial temporal and spatial variations. As a result, they behave nonconservatively in the ocean.

E. Organic Compounds

Seawater also contains small amounts ($<0.01\%$ of the total amount of salts) of organic matter, which consists of a wide array of dissolved and particulate compounds. Although not all of the individual compounds have been characterized, six major classes of naturally occurring organic compounds exist in seawater. These include carbohydrates, hydrocarbons, humic substances, amino acids and proteins, carboxylic acids, and steroids.

Most organic compounds in the ocean derive from primary production of marine phytoplankton. Other sources are riverine influx, atmospheric deposition, resuspension of organic matter from seafloor sediments, excretion by marine organisms, and oil spills.[1] In estuaries and coastal marine waters, various anthropogenic activities can contribute additional quantities of organic matter. Releases of organic compounds from motorized vehicles, marinas, and stormwater outfalls provide examples.

F. Dissolved Constituent Behavior

The dissolved components of seawater may behave conservatively or nonconservatively as alluded to above. Most major constituents of seawater are conservative, showing little or no variation in concentration throughout the ocean as a consequence of their nonreactivity or low crustal abundance. In contrast, nonconservative constituents, such as minor and trace elements, have concentrations that differ from place to place in the oceans because of variable reactivity or inputs (e.g., emissions from hydrothermal vents and organismal uptake). In certain marine environments, major constituents of seawater also may behave nonconservatively. For example, in estuaries, anoxic basins and sediments, evaporated basins, and deep-sea hydrothermal vent systems, various processes (e.g., dissolution, evaporation, precipitation, freezing, and oxidation) can cause the concentration of some major seawater components to change.[1] Estuaries are notable environments where changes in constituent concentrations occur during the mixing of river water and seawater because of chemical and biological processes that facilitate the addition, removal, or alteration of dissolved components in the system.[27,28]

G. Vertical Profiles

Millero (p. 107)[1] has examined the vertical distribution of major, minor, trace, and nutrient elements in the oceans. Several major types of element profiles have been described.

1. Conservative Profile

A constant ratio of the concentration of an element to chlorinity or salinity is found for some elements because of low reactivity. Along with the major components of seawater, trace metals such as rubidium (Rb^+) and cesium (Cs^+) and anions such as molybdenum (MoO_4^{2-}) and tungsten (WO_4^{2-}) exhibit this type of behavior. One might also expect elements such as gold and silver to be conservative in seawater.

2. Nutrient-Type Profile

The depletion of an element in surface waters and its enrichment at depth is a nutrient-type profile. The element is removed from the surface waters by plankton or biologically produced particulate matter. It is regenerated in deep waters when the biologically produced particulate matter is oxidized by bacteria. Three nutrient-type profiles have been delineated:

a. In this profile, shallow water regeneration accounts for maximum elemental concentrations near 1 km depth, similar to the nutrients phosphate PO_4^{3-} and nitrate NO_3^-. Cadmium (Cd^{2+}) provides a good example of this type of nutrient behavior.
b. Some metals follow a deep regeneration cycle and hence display maximum concentrations in deep waters similar to the distributions of silica and total alkalinity. Examples of this type of profile include the elements barium, zinc, and germanium.
c. The combination of shallow and deep generation is inferred from the nutrient-type profiles of nickel and selenium.

3. Surface Enrichment and Depletion at Depth

Elements following this type of profile enter surface waters via river inflow, other land sources, and atmospheric deposition. Because they are rapidly removed from seawater, their residence times are very short. Lead is a good example of an element entering the oceans via the atmosphere. The mechanism for the scavenging process for lead is not well characterized. Manganese (Mn^{2+}) enters surface waters from river discharges or upon being released from shelf sediments. Elements that can occupy different oxidation states may also exhibit this type of profile. The reduction of metals in surface waters can be caused by biological and photochemical processes. Subsequent oxidation can produce an oxidized form that is insoluble in seawater. The elements chromium (Cr^{3+}), arsenic (As^{3+}), and iodine (I^-) fall into this category.

4. Mid-Depth Minima

A mid-depth minimum can result from the surface input and regeneration of an element at or near the bottom, or scavenging throughout the water column. Copper (Cu^{2+}), tin (Sn), and aluminum (Al^{3+}) show this type of profile. Their input into surface waters far at sea generally derives from the fallout of atmospheric dust particles that originate on the continents. They may be quickly scavenged by adsorption or uptake by plant material. The particles gradually settle through the water column and are ultimately deposited in seafloor sediments. The resuspension and flux of aluminum from the sediments increases its concentration in bottom waters.

5. Mid-Depth Maxima

This type of profile can arise from hydrothermal emissions along a mid-ocean ridge system. Manganese (Mn^{2+}) and helium (^3He) are examples. Fluxes from hydrothermal plumes have been used to trace these elements in deep ocean waters.

6. Mid-Depth Maxima or Minima in the Suboxic Layer (~1 km)

A large suboxic layer exists in some regions of the Pacific and Indian Oceans. Reduction and oxidation processes in the water column or adjacent sediments can yield maxima of the reduced form (Mn^{2+} and Fe^{2+}) and minima of the reduced form if it is insoluble or scavenged by solid phases (Cr^{3+}).

7. Maxima and Minima in Anoxic Waters

In areas of restricted circulation such as the Black Sea, Cariaco Trench, and fjords, bottom waters often become anoxic (devoid of O_2) with the production of H_2S. Near certain interfaces redox processes can cause maxima and minima of some forms due to solubility changes of the various species. Manganese (Mn^{2+}) and iron (Fe^{2+}), for example, exhibit maxima in these waters because of the increased solubility of the reduced forms.

H. Salinity

Salinity is defined as the weight in grams of dissolved inorganic salts in 1 kg of seawater after all the bromine and iodine have been replaced by an equivalent quantity of chlorine, all carbonate has been converted to an equivalent quantity of oxide, and all organic compounds have been oxidized at a temperature of 480°C. Salinity is usually reported in parts per thousand (‰), but may also be expressed in milligrams per liter (mg/l), milliequivalents per liter (meq/l), grams per kilogram (g/kg), or percent (%). The constancy of the proportions of the major dissolved constituents in seawater has enabled salinity to be determined by the concentration of the chloride ion, which represents an index of salinity. By titrating a seawater sample with silver nitrate (Knudsen method), the amount of chloride, plus a chloride equivalent of bromide and iodide, can be obtained. Chlorinity (Cl)—the total weight of these elements in a 1-kg seawater sample—is related to salinity (S) according to the following equation:

$$S(‰) = 1.80655 \times Cl\ (‰)$$

Because seawater is an electrolyte and conducts an electric current, salinity can be measured by electrical conductivity. Chlorinity measurements are less accurate (0.02% for 32 to 38‰) than conductivity techniques (0.003‰) of salinity estimation.[29] Therefore, salinity is now more frequently measured by a salinometer, which records conductivity, rather than by volumetric titration of chlorinity. Measurements of density (with a hydrometer), the refractive index of seawater (with a refractometer), and freezing-point depression provide alternate methods of estimating salinity.

In 1978, the Joint Panel for Oceanographic Tables and Standards recommended the Practical Salinity Scale as a conductivity standard for salinity. The Practical Salinity Scale is the salinity defined by the conductivity ratio of seawater relative to a given mass of KCl. As stated by Millero (p. 72),[1] "A standard seawater of salinity D = 35.000 [no units or ‰ are needed] has, by definition, a conductivity ratio of 1.0 at 15°C with a KCl solution containing a mass of 32.4356 g of KCl in a mass of 1 kg of solution." Today, seawater salinity is most commonly determined by conductivity using the Practical Salinity Scale.

Salinity in the ocean ranges from ~33 to 38‰ and averages ~35‰. Deviations from average salinity occur locally at ocean boundaries due to dilution from river runoff and at deep-sea hydrothermal vents along seafloor spreading centers due to the emission of mineral-laden water from the oceanic crust. On a regional scale, deviations from average salinity arise from differences in evaporation and precipitation. Evaporation and the formation of sea ice raise salinity in the ocean, whereas large rates of precipitation and river runoff reduce salinity. High rates of precipitation associated with tropical rains depress salinity in equatorial seas, as does the melting of ice caps in polar regions. At high latitudes greater than 40°N and 40°S latitude, precipitation in excess of evaporation also lowers salinity. Salinity maxima develop between 20° and 30° north and south latitudes.[23,30]

Estuaries have much more variable salinity than the oceans. In these coastal ecotones, salinity ranges from ~0.5 to more than 35‰. It changes temporally over annual, seasonal, daily, and tidal cycles, and spatially along longitudinal, transverse, and vertical planes. The mixing of river water and seawater, solutions significantly different with respect to physical–chemical properties and chemical composition, is responsible for the variable salinity observed in estuaries.[27,31] Numerous chemical and biogenic reactions may take place independently and simultaneously; chemical constituents are thus readily added, removed, or altered in these systems. Anthropogenic input of large volumes of wastewater discharges, stormwater runoff, and other pollutant components can mask contributions of certain chemical constituents from natural waters.[5,22]

Dissolved constituents in seawater originate principally from the weathering of rocks and leaching of soils on land. Solution, oxidation–reduction, the action of hydrogen ions, and the formation of complexes are weathering processes that control the supply of ions. Fluvial and glacial processes continually deliver most dissolved material to the ocean; of secondary importance are materials supplied by terrestrial and cosmic sources (e.g., aeolian, meteortic, and volcanic particles), atmospheric processes, and hydrothermal activity at mid-ocean ridges. Equally important are biological, chemical, and sedimentological processes operating in the oceans, which remove dissolved components from seawater, tending to maintain equilibrium among the constituents. Authigenesis involving inorganic reactions in seawater and biogenesis involving organic reactions that use seawater constituents are two of the most important processes for removing dissolved constituents.[32] Biogenic processes (e.g., growth of skeletal material) are important in the removal of certain chemical species, such as silica (SiO_2), calcium (Ca^{2+}), and bicarbonate (HCO_3^-). Bicarbonate, as part of the carbon dioxide–carbonate–bicarbonate chemical system, plays a role in regulating the pH of seawater. The removal of carbon dioxide by autotrophs during photosynthesis contributes substantially to diurnal and seasonal variations in the carbonate system.[1]

REFERENCES

1. Millero, F. J., *Chemical Oceanography,* 2nd ed., CRC Press, Boca Raton, FL, 1996.
2. Bruland, K. W., Trace elements in seawater, in *Chemical Oceanography,* Riley, J. P. and Chester, R., Eds., Academic Press, London, 1983, 157.
3. Wong, C. S., Boyle, E., Bruland, K. W., Burton, J. D., and Goldberg, E. D., Eds., *Trace Metals in Sea Water,* Plenum Press, New York, 1983.
4. Cutter, G. A., Trace elements in estuarine and coastal waters, *Rev. Geophys. (Suppl.), Contrib. Oceanogr.,* 639, 1991.
5. Kennish, M. J., *Ecology of Estuaries: Anthropogenic Effects,* CRC Press, Boca Raton, FL, 1992.
6. Kennish, M. J. and Lutz, R. A., Ocean crust formation, *Rev. Aquat. Sci.,* 6, 493, 1992.
7. Cann, J. R., Elderfield, H., and Laughton, A. S., Eds., *Mid-Ocean Ridges: Dynamics of Processes Associated with Creation of New Ocean Crust,* Cambridge University Press, New York, 1998.
8. Wells, M. L., Zorkin, N. G., and Lewis, A. G., The role of colloid chemistry in providing a source of iron to phytoplankton, *J. Mar. Res.,* 41, 731, 1983.
9. Spaargaren, D. H. and Ceccaldi, H. J., Some relations between the elementary chemical composition of marine organisms and that of seawater, *Oceanol. Acta,* 7, 63, 1984.
10. Abel, P. D., *Water Pollution Biology,* Ellis Horwood, Chichester, 1989.
11. Rainbow, P. S., The significance of trace metal concentrations in marine invertebrates, in *Ecotoxicology of Metals in Invertebrates,* Dallinger, R. and Rainbow, P. S., Eds., Lewis Publishers, Ann Arbor, MI, 1993, 3.
12. Viarengo, A., Biochemical effects of trace metals, *Mar. Pollut. Bull.,* 16, 153, 1985.
13. Viarengo, A., Heavy metals in marine invertebrates: mechanisms of regulation and toxicity at the cellular level, *Rev. Aquat. Sci.,* 1, 295, 1989.
14. Roesijadi, G., Behavior of metallothionein-bound metals in a natural population of an estuarine mollusc, *Mar. Environ. Res.,* 38, 147, 1994.
15. Schlenk, D., Ringwood, A. H., Brouwer-Hoexum, T., and Brouwer, M., Crustaceans as models for metal metabolism. II. Induction and characterization of metallothionein isoforms from the blue crab (*Callinectes sapidus*), *Mar. Environ. Res.,* 35, 7, 1993.
16. Viarengo, A., Moore, M. N., Mancinelli, G., Mazzucotelli, A., Pipe, R. K., and Farrar, S. V., Metallothioneins and lysosomes in metal toxicity and accumulation in marine mussels: the effect of cadmium in the presence and absence of phenanthrene, *Mar. Biol.,* 94, 251, 1987.
17. Engel, D. W. and Brouwer, W., Trace metal-binding proteins in marine molluscs and crustaceans, *Mar. Environ. Res.,* 13, 177, 1984.
18. Clark, R. B., *Marine Pollution,* 3rd ed., Clarendon Press, Oxford, 1992.
19. Von Damm, K. L., Seafloor hydrothermal activity: black smoker chemistry and chimneys, *Annu. Rev. Earth Planet. Sci.,* 18, 173, 1990.

20. Valiela, I., *Marine Ecological Processes,* 2nd ed., Springer-Verlag, New York, 1995.
21. Alongi, D. M., *Coastal Ecosystem Processes,* CRC Press, Boca Raton, FL, 1998.
22. Kennish, M. J., *Practical Handbook of Estuarine and Marine Pollution,* CRC Press, Boca Raton, FL, 1997.
23. Pinet, P. R., *Invitation to Oceanography,* Jones and Bartlett Publishers, Sudbury, MA, 1998.
24. Riley, J. P. and Chester, R., *Introduction to Marine Chemistry,* Academic Press, London, 1971.
25. Kennish, M. J., *Ecology of Estuaries: Biological Aspects,* CRC Press, Boca Raton, FL, 1990.
26. Helgeson, H. C. and Mackenzie, F. T., Silicate–seawater equilibria in the ocean system, *Deep-Sea Res.,* 17, 877, 1970.
27. Kennish, M. J., *Ecology of Estuaries: Physical and Chemical Aspects,* CRC Press, Boca Raton, FL, 1986.
28. Takayanagi, K. and Wong, G. T. F., Total selenium and selenium (IV) in the James River estuary and southern Chesapeake Bay, *Estuarine Coastal Shelf Sci.,* 18, 113, 1984.
29. Levinton, J. S., *Marine Ecology,* Prentice-Hall, Englewood Cliffs, NJ, 1982.
30. Sverdrup, H. U., Johnson, M. W., and Fleming, R. H., *The Oceans, Their Physics, Chemistry, and General Biology,* Prentice-Hall, Englewood Cliffs, NJ, 1942.
31. Kjerfve, B., *Hydrodynamics of Estuaries,* CRC Press, Boca Raton, FL, 1988.
32. Smith, D. G., Ed., *The Cambridge Encyclopedia of Earth Sciences,* Cambridge University Press, Cambridge, 1981.

2.1 PERIODIC TABLE

Table 2.1–1   Periodic Table of the Elements

Note: The new IUPAC format numbers the groups from 1 to 18. The previous IUPAC numbering system and the system used by Chemical Abstracts Service (CAS) are also shown. For radioactive elements that do not occur in nature, the mass number of the most stable isotope is given in parentheses.

REFERENCES

1. G. J. Leigh, Editor, *Nomenclature of Inorganic Chemistry*, Blackwell Scientific Publications, Oxford, 1990.
2. *Chemical and Engineering News*, 63(5), 27, 1985.
3. Atomic Weights of the Elements, 1995, *Pure Appl. Chem.*, 68, 2339, 1996.

Source: Lide, D. R. and Frederikse, H. P. R., Eds., *CRC Handbook of Chemistry and Physics*, 79th ed., CRC Press, Boca Raton, FL, 1998, 14–10. With permission.

Table 2.1–2   Atomic Numbers and Atomic Weights

Actinium	Ac	89	227.0278	Mercury	Hg	80	200.59
Aluminum	Al	13	26.98154	Molybdenum	Mo	42	95.94
Americium	Am	95	(243)	Neodymium	Nd	60	144.24
Antimony	Sb	51	121.75	Neon	Ne	10	20.179
Argon	Ar	18	39.948	Neptunium	Np	93	237.0482
Arsenic	As	33	74.9216	Nickel	Ni	28	58.69
Astatine	At	85	(210)	Niobium	Nb	41	92.9064
Barium	Ba	56	137.33	Nitrogen	N	7	14.0067
Berkelium	Bk	97	(247)	Nobelium	No	102	(259)
Beryllium	Be	4	9.01218	Osmium	Os	76	190.2
Bismuth	Bi	83	208.9804	Oxygen	O	8	15.9994
Boron	B	5	10.81	Palladium	Pd	46	106.42
Bromine	Br	35	79.904	Phosphorus	P	15	30.97376
Cadmium	Cd	48	112.41	Platinum	Pt	78	195.08
Calcium	Ca	20	40.08	Plutonium	Pu	94	(244)
Californium	Cf	98	(251)	Polonium	Po	84	(209)
Carbon	C	6	12.011	Potassium	K	19	39.0983
Cerium	Ce	58	140.12	Praseodymium	Pr	59	140.9077
Cesium	Cs	55	132.9054	Promethium	Pm	61	(145)
Chlorine	Cl	17	35.453	Protactinium	Pa	91	231.0359
Chromium	Cr	24	51.996	Radium	Ra	88	226.0254
Cobalt	Co	27	58.9332	Radon	Rn	86	(222)
Copper	Cu	29	63.546	Rhenium	Re	75	186.207
Curium	Cm	96	(247)	Rhodium	Rh	45	102.9055
Dysprosium	Dy	66	162.50	Rubidium	Rb	37	85.4678
Einsteinium	Es	99	(252)	Ruthenium	Ru	44	101.07
Erbium	Er	68	167.26	Samarium	Sm	62	150.36
Europium	Eu	63	151.96	Scandium	Sc	21	44.9559
Fermium	Fm	100	(257)	Selenium	Se	34	78.96
Fluorine	F	9	18.998403	Silicon	Si	14	28.0855
Francium	Fr	87	(223)	Silver	Ag	47	107.868
Gadolinium	Gd	64	157.25	Sodium	Na	11	22.98977
Gallium	Ga	31	69.72	Strontium	Sr	38	87.62
Germanium	Ge	32	72.59	Sulfur	S	16	32.06
Gold	Au	79	196.9665	Tantalum	Ta	73	180.9479
Hafnium	Hf	72	178.49	Technetium	Tc	43	(98)
Helium	He	2	4.00260	Tellurium	Te	52	127.60
Holmium	Ho	67	164.9304	Terbium	Tb	65	158.9254
Hydrogen	H	1	1.0079	Thallium	Tl	81	204.383
Indium	In	49	114.82	Thorium	Th	90	232.0381
Iodine	I	53	126.9045	Thulium	Tm	69	168.9342
Iridium	Ir	77	192.22	Tin	Sn	50	118.69
Iron	Fe	26	55.847	Titanium	Ti	22	47.88
Krypton	Kr	36	83.80	Tungsten	W	74	183.85
Lanthanum	La	57	138.9055	Uranium	U	92	238.0289
Lawrencium	Lr	103	(260)	Vanadium	V	23	50.9415
Lead	Pb	82	207.2	Xenon	Xe	54	131.29
Lithium	Li	3	6.941	Ytterbium	Yb	70	173.04
Lutetium	Lu	71	174.967	Yttrium	Y	39	88.9059
Magnesium	Mg	12	24.305	Zinc	Zn	30	65.38
Manganese	Mn	25	54.9380	Zirconium	Zr	40	91.22
Mendelevium	Md	101	(258)				

Source: Pankow, J. F., *Aquatic Chemistry Concepts*, Lewis Publishers, Chelsea, MI, 1991. With permission.

Table 2.1–3   Atomic Masses and Abundances

Z	Isotope	Mass in u	Abundance in %	Z	Isotope	Mass in u	Abundance in %
1	^{1}H	1.007 825 032(1)	99.985(1)	18	^{38}Ar	37.962 732 2(5)	0.063(1)
1	^{2}H	2.014 101 778(1)	0.015(1)	18	^{39}Ar	38.964 313(5)	*
1	^{3}H	3.016 049 268(1)	*	18	^{40}Ar	39.962 383 123(3)	99.600(3)
2	^{3}He	3.016 029 310(1)	0.000137(3)	19	^{39}K	38.963 706 8(3)	93.2581(44)
2	^{4}He	4.002 603 250(1)	99.999863(3)	19	^{40}K	39.963 998 7(3)	0.0117(1)
3	^{6}Li	6.015 122 3(5)	7.5(2)	19	^{41}K	40.961 826 0(3)	6.7302(44)
3	^{7}Li	7.016 004 0(5)	92.5(2)	20	^{40}Ca	39.962 591 1(3)	96.941(18)
4	^{9}Be	9.012 182 1(4)	100	20	^{42}Ca	41.958 618 3(4)	0.647(9)
5	^{10}B	10.012 937 0(3)	19.9(2)	20	^{43}Ca	42.958 766 8(5)	0.135(6)
5	^{11}B	11.009 305 5(5)	80.1(2)	20	^{44}Ca	43.955 481 1(9)	2.086(12)
6	^{12}C	12 by definition	98.90(3)	20	^{46}Ca	45.953 693(3)	0.004(3)
6	^{13}C	13.003 354 838(1)	1.10(3)	20	^{48}Ca	47.952 534(4)	0.187(4)
6	^{14}C	14.003 241 988(4)	*	21	^{45}Sc	44.955 910(1)	100
7	^{14}N	14.003 074 005(1)	99.634(9)	22	^{46}Ti	45.952 629(1)	8.0(1)
7	^{15}N	15.000 108 898(1)	0.366(9)	22	^{47}Ti	46.951 764(1)	7.3(1)
8	^{16}O	15.994 914 622(2)	99.762(15)	22	^{48}Ti	47.947 947(1)	73.8(1)
8	^{17}O	16.999 131 5(2)	0.038(3)	22	^{49}Ti	48.947 871(1)	5.5(1)
8	^{18}O	17.999 160 4(9)	0.200(12)	22	^{50}Ti	49.944 792(1)	5.4(1)
9	^{19}F	18.998 403 21(8)	100	23	^{50}V	49.947 163(1)	0.250(2)
10	^{20}Ne	19.992 440 176(2)	90.48(3)	23	^{51}V	50.943 964(1)	99.750(2)
10	^{21}Ne	20.993 846 74(4)	0.27(1)	24	^{50}Cr	49.946 050(1)	4.345(13)
10	^{22}Ne	21.991 385 5(2)	9.25(3)	24	^{52}Cr	51.940 512(2)	83.789(18)
11	^{23}Na	22.989 769 7(2)	100	24	^{53}Cr	52.940 654(2)	9.501(17)
12	^{24}Mg	23.985 041 9(2)	78.99(3)	24	^{54}Cr	53.938 885(1)	2.365(7)
12	^{25}Mg	24.985 837 0(2)	10.00(1)	25	^{55}Mn	54.938 050(1)	100
12	^{26}Mg	25.982 593 0(2)	11.01(2)	26	^{54}Fe	53.939 615(1)	5.8(1)
13	^{27}Al	26.981 538 4(1)	100	26	^{56}Fe	55.934 942(1)	91.72(30)
14	^{28}Si	27.976 926 533(2)	92.23(1)	26	^{57}Fe	56.935 399(1)	2.1(1)
14	^{29}Si	28.976 494 72(3)	4.67(1)	26	^{58}Fe	57.933 280(1)	0.28(1)
14	^{30}Si	29.973 770 22(5)	3.10(1)	27	^{59}Co	58.933 200(2)	100
15	^{31}P	30.973 761 5(2)	100	28	^{58}Ni	57.935 348(2)	68.077(9)
16	^{32}S	31.972 070 7(1)	95.02(9)	28	^{60}Ni	59.930 791(2)	26.223(8)
16	^{33}S	32.971 458 5(1)	0.75(4)	28	^{61}Ni	60.931 060(2)	1.140(1)
16	^{34}S	33.967 866 8(1)	4.21(8)	28	^{62}Ni	61.928 349(2)	3.634(2)
16	^{36}S	35.967 080 9(3)	0.02(1)	28	^{64}Ni	63.927 970(2)	0.926(1)
17	^{35}Cl	34.968 852 71(4)	75.77(7)	29	^{63}Cu	62.929 601(2)	69.17(3)
17	^{37}Cl	36.965 902 60(5)	24.23(7)	29	^{65}Cu	64.927 794(2)	30.83(3)
18	^{36}Ar	35.967 546 3(3)	0.337(3)	30	^{64}Zn	63.929 147(2)	48.6(3)
18	^{37}Ar	36.966 775 9(3)	*	30	^{66}Zn	65.926 037(2)	27.9(2)
30	^{67}Zn	66.927 131(2)	4.1(1)	46	^{108}Pd	107.903 894(4)	26.46(9)
30	^{68}Zn	67.924 848(2)	18.8(4)	46	^{110}Pd	109.905 15(1)	11.72(9)
30	^{70}Zn	69.925 325(4)	0.6(1)	47	^{107}Ag	106.905 093(6)	51.839(7)
31	^{69}Ga	68.925 581(3)	60.108(9)	47	^{109}Ag	108.904 756(3)	48.161(7)
31	^{71}Ga	70.924 705(2)	39.892(9)	48	^{106}Cd	105.906 458(6)	1.25(4)
32	^{70}Ge	69.924 250(2)	21.23(4)	48	^{108}Cd	107.904 183(6)	0.89(2)
32	^{72}Ge	71.922 076(2)	27.66(3)	48	^{110}Cd	109.903 006(3)	12.49(12)
32	^{73}Ge	72.923 459(2)	7.73(1)	48	^{111}Cd	110.904 182(3)	12.80(8)
32	^{74}Ge	73.921 178(2)	35.94(2)	48	^{112}Cd	111.902 757(3)	24.13(14)
32	^{76}Ge	75.921 403(2)	7.44(2)	48	^{113}Cd	112.904 401(3)	12.22(8)
33	^{75}As	74.921 596(2)	100	48	^{114}Cd	113.903 358(3)	28.73(28)
34	^{74}Se	73.922 477(2)	0.89(2)	48	^{116}Cd	115.904 755(3)	7.49(12)
34	^{76}Se	75.919 214(2)	9.36(11)	49	^{113}In	112.904 061(4)	4.3(2)
34	^{77}Se	76.919 915(2)	7.63(6)	49	^{115}In	114.903 878(5)	95.7(2)
34	^{78}Se	77.917 310(2)	23.78(9)	50	^{112}Sn	111.904 821(5)	0.97(1)
34	^{80}Se	79.916 522(2)	49.61(10)	50	^{114}Sn	113.902 782(3)	0.65(1)
34	^{82}Se	81.916 700(2)	8.73(6)	50	^{115}Sn	114.903 346(3)	0.34(1)

Table 2.1–3 Atomic Masses and Abundances (continued)

Z	Isotope	Mass in u	Abundance in %	Z	Isotope	Mass in u	Abundance in %
35	^{79}Br	78.918 338(2)	50.69(7)	50	^{116}Sn	115.901 744(3)	14.53(1)
35	^{81}Br	80.916 291(3)	49.31(7)	50	^{117}Sn	116.902 954(3)	7.68(7)
36	^{78}Kr	77.920 386(7)	0.35(2)	50	^{118}Sn	117.901 606(3)	24.23(11)
36	^{80}Kr	79.916 378(4)	2.25(2)	50	^{119}Sn	118.903 309(3)	8.59(4)
36	^{82}Kr	81.913 485(3)	11.6(1)	50	^{120}Sn	119.902 197(3)	32.59(10)
36	^{83}Kr	82.914 136(3)	11.5(1)	50	^{122}Sn	121.903 440(3)	4.63(3)
36	^{84}Kr	83.911 507(3)	57.0(3)	50	^{124}Sn	123.905 275(1)	5.79(5)
36	^{86}Kr	85.910 610(1)	17.3(2)	51	^{121}Sb	120.903 818(2)	57.36(8)
37	^{85}Rb	84.911 789(3)	72.165(20)	51	^{123}Sb	122.904 216(2)	42.64(8)
37	^{87}Rb	86.909 183(3)	27.835(20)	52	^{120}Te	119.904 02(1)	0.096(2)
38	^{84}Sr	83.913 425(4)	0.56(1)	52	^{122}Te	121.903 047(2)	2.603(4)
38	^{86}Sr	85.909 262(2)	9.86(1)	52	^{123}Te	122.904 273(2)	0.908(2)
38	^{87}Sr	86.908 879(2)	7.00(1)	52	^{124}Te	123.902 819(2)	4.816(6)
38	^{88}Sr	87.905 614(2)	82.58(1)	52	^{125}Te	124.904 425(2)	7.139(6)
39	^{89}Y	88.905 848(3)	100	52	^{126}Te	125.903 306(2)	18.95(1)
40	^{90}Zr	89.904 704(2)	51.45(3)	52	^{128}Te	127.904 461(2)	31.69(1)
40	^{91}Zr	90.905 645(2)	11.22(4)	52	^{130}Te	129.906 223(2)	33.80(1)
40	^{92}Zr	91.905 040(2)	17.15(2)	53	^{127}I	126.904 468(4)	100
40	^{94}Zr	93.906 316(3)	17.38(4)	54	^{124}Xe	123.905 896(2)	0.10(1)
40	^{96}Zr	95.908 276(3)	2.80(2)	54	^{126}Xe	125.904 269(7)	0.09(1)
41	^{93}Nb	92.906 378(2)	100	54	^{128}Xe	127.903 530(2)	1.91(3)
42	^{92}Mo	91.906 810(4)	14.84(4)	54	^{129}Xe	128.904 779 4(9)	26.4(6)
42	^{94}Mo	93.905 088(2)	9.25(3)	54	^{130}Xe	129.903 508(1)	4.1(1)
42	^{95}Mo	94.905 841(2)	15.92(5)	54	^{131}Xe	130.905 082(1)	21.2(4)
42	^{96}Mo	95.904 679(2)	16.68(5)	54	^{132}Xe	131.904 154(1)	26.9(5)
42	^{97}Mo	96.906 021(2)	9.55(3)	54	^{134}Xe	133.905 394 5(9)	10.4(2)
42	^{98}Mo	97.905 408(2)	24.13(7)	54	^{136}Xe	135.907 220(8)	8.9(1)
42	^{100}Mo	99.907 477(6)	9.63(3)	55	^{133}Cs	132.905 447(3)	100
43	^{98}Tc	97.907 216(4)	*	56	^{130}Ba	129.906 310(7)	0.106(2)
43	^{99}Tc	98.906 255(5)	*	56	^{132}Ba	131.905 056(3)	0.101(2)
44	^{96}Ru	95.907 598(8)	5.52(6)	56	^{134}Ba	133.904 503(3)	2.417(27)
44	^{98}Ru	97.905 287(7)	1.88(6)	56	^{135}Ba	134.905 683(3)	6.592(18)
44	^{99}Ru	98.905 939(2)	12.7(1)	56	^{136}Ba	135.904 570(3)	7.854(36)
44	^{100}Ru	99.904 220(2)	12.6(1)	56	^{137}Ba	136.905 821(3)	11.23(4)
44	^{101}Ru	100.905 582(2)	17.0(1)	56	^{138}Ba	137.905 241(3)	71.70(7)
44	^{102}Ru	101.904 350(2)	31.6(2)	57	^{138}La	137.907 107(4)	0.0902(2)
44	^{104}Ru	103.905 430(4)	18.7(2)	57	^{139}La	138.906 348(3)	99.9098(2)
45	^{103}Rh	102.905 504(3)	100	58	^{136}Ce	135.907 14(5)	0.19(1)
46	^{102}Pd	101.905 608(3)	1.02(1)	58	^{137}Ce	136.907 78(5)	*
46	^{104}Pd	103.904 035(5)	11.14(8)	58	^{138}Ce	137.905 99(1)	0.25(1)
46	^{105}Pd	104.905 084(5)	22.33(8)	58	^{139}Ce	138.906 647(8)	*
46	^{106}Pd	105.903 483(5)	27.33(3)	58	^{140}Ce	139.905 434(3)	88.48(10)
58	^{141}Ce	140.908 271(3)	*	72	^{180}Hf	179.946 549(3)	35.100(7)
58	^{142}Ce	141.909 240(4)	11.08(10)	73	^{180}Ta	179.947 466(3)	0.012(2)
59	^{141}Pr	140.907 648(3)	100	73	^{181}Ta	180.947 996(3)	99.988(2)
60	^{142}Nd	141.907 719(3)	27.13(12)	74	^{180}W	179.946 706(5)	0.13(4)
60	^{143}Nd	142.909 810(3)	12.18(6)	74	^{182}W	181.948 206(3)	26.3(2)
60	^{144}Nd	143.910 083(3)	23.80(12)	74	^{183}W	182.950 224(3)	14.3(1)
60	^{145}Nd	144.912 569(3)	8.30(6)	74	^{184}W	183.950 933(3)	30.67(15)
60	^{146}Nd	145.913 112(3)	17.19(9)	74	^{186}W	185.954 362(3)	28.6(2)
60	^{148}Nd	147.916 889(3)	5.76(3)	75	^{185}Re	184.952 956(3)	37.40(2)
60	^{150}Nd	149.920 887(4)	5.64(3)	75	^{187}Re	186.955 751(3)	62.60(2)
61	^{143}Pm	142.910 928(4)	*	76	^{184}Os	183.952 491(3)	0.02(1)
61	^{145}Pm	144.912 744(4)	*	76	^{186}Os	185.953 838(3)	1.58(30)
61	^{147}Pm	146.915 134(3)	*	76	^{187}Os	186.955 748(3)	1.6(3)
62	^{144}Sm	143.911 995(4)	3.1(1)	76	^{188}Os	187.955 836(3)	13.3(7)

Table 2.1–3   Atomic Masses and Abundances (continued)

Z	Isotope	Mass in u	Abundance in %	Z	Isotope	Mass in u	Abundance in %
62	[147]Sm	146.914 893(3)	15.0(2)	76	[189]Os	188.958 145(3)	16.1(8)
62	[148]Sm	147.914 818(3)	11.3(1)	76	[190]Os	189.958 445(3)	26.4(12)
62	[149]Sm	148.917 180(3)	13.8(1)	76	[192]Os	191.961 479(4)	41.0(8)
62	[150]Sm	149.917 271(3)	7.4(1)	77	[191]Ir	190.960 591(3)	37.3(5)
62	[152]Sm	151.919 728(3)	26.7(2)	77	[193]Ir	192.962 924(3)	62.7(5)
62	[154]Sm	153.922 205(3)	22.7(2)	78	[190]Pt	189.959 930(7)	0.01(1)
63	[151]Eu	150.919 846(3)	47.8(15)	78	[192]Pt	191.961 035(4)	0.79(6)
63	[153]Eu	152.921 226(3)	52.2(15)	78	[194]Pt	193.962 664(3)	32.9(6)
64	[152]Gd	151.919 788(3)	0.20(1)	78	[195]Pt	194.964 774(3)	33.8(6)
64	[154]Gd	153.920 862(3)	2.18(3)	78	[196]Pt	195.964 935(3)	25.3(6)
64	[155]Gd	154.922 619(3)	14.80(5)	78	[198]Pt	197.967 876(5)	7.2(2)
64	[156]Gd	155.922 120(3)	20.47(4)	79	[197]Au	196.966 552(3)	100
64	[157]Gd	156.923 957(3)	15.65(3)	80	[196]Hg	195.965 815(4)	0.15(1)
64	[158]Gd	157.924 101(3)	24.84(12)	80	[198]Hg	197.966 752(3)	9.97(8)
64	[160]Gd	159.927 051(3)	21.86(4)	80	[199]Hg	198.968 262(3)	16.87(10)
65	[159]Tb	158.925 343(3)	100	80	[200]Hg	199.968 309(3)	23.10(16)
66	[156]Dy	155.924 278(7)	0.06(1)	80	[201]Hg	200.970 285(3)	13.18(8)
66	[158]Dy	157.924 405(4)	0.10(1)	80	[202]Hg	201.970 626(3)	29.86(20)
66	[160]Dy	159.925 194(3)	2.34(6)	80	[204]Hg	203.973 476(3)	6.87(4)
66	[161]Dy	160.926 930(3)	18.9(2)	81	[203]Tl	202.972 329(3)	29.524(14)
66	[162]Dy	161.926 795(3)	25.5(2)	81	[205]Tl	204.974 412(3)	70.476(14)
66	[163]Dy	162.928 728(3)	24.9(2)	82	[204]Pb	203.973 029(3)	1.4(1)
66	[164]Dy	163.929 171(3)	28.2(2)	82	[206]Pb	205.974 449(3)	24.1(1)
67	[165]Ho	164.930 319(3)	100	82	[207]Pb	206.975 881(3)	22.1(1)
68	[162]Er	161.928 775(4)	0.14(1)	82	[208]Pb	207.976 636(3)	52.4(1)
68	[164]Er	163.929 197(4)	1.61(2)	83	[209]Bi	208.980 383(3)	100
68	[166]Er	165.930 290(3)	33.6(2)	84	[209]Po	208.982 416(3)	*
68	[167]Er	166.932 045(3)	22.95(15)	85	[210]At	209.987 131(9)	*
68	[168]Er	167.932 368(3)	26.8(2)	86	[211]Rn	210.990 585(8)	*
68	[170]Er	169.935 460(3)	14.9(2)	86	[222]Rn	222.017 570(3)	*
69	[169]Tm	168.934 211(3)	100	87	[223]Fr	223.019 731(3)	*
70	[168]Yb	167.933 894(5)	0.13(1)	88	[223]Ra	223.018 497(3)	*
70	[170]Yb	169.934 759(3)	3.05(6)	88	[225]Ra	225.023 604(3)	*
70	[171]Yb	170.936 322(3)	14.3(2)	88	[226]Ra	226.025 403(3)	*
70	[172]Yb	171.936 378(3)	21.9(3)	89	[227]Ac	227.027 747(3)	*
70	[173]Yb	172.938 207(3)	16.12(21)	90	[229]Th	229.031 755(3)	*
70	[174]Yb	173.938 858(3)	31.8(4)	90	[232]Th	232.038 050(2)	100
70	[176]Yb	175.942 568(3)	12.7(2)	91	[231]Pa	231.035 879(3)	*
71	[175]Lu	174.940 768(3)	97.41(2)	92	[234]U	234.040 946(2)	0.0055(5)
71	[176]Lu	175.942 682(3)	2.59(2)	92	[235]U	235.043 923(2)	0.7200(12)
72	[174]Hf	173.940 040(3)	0.162(3)	92	[238]U	238.050 783(2)	99.2745(60)
72	[176]Hf	175.941 402(3)	5.206(5)	93	[237]Np	237.048 167(2)	*
72	[177]Hf	176.943 220(3)	18.606(4)	94	[239]Pu	239.052 157(2)	*
72	[178]Hf	177.943 698(3)	27.297(4)	94	[244]Pu	244.064 198(5)	*
72	[179]Hf	178.945 815(3)	13.629(6)				

Note: This table lists the mass (in atomic mass units, symbol u) and the natural abundance (in percent) of the stable nuclides and a few important radioactive nuclides.

The atomic masses were taken from the 1995 evaluation of Audi and Wapstra (References 2, 3). Mass values were rounded in accordance with the stated uncertainty. The number in parentheses following the mass value is the uncertainty (specifically, the rounded value of the estimated standard deviation) in the last digit given.

Natural abundance values are also followed by uncertainties in the last digit(s) of the stated values. This uncertainty includes both the estimated measurement uncertainty and the reported range of variation in different terrestrial sources of the element (see Reference 4 for more details). An asterisk in the abundance column indicates a radioactive nuclide not present in nature or an element whose isotopic composition varies so widely that a meaningful natural abundance cannot be defined.

Table 2.1–3   Atomic Masses and Abundances (continued)

Source: Lide, D. R. and Frederikse, H. P. R., Eds., *CRC Handbook of Chemistry and Physics,* 79th ed., CRC Press, Boca Raton, FL, 1998, 1–10. With permission.

REFERENCES

1. Holden, N. E., Table of the isotopes, in Lide, D. R., Ed., *CRC Handbook of Chemistry and Physics,* 79th ed., CRC Press, Boca Raton, FL, 1998.
2. Audi, G. and Wapstra, A. H., *Nucl. Phys.,* A595, 409, 1995.
3. Audi, G. and Wapstra, A. H., Atomic Mass Data Center, World Wide Web site, available at http://csnwww.in2p3.fr. Note that the print version in Reference 2 gives mass excess in energy units but not the actual atomic mass. The latter is included in the electronic version.
4. IUPAC Commission on Atomic Weights and Isotopic Abundances, *Pure Appl. Chem.,* 63, 991, 1991.

2.2   PROPERTIES OF SEAWATER

In addition to the dependence on temperature and pressure, the physical properties of seawater vary with the concentration of the dissolved constituents. A convenient parameter for describing the composition is the salinity, S, which is defined in terms of the electrical conductivity of the seawater sample. The defining equation for the practical salinity is:

$$S = a_0 + a_1 K^{1/2} + a_2 K + a_3 K^{3/2} + a_4 K^2 + a_5 K^{5/2}$$

where K is the ratio of the conductivity of the seawater sample at 15°C and atmospheric pressure to the conductivity of a potassium chloride solution in which the mass fraction of KCl is 0.0324356, at the same temperature and pressure. The values of the coefficients are

$$a_0 = 0.080 \qquad a_3 = 14.0941$$
$$a_1 = -0.1692 \qquad a_4 = -7.0261$$
$$a_2 = 25.3851 \qquad a_5 = 2.7081$$
$$\Sigma a_i = 35.0000$$

Thus when $K = 1$, $S = 35$ exactly (S is normally quoted in units of ‰ i.e., parts per thousand). The value of S can be roughly equated with the mass of dissolved material in grams per kilogram of seawater. Salinity values in the oceans at mid-latitudes typically fall between 34 and 36.

The freezing point of seawater at normal atmospheric pressure varies with salinity as follows:

S	0	5	10	15	20	25	30	35	40
$t_f/°C$	0.000	−0.274	−0.542	−0.812	−1.083	−1.358	−1.636	−1.922	−2.212

The first section of Table 2.2–1 below gives several properties of seawater as a function of temperature for a salinity of 35. The second section gives electrical conductivity as a function of salinity at several temperatures, and the third section lists typical concentrations of the main constituents of seawater as a function of salinity.

REFERENCES

1. *The Practical Salinity Scale 1978 and the International Equation of State of Seawater 1980,* UNESCO Technical Papers in Marine Science No. 36, UNESCO, Paris, 1981; sections No. 37, 38, 39, and 40 in this series give background papers and detailed tables.
2. Kennish, M. J., *Practical Handbook of Marine Science,* CRC Press, Boca Raton, FL, 1989.
3. Poisson, A., *IEEE J. Ocean. Eng.,* OE-5, 50, 1981.

Table 2.2–1   Properties of Seawater

Properties of Seawater as a Function of Temperature at Salinity $S = 35$ and Normal Atmospheric Pressure (100 kPa)

ρ = density in g/cm^3
β = $(1/\rho)$ $(d\rho/dS)$ = fractional change in density per unit change in salinity
α = $(1/\rho)$ $(d\rho/dt)$ = fractional change in density per unit change in temperature ($°C^{-1}$)
κ = electrical conductivity in S/cm
η = viscosity in mPa s (equal to c_p)
c_p = specific heat in J/kg
v = speed of sound in m/s

$t/°C$	$\rho/g\ cm^{-3}$	$10^7 \cdot \beta$	$10^7 \cdot \alpha/°C$	$\kappa/S/cm$	$\eta/mPa\ s$	$c_p/J/kg/°C$	$v/m/s$
0	1.028106	7854	526	0.029048	1.892	3986.5	1449.1
5	1.027675	7717	1136	0.033468	1.610		
10	1.026952	7606	1668	0.038103	1.388	3986.3	1489.8
15	1.025973	7516	2141	0.042933	1.221		
20	1.024763	7444	2572	0.047934	1.085	3993.9	1521.5
25	1.023343	7385	2970	0.053088	0.966		
30	1.021729	7338	3341	0.058373	0.871	4000.7	1545.6
35	1.019934	7300	3687				
40		7270	4004			4003.5	1563.2

Electrical Conductivity of Seawater in S/cm as a Function of Temperature and Salinity

$t/°C$	$S = 5$	$S = 10$	$S = 15$	$S = 20$	$S = 25$	$S = 30$	$S = 35$	$S = 40$
0	0.004808	0.009171	0.013357	0.017421	0.021385	0.025257	0.029048	0.032775
5	0.005570	0.010616	0.015441	0.020118	0.024674	0.029120	0.033468	0.037734
10	0.006370	0.012131	0.017627	0.022947	0.028123	0.033171	0.038103	0.042935
15	0.007204	0.013709	0.019905	0.025894	0.031716	0.037391	0.042933	0.048355
20	0.008068	0.015346	0.022267	0.028948	0.035438	0.041762	0.047934	0.053968
25	0.008960	0.017035	0.024703	0.032097	0.039276	0.046267	0.053088	0.059751
30	0.009877	0.018771	0.027204	0.035330	0.043213	0.050888	0.058373	0.065683

Composition of Seawater and Ionic Strength at Various Salinities[a]

Constituent	$S = 30$	$S = 35$	$S = 40$
Cl^-	0.482	0.562	0.650
Br^-	0.00074	0.00087	0.00100
F^-		0.00007	
SO_4^{2-}	0.0104	0.0114	0.0122
HCO_3^-	0.00131	0.00143	0.00100
$NaSO_4^-$	0.0085	0.0108	0.0139
KSO_4^-	0.00010	0.00012	0.00015
Na^+	0.405	0.472	0.544
K^+	0.00892	0.01039	0.01200
Mg^{2+}	0.0413	0.0483	0.0561
Ca^{2+}	0.00131	0.00143	0.00154
Sr^+	0.00008	0.00009	0.00011
$MgHCO_3^+$	0.00028	0.00036	0.00045
$MgSO_4$	0.00498	0.00561	0.00614
$CaSO_4$	0.00102	0.00115	0.00126
$NaHCO_3$	0.00015	0.00020	0.00024
H_3BO_3	0.00032	0.00037	0.00042
Ionic strength (mol/kg)	0.5736	0.6675	0.7701

[a] Concentration of major constituents expressed as molality (moles per kilogram of H_2O).

Source: Lide, D. R. and Frederikse, H. P. R., Eds., *CRC Handbook of Chemistry and Physics,* 79th ed., CRC Press, Boca Raton, FL, 1998, 14–12. With permission.

Table 2.2–2   The Volumetric Properties of Seawater ($S = 35$) at 1 atm

Temperature,°C	ρ (g/cm)	v (m³/kg)	$10^6\alpha$ (deg⁻¹)	$10^6\beta$ (bar⁻¹)	$10^6\beta s$ (bar⁻¹)
0	1.028106	972.662	52.55	46.334	46.316
5	1.027675	973.070	113.61	45.074	44.985
10	1.026952	973.755	116.79	44.062	43.870
15	1.025972	974.685	214.14	43.258	42.936
20	1.024763	975.835	257.18	42.627	42.154
25	1.023343	977.189	296.98	42.147	41.504
30	1.021729	978.733	334.13	41.799	40.969
35	1.019934	980.456	368.73	41.568	40.537
40	1.017973	982.344	400.38	41.445	40.199

Source: Millero, F. J., *Chemical Oceanography,* 2nd ed., CRC Press, Boca Raton, FL, 1996, 432. With permission.

Table 2.2–3   The Effect of Pressure on the Volumetric Properties of Seawater ($S = 35$)

Pressure, bar	0°C			25°C		
	v (m³/kg)	$10^6\alpha$ (deg⁻¹)	$10^6\beta$ (bar⁻¹)	v (m³/kg)	$10^6\alpha$ (deg⁻¹)	$10^6\beta$ (bar⁻¹)
0	972.662	52.55	46.334	977.189	296.98	42.147
100	968.224	79.90	45.128	973.129	305.07	41.132
200	963.921	105.81	43.962	969.182	312.86	40.157
300	959.748	130.30	42.835	965.344	320.36	39.220
400	955.698	153.38	41.746	961.609	327.57	38.318
500	951.767	175.08	40.692	957.973	334.50	37.451
600	947.950	195.40	39.673	954.432	341.15	36.615
700	944.244	214.36	38.686	950.982	347.52	35.811
800	940.643	231.99	37.731	947.620	353.62	35.035
900	937.144	248.30	36.806	944.341	359.46	34.288
1000	933.743	263.31	35.910	941.143	365.04	33.566

Source: Millero, F. J., *Chemical Oceanography,* 2nd ed., CRC Press, Boca Raton, FL, 1996, 432. With permission.

Table 2.2–4   Ocean Pressure as a Function of Depth and Latitude

Depth (meters)	Pressure in MPa at the Specified Latitude						
	0°	15°	30°	45°	60°	75°	90°
0	0.0000	0.0000	0.0000	0.0000	0.0000	0.0000	0.0000
500	5.0338	5.0355	5.0404	5.0471	5.0537	5.0586	5.0605
1000	10.0796	10.0832	10.0930	10.1064	10.1198	10.1296	10.1333
1500	15.1376	15.1431	15.1577	15.1778	15.1980	15.2127	15.2182
2000	20.2076	20.2148	20.2344	20.2613	20.2882	20.3080	20.3153
2500	25.2895	25.2985	25.3231	25.3568	25.3905	25.4153	25.4244
3000	30.3831	30.3940	30.4236	30.4641	30.5047	30.5345	30.5453
3500	35.4886	35.5012	35.5358	35.5832	35.6307	35.6654	35.6782
4000	40.6056	40.6201	40.6598	40.7140	40.7683	40.8082	40.8229
4500	45.7342	45.7505	45.7952	45.8564	45.9176	45.9626	45.9791
5000	50.8742	50.8924	50.9421	51.0102	51.0785	51.1285	51.1469
5500	56.0255	56.0456	56.1004	56.1755	56.2508	56.3059	56.3262
6000	61.1882	61.2100	61.2700	61.3521	61.4344	61.4947	61.5168
6500	66.3619	66.3857	66.4508	66.5399	66.6292	66.6947	66.7187
7000	71.5467	71.5724	71.6427	71.7388	71.8352	71.9059	71.9318
7500	76.7426	76.7701	76.8456	76.9488	77.0523	77.1282	77.1560

Table 2.2–4   Ocean Pressure as a Function of Depth and Latitude (continued)

Depth	Pressure in MPa at the Specified Latitude						
(meters)	0°	15°	30°	45°	60°	75°	90°
8000	81.9493	81.9788	82.0594	82.1697	82.2804	82.3614	82.3911
8500	87.1669	87.1983	87.2841	87.4016	87.5193	87.6057	87.6373
9000	92.3950	92.4284	92.5194	92.6440	92.7689	92.8606	92.8941
9500	97.6346	97.6698	97.7661	97.8978	98.0300	98.1269	98.1624
10000	102.8800	102.9170	103.0185	103.1572	103.2961	103.3981	103.4355

Note: This table is based upon an ocean model that takes into account the equation of state of standard seawater and the dependence on latitude of the acceleration of gravity. The tabulated pressure value is the excess pressure over the ambient atmospheric pressure at the surface.

Source: Lide, D. R. and Frederikse, H. P. R., Eds., CRC Handbook of Chemistry and Physics, 79th ed., CRC Press, Boca Raton, FL, 1998, 14–11. With permission.

REFERENCES

1. International Oceanographic Tables, Volume 4, UNESCO Technical Papers in Marine Science No. 40, UNESCO, Paris, 1987.
2. Saunders, P. M. and Fofonoff, N. P., Deep-Sea Res., 23, 109–111, 1976.

Table 2.2–5   The Thermochemical Properties of Seawater ($S = 35$) at 1 atm

Temperature,°C	C_p(J/g/K)	C_V(J/g/K)	$10^3 h$(J/g)	$-g$(J/g)	$10^6 s$(J/g/K)
0	3.9865	3.9848	−30.59	5.01	1.718
5	3.9842	3.9772	−17.18	5.10	1.735
10	3.9861	3.9691	−4.73	5.19	1.815
15	3.9912	3.9599	6.88	5.28	1.858
20	3.9937	3.9492	17.86	5.37	1.897
25	3.9962	3.9366	28.30	5.47	1.933
30	4.0011	3.9215	38.42	5.57	1.968
35	4.0031	3.9037	48.35	5.66	2.001
40	4.0039	3.8829	58.27	5.76	2.035

Note: C_p = specific heat capacity at constant pressure; C_V = specific heat capacity at constant volume; h = specific enthalpy; g = specific free energy; s = specific entropy.

Source: Millero, F. J., Chemical Oceanography, 2nd ed., CRC Press, Boca Raton, FL, 1996, 433. With permission.

Table 2.2–6   The Colligative Properties of Seawater ($S = 35$) at 1 atm

Salinity	$-T_f$(°C)	Temp. (°C)	φ	a	p(mm Hg)	π (bar)
0	0	0	0.8925	0.9815	4.496	23.54
5	0.274	5	0.8954	0.9814	6.419	24.05
10	0.542	10	0.8978	0.9814	9.036	24.54
15	0.811	15	0.8996	0.9814	12.551	25.02
20	1.083	20	0.9009	0.9813	17.213	25.46
25	1.358	25	0.9017	0.9813	23.323	25.90
30	1.637	30	0.9023	0.9813	31.245	26.31
35	1.922	35	0.9025	0.9813	41.412	26.70
40	2.211	40	0.9025	0.9813	54.329	27.09

Note: T_f = freezing point; φ = osmotic coefficient; a = activity of water; p = vapor pressure; π = osmotic pressure.

Source: Millero, F. J., Chemical Oceanography, 2nd ed., CRC Press, Boca Raton, FL, 1996, 433. With permission.

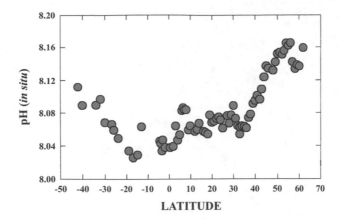

Figure 2.2–1 The pH of surface waters in the Atlantic Ocean. (From Millero, F. J., *Chemical Oceanography,* 2nd ed., CRC Press, Boca Raton, FL, 1996, 265. With permission.)

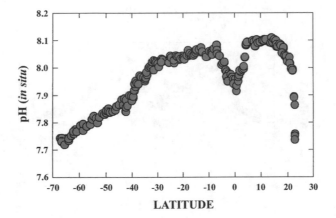

Figure 2.2–2 The pH of surface waters in the Pacific Ocean. (From Millero, F. J., *Chemical Oceanography,* 2nd ed., CRC Press, Boca Raton, FL, 1996, 265. With permission.)

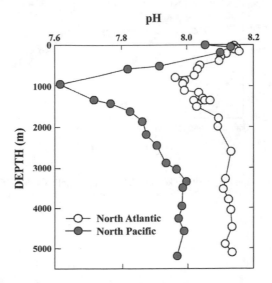

Figure 2.2–3 Depth profile of pH in the Atlantic and Pacific Oceans. (From Millero, F. J., *Chemical Oceanography,* 2nd ed., CRC Press, Boca Raton, FL, 1996, 265. With permission.)

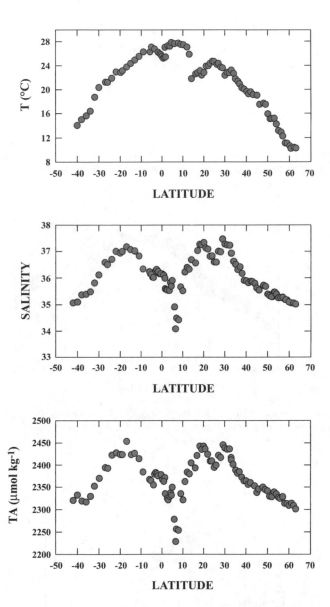

Figure 2.2–4   Temperature, salinity, and total alkalinity (TA) in the Atlantic Ocean. (From Millero, F. J., *Chemical Oceanography,* 2nd ed., CRC Press, Boca Raton, FL, 1996, 266. With permission.)

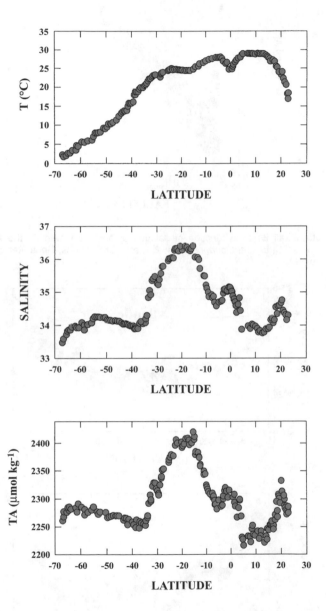

Figure 2.2–5 Temperature, salinity, and total alkalinity (TA) in the Pacific Ocean. (From Millero, F. J., *Chemical Oceanography,* 2nd ed., CRC Press, Boca Raton, FL, 1996, 267. With permission.)

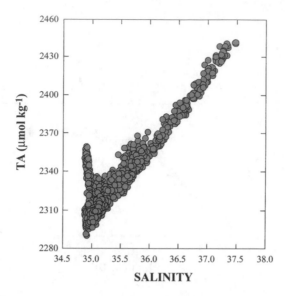

Figure 2.2–6 The total alkalinity as a function of salinity for surface waters in the Atlantic Ocean. (From Millero, F. J., *Chemical Oceanography*, 2nd ed., CRC Press, Boca Raton, FL, 1996, 268. With permission.)

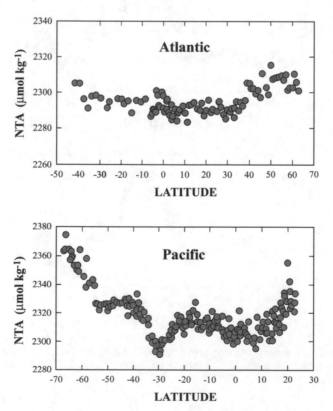

Figure 2.2–7 Normalized total alkalinity (NTA) in the Atlantic and Pacific Oceans. (From Millero, F. J., *Chemical Oceanography,* 2nd ed., CRC Press, Boca Raton, FL, 1996, 268. With permission.)

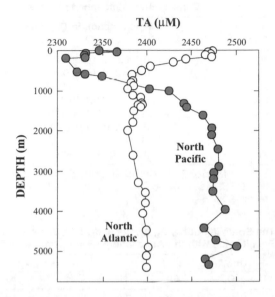

Figure 2.2–8 The total alkalinity as a function of depth in the Atlantic and Pacific Oceans. (From Millero, F. J., *Chemical Oceanography,* 2nd ed., CRC Press, Boca Raton, FL, 1996, 269. With permission.)

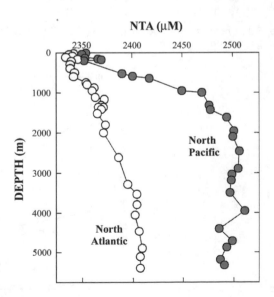

Figure 2.2–9 Normalized total alkalinity (NTA) as a function of depth in the Atlantic and Pacific Oceans. (From Millero, F. J., *Chemical Oceanography,* 2nd ed., CRC Press, Boca Raton, FL, 1996, 269. With permission.)

2.3 ATMOSPHERIC AND FLUVIAL FLUXES

Table 2.3–1   Abundance of the Major Conservative Atmospheric Gases

Gas	Mole Fraction in Dry Air (X_i)
N_2	0.78084 ± 0.00004
O_2	0.20946 ± 0.00002
Ar	$(9.34 \pm 0.01) \times 10^{-3}$
CO_2	$(3.5 \pm 0.1) \times 10^{-4}$
Ne	$(1.818 \pm 0.004) \times 10^{-5}$
He	$(5.24 \pm 0.004) \times 10^{-6}$
Kr	$(1.14 \pm 0.01) \times 10^{-6}$
Xe	$(8.7 \pm 0.1) \times 10^{-8}$

Source: Millero, F. J., *Chemical Oceanography,* 2nd ed., CRC Press, Boca Raton, FL, 1996, 177. With permission.

Table 2.3–2   The Solubility of N_2, O_2, Ar, Ne, and He in Seawater $S = 35$ Equilibrated with the Atmosphere ($P = 1$ atm) at 100% Humidity

	μmol/kg			nmol/kg			
	N_2	O_2	Ar	Ne	He	Kr	Xe
0°C	616.4	349.5	16.98	7.88	1.77	4.1	0.66
5	549.6	308.1	15.01	7.55	1.73	3.6	0.56
10	495.6	274.8	13.42	7.26	1.70	3.1	0.46
15	451.3	247.7	12.11	7.00	1.68	2.7	0.39
20	414.4	225.2	11.03	6.77	1.66	2.4	0.33
25	383.4	206.3	10.11	6.56	1.65	2.2	0.29
30	356.8	190.3	9.33	6.36	1.64	2.0	0.25

Source: Millero, F. J., *Chemical Oceanography,* 2nd ed., CRC Press, Boca Raton, FL, 1996, 209. With permission.

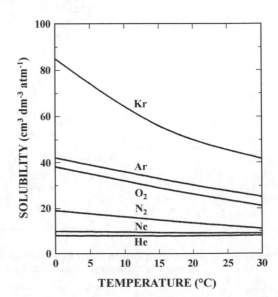

Figure 2.3–1   The effect of temperature on the solubility of gases in seawater. (From Millero, F.J., *Chemical Oceanography,* 2nd ed., CRC Press, Boca Raton, FL, 1996, 216. With permission.)

Table 2.3–3   The Composition of Minor Gases in the Atmosphere

Species	X_i Actual	Reliability	Source	Sink
CH_4	1.7×10^{-6}	High	Biog.	Photochem.
CO	$0.5 - 2 \times 10^{-7}$	Fair	Photo., anthr.	Photochem.
O_3	5×10^{-8} (clean)	Fair	Photo	Photochem.
	4×10^{-7} (polluted)			
	10^{-7} to 6×10^{-6} (stratosphere)			
$NO + NO_2$	10^{-12}–10^{-8}	Low	Lightn., anthr. photo.	Photochem.
HNO_3	10^{-11}–10^{-9}	Low	Photo.	Rainout
NH_3	10^{-10}–10^{-9}	Low	Biog.	Photo., rainout
N_2O	3×10^{-7}	High	Biog.	Photo.
H_2	5×10^{-7}	High	Biog., photo.	Photo.
OH	10^{-15}–10^{-12}	Very low	Photo.	Photo.
HO_2	10^{-11}–10^{-13}	Very low	Photo.	Photo.
H_2O_2	10^{-10}–10^{-18}	Very low	Photo.	Rainout
H_2CO	10^{-10}–10^{-9}	Low	Photo.	Photo.
SO_2	10^{-11}–10^{-10}	Fair	Anth., photo.	Photo., volcanic
CS_2	10^{-11}–10^{-10}	Low	Anthr., biol.	Photo.
OCS	5×10^{-10}	Fair	Anthr., biol., photo.	Photo.
CH_3CCl_3	$0.7 - 2 \times 10^{-10}$	Fair	Anthropogenic	Photo.

Source: Millero, F.J., *Chemical Oceanography,* 2nd ed., CRC Press, Boca Raton, FL, 1996, 178. With permission.

Table 2.3–4   Concentrations of Elements in Precipitation over Marine Regions (units, μg/l)

Element	North Sea	Bay of Bengal	Bermuda	North Pacific (Enewetak)	South Pacific (Samoa)
Na	19,000	14,000	3,400	1,100	2,500
Mg	—	1,600	490	170	270
K	—	1,400	160	39	88
Ca	—	1,700	310	50	75
Cl	32,000	25,000	6,800	2,000	4,700
Br	—	—	—	7.1	12
Al	105	—	—	2.1	16
Fe	84	30	4.8	1.0	0.42
Mn	<12	—	0.27	0.012	0.020
Sc	0.010	0.016	—	0.00023	—
Th	—	0.12	—	0.00091	—
Co	0.17	0.95	—	—	—
I	—	—	—	1.2	0.021
V	3.7	—	—	0.018	—
Zn	35	100	1.15	0.052	1.6
Cd	—	—	0.06	0.0021	—
Cu	15	—	0.66	0.013	0.021
Pb	—	—	0.77	0.035	0.014
Ag	—	—	—	0.0056	—
Se	—	—	—	0.021	0.026

Source: Furness, R. W. and Rainbow, P. S., Eds., *Heavy Metals in the Marine Environment,* CRC Press, Boca Raton, FL, 1990, 36. With permission.

Table 2.3–5 Atmospheric Deposition Fluxes to the Sea Surface over a
Number of Marine Regions (μg/cm^2/year)

Element	North Atlantic[a]	North Pacific[b,d]	South Pacific[c,d]
Al	5.0	1.8	0.13
Fe	3.2	0.82	0.047
Mn	0.07	0.0113	0.0036
Ni	0.02	—	—
Co	0.0027	0.00016	0.000025
Cr	0.014	—	—
V	0.017	0.0069	—
Cu	0.025	0.0071	0.0044
Pb	0.31	0.0074	0.0014
Zn	0.13	0.065	0.0058
Cd	0.002	0.00029	—

[a] Data for the tropical North Atlantic. Deposition fluxes for all elements calculated
on the basis of a global total deposition (wet and dry) rate of 1 cm/s.
[b] Data for Enewetak, North Pacific.
[c] Data for Samoa, South Pacific.
[d] Wet deposition fluxes were calculated using two methods: (1) those based on
the concentration of elements in rain and the total rainfall at the two islands and
(2) those based on scavenging ratios. To avoid any bias which may be introduced
by unrepresentative total rainfall over the islands compared to the open-ocean
regions, wet deposition rates for both Enewetak and Samoa have been estimated
from scavenging ratios only; both wet and dry deposition rates have been
adjusted to take account of sea-surface recycling.

Source: Furness, R. W. and Rainbow, P. S., Eds., *Heavy Metals in the Marine
Environment,* CRC Press, Boca Raton, FL, 1990, 45. With permission.

Table 2.3–6 Atmospheric Dissolved vs. Fluvial Dissolved Inputs to Regions of the World
Ocean (μg/cm^2/year)

	Oceanic Region					
	North Atlantic		North Pacific		South Pacific	
Element	Fluvial Dissolved Flux[a]	Atmospheric Dissolved Flux	Fluvial Dissolved Flux	Atmospheric Dissolved Flux	Fluvial Dissolved Flux	Atmospheric Dissolved Flux
Al	0.56	0.25	0.24	0.09	0.12	0.0066
Fe	0.18	0.24	0.076	0.062	0.038	0.0035
Mn	0.18	0.025	0.076	0.004	0.038	0.0013
Ni	0.017	0.008	0.0074	—	0.0037	—
Co	0.0044	0.00061	0.0019	0.00004	0.0009	0.000006
Cr	0.011	0.0014	0.0047	—	0.0023	—
V	0.022	0.0043	0.0094	0.0017	0.0047	—
Cu	0.037	0.0075	0.016	0.0021	0.0079	0.0013
Pb	0.0022	0.093	0.0009	0.0022	0.0005	0.0004
Zn	0.016	0.059	0.0071	0.029	0.0035	0.0026
Cd	0.0018	0.0016	0.0008	0.00023	0.0004	—

[a] No attempt has been made to adjust the North Atlantic fluvial flux for European and North American
anthropogenic inputs.

Source: Chester, R. and Murphy, K. J. T., in *Heavy Metals in the Marine Environment,* Rainbow, P. S. and
Furness, R. W., Eds., CRC Press, Boca Raton, FL, 1990, 27. With permission.

Table 2.3–7   Net Fluvial and Atmospheric Fluxes to the Global Sea Surface (μg/cm²/year)

Element	Net Global Fluvial Dissolved Flux		Net Global Total Atmospheric Input		Estimated Average Seawater Solubility from Aerosols (%)	Net Global Dissolved Atmospheric Flux[d]
	Present Work[a]	Collier and Edmond[b]	Present Work[c]	Collier and Edmond[b]		
Al	0.27	0.13–0.67	1.85	0.27–5.4	5	0.088
Fe	0.085	<0.73	1.0	1.12–3.35	7.5	0.075
Mn	0.085	0.093	0.021	0.016–0.044	35	0.0074
Ni	0.0082	0.0035	<0.02	0.003	40	<0.008
Co	0.0021	—	0.00068	—	22.5	0.00015
Cr	0.0052	—	<0.014	—	10	<0.0014
V	0.010	—	<0.010	—	25	<0.0028
Cu	0.018	0.019	0.010	0.00064–0.0095	30	0.0033
Pb	0.001	—	0.074	—	30	0.022
Zn	0.0079	<0.0065	0.055	0.022–0.013	45	0.025
Cd	0.00088	0.00022	<0.00096	0.0023	80	<0.00077

[a] Calculated from net fluvial flux data assuming a global ocean area of 352.6 ×10⁶ km²; at a global river discharge of 37,400 km³/year, which is included in the net flux estimates, this would yield a layer 10.6 cm spread evenly over the whole ocean surface.

[b] Collier, R. and Edmond, J., The trace element geochemistry of marine biogenic particulate matter, *Prog. Oceanogr.*, 13, 113, 1984.

[c] Calculated from the tropical North Atlantic (column 1), North Pacific (column 2), and South Pacific (column 3) atmospheric deposition fluxes by scaling them to global fluxes weighted on an areal basis for each oceanic region.

[d] Net total atmospheric flux adjusted for the average seawater solubility of the individual elements.

Source: From Furness, R. W. and Rainbow, P. S., Eds., *Heavy Metals in the Marine Environment,* CRC Press, Boca Raton, FL, 1990, 46. With permission.

Table 2.3–8   Relative Fluvial and Atmospheric Fluxes to Some Coastal Oceanic Regions

A. Ratio of Total Atmospheric Flux to Fluvial Flux

Element	South Atlantic Bight	New York Bight	North Sea	Western Mediterranean
Fe	5.8	6.4	1.7	—
Mn	0.6	—	0.8	—
Cu	1.9	—	1.9	—
Ni	1.7	—	1.3	—
Pb	9.5	20	6.8	6.2
Zn	2.3	3.1	1.9	0.8
Cd	2.7	3.1	1.1	—
As	2.1	1.0	1.7	—
Hg	22	—	2.1	0.8

B. Estimated Fluvial and "Soluble" Atmospheric Fluxes to the South Atlantic Bight (units, g $\times$ 10^6)

Element	Fluvial Flux	Total Atmospheric Flux	Soluble Atmospheric Flux	Ratio: Soluble Atmospheric Flux to Fluvial Flux
Fe	950	5500	413	0.43
Mn	91	57	20	0.29
Cu	110	210	63	0.57
Ni	220	370	148	0.85
Pb	65	620	186	4.8
Zn	310	710	320	1.6
Cd	3	8	6.4	2.3

Source: Furness, R. W. and Rainbow, P. S. Eds., *Heavy Metals in the Marine Environment,* CRC Press, Boca Raton, FL, 1990, 42. With permission.

Table 2.3–9   Composition of Average World River Water (1 l)[a]

Species	$10^6 g_i$	$10^3 n_i$	$10^3 e_i$	$10^3 l_i$
Na^+	6.5	0.283	0.283	0.283
Mg^{2+}	4.1	0.169	0.337	0.674
Ca^{2+}	15.0	0.374	0.749	1.496
K^+	2.3	0.059	0.059	0.059
Cl^-	7.8	0.220	0.220	0.220
SO_4^{2-}	11.2	0.117	0.233	0.466
HCO_3^-	58.4	0.950	0.950	0.950
CO_3^{2-}	—	0.002	0.004	0.008
NO_3^-	1.0	0.016	0.016	0.016
$Si(OH)_3O^-$	—	0.005	0.005	0.005
1/2 Σ		1.086	1.428	2.089
$Si(OH)_4$	21.5	0.213	0.213	—
	$g_T = 126.8$	$n_T = 1.299$	$e_T = 1.641$	$l_T = 2.089$

[a] To convert to molar units multiply by the density. To convert to molal units divide by $X_{H_2O} = 0.96483$.

Source: Millero, F. J., *Chemical Oceanography,* 2nd ed., CRC Press, Boca Raton, FL, 1996, 84. With permission.

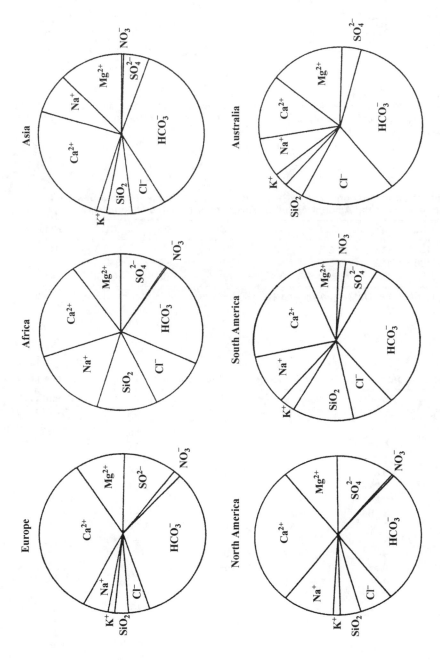

Figure 2.3–2 Major components of various continental rivers of the world. (From Millero, F. J., *Chemical Oceanography*, 2nd ed., CRC Press, Boca Raton, FL, 1996, 77. With permission.)

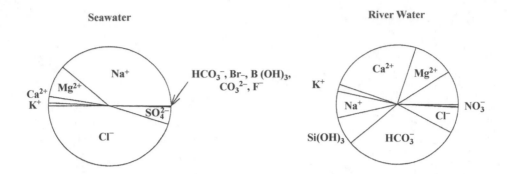

Figure 2.3–3   Equivalent fractions of the major constituents of the average world river and seawater. (From Millero, F. J., *Chemical Oceanography,* 2nd ed., CRC Press, Boca Raton, FL, 1996, 78. With permission.)

Table 2.3–10   Present Concentration (mol/cm² of surface) of Major Components of Seawater Compared with the Amount Added by Rivers over Last 100 Million Years

	Na	Mg	Ca	K	Cl	SO₄	CO₃	NO₃
Present	129	15	2.8	2.7	150	8	0.3	0.01
Added by rivers	196	122	268	42	157	84	342	11
Excess added	67	107	265	39	7	76	342	11

Source: Millero, F. J., *Chemical Oceanography,* 2nd ed., CRC Press, Boca Raton, FL, 1996, 119. With permission.

Table 2.3–11   The Concentration of Various Components of the Geochemical Cycle of the Oceans

Component	Rock	Volatile	Air	SW	Sediments
H_2O	—	54.90	—	54.90	—
Cl(HCl)	—	0.94	—	0.55	0.40
Na(NaO$_{.5}$, NaOH)	1.47	—	—	0.47	1.00
Ca(CaO, Ca(OH)$_2$)	1.09	—	—	0.01	1.08
Mg(MgO, Mg(OH)$_2$)	0.87	—	—	0.05	0.82
K(KO$_{.5}$, KOH)	0.79	—	—	0.01	0.78
Si(SiO$_2$)	12.25	—	—	—	12.25
Al(AlO$_{1.5}$, Al(OH)$_3$)	3.55	—	—	—	3.55
C(CO$_2$)	0.03	1.05	—	0.002	1.08
C(s)	—	1.01	—	—	1.01
O_2	—	0.022	0.022	—	—
Fe(FeO, Fe(OH)$_2$)	0.52	—	—	—	0.53
(FeO$_{1.5}$, FeOH)	0.38	—	—	—	0.38
Ti(TiO$_2$)	0.12	—	—	—	0.12
S	0.02	0.06	—	0.03	0.05
F(HF)	0.05	—	—	—	0.05
P(PO$_{2.5}$,H$_3$PO$_4$)	0.04	—	—	—	0.04
Mn(MnO$_{1 \text{ to } 2}$)	0.05	—	—	—	0.05
N_2	—	0.082	0.082	—	—

Source: Millero, F. J., *Chemical Oceanography,* 2nd ed., CRC Press, Boca Raton, FL, 1996, 121. With permission.

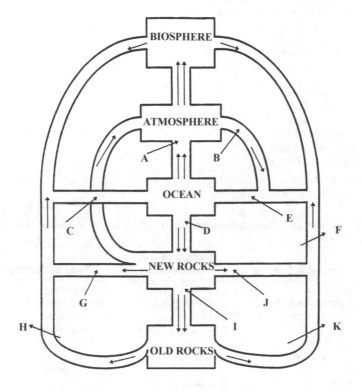

A. Ocean to Atmosphere
B. Atmosphere to Streams
C. Total Suspended Load of Streams
D. Total Sediment Flux
E. Dissolved Load of Streams
F. Total Flux Dissolved from Rocks

G. Suspended Load from New Rocks
H. Suspended Load from Old Rocks
I. Flux from New Rocks to Old Rocks
J. Dissolved Load from New Rocks
K. Dissolved Load from Old Rocks

NEW ROCKS
4800×10^{20} g of
elements (150×10^6 y)

BIOSPHERE
0.01×10^{20} gC Living
0.04×10^{20} gC Dead

OLD ROCKS
8450×10^{20} g of
elements (663×106 y)

ATMOSPHERE
0.065×10^{20} g C
as CO_2

OCEAN
500×10^{20} g of
elements

Figure 2.3–4 The cycle of elements. (From Millero, F. J., *Chemical Oceanography*, 2nd ed., CRC Press, Boca Raton, FL, 1996, 120. With permission.)

2.4 COMPOSITION OF SEAWATER

Table 2.4–1   Major Components of Seawater

1. Solids (material that does not pass through a 0.45-μm filter)
   a. Particulate organic material (plant detritus)
   b. Particulate inorganic material (minerals)
2. Gases
   a. Conservative (N_2, Ar, Xe)
   b. Nonconservative (O_2 and CO_2)
3. Colloids (passes through 0.45-μm filter, but is not dissolved)
   a. Organic
   b. Inorganic
4. Dissolved Solutes
   a. Inorganic solutes
      1. Major (>1ppm)
      2. Minor (<1ppm)
   b. Organic solutes

Source: Millero, F. J. and Sohn, M. L., *Chemical Oceanography,* CRC Press, Boca Raton, FL, 1992, 59. With permission.

Table 2.4–2   Speciation, Concentration, and Distribution Types of Elements in Ocean Waters

Element	Probable Species	Range and Avg. Concentration	Type of Distribution
Li	Li^+	25 μM	Conservative
Be	$BeOH^+$, $Be(OH)_2$	4–30 pM, 20 pM	Nutrient type
B	$B(OH)_3$, $B(OH)_4$	0.416 mM	Conservative
C	HCO_3^-, CO_3^{2-}	2.0–2.5 mM, 2.3 mM	Nutrient type
N	NO_3^-, (N_2)	0–45 μM	Nutrient type
F	F^-, MgF^+, CaF^+	68 μM	Conservative
Na	Na^+	0.468 M	Conservative
Mg	Mg^{2+}	53.2 mM	Conservative
Al	$Al(OH)_4^-$, $Al(OH)_3$	5–40 nM, 2 nM	Mid-depth-min.
Si	$Si(OH)_4$	0–180 μM	Nutrient type
P	HPO_4^{2-}, $MgHPO_4$	0–3.2 μM	Nutrient type
S	SO_4^{2-}, $NaSO_4^-$, $MgSO_4$	28.2 mM	Conservative
Cl	Cl^-	0.546 M	Conservative
K	K^+	10.2 mM	Conservative
Ca	Ca^{2+}	10.3 mM	Conservative
Sc	$Sc(OH)_3$	8–20 pM, 15 pM	Surface depletion
Ti	$Ti(OH)_4$	Few pM	?
V	HVO_4^{2-}, $H_2VO_4^-$	20–35 nM	Surface depletion
Cr	CrO_4^{2-}	2–5 nM, 4 nM	Nutrient type
Mn	Mn^{2+}	0.2–3 nM, 0.5 nM	Depletion at depth
Fe	$Fe(OH)_3$	0.1–2.5 nM, 1 nM	Surface and depth depletion
Co	Co^{2+}, $CoCO_3$	0.01–0.1 nM, 0.02 nM	Surface and depth depletion
Ni	$NiCO_3$	2–12 nM, 8 nM	Nutrient type
Cu	$CuCO_3$	0.5–6 nM, 4 nM	Nutrient type, scavenging
Zn	Zn^{2+}, $ZnOH^+$	0.05–9 nM, 6 nM	Nutrient type
Ga	$Ga(OH)_4^-$	5–30 pM	?
As	$HAsO_4^{2-}$	15–25 nM, 23 nM	Nutrient type
Se	SeO_4^{2-}, SeO_3^{2-}	0.5–2.3 nM, 1.7 nM	Nutrient type
Br	Br^-	0.84 nM	Conservative
Rb	Rb^+	1.4 μM	Conservative
Sr	Sr^{2+}	90 μM	Conservative
Y	YCO_3^+	0.15 nM	Nutrient type
Zr	$Zr(OH)_4$	0.3 nM	?

Table 2.4–2 Speciation, Concentration, and Distribution Types of Elements in Ocean Waters (continued)

Element	Probable Species	Range and Avg. Concentration	Type of Distribution
Nb	$NbCO_3^+$	50 pM	Nutrient type(?)
Mo	MoO_4^{2-}	0.11 μM	Conservative
Tc	TcO_4^-	No stable isotope	?
Ru	?	<0.05 pM	?
Rh	?	?	?
Pd	?	0.2 pM	?
Ag	$AgCl_2^-$	0.5–35 pM, 25 pM	Nutrient type
Cd	$CdCl_2^-$	0.001–1.1 nM, 0.7 nM	Nutrient type
In	$In(OH)_3$	1 pM	?
Sn	$Sn(OH)_4$	1–12 pM, 4 pM	Surface input
Sb	$Sb(OH)_6^-$	1.2 nM	?
Te	TeO_3^{2-}, $HTeO_3^-$	?	?
I	IO_3^-	0.2–0.5 μM, 0.4 μM	Nutrient type
Cs	Cs^+	2.2 nM	Conservative
Ba	Ba^{2+}	32–150 nM, 100 nM	Nutrient type
La	$LaCO_3^+$	13–37 pM, 30 pM	Surface depletion
Ce	$CeCO_3^+$	16–26 pM, 20 pM	Surface depletion
Pr	$PrCO_3^+$	4 pM	Surface depletion
Nd	$NdCO_3^+$	12–25 pM, 10 pM	Surface depletion
Sm	$SmCO_3^+$	3–5 pM, 4 pM	Surface depletion
Eu	$EuCO_3^+$	0.6–1 pM, 0.9 pM	Surface depletion
Gd	$GdCO_3^+$	3–7 pM, 6 pM	Surface depletion
Tb	$TbCO_3^+$	0.9 pM	Surface depletion
Dy	$DyCO_3^+$	5–6 pM, 6 pM	Surface depletion
Ho	$HoCO_3^+$	1.9 pM	Surface depletion
Er	$ErCO_3^+$	4–5 pM, 5 pM	Surface depletion
Tm	$TmCO_3^+$	0.8 pM	Surface depletion
Yb	$YbCO_3^+$	3–5 pM, 5 pM	Surface depletion
Lu	$LuCO_3^+$	0.9 pM	Surface depletion
Hf	$Hf(OH)_4$	<40 pM	?
Ta	$Ta(OH)_5$	<14 pM	?
W	WO_4^{2-}	0.5 nM	Conservative
Re	ReO_4^-	14–30 pM, 20 pM	Conservative
Os	?	?	?
Ir	?	0.01 pM	?
Pt	$PtCl_4^{2-}$	0.5 pM	?
Au	$AuCl_2^-$	0.1–0.2 pM	?
Hg	$HgCl_4^{2-}$	2–10 pM, 5 pM	?
Tl	Tl^+, TCl	60 pM	Conservative
Pb	$PbCO_3$	5–175 pM, 10 pM	Surface input, depletion at depth
Bi	BiO^+, $Bi(OH)_2^+$	<0.015–0.24 pM	Depletion at depth

Source: Millero, F. J., *Chemical Oceanography,* 2nd ed., CRC Press, Boca Raton, FL, 1996, 98. With permission.

Dissolved Elements in Seawater

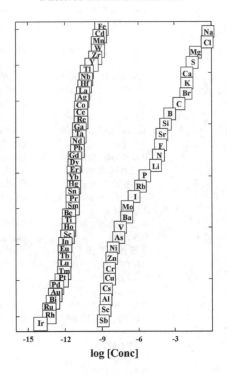

Figure 2.4–1   Range of concentrations of elements in seawater. (From Millero, F. J., *Chemical Oceanography,* 2nd ed., CRC Press, Boca Raton, FL, 1996, 100. With permission.)

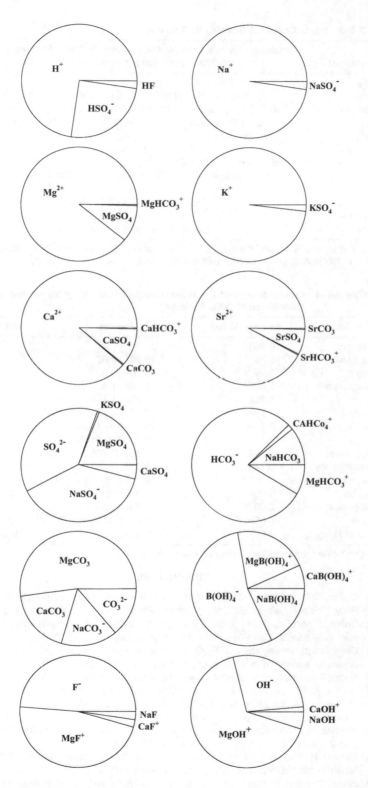

Figure 2.4–2 The speciation of cations (above) and anions (below) in seawater. (From Millero, F. J., *Chemical Oceanography*, 2nd ed., CRC Press, Boca Raton, FL, 1996, 157. With permission).

Table 2.4–3   Major Chemical Species in Seawater

Constituent	Percentage of the Constituent Present in Each Species at 25°C, 19.375‰ chlorinity, 1 atm, and pH 8.0
Chloride	—
Sodium	Na^+ (97.7%); $NaSO_4^-$ (2.2%); $NaHCO_3^0$ (0.03%)
Magnesium	Mg^{2+} (89%); $MgSO_4^0$ (10%); $MgHCO_3^+$ (0.6%); $MgCO_3^0$ (0.1%)
Sulfate	SO_4^{2-} (39%); $NaSO_4^-$ (37%); $MgSO_4^0$ (19%); $CaSO_4^0$ (4%)
Calcium	Ca^{2+} (88%); $CaSO_4^0$ (11%); $CaHCO_3^+$ (0.6%); $CaCO_3^0$ (0.1%)
Potassium	K^+ (98.8%); KSO_4^- (1.2%)
Bicarbonate	HCO_3^- (64%); $MgHCO_3^+$ (16%); $NaHCO_3^0$ (8%); $CaHCO_3^+$ (3%); CO_3^{2-} (0.8%); $MgCO_3^0$ (6%); $NaCO_3$(1%); $CaCO_3^0$ (0.5%)
Bromide	—
Boron	$B(OH)_3$ (84%); $B(OH)_4^-$(16%)
Strontium	—
Fluoride	F^- (50–80%); MgF^+ (20–50%)

Source: Pytkowicz, R. M. and Kester, D. R., in *Oceanogr. Mar. Biol. Ann. Rev.,* Barnes, H., Ed., 9, 11, 1971. With permission of George Allen and Unwin, Ltd., London.

Table 2.4–4   Concentration of the Major Ions in Seawater (g/kg Seawater) Normalized to 35‰ Salinity

Ion	Average Value	Range	Ref.
Chloride	19.353		
Sodium	10.76	10.72–10.80	2
	10.79	10.76–10.80	9
Magnesium	1.297	1.292–1.301	2
	1.292	1.296–1.287	9
Sulfate	2.712	2.701–2.724	5
Calcium	0.4119	0.4098–0.4134	2
	0.4123	0.4088–0.4165	9
Potassium	0.399	0.393–0.405	2, 9
Bicarbonate[a]	0.145	0.137–0.153	4, 7, 6
Bromide	0.0673	0.0666–0.0680	5
Boron	0.0046	0.0043–0.0051	1
Strontium	0.0078	0.0074–0.0079	2
	0.0081	0.0078–0.0085	9
Fluoride	0.0013	0.0012–0.0017	3, 8

[a] The values reported for bicarbonate are actually titration alkalinities.

REFERENCES

1. Culkin, F., The major constituents of sea water, in *Chemical Oceanography,* Vol. 1, Riley, J. P. and Skirrow, G., Eds., Academic Press, London, 1965, 121–161.
2. Culkin, F. and Cox, R. A., Sodium, potassium, magnesium, calcium, and strontium in seawater, *Deep-Sea Res. Oceanogr. Abstr.,* 13, 789, 1966.
3. Greenhalgh, R. and Riley, J. P., Occurrence of abnormally high fluoride concentrations at depth in the oceans, *Nature,* 197, 371, 1963.
4. Koczy, F. F., The specific alkalinity, *Deep-Sea Res. Oceanogr. Abstr.,* 3, 279, 1956.
5. Morris, A. W. and Riley, J. P., The bromide/chlorinity and sulphate/chlorinity ratio in sea water, *Deep-Sea Res. Oceanogr. Abstr.* 3, 699, 1966.
6. Park, Kilho, Deep-sea pH, *Science,* 154, 1540, 1966.
7. Postma, H., The exchange of oxygen and carbon dioxide between the ocean and the atmosphere, *Neth. J. Sea Res.,* 2, 258, 1964.
8. Riley, J. P., The occurrence of anomalously high fluoride concentrations in the North Atlantic, *Deep-Sea Res. Oceanogr. Abstr.,* 12, 219, 1965.
9. Riley, J. P. and Tongudai, M., The major cation/chlorinity ratios in sea water, *Chem. Geol.,* 2, 263, 1967.

Compiled by D. R. Kester, Graduate School of Oceanography, University of Rhode Island.

Table 2.4–5   Relative Composition of Major Components of Seawater (pH_{SWS} = 8.1, S = 35, and 25°C)

	$g_i/Cl(‰)$			
Solute	A	B	C	D
Na^+	0.5556	0.5555	0.5567	0.55661
Mg^{2+}	0.06695	0.06692	0.06667	0.06626
Ca^{2+}	0.02106	0.02126	0.02128	0.02127
K^+	0.0200	0.0206	0.0206	0.02060
Sr^{2+}	0.00070	0.00040	0.00042	0.00041
Cl^-	0.99894	—	—	0.99891
SO_4^{2-}	0.1394	—	0.1400	0.14000
HCO_3^-	0.00735	—	—	0.00552
Br^-	0.00340	—	0.003473	0.00347
CO_3^{2-}	—	—	—	0.00083
$B(OH)_4^-$	—	—	—	0.000415
F^-	—	—	—	0.000067
$B(OH)_3$	0.00137	—	—	0.001002
$\Sigma =$	1.81484			1.81540

Note:

Column A: Dittmar as recalculated by Lyman and Flemming (*J. Mar. Res.,* 3, 134, 1940).

Column B: Cox and Culkin (*Deep-Sea Res.,* 13, 789, 1966).

Column C: Riley and Tongadai (*Chem. Geol.,* 2, 263, 1967); Morris and Riley (*Deep-Sea Res.,* 13, 699, 1966).

Column D: Update of earlier calculations of Millero (*Ocean Sci. Eng.,* 7, 403, 1982) using new dissociation constants for carbonic (Roy et al., *Mar. Chem.,* 44, 249, 1994) and boric acids (Dickson. *Deep-Sea Res.,* 37, 755, 1992) and 1992 Atomic Weights (Appendix I). The values of TA/Cl(‰) = 123.88 mol kg^{-1} (Millero, 1995) and B/Cl(‰) = 0.000232 (Uppström, *Deep-Sea Res.,* 21, 161, 1974) were also used to determine total carbonate and borate.

Source: Millero, F. J., *Chemical Oceanography,* 2nd ed., CRC Press, Boca Raton, FL, 1996, 61. With permission.

Table 2.4–6   Composition of Natural Seawater (1 kg) as a Function of Chlorinity[a]

Species	g_i/Cl	M_i	n_i/Cl	e_i/Cl	$n_iZ_i^2 2/Cl$
Na^+	0.556614	22.9898	0.024211	0.024211	0.024211
Mg^{2+}	0.066260	24.3050	0.002726	0.005452	0.010905
Ca^{2+}	0.021270	40.0780	0.000531	0.001061	0.002123
K^+	0.020600	39.0983	0.000527	0.000527	0.000527
Sr^{2+}	0.000410	87.6200	0.000005	0.000009	0.000018
Cl^-	0.998910	35.4527	0.028176	0.028176	0.028176
SO_4^{2-}	0.140000	96.0636	0.001457	0.002915	0.005830
HCO_3^-	0.005524	61.0171	0.000091	0.000091	0.000091
Br^-	0.003470	79.9040	0.000043	0.000043	0.000043
CO_3^{2-}	0.000830	60.0092	0.000014	0.000028	0.000055
$B(OH)_4^-$	0.000407	78.8404	0.000005	0.000005	0.000005
F^-	0.000067	18.9984	0.000004	0.000004	0.000004
OH^-	0.000007	17.0034	0.0000004	0.0000004	0.0000004
$\frac{1}{2}\Sigma$			0.028895	0.031261	0.035994
$B(OH)_3$	0.000996	61.8322	0.000016	0.000016	
Σ	1.815402		0.028903	0.031261	

[a] For average seawater S = 35, Cl = 19.374, pH_{SWS} = 8.1, TA = 2400 mmol kg^{-1}, and t = 25°C.

Source: Millero, F. J., *Chemical Oceanography,* 2nd ed., CRC Press, Boca Raton, FL, 1996, 64. With permission.

Table 2.4–7   Minor Constituents of Seawater Excluding the Dissolved Gases[a]

Element	Concentration µg/l		Ref. on the Distribution in the Oceans
	Average	Range	
Lithium	185 (2, 18, 29, 52, 62)	180–195 (2, 18, 29, 52, 62)	(2, 18, 29, 52, 62)
Beryllium	5.7×10^{-4} (48)		
Nitrogen	280 (82)	0–560 (82)	(82)
Aluminum	2 (65, 70, 10, 31)	0–7 (65, 70)	(65)
Silicon	2000 (5)	0–4900 (5)	(5)
Phosphorus	30 (4)	0–90(4)	(4)
Scandium	0.04 (31)		
	<0.004 (68)	$0.1–18 \times 10^{-4}$ (37)	
	9.6×10^{-4} (37)		
Titanium	1 (34)		
Vanadium	2.5 (11, 12)	2.0–3.0 (11, 12)	(11)
Chromium	0.3 (30)	0.23–0.43 (30)	
	0.05 (41)	0.04–0.07 (41)	
Manganese	1.5 (64)	0.2–8.6 (64)	(64, 78)
	0.9 (78)	0.7–1.3 (78)	
		3.0–4.4 (28)	
Iron	6.6 (78)	0.1–62 (78)	(3, 7, 22, 78)
	2.6 (70)	8–13 (22)	
	0.2 (7)	0–7 (28, 70)	
		0.03–2.56 (7)	
Cobalt	0.27 (68)	0.035–4.1 (68)	(68, 69, 78)
	0.032 (78)	<0.005–0.092 (78)	
Nickel	5.4 (68)	0.43–43 (22, 68)	(68, 69, 73)
	1.7 (73)	0.8–2.4 (73)	
		0.13–0.37 (28)	
Copper	2 (28, 80)	0.2–4 (28, 73, 78, 80)	(1, 73, 78, 80)
	1.2 (78)	0.5–27 (1, 9, 22, 38)	
	0.7 (73)		
Zinc	12.3 (78)	3.9–48.4 (78)	(64, 73, 78, 80)
	6.5 (64, 80)	2–18 (64, 80)	
	2 (73)	1–8 (73)	
		29–50 (9)	
Gallium	0.03 (23)	0.023–0.037 (23)	
Germanium	0.05 (12, 27)	0.05–0.06 (12)	
Arsenic	4 (39)	3–6 (39)	
	0.46 (41)	2–35 (61)	
Selenium	0.2 (15, 68)	0.34–0.50 (15)	(68)
		0.052–0.12 (68)	
Rubidium	120 (8, 52, 63, 71)	112–134 (8, 52, 63, 71)	(8, 29, 52, 71)
		86–119 (29)	
Yttrium	0.03 (31)	0.0112–0.0163 (37)	
	0.0133 (37)		
Zirconium	2.6×10^{-2} (88)		
Niobium	0.01 (13)	0.01–0.02 (13)	
Molybdenum	10 (41)	0.24–12.2 (9, 85)	
	1 (9)		
Technetium			
Ruthenium	0.0007 (88)		
Rhodium			
Palladium			
Silver	0.29 (68)	0.055–1.5 (68)	(68, 69)
	0.04 (31)		
Cadmium	0.113 (53)	0.02–0.25 (53)	
Indium	<20 (31)		
Tin	0.8 (31)		

Table 2.4–7 Minor Constituents of Seawater Excluding the Dissolved Gases[a] (continued)

| Element | Concentration μg/l | | Ref. on the Distribution in the Oceans |
	Average	Range	
Antimony	0.33 (68)	0.18–1.1 (68)	(68)
Tellurium			
Iodine	63 (6)	48–60 (6)	(6)
	44 (41)		
Cesium	0.4 (8, 63)	0.27–0.33 (8)	8
		0.48–0.58 (63)	
Barium	20 (8, 17, 19, 81)	5–93 (8, 17, 19, 81)	(19, 81)
Lanthanum	3×10^{-3} (33, 37)	$1–6 \times 10^{-3}$ (37)	(35–37)
Cerium	14×10^{-3} (14)	$4–850 \times 10^{-3}$ (14)	(35–37)
	1×10^{-3} (37)	$0.6–2.8 \times 10^{-3}$ (37)	
Praseodymium	6.4×10^{-4} (33, 37)	$4.1–15.8 \times 10^{-4}$ (37)	(35–37)
Neodymium	23×10^{-4} (33)	$13–65 \times 10^{-4}$ (37)	(35–37)
	28×10^{-4} (37)		
Promethium			
Samarium	4.2×10^{-4} (33)	$2.6–10 \times 10^{-4}$ (37)	(35–37)
	4.5×10^{-4} (37)		
Europium	1.14×10^{-4}	$0.9–7.9 \times 10^{-4}$ (37)	(35–37)
	1.3×10^{-4} (37)		
Gadolinium	6.0×10^{-4} (33)	$5.2–11.5 \times 10^{-4}$ (37)	(35–37)
	7.0×10^{-4} (37)		
Terbium	1.4×10^{-4} (37)	$0.6–3.6 \times 10^{-4}$ (37)	(35–37)
Dysprosium	7.3×10^{-4} (33)	$5.2–14.0 \times 10^{-4}$ (37)	(35–37)
	9.1×10^{-4} (37)		
Holmium	2.2×10^{-4} (33, 37)	$1.2–7.2 \times 10^{-4}$ (33, 37)	(33, 35–37)
Erbium	6.1×10^{-4} (33)	$6.6–12.4 \times 10^{-4}$ (37)	(35–37)
	8.7×10^{-4} (37)		
Thulium	1.3×10^{-4} (33)	$0.9–3.7 \times 10^{-4}$ (37)	(35–37)
	1.7×10^{-4} (37)		
Ytterbium	5.2×10^{-4} (33)	$4.8–28 \times 10^{-4}$ (33, 37)	(33, 35–37)
	8.2×10^{-4} (37)		
Lutetium	2.0×10^{-4} (33)	$1.2–7.5 \times 10^{-4}$ (33, 37)	(33, 35–37)
	1.5×10^{-4} (37)		
Hafnium	80×10^{-4} (68)		
Tantalum	25×10^{-4} (68)		
Tungsten	0.1 (41)		
Rhenium	8.4×10^{-3} (66)		
Osmium			
Indium	1×10^{-4} (88)		
Platinum			
Gold	0.068 (86)	0.004–0.027 (68)	
Mercury	0.03 (31)		
Thallium	<0.01 (31)		
Lead	0.05 (19)	0.02–0.4 (19, 76, 77)	(19, 76, 77)
Bismuth	0.02 (56)	0.015–0.033 (56)	
Polonium			
Astatine			
Francium			
Radium	8×10^{-8} (55)	$4–15 \times 10^{-8}$ (45, 49, 55)	(45, 49, 75)
Actinium			
Thorium	0.05 (31)	$2–40 \times 10^{-4}$ (51, 72)	
	0.02 (55)		
	6×10^{-4} (51, 72)		
	$<7 \times 10^{-5}$ (42)		

Table 2.4–7   Minor Constituents of Seawater Excluding the Dissolved Gases[a] (continued)

Element	Concentration µg/l		Ref. on the Distribution in the Oceans
	Average	Range	
Protactinium	2×10^{-6} (31)		
	5×10^{-8} (55)		
Uranium	3 (50, 55, 79)	2–4.7 (50, 55, 79)	(50, 55, 79)

[a] The numbers in parentheses refer to the citations listed after the table. The concentrations represent the dissolved and particulate forms of the elements.

Based on compilations of Pytkowicz, R. M. and Kester, D. R., in *Oceanogr. Mar. Biol. Ann. Rev.,* Barnes, H., Ed., 9, 11, 1971. With permission of George Allen and Unwin, Ltd., London.

REFERENCES

1. Alexander, J. E. and Corcoran, E. F., *Limnol. Oceanogr.,* 12, 236, 1967.
2. Angino, E. E. and Billings, G. K., *Geochim. Cosmochim. Acta,* 30, 153, 1966.
3. Armstrong, F. A. J., *J. Mar. Biol. Assoc. U.K.,* 36, 509, 1957.
4. Armstrong, F. A. J., in *Chemical Oceanography.* Vol. 1, Riley, J. P. and Skirrow, G., Eds., Academic Press, London, 1965, 323–364.
5. Armstrong, F. A. J., in *Chemical Oceanography.* Vol. 1, Riley, J. P. and Skirrow, G., Eds., Academic Press, London, 1965, 409–432.
6. Barkley, R. A. and Thompson, T. G., *Deep Sea Res.,* 7, 24, 1960.
7. Betzer, P. and Pilson, M. E. Q., *J. Mar. Res.,* 28, 251, 1970.
8. Bolter, E., Turekian, K. K., and Schutz, D. F., *Geochim. Cosmochim. Acta,* 28, 1459, 1964.
9. Brooks, R. R., *Geochim. Cosmochim. Acta,* 29, 1369, 1965.
10. Burton, J. D., *Nature,* 212, 976, 1966.
11. Burton, J. D. and Krishnamurty, K., *Rep. Challenger Soc.,* 3, 24, 1967.
12. Burton, J. D. and Riley, J. P., *Nature,* 181, 179, 1958.
13. Carlisle, D. B. and Hummerstone, L. G., *Nature,* 181, 1002, 1958.
14. Carpenter, J. H. and Grant, V. E., *J. Mar. Res.,* 25, 228, 1967.
15. Chau, Y. K. and Riley, J. P., *Anal. Chim. Acta,* 33, 36, 1965.
16. Chester, R., *Nature,* 206, 884, 1965.
17. Chow, T. J. and Goldberg, E. D., *Geochim. Cosmochim. Acta,* 20, 192, 1960.
18. Chow, T. J. and Goldberg, E. D., *J. Mar. Res.,* 20, 163, 1962.
19. Chow, T. J. And Patterson, C. C., *Earth Planet. Sci. Lett.,* 1, 397, 1966.
20. Chow, T. J. and Tatsumoto, M., in *Recent Researches in the Fields of Hydrosphere, Atmosphere and Nuclear Geochemistry,* Miyake, Y. and Koyama, T., Eds., Maruzen Co., Tokyo, 1964, 179–183.
21. Chuecas, L. and Riley, J. P., *Anal. Chim. Acta,* 35, 240, 1966.
22. Corcoran, E. F. and Alexander, J. E., *Bull. Mar. Sci. Gulf Caribbean,* 14, 594, 1964.
23. Culkin, F. and Riley, J. P., *Nature,* 181, 180, 1958.
24. Curl, H., Cutshall, N., and Osterberg, C., *Nature,* 205, 275, 1965.
25. Cutshall, N., Johnson, V., and Osterberg, C., *Science,* 152, 202, 1966.
26. Duursma, E. K. and Sevenhuysen, W., *Neth. J. Sea Res.,* 3, 95, 1966.
27. El Wardani, S. A., *Geochim. Cosmochim. Acta,* 15, 237, 1958.
28. Fabricand, B. P., Sawyer, R. R., Ungar, S. G., and Adler, S., *Geochim. Cosmochim. Acta,* 26, 1023, 1962.
29. Fabricand, B. P., Imbimbo, E. S., Brey, M. E., and Weston, J. A., *J. Geophys. Res.,* 71, 3917, 1966.
30. Fukai, R., *Nature,* 213, 901, 1967.
31. Goldberg, E. D., in *Chemical Oceanography,* Vol. 1, Riley, J. P. and Skirrow, G., Eds., Academic Press, London, 1965, 163–196.
32. Goldberg, E. D. and Arrhenius, G. S., *Geochim. Cosmochim. Acta,* 13, 153, 1958.
33. Goldberg, E. D., Koide, M., Schmitt, R. A., and Smith, R. H., *J. Geophys. Res.,* 68, 4209, 1963.
34. Griel, J. V. and Robinson, R. J., *J. Mar. Res.,* 11, 173, 1952.
35. Høgdahl, O., Semi Annual Progress Report No. 5, NATO Scientific Affairs Div., Brussels, 1967.
36. Høgdahl, O., Semi Annual Progress Report No. 6, NATO Scientific Affairs Div., Brussels, 1968.
37. Høgdahl, O., Melsom, S., and Bowen, V. T., Trace inorganics in water, in *Advances in Chemistry Series,* No. 73, American Chemical Society, Washington, D.C., 1968, 308–325.

38. Hood, D. W., in *Oceanogr. Mar. Biol. Annu. Rev.*, Vol. 1, Barnes, H., Ed., George Allen and Unwin, Ltd., London, 1963, 129–155.
39. Ishibashi, M., *Rec. Oceanogr. Works Jpn.*, 1, 88, 1953.
40. Johnson, V., Cutshall, N., and Osterberg, C., *Water Resour. Res.*, 3, 99, 1967.
41. Kappanna, A. N., Gadre, G. T., Bhavnagary, H. M., and Joshi, J. M., *Curr. Sci (India)*, 31, 273, 1962.
42. Kaufman, A., *Geochim. Cosmochim. Acta*, 33, 717, 1969.
43. Kester, D. R. and Pytkowicz, R. M., *Limnol. Oceanogr.*, 12, 243, 1967.
44. Kharkar, D. P., Turekian, K. K., and Bertine, K. K., *Geochim. Cosmochim. Acta*, 32, 285, 1968.
45. Koczy, F. F., *Proc. Second U.N. Int. Conf. Peaceful Uses Atomic Energy*, 18, 351, 1958.
46. Krauskopf, K. B., *Geochim. Cosmochim. Acta.*, 9, 1, 1956.
47. Menzel, D. W. and Ryther, J. H., *Deep Sea Res.*, 7, 276, 1961.
48. Merrill, J. R., Lyden, E. F. X., Honda, M., and Arnold, J., *Geochim. Cosmochim. Acta*, 18, 108, 1960.
49. Miyake, Y. and Sugimura, Y., in *Studies on Oceanography*, Yoshida, K., Ed., University of Washington Press, Seattle, 1964, 274.
50. Miyake, Y., Sugimura, Y., and Uchida, T., *J. Geophys. Res.*, 71, 3083, 1966.
51. Moore, W. S. and Sackett, W. M., *J. Geophys. Res.*, 69, 5401, 1964.
52. Morozov, N. P., *Oceanology*, 8, 169, 1968.
53. Mullin, J. B. and Riley, J. P., *J. Mar. Res.*, 15, 103, 1956.
54. Peshchevitskiy, B. I., Anoshin, G. N., and Yereburg, A. M., *Dokl. Earth Sci. Sect.*, 162, 205, 1965.
55. Picciotto, E. E., in *Oceanography*, Sears, M., Ed., Amercan Association for the Advancement Science, Washington, D.C., 1961, 367.
56. Portmann, J. E. and Riley, J. P., *Anal. Chim. Acta*, 34, 201, 1966.
57. Putnam, G. L., *J. Chem. Educ.*, 30, 576, 1953.
58. Pytkowicz, R. M., *J. Oceanogr. Soc. Jpn.*, 24, 21, 1968.
59. Pytkowicz, R. M. and Kester, D. R., *Deep Sea Res.*, 13, 373, 1966.
60. Pytkowicz, R. M. and Kester, D. R., *Limnol. Oceanogr.*, 12, 714, 1967.
61. Richards, F. A., in *Physics and Chemistry of the Earth*, Vol. 2, Ahrens, L. H., Press, F., Rankama, K., and Runcorn, S. K., Eds., Pergamon Press, New York, 1957, 77–128.
62. Riley, J. P. and Tongudai, M., *Deep Sea Res.*, 11, 563, 1964.
63. Riley, J. P. and Tongudai, M., *Chem. Geol.*, 1, 291, 1966.
64. Rona, E., Hood, D. W., Muse, L., and Buglio, B., *Limnol. Oceanogr.*, 7, 201, 1962.
65. Sackett, W. and Arrhenius, G., *Geochim. Cosmochim. Acta*, 26, 955, 1962.
66. Scadden, E. M., *Geochim. Cosmochim. Acta*, 33, 633, 1969.
67. Schink, D. R., *Geochim. Cosmochim. Acta*, 31, 987, 1967.
68. Schutz, D. F. and Turekian, K. K., *Geochim. Cosmochim. Acta*, 29, 259, 1965.
69. Schutz, D. F. and Turekian, K. K., *J. Geophys. Res.*, 70, 5519, 1965.
70. Simmons, L. H., Monaghan, P. H., and Taggart, M. S., *Anal. Chem.*, 25, 989, 1953.
71. Smith, R. C., Pillai, K. C., Chow, T. J., and Folson, T. R., *Limnol. Oceanogr.*, 10, 226, 1965.
72. Somayajulu, B. L. K. and Goldberg, E. D., *Earth Planet. Sci. Lett.*, 1, 102, 1966.
73. Spencer, D. W. and Brewer, P. G., *Geochim. Cosmochim. Acta*, 33, 325, 1969.
74. Sugawara, K. and Terada, K., *Nature*, 182, 250, 1958.
75. Szabo, B. J., *Geochim. Cosmochim. Acta*, 31, 1321, 1967.
76. Tatsumoto, M. and Patterson, C. C., *Nature*, 199, 350, 1963.
77. Tatsumoto, M. and Patterson, C. C., in *Earth Sciences and Meteoritics*, Geiss, J. and Goldberg, E. D., Compilers, North Holland Publ. Co., Amsterdam, 1963, 74–89.
78. Topping, G., *J. Mar. Res.*, 27, 318, 1969.
79. Torii, T. and Murata, S., in *Recent Researches in the Fields of Hydrosphere, Atmosphere, and Nuclear Geochemistry*, Miyake, Y. and Koyama, T., Eds., Maruzen Co., Tokyo, 1964.
80. Torri, T. and Murata, S., *J. Oceanogr. Soc. Jpn.*, 22, 56, 1966.
81. Turekian, K. K. and Johnson, D. G., *Geochim. Cosmochim. Acta*, 30, 1153, 1966.
82. Vaccaro, R. F., in *Chemical Oceanography*, Vol. 1, Riley, J. P. and Skirrow, G., Eds., Academic Press, London, 1965, 365–408.
83. Veeh, H. H., *Earth Planet. Sci. Lett.*, 3, 145, 1967.
84. Wangersky, P. J. and Gordon, D. C., Jr., *Limnol. Oceanogr.*, 10, 544, 1965.
85. Weiss, H. V. and Lai, M. G., *Talanta*, 8, 72, 1961.
86. Weiss, H. V. and Lai, M. G., *Anal. Chim. Acta*, 28, 242, 1963.
87. Williams, P. M., *Limnol. Oceanogr.*, 14, 156, 1969.
88. Riley, J. P. and Chester, R., *Introduction to Marine Chemistry*, Academic Press, London, 1971.

Table 2.4–8   Residence Times of Elements in Seawater

Element	Residence Time (million years)	
	River Input	Sedimentation
Na	210	260
Mg	22	45
Ca	1	8
K	10	11
Sr	10	19
Si	0.935	0.01
Li	12	19
Rb	6.1	0.27
Ba	0.05	0.084
Al	0.0031	0.0001
Mo	2.15	0.5
Cu	0.043	0.05
Ni	0.015	0.018
Ag	0.25	2.1
Pb	0.00056	0.002

Source: Millero, F.J., *Chemical Oceanography,* 2nd ed., CRC Press, Boca Raton, FL, 1996, 104. With permission.

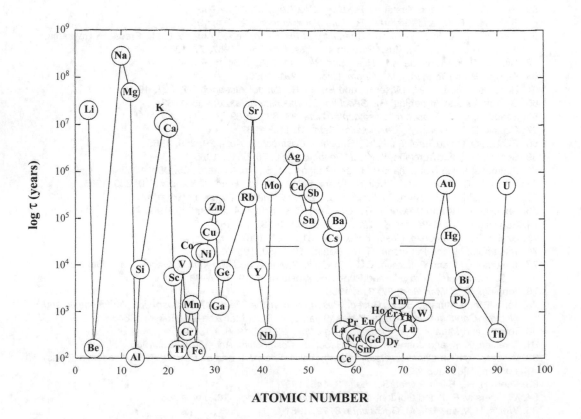

Figure 2.4–3   Residence time of elements in seawater plotted vs. atomic number. (From Millero, F. J., *Chemical Oceanography,* 2nd ed., CRC Press, Boca Raton, FL, 1996, 105. With permission.)

Table 2.4–9 Calculation of the Salinity for Average Seawater from Composition Data

	Before Evaporation (g)		After Evaporation (g)
HCO_3^-	0.1070		0.0137
CO_3^{2-}	0.0161		0.0043
CO_2	0.0005		0.0000
Br^-	0.0672	Cl	0.0298
	0.1908		0.0478

Salt loss from HCO_3^-, CO_3^{2-} and Br^-; $0.1908 - 0.0478 = 0.1430$ g
$B(OH)_3$ lost $= 0.0275$ g
Total salts lost $= 0.1705$ g

g_τ	35.1716
Loss of salts	−0.1705
	35.0011

Source: Millero, F. J., *Chemical Oceanography,* 2nd ed., CRC Press, Boca Raton, FL, 1996, 68. With permission.

Table 2.4–10 Precision in Salinity Determined by Various Methods

1. Composition studies of major components	±0.01
2. Evaporation to dryness	±0.01
3. Chlorinity	±0.002
4. Density	±0.004
5. Conductivity	±0.001
6. Sound speeds	±0.03
7. Refractive index	±0.05

Source: Millero, F. J., *Chemical Oceanography,* 2nd ed., CRC Press, Boca Raton, FL, 1996, 76. With permission.

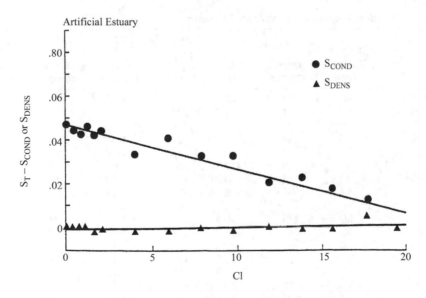

Figure 2.4–4 Comparison of the salinity determined from conductivity and density for estuarine waters. (From Millero, F. J., *Chemical Oceanography,* 2nd ed., CRC Press, Boca Raton, FL, 1996, 85. With permission.)

2.5 TRACE ELEMENTS

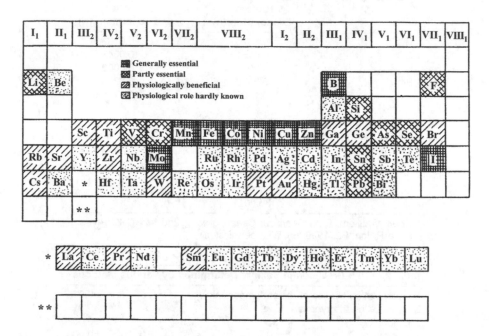

Figure 2.5–1   Periodic Table identification of trace elements and micronutrients. (From Pais, I. and Jones, J. B., Jr., *The Handbook of Trace Elements,* St. Lucie Press, Boca Raton, FL, 1997, 3. With permission.)

Table 2.5–1   Chemical Properties of the Trace Elements

Element	Group in Periodic Table	Atomic Number	Atomic Weight	Ions	Ionic Radius[a]	Electro-negativity[b]	Ion Potential (charge/radius)
Aluminum (Al)	IIIA	13	26.98	Al^{3+}	0.57	1.6	—
Antimony (Sb)	VA	51	121.75	Sb^{3+}	0.76	—	—
				Sb^{5+}	0.62	1.9	—
Arsenic (As)	VA	33	74.92	As^{3+}	0.58	—	—
				As^{5+}	0.46	—	—
Barium (Ba)	IIA	56	137.32	Ba^{2+}	1.43	0.9	—
Beryllium (Be)	IIA	4	9.01	Be^{2+}	0.34	1.6	—
Bismuth (Bi)	VA	83	208.98	Bi^{5+}	0.74	2.0	—
				Bi^{3+}	0.96	—	—
Boron (B)	IIIA	5	10.81	B^{3+}	0.23	2.0	—
Cadmium (Cd)	IIB	48	122.40	Cd^{2+}	0.97	1.7	—
Cesium (Cs)	IA	55	132.91	Cs^+	1.65	0.8	—
Chromium (Cr)	VIB	24	52.00	Cr^{3+}	0.63	1.6	4.3
				Cr^{6+}	0.52	—	16.0
Cobalt (Co)	VIII	27	58.93	Co^{2+}	0.72	1.8	2.6
Copper (Cu)	IB	29	63.54	Cu^+	0.96	1.9	—
				Cu^{2+}	0.72	2.0	2.5
Fluorine (F)	VIIA	9	18.99	F^-	1.33	4.1	—
Gallium (Ga)	IIIA	31	69.72	Ga^{3+}	0.62	1.8	—
				Ga^+	1.13	—	—

Table 2.5–1 Chemical Properties of the Trace Elements (continued)

Element	Group in Periodic Table	Atomic Number	Atomic Weight	Ions	Ionic Radius[a]	Electro-negativity[b]	Ion Potential (charge/radius)
Germanium (Ge)	IVA	32	72.61	Ge^{2+}	0.90	2.0	—
Gold (Au)	IB	79	196.97	Au^+	1.37	2.4	—
Hafnium (Hf)	IVB	72	178.49	Hf^{4+}	0.84	1.3	—
Indium (In)	IIIA	49	114.82	In^{3+}	0.92	1.8	—
				In^+	1.32	—	—
Iodine (I)	VIIA	53	126.90	I^-	2.20	2.7	—
Iron (Fe)	VIII	26	55.85	Fe^{2+}	0.82	1.8	—
				Fe^{3+}	0.67	—	—
Lead (Pb)	IVA	82	207.19	Pb^{2+}	1.20	1.8	1.9
Lithium (Li)	IA	3	6.94	Li^+	0.78	1.0	—
Manganese (Mn)	VIIB	25	54.94	Mn^{2+}	0.80	1.5	—
				Mn^{3+}	0.66	—	—
				Mn^{4+}	0.60	—	6.5
Mercury (Hg)	IIB	80	200.59	Hg^{2+}	1.10	1.9	—
Molybdenum (Mo)	VIB	42	95.94	Mo^{4+}	0.70	—	—
				Mo^{6+}	0.62	1.8	12.0
Nickel (Ni)	VIII	28	59.71	Ni^{2+}	0.69	1.8	2.6
Niobium (Nb)	VB	41	92.91	Nb^{4+}	0.74	1.6	—
Platinum (Pt)	VIII	78	195.08	Pt^{2+}	0.85	2.2	—
Rubidium (Rb)	IA	37	85.47	Rb^+	1.49	0.8	—
Selenium (Se)	VIA	34	78.96	Se^{2-}	12.001	2.4	3.7
				Se^{4+}	0.42	—	—
Silicon (Si)	IVA	14	28.09	Si^{4+}	0.26	1.9	—
				Si^{4-}	2.71	—	—
Silver (Ag)	IB	47	107.87	Ag^+	1.26	1.9	—
Strontium (Sr)	IIA	38	87.62	Sr^{2+}	1.27	0.9	—
Thallium (Tl)	IIIA	81	204.37	Tl^+	1.47	—	—
				Tl^{3+}	0.95	1.8	—
Tin (Sn)	IVA	50	118.69	Sn^{2+}	0.93	1.8	1.5
				Sn^{4+}	0.71	1.9	—
Titanium (Ti)	IVB	22	47.88	Ti^{4+}	0.80	1.5	—
				Ti^{3+}	0.69	—	—
Tungsten (W)	VIB	74	183.85	W^{6+}	0.62	1.7	—
Uranium (U)	Actinide series	92	238.04	U^{4+}	0.97	—	—
				U^{6+}	0.80	1.7	—
Vanadium (V)	VB	23	50.94	V^{3+}	10.651	—	—
				V^{4+}	(0.65)	—	—
				V^{5+}	0.59	—	11.0
Zinc (Zn)	IIB	30	65.37	Zn^{2+}	0.74	1.7	2.6
Zirconium (Zr)	IVB	40	91.22	Zr^{2+}	1.09	1.3	—
				Zr^{4+}	0.87	—	—

[a] Ionic radius is for 6-coordination.
[b] Electronegativity values for elements: S = 2.5, O = 3.5, I = 2.5, Cl = 3.0, F = 4.0. From these numbers, it can be generalized that a bond between any two atoms will be largely covalent if the electronegativities are similar and mainly ionic if they are very different.

Source: Pais. I. and Jones, J. B., Jr., *The Handbook of Trace Elements,* St. Lucie Press, Boca Raton, FL, 1997, 3. With permission.

Table 2.5–2   Trace Element Content in Seawater

Trace Element	Mg/l Atlantic Surface	Atlantic Deep	Pacific Surface	Pacific Deep	Other
Aluminum (Al)	9.7×10^{-4}	5.2×10^{-4}	1.3×10^{-4}	0.13×10^{-4}	—
Antimony (Sb)	—	—	—	—	0.3×10^{-3}
Arsenic (As)	1.45×10^{-3}	1.53×10^{-3}	1.45×10^{-3}	1.75×10^{-3}	—
Barium (Ba)	4.7×10^{-3}	9.3×10^{-3}	4.7×10^{-3}	20.0×10^{-3}	—
Beryllium (Be)	8.8×10^{-8}	17.5×10^{-8}	3.5×10^{-8}	22.0×10^{-8}	—
Bismuth (Bi)	5.1×10^{-8}	—	4.0×10^{-8}	0.4×10^{-8}	—
Boron (B)	—	—	—	—	4.41
Cadmium (Cd)	1.1×10^{-6}	38.0×10^{-6}	1.1×10^{-6}	100.0×10^{-6}	—
Cesium (Cs)	—	—	—	—	3.0×10^{-4}
Chromium (Cr)	1.8×10^{-4}	2.3×10^{-4}	1.5×10^{-4}	2.5×10^{-4}	—
Cobalt (Co)	—	—	6.9×10^{-6}	1.1×10^{-6}	—
Copper (Cu)	8.0×10^{-5}	12.0×10^{-5}	8.0×10^{-5}	28.0×10^{-5}	—
Fluorine (F)	—	—	—	—	1.3
Gallium (Ga)	—	—	—	—	3.0×10^{-5}
Germanium (Ge)	0.07×10^{-6}	0.14×10^{-6}	0.35×10^{-6}	7.0×10^{-6}	—
Gold (Au)	—	—	—	—	1.0×10^{-5}
Hafnium (Hf)	—	—	—	—	7.0×10^{-6}
Indium (In)	—	—	—	—	1.0×10^{-7}
Iodine (I)	0.049	0.0056	0.043	0.058	—
Iron (Fe)	1.0×10^{-4}	4.0×10^{-4}	0.1×10^{-4}	1.0×10^{-4}	—
Lead (Pb)	30.0×10^{-6}	4.0×10^{-6}	10.0×10^{-6}	1.0×10^{-6}	—
Lithium (Li)	—	—	—	—	0.17
Manganese (Mn)	1.0×10^{-4}	0.96×10^{-4}	1.0×10^{-4}	0.4×10^{-4}	—
Mercury (Hg)	4.9×10^{-7}	4.9×10^{-7}	3.3×10^{-7}	3.3×10^{-7}	—
Molybdenum (Mo)	—	—	—	—	0.01
Nickel (Ni)	1.0×10^{-4}	4.0×10^{-4}	1.0×10^{-4}	5.7×10^{-4}	—
Platinum (Pt)	—	—	1.1×10^{-7}	2.7×10^{-7}	—
Rubidium (Rb)	—	—	—	—	0.12
Selenium (Se)	0.46×10^{-7}	1.8×10^{-7}	0.15×10^{-7}	1.65×10^{-7}	—
Silicon (Si)	0.03	0.82	0.03	4.09	—
Silver (Ag)	—	—	1.0×10^{-7}	24.0×10^{-7}	—
Strontium (Sr)	7.6	7.7	7.6	7.7	—
Thallium (Tl)	—	—	—	—	1.4×10^{-5}
Tin (Sn)	2.3×10^{-6}	5.8×10^{-6}	—	—	—
Titanium (Ti)	—	—	—	—	4.8×10^{-4}
Uranium (U)	—	—	—	—	3.13×10^{-3}
Vanadium (V)	1.1×10^{-3}	—	1.6×10^{-3}	1.8×10^{-3}	—
Zinc (Zn)	0.05×10^{-4}	1.0×10^{-4}	0.5×10^{-4}	5.2×10^{-4}	—

Source: Pais, I. and Jones, J. B., Jr., *The Handbook of Trace Elements,* St. Lucie Press, Boca Raton, FL, 1997, 19. With permission.

Table 2.5–3   Metals on Shelf vs. Open-Sea Surface Waters

Metal	Shelf	Open
Mn	21 nM	2.4 nM
Ni	5.9 nM	2.3 nM
Cu	4.0 nM	1.2 nM
Zn	2.4 nM	0.06 nM
Cd	200 pM	2 pM

Source: Millero, F. J., *Chemical Oceanography,* 2nd ed., CRC Press, Boca Raton, FL, 1996, 114. With permission.

Table 2.5–4   Metals in Central Gryes

Metal	Atlantic	Pacific
Mn	2.4 nM	1.0 nM
Cu	1.2 nM	0.5 nM
Ni	2.1 nM	2.4 nM
Zn	0.06 nM	0.06 nM
Cd	2 pM	2 pM

Source: Millero, F. J., *Chemical Oceanography,* 2nd ed., CRC Press, Boca Raton, FL, 1996, 115. With permission.

Table 2.5–5   The Fraction of Free Metals and the Dominant Forms in Seawater at pH = 8.2

Cation	Free	OH	F	Cl	SO$_4$	CO$_3$	Log α
Ag$^+$	*	*	*	100	*	*	5.26
Al^{3+}	*	100	*	—	—	*	9.22
Au$^+$	*	—	—	100	—	—	12.86
Au^{3+}	*	100	*	9	5	*	27.30
Ba^{2+}	86	*	*	9	5	*	0.07
Be^{2+}	*	99	2	*	*	*	2.74
Bi^{3+}	*	100	*	*	*	—	14.79
Cd^{2+}	3	*	*	97	*	*	1.57
Ce^{3+}	21	5	1	12	10	51	0.68
Co^{2+}	58	1	*	30	5	6	0.24
Cr^{3+}	*	100	*	*	*	—	5.82
Ca^{2+}	93	—	—	7	—	—	0.03
Cu$^+$	*	—	—	100	—	—	5.18
Cu^{2+}	9	8	*	3	1	79	1.03
Dy^{3+}	11	8	1	5	6	68	0.94
Er^{3+}	8	12	1	4	4	70	1.08
Eu^{3+}	18	13	1	10	9	50	0.74
Fe^{2+}	69	2	*	20	4	5	0.16
Fe^{3+}	*	100	*	*	*	*	11.98
Ga^{3+}	*	100	*	*	—	*	15.35
Gd^{3+}	9	5	1	4	6	74	1.02
Hf^{4+}	*	100	*	*	*	—	22.77
Hg^{2+}	*	*	*	100	*	*	14.24
Ho^{3+}	10	8	1	5	5	70	0.99
In^{3+}	*	100	*	*	*	*	11.48
La^{3+}	38	5	1	18	16	22	0.42
Li^{3+}	99	*	—	—	1	—	0.00
Lu^{3+}	5	21	1	1	1	71	1.32
Mn^{2+}	58	*	*	37	4	1	0.23
Nd^{3+}	22	8	1	19	12	45	0.66
Ni^{2+}	47	1	*	34	4	14	0.33
Pb^{2+}	3	9	*	47	1	41	1.51
Pr^{3+}	25	8	1	12	13	41	0.61
Rb$^+$	95	—	—	5	—	—	0.02
Sc^{3+}	*	100	*	*	*	*	7.41
Sm^{3+}	18	10	1	8	11	52	0.75
Sn^{4+}	*	100	—	—	—	—	32.05
Tb^{3+}	16	11	1	8	9	55	0.80
Th^{4+}	*	100	*	*	*	*	0.80
TiO^{2+} a	*	100	—	—	*	—	11.14
Tl$^+$	53	*	*	45	2	—	0.28
Tl^{3+}	*	100	—	*	*	—	20.49
Tm^{3+}	11	21	1	5	6	55	0.94
U^{4+}	*	100	*	*	*	—	23.65
UO$_2^{2+}$ a	*	*	*	*	*	100	6.83
Y^{3+}	15	14	3	7	6	54	0.81
Yb^{3+}	5	9	1	2	3	81	1.30
Zn^{2+}	46	12	*	35	4	3	0.34
Zr^{4+}	*	100	*	*	*	—	23.96

— Indicates ligand not considered.
* Indicates calculated abundance <1%.
a Classified as fully hydrolyzed oxidation states.

Source: Millero, F. J., *Chemical Oceanography,* 2nd ed., CRC Press, Boca Raton, FL, 1996, 160. With permission.

Table 2.5-6 Fluxes of Trace Metals to the Sea Surface (units ng/cm²/year)

	New York Bight	North Sea	Western Med.	South Atlantic Bight	Bermuda	North Atlantic Northeast Trades	Tropical North Atlantic	Total Net Deposition — Tropical North Pacific	South Pacific	Westerlies — North Atlantic	North Pacific	South Pacific
Al	6,000	30,000	5,000	2,900	3,900	97,000	5,000	1,200	132–1,800	—	—	—
Sc	—	5	1	—	0.6	—	1.1	0.18	0.06	—	—	—
V	—	480	—	—	5	—	17	7.8	—	—	—	—
Cr	—	210	49	—	9	111	14	—	—	—	—	—
Mn	—	920	—	60	45	570	70	9.0	3.6	—	—	—
Fe	5,700	25,500	5,100	5,900	3,000	48,000	3,200	560	47–337	—	—	—
Co	—	39	3.5	—	1.2	12	2.7	—	0.25	—	—	—
Ni	—	260	—	390	3	67	20	—	—	—	—	—
Cu	—	1,300	96	220	30	48	25	8.9	4.4–7.9	—	—	—
Zn	1,400	8,950	1,080	750	75	152	130	67	2.4–5.8	—	—	—
As	—	280	54	45	3	—	—	—	—	—	—	—
Se	—	22	48	—	3	—	14	4.2	0.8	—	—	—
Ag	—	3	3	—	—	—	0.9	—	—	—	—	—
Cd	30	43	13	9	4.5	—	5	0.35	—	—	—	—
Sb	—	58	48	—	1.0	—	3.5	—	—	—	—	—
Au	—	—	0.05	—	—	—	0.1	—	—	—	—	—
Hg	—	5	5	24	—	—	2.1	—	—	—	—	—
Pb	3,900	2,650	1,050	660	100	32	310	7.0	1.4–2.8	170	50	3
Th	—	4	1.2	—	—	—	0.9	0.61	0.036	—	—	—

Source: Furness, R. W. and Rainbow, P. S., Eds., *Heavy Metals in the Marine Environment*, CRC Press, Boca Raton, FL, 1990, 37. With permission.

Table 2.5–7 Survey of Selected Trace Element Concentrations in Precipitation (volume-weight, µg/l) and in Wet Deposition Flux (values in parentheses, mg/m²/year)

Location[a]	As	Cd	Cr	Cu	Pb	Hg	Ni	V	Zn
Rural/Remote									
Canada									
Eastern[63]	0.23 (0.25)	<0.2	<1.0	1.3	15.4 (16.92)	—	—	0.41 (0.45)	4.7 (5.16)
Ontario (north)[35]	—	0.12	—	1.5	4.8	—	—	—	—
Ontario (south)[35]	—	0.12	—	1.6	7.0	—	—	—	4.9–10
Great Lakes[54]	—	0.2–0.8 (0.1–0.6)	—	—	2–11 (2–10)	—	—	—	—
United States									
Eastern[50]	—	0.12 (0.24)	—	—	1.1 (2.16)	—	—	—	3.1 (6.0)
Eastern[51]	0.096	0.31	0.14	0.95	4.5	—	0.75	1.1	3.7
Eastern[65]	0.141	—	—	—	—	—	—	1.27	8.3
Central[28]	0.15	0.23	—	2.0	2.0	—	—	—	3.0
Central[55]	—	—	0.6	2.9	1.9	—	1.0	<0.1	9.3
Northern[86]	—	—	—	—	—	0.0187 (0.0126)	—	—	—
Northern[87]	—	—	—	—	—	0.0105 (0.0045)	—	—	—
Sweden									
North[19]	—	0.041 (0.024)	0.064 (0.048)	—	1.84 (1.32)	—	0.13 (0.10)	0.27 (0.24)	4.1 (2.88)
South[19]	—	0.125 (0.072)	0.160 (0.084)	—	3.75 (1.92)	—	0.44 (0.24)	1.16 (0.60)	10.2 (5.4)
Holland[56]	0.5 (0.29)	0.5 (0.24)	1.4 (0.66)	6.7 (4.54)	13 (7.4)	—	2.4 (0.98)	6.7 (4.27)	21 (13.7)
Central Italy[88]	—	—	—	—	—	0.01	—	—	—
Germany[89]	—	0.10–0.30	—	0.74–3.40	4.27–14.1	—	—	—	6.93–33.4
Northern India[64]	0.27	—	5.40	—	—	—	—	—	14.22

Marine								
Atlantic								
Ireland[57]	—	0.04	—	0.86	0.51	—	—	—	8.05
Scotland[52]	—	0.68	—	2.3	4.0	—	—	—	13
	—	(0.39)	—	(1.3)	(2.3)	—	—	—	(7.6)
Bermuda[48]	—	0.062	—	0.32	0.722	—	0.167	0.096	1.53
East U.S.[90]	—	0.10	—	0.76	1.9	—	1.12	—	5.16
	—	(0.108)	—	(0.836)	(2.09)	—	(1.23)	—	(5.68)
Cruise[91]	—	0.07–0.95	—	0.13–0.67	0.13–3.65	—	—	—	1.23–4.75
Cruise[92]	—	0.031	—	0.5	0.471	—	—	—	2.1
Pacific								
Northeast[93]	—	—	—	—	—	0.009	—	—	—
Samoa[49]	—	—	<0.010	0.042	0.040	—	—	<0.050	0.96
	—	—	—	(0.04)	(0.03)	—	—	—	(0.3)
Enewetak[49]	—	—	<0.010	0.020	0.038	—	—	<0.050	0.088
Enewetak[1]	—	0.002	—	0.013	<0.04	—	—	0.018	0.05
NW U.S.[94]	—	0.012	—	0.14	0.15	—	—	—	0.99
Polar								
E. Arctic Ocean[41]	—	0.005	—	0.097	0.185	—	0.197	—	—
W. Can. Arctic[53]	—	0.012	—	0.26	0.63	—	0.18	0.72	0.89
Antarctic[95]	—	—	—	—	—	<0.00096	—	—	—

[a] See original source for reference citations.

Source: Ross, H. B. and Vermette, S. J., Precipitation, in *Trace Elements in Natural Waters*, Salbu, B. and Steinnes, E., Eds., CRC Press, Boca Raton, FL, 1995, 110. With permission.

Table 2.5–8 The Solubility of Atmospheric Trace Metals in Seawater

Atmospheric Population	Al	Fe	Mn	Ni	Co	Cr	V	Ag	Zn	Cu	Pb	Cd
Anthropogenic-rich[a]												
Mean concentration (ng/m³ of air)	900	610	11	11	<0.4	<1.6	13	<0.05	25	16	560	0.25
Mean % soluble in seawater	0.56	1.1	47	47	25	12.5	31	80	68	28	39	84
Mean EF$_{crust}$	1.0	0.99	1.0	14	1.5	1.5	9.0	—	33	26	4148	16
Dust-rich[b]												
Mean concentration (ng/m³ of air)	3380	2100	46	2.8	<0.5	4.0	<11	<0.03	18	20	150	0.20
Mean % soluble in seawater	0.09	0.19	34	28	20	10	18	33	24	14.5	13	80
Mean EF$_{crust}$	1.0	0.91	1.2	0.92	0.49	0.99	2.0	—	6.2	8.4	296	25

[a] Southern California.
[b] Baja, California.

Source: Furness, R. W. and Rainbow, P. S., Eds., Heavy Metals in the Marine Environment, CRC Press, Boca Raton, FL, 1990, 41. With permission.

Table 2.5–9 Dissolved Trace Element Concentrations in Uncontaminated Rivers

Element	System	Form, μm	Concentration, ng/1	Ref.
Cd	Mountain stream, California	NR[a]	0.3–2.1 (1.2)[b]	63
	Manuherikia River, New Zealand	<0.4	8	62
	Amazon River	<0.4	10–90	71
	Background	NR	20	38
Cu	Manuherikia River, New Zealand	<0.4	150	62
	Niger River	<0.4	140	70
	Negros River	<0.4	320	71
	Amazon River	<0.4	1,200–1,400	70
		<0.4	2,200	71
	Orinoco River	<0.4	1,200	70
	Background	NR	1,000	38
	"World average"	NR	10,000	64
Pb	Mountain stream, California	<0.4	1–24(9)[b]	63
	Manuherikia River, New Zealand	NR	20–30	62
	Southeastern U.S. rivers	<0.4	20–500	72
	Background	NR	200	38
	"World average"	NR	1,000	64
Zn	Magela Creek, Australia	<0.4	87–121	67
	Manuherikia River, New Zealand	<0.4	150–200	62
	Amazon Basin, Orinoco Basin, Yangtze	<0.4	20–1,800 (~200)[b]	65
	Background	NR	500–10,000	38
	"World average"	NR	30,000	64
Ni	Manuherikia River, New Zealand	<0.4	100–150	62
	Yangtze River	<0.4	130	70
	Orinoco River	<0.4	290	70
	Amazon River	<0.4	320	70
	Background	NR	300	38
	"World average"	NR	2,200	64
Hg	Manuherikia River, New Zealand	c	0.3	62
	Eight rural lakes, Wisconsin	c	0.2–0.5	134
	Background	NR	10	38

Note: See original source for reference citations.

[a] NR, not recorded.
[b] Range (mean).
[c] Sample not filtered. Total Hg released by borohydride reduction.

Source: Hart, B.T. and Hines, T., in Trace Elements in Natural Waters, Salbu, B. and Steinnes, E., Eds., CRC Press, Boca Raton, FL, 1995, 210. With permission.

Table 2.5–10 Field Observations of the Behavior of Dissolved Trace Metals in the Estuaries

Estuary	Metals	Trace Metal Behavior	Ref.
Amazon	Cu, Ni, V[a]	Conservative	5, 82
	[9]Be, Zn	Removal	67, 81
	Cd	Possible addition	5
St. Lawrence	As,[a] Co,[a] Cu	Conservative	61, 84
	Ni, Mn		
	Fe, Cr	Removal at low salinity	77, 84
	Hg	Removal	69
Savannah	As[a]	Conservative	54, 59
	U[a]	Conservative, removal at low Q_w	92
	Fe	Removal at low salinity	80
	Pb, Sb, Zn	Removal	59, 78, 80
	Cu, Cd, Ni	Addition	76, 80
Scheldt	Mo,[a] Ni, Se	Conservative	74, 88
	As, V, Sb	Removal at low salinity	88
	As (tot.)	Possible addition at low salinity	58

Table 2.5–10   Field Observations of the Behavior of Dissolved Trace Metals in the Estuaries
(continued)

Estuary	Metals	Trace Metal Behavior	Ref.
Scheldt	Cd, Cu Zn	Removal at low salinity; midestuary addition	74, 85
	Mn	Addition at low salinity; midestuary removal	74
Tamar	U[a]	Conservative	91
	Al, Cd, Cu, Mn, Ni, Zn	Removal at low salinity; midestuary addition	17, 53, 73, 79
	As, Sn	Addition	63

Note: See original source for reference citations.

[a] Trace metals displaying a positive correlation with salinity.

Source: Millward, G. E. and Turner, A., in *Trace Elements in Natural Waters,* Salbu, B. and Steinnes, E., Eds., CRC Press, Boca Raton, FL, 1995, 231. With permission.

2.6 DEEP-SEA HYDROTHERMAL VENT CHEMISTRY

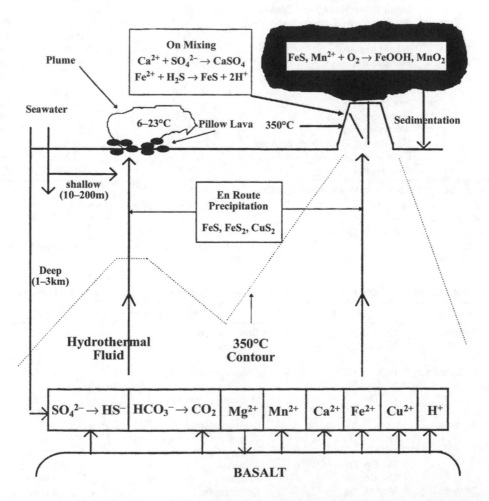

Figure 2.6–1   Schematic diagram showing the inorganic processes occurring at a vent site. (From Millero, F. J., *Chemical Oceanography,* 2nd ed., CRC Press, Boca Raton, FL, 1996, 377. With permission.)

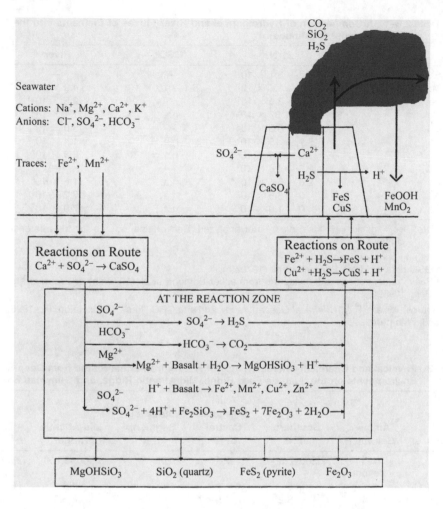

Figure 2.6–2 Hydrothermal chemical reactions. (From Millero, F. J., *Chemical Oceanography,* 2nd ed., CRC Press, Boca Raton, FL, 1996, 378. With permission.)

Table 2.6–1 A Comparison of Hydrothermal and River Fluxes of Elements into the Oceans[a]

	21° N	GSC[b]	River[c]
Li	$1.2 \rightarrow 1.9 \times 10^{11}$	$9.5 \rightarrow 16 \times 10^{10}$	1.4×10^{10}
Na	$-8.6 \rightarrow 1.9 \times 10^{12}$	$+, - 4$	6.9×10^{12}
K	$1.9 \rightarrow 2.3 \times 10^{12}$	1.3×10^{12}	1.9×10^{12}
Rb	$3.7 \rightarrow 4.6 \times 10^{9}$	$1.7 \rightarrow 2.8 \times 10^{9}$	5×10^{6}
Be	$1.4 \rightarrow 5.3 \times 10^{6}$	$1.6 \rightarrow 5.3 \times 10^{6}$	3.3×10^{7}
Mg	-7.5×10^{12}	-7.7×10^{12}	5.3×10^{12}
Ca	$2.4 \rightarrow 15 \times 10^{11}$	$2.1 \rightarrow 4.3 \times 10^{12}$	1.2×10^{13}
Sr	$-3.1 \rightarrow 1.4 \times 10^{9}$	0	2.2×10^{10}
Ba	$1.1 \rightarrow 2.3 \times 10^{9}$	$2.5 \rightarrow 6.1 \times 10^{9}$	1.0×10^{10}
F	$-1.0 \rightarrow 2.3 \times 10^{9}$	$2.5 \rightarrow 6.1 \times 10^{9}$	1.0×10^{10}
Cl	$0 \rightarrow 1.2 \times 10^{13}$	$-31 \rightarrow 7.8 \times 10^{12}$	6.9×10^{12}
SiO_2	$2.2 \rightarrow 2.8 \times 10^{12}$	3.1×10^{12}	6.4×10^{12}
Al	$5.7 \rightarrow 7.4 \times 10^{8}$	n.a.	6.0×10^{10}
SO_4	-4.0×10^{12}	-3.8×10^{12}	3.7×10^{12}
H_2S	$9.4 \rightarrow 12 \times 10^{11}$	$+$	

Table 2.6–1   A Comparison of Hydrothermal and River Fluxes of Elements into the Oceans[a] (continued)

	21°N	GSC[b]	River[c]
S	$-2.8 \rightarrow 3.1 \times 10^{11}$	—	—
Mn	$1.0 \rightarrow 1.4 \times 10^{11}$	$5.1 \rightarrow 16 \times 10^{10}$	4.9×10^9
Fe	$1.1 \rightarrow 3.5 \times 10^{11}$	$+$	2.3×10^{10}
Co	$3.1 \rightarrow 32 \times 10^6$	n.a.	1.1×10^8
Cu	$0 \rightarrow 6.3 \times 10^9$	—	5.0×10^9
Zn	$5.7 \rightarrow 15 \times 10^9$	n.a.	1.4×10^{10}
Ag	$0 \rightarrow 5.4 \times 10^6$	n.a.	8.8×10^7
Cd	$2.3 \rightarrow 26 \times 10^6$	—	—
Pb	$2.6 \rightarrow 5.1 \times 10^7$	n.a.	1.5×10^8
As	$0 \rightarrow 6.5 \times 10^7$	n.a.	7.2×10^8
Se	$0 \rightarrow 1.0 \times 10^7$	n.a.	7.9×10^7

Note: + = gain; − = loss; n.a. = not analyzed. See original source for full reference citations.

[a] All numbers are in mol/year.
[b] GSC data are from Edmond et al. (1979a,b).
[c] River concentrations and fluxes are from either Edmond et al. (1979a,b) or Broecker and Peng (1982).

Source: Millero, F. J., Chemical Oceanography, 2nd ed., CRC Press, Boca Raton, FL, 1996, 386. With permission.

Table 2.6–2   Physical and Chemical Characteristics of Hot Hydrothermal Fluids from Deep-Sea Vent Environments on the Juan de Fuca Ridge, Mid-Atlantic Ridge, and Guaymas Basin

Parameters	Ambient Seawater[a]	Southern JFR[b]	Axial Segment, Central JFR[c]	Endeavour Segment, JFR[c]	Mid-Atlantic Ridge[d]	Guaymas Basin[e]
Depth (m)	Near bottom	2,200–2,300	1,542	2,200	3,500–3,700	2,003
Max. temp. (°C)	5	285	328	400	350	355
pH	7.8	3.2	3.5–4.4	4.2	3.9	5.9
Gases						
CO_2 (mM)	2.3	3.9–4.5	50–285	11.6–18.2	n.d.	16–24
NH_4^+ (mM)	<0.01	n.d.	n.d.	0.64–0.95	n.d.	10–16
H_2S (mM)	0	3.5	7.0	3.2–6.8	5.9	3.8–6.0
CH_4 (μM)	0.0003	82–118	25–45	1,800–3,400	n.d.	2,000–6,800
H_2O (μM)	0.0003	335	25–80	160–420	n.d.	n.d.
CO (nM)	0.3	n.d.	n.d.	n.d.	n.d.	n.d.
Elements						
Mg (mM)	53.2	0	0	0	0	0
Cl (mM)	540	1,087	176–624	255–506	559–659	581–606
Si (mM)	0.16	23.3	13.5–15.1	5–16.9	18.2–22.0	9.3–13.8
Li (μM)	28	1,718	184–636	178–455	411–843	630–1,054
Fe (μM)	<0.001	18,739	12–1,065	500–1,013	1,640–2,180	17–180
Mn (μM)	<0.001	3,585	160–1,080	195–297	490–1,000	128–148

Note: n.d. = not determined; JFR = Juan de Fuca Ridge. See original source for reference citations.
[a] Shown for comparative purposes.
[b] Evans et al. (1988); Von Damm and Bischoff (1987); Von Damm (1988).
[c] Butterfield (1990); Lilley et al. (1993).
[d] Campbell et al. (1988).
[e] Von Damm et al. (1985b).

Source: Modified from Baross, J. A. and Deming, J. W., in The Microbiology of Deep-Sea Hydrothermal Vents, Karl, D. M., Ed., CRC Press, Boca Raton, FL, 1995, 179. With permission.

Table 2.6-3 Summary of Chemical Data for Seafloor Hydrothermal Solutions. Alkali and Alkaline Earth Metals, Ammonium, and Silica[a]

Vent	Temp. (°C)	Li (μmol/kg)	Na (mmol/kg)	K (mmol/kg)	Rb (μmol/kg)	Cs (nmol/kg)	NH4 (mmol/kg)	Be (nmol/kg)	Mg (mmol/kg)	Ca (mmol/kg)	Sr (μmol/kg)	$^{87}Sr/^{86}Sr$	Ba (μmol/kg)	SiO2 (mmol/kg)
Galapagos spreading center								11–37						
CB	<13	1142	487	18.7	20.3				0	40.2	87		>42.6	21.9
GE	<13	1142	451	18.8	21.2				0	34.3	87		>17.2	21.9
DL	<13	1142	313	18.8	17.3				0	34.3	87		>17.2	21.9
OB	<13	689	259	18.8	13.4				0	24.6	87		>17.2	21.9
21°N EPR														
NGS	273	1033	510	25.8	31.0		<0.01	37	0	20.8	97	0.7030	>16	19.5
OBS	350	891	432	23.2	28.0	202	<0.01	15	0	15.6	81	0.7031	>8	17.6
SW	355	899	439	23.2	27.0		<0.01	10	0	16.6	83	0.7033	>10	17.3
HG	351	1322	443	23.9	33.0		<0.01	13	0	11.7	65	0.7030	>11	15.6
Guaymas Basin														
1	291	1054	489	48.5	85.0		15.6	12	0	29.0	202		>12	12.9
2	291	954	478	46.3	77.0		15.3	18	0	28.7	184		>20	12.5
3	285	720	513	37.1	57.0		10.3	42	0	41.5	253	0.7052	>15	13.5
4	315	873	485	40.1	66.0		12.9	29	0	34.0	226	0.7052	>54	13.8
5	287	933	488	43.1	74.0		14.5	29	0	30.9	211		>13	12.4
6	264	896	475	45.1	74.0		14.5	60	0	26.6	172	0.7059	>16	10.8
7	300	1076	490	49.2	86.0		15.2	17	0	29.5	212		>24	12.8
9	100	630	480	32.5	57.0		10.7	91	0	30.2	160			9.3
Southern Juan de Fuca														
Plume	224	1718	796	51.6	37.0			95	0	96.4	312	0.7034		23.3
Vent 1	285	1108	661	37.3	28.0			150	0	84.7	230			22.8
Vent 3		1808	784	45.6	32.0			150	0	77.3	267			22.7
11–13°N EPR														
N & S −13°N	317	688	560	29.6	14.1				0	55.0	175	0.7041		22.0
1–13°N		614	587	29.8	18.0				0	44.6	171			21.9
2–13°N	354	592	551	27.5	19.0				0	53.7	182			19.4
3–13°N	(380)	591	596	28.8	20.0				0	54.8	168			17.9
4–11°N	347	884	472	32.0	24.0				0	22.5	80			18.8
5–11°N		623	577	32.9	25.0				0	35.2	135			20.6
6–11°N		484	290	18.7	15.0				0	10.6	38			14.3

(continued)

Table 2.6–3 Summary of Chemical Data for Seafloor Hydrothermal Solutions. Alkali and Alkaline Earth Metals, Ammonium, and Silica[a] (continued)

Vent	Temp. (°C)	Li (μmol/ kg)	Na (mmol/ kg)	K (mmol/ kg)	Rb (μmol/ kg)	Cs (nmol/ kg)	NH$_4$ (mmol/ kg)	Be (nmol/ kg)	Mg (mmol/ kg)	Ca (mmol/ kg)	Sr (μmol/ kg)	^{87}Sr ^{86}Sr	Ba (μmol/ kg)	SiO$_2$ (mmol/ kg)
Mid-Atlantic Ridge														
TAG	290–3 21	411	584	17.0	10.0	100			0	26.0	99	0.7029		22.0
MARK-1	350	843	510	23.6	10.5	177		38.5	0	9.9	50	0.7028		18.2
MARK-2	335	849	509	23.9	10.8	181		38.0	0	10.5	51	0.7028		18.3
Axial Volcano														
HE, HI, MR	136–3 23	512	415	22.0					0	37.3				15.1
Inferno	149–3 28	637	500	27.5					0	46.8				15.1
VM, Crack	5–299	204	159	7.6					0	10.2				13.5
Seawater	2	26	464	9.8	1.3	2.0	<0.01	0.0	52.7	10.2	87	0.7090	0.14	0.16

[a] Data references listed at end of Table 2.6–5.

Source: Annual Review of Earth Planetary Science, Vol. 18, ©1990 by Annual Reviews, Inc. With permission.

Table 2.6-4 Summary of Chemical Data for Seafloor Hydrothermal Solutions. pH, Carbon, and Sulfur Systems, Halogens, Boron, Aluminum, and Water Isotopes[a]

Vent	pH	Alk$_T$ (meq/kg)	Total CO$_2$ (mmol/kg)	SO$_4$ (mmol/kg)	H$_2$S (mmol/kg)	δ^{34}S	As (nmol/kg)	Se (nmol/kg)	Cl (mmol/kg)	Br (μmol/kg)	B (μmol/kg)	δ^{11}B	Al (μmol/kg)	δ^{18}O	δD
Galapagos spreading center			9.3–11.3							832–835					
CB	—	<0		0	+				595						
GE	—	<0		0	+				543						
DL	—	<0		0	+				395						
OB	—	<0		0	+				322						
21°N EPR			5.72												
NGS	3.8	−0.19		0.00	6.6	3.4	30	<0.6	579	929	507	32.7	4.0	1.6–2.0	2.5
OBS	3.4	−0.40		0.50	7.3	1.3–1.5	247	72	489	802	505	32.2	5.2		
SW	3.6	−0.30		0.60	7.5	2.7–5.5	214	70	496	877	500	31.5	4.7		
HG	3.3	−0.50		0.40	8.4	2.3–3.2	452	61	496	855	548	30.0	4.5		
Guaymas Basin															
1	5.9	10.60		−0.15	5.8		283	82	601	1054–1117			0.9		
2	5.9	9.60		−0.09	4.0		732	87	589		1630	17.4	1.2		
3	5.9	6.50		−0.34	5.2		1071	38	637				6.7		
4	5.9	8.10		0.06	4.8		1074	103	599	1063	1570	23.2	3.7		
5	5.9	9.70		−0.07	4.1		516		599				3.0		
6	5.9	7.30		−0.32	3.8		669	49	582				3.9		
7	5.9	10.50		−0.06	6.0		711	92	606	1054–1117	1730	19.6	1.0		
9	5.9	2.80		−4.20	4.6		577		581				7.9		
Southern Juan de Fuca			3.92–4.46												−2.5 – +0.5
Plume	3.2			−0.50	3.5	4.2–7.3		<1	1087	1832	496	34.2		0.65	
Vent 1	3.2			−1.30	(3.0)	4.0–6.4		<1	896	1580				0.60	
Vent 3	3.2			−1.70	(4.4)	7.2–7.4		<1	951	1422				0.80	
11–13°N EPR			10.8–16.7												
N & S −13°N	3.2			0.00					740	1163				0.39–0.69	0.62–1.49
1–13°N	3.2	−0.64			2.9				718	1131			13.3		
2–13°N	3.1	−0.74			8.2	4.7			712	1158	467	34.9	19.8		

(continued)

Table 2.6–4   Summary of Chemical Data for Seafloor Hydrothermal Solutions. pH, Carbon, and Sulfur Systems, Halogens, Boron, Aluminum, and Water Isotopes[a] (continued)

Vent	pH	Alk_T (meq/ kg)	Total CO_2 (mmol/ kg)	SO_4 (mmol/ kg)	H_2S (mmol/ kg)	$\delta^{14}S$	As (nmol/ kg)	Se (nmol/ kg)	Cl (mmol/ kg)	Br (μmol/kg)	B (μmol/ kg)	$\delta^{11}B$	Al (μmol/ kg)	$\delta^{18}O$	δD
3–13°N	3.3	−0.40			4.5	2.3–3.5			760	1242			20.0		
4–11°N	3.1	−1.02			8.0	4.6			563	940			12.9		
5–11°N	3.7	−0.28			4.4	4.7–4.9			686	1105	451	36.8	13.5		
6–11°N	3.1	−0.88			12.2	4.1–5.2			338	533	493	31.5	13.6		
Mid-Atlantic Ridge															
TAG									659		476	34.7			
MARK-1	3.9	−0.06			5.9	4.9			559	847	518	26.8	5.3	2.37	
MARK-2	3.7	−0.24			5.9	5.0			559	847	530	26.5	5.0	2.37	
Axial Volcano															
HE, HI, MR	3.5	−0.52			8.1				515	760	565			0.8–1.1	
Inferno	3.5	−0.45			7.0				625	950	565			0.7–1.1	
VM, Crack	4.4	0.58	150–170		(19.5)				188	240	503			0.7–0.9	
Seawater	7.8	2.3	2.3	27.9	0.0		27	2.5	541	840	416	39.5	0.020	0.0	0.0

[a] Data references listed at end of Table 2.6–5.

Source: Annual Review of Earth Planetary Science, Vol. 18, ©1990 by Annual Reviews, Inc. With permission.

Table 2.6-5 Summary of Chemical Data for Seafloor Hydrothermal Solutions. Trace Metals[a]

Vent	Mn (μmol/kg)	Fe (μmol/kg)	Co (nmol/kg)	Cu (μmol/kg)	Zn (μmol/kg)	Ag (nmol/kg)	Cd (nmol/kg)	Pb (nmol/kg)
Galapagos spreading center								
CB	1,140	+		0			0	
GE	390	+		0			0	
DL	480	+		0			0	
OB	360	+		0			0	
21°N EPR								
NGS	1,002	871	22	<0.02	40	<1	17	183
OBS	960	1,664	213	35.00	106	38	155	308
SW	699	750	66	9.70	89	26	144	194
HG	878	2,429	227	44.00	104	37	180	359
Guaymas Basin								
1	139	56	<5	<0.02	4.2	230	<10	265
2	222	49	<5	<0.02	1.8	<1	<10	304
3	236	180	<5	<0.02	40.0	24	46	652
4	139	77	<5	1.10	19.0	2	27	230
5	128	33	<5	0.10	2.2	<1	<10	<20
6	148	17	<5	<0.02	0.1	<1	<10	<20
7	139	37	<5	<0.02	2.2	<1	<10	<20
9	132	83	<5	<0.02	21.0	<1	<10	<20
Southern Juan de Fuca								
Plume 1	3,585	18,739		<2	900			900
Vent 1	2,611	10,349		<2	<600			
Vent 3	4,480	17,770		<2				

(continued)

Table 2.6-5 Summary of Chemical Data for Seafloor Hydrothermal Solutions. Trace Metals[a] (continued)

Vent	Mn (μmol/kg)	Fe (μmol/kg)	Co (nmol/kg)	Cu (μmol/kg)	Zn (μmol/kg)	Ag (nmol/kg)	Cd (nmol/kg)	Pb (nmol/kg)
11–13°N EPR								
N & S–13°N	1,000	1,450						
1–13°N	1,689	3,980			102.0		55	135
2–13°N	2,932	10,370			2.0		70	27
3–13°N	2,035	10,760			2.0		65	14
4–11°N	766	6,470			105.0		30	50
5–11°N	742	1,640			73.0		43	270
6–11°N	925	2,640			44.0		1	9
Mid-Atlantic Ridge								
TAG	1,000	1,640						
MARK-1	491	2,180		17.0	50.0			
MARK-2	493	1,832		10.0	47.0			
Axial Volcano								
HE, HI, MR	1,081	1,006		12.0	113.0			302
Inferno	1,081	1,006		12.0	113.0			302
VM, Crack	162	9		0.70	2.3			101
Seawater	<0.001	<0.001	0.03	0.007	0.01	0.02	1.0	0.0100

[a] Data from the following references (full citations in original source):

Galapagos: Edmond et al., 1979a,b; Welhan, 1981.

21°N EPR: Craig et al., 1980; Welhan, 1981; Von Damm et al., 1985a; Spivack and Edmond, 1987; Woodruff and Shanks, 1988: Campbell and Edmond, 1989.

Guaymas: Von Damm et al., 1985b; Spivack et al., 1987; Campbell and Edmond, 1989.

Juan de Fuca: Von Damm and Bischoff, 1987; Shanks and Seyfried, 1987; Hinkley and Tatsumoto, 1987; Evans et al., 1988; Campbell and Edmond, 1989.

11–13°N EPR: Michard et al., 1984; Merlivat et al., 1987; Bowers et al., 1988; Bluth and Ohmoto, 1988; Campbell and Edmond, 1989.

Mid-Atlantic Ridge: Campbell et al., 1988b. Campbell and Edmond, 1989.

Axial: Butterfield et al., 1988; Massoth et al., 1989.

Source: Annual Review of Earth Planetary Science, Vol. 18, ©1990 by Annual Reviews, Inc. With permission.

2.7 ORGANIC MATTER

Table 2.7–1 Levels of Dissolved and Particulate Organic Material in Natural Waters

Source	Dissolved, μM	Particulate, μM
Seawater		
Surface	75–150	1–17
Deep	4–75	0.2–1.3
Coastal	60–210	4–83
Estuarine	8–833	8–833
Drinking water	17	
Groundwater	58	
Precipitation	92	
Oligotrophic lake	183	80
River	420	170
Eutrophic lake	830–4,170	170
Marsh	1,250	170
Bog	2,500	250

Source: Millero, F. J., *Chemical Oceanography,* 2nd ed., CRC Press, Boca Raton, FL, 1996, 347. With permission.

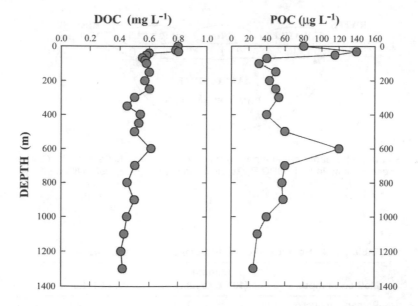

Figure 2.7–1 Profile of dissolved and particulate organic carbon in the oceans. (From Millero, F. J., *Chemical Oceanography,* 2nd ed., CRC Press, Boca Raton, FL, 1996, 347. With permission.)

Table 2.7–2 Comparison of the Dissolved and Particulate Organic Carbon, Nitrogen, and Phosphorus

Location	Carbon (μM C)		Nitrogen (μM N)		Phosphorus (μM P)	
	Diss	Part	Diss	Part	Diss	Part
Surface	75–150	12.5	4–10	2.1	0.1–0.6	0.06
Oxygen min.	—	10	3–5	0.6	—	0.03
Deep	4–75	—	1.7	0.9	—	0.03
Coastal	60–210	4–83	4–60	—	0.6–1.6	

Source: Millero, F. J., *Chemical Oceanography,* 2nd ed., CRC Press, Boca Raton, FL, 1996, 348. With permission.

Table 2.7–3   Some Organic Compounds
              Found in the Oceans

Compound	Formula
Carbohydrates	$C_n(H_2O)_m$
Amino acids	R–CH–COOH
	$\vert$
	NH_2
Hydrocarbons	C_nH_m
Carboxylic acids	R—COOH
Humic substances	Phenolic?
Steroids	

Source: Millero, F. J., *Chemical Oceanography,* 2nd ed., CRC Press, Boca Raton, FL, 1996, 350. With permission.

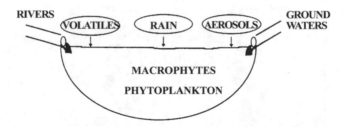

Figure 2.7–2   Sources of organic compounds to the oceans. (From Millero, F. J., *Chemical Oceanography,* 2nd ed., CRC Press, Boca Raton, FL, 1996, 308. With permission.)

Table 2.7–4   Sources of Organic Compounds to the Oceans

Method of Input	Amount (10^{15} g C/year)	% of Total	
Primary production			
Phytoplankton	23.1	84.4	
Macrophytes	1.7	6.2	90.6
Liquid input			
Rivers	1.0	3.65	
Groundwaters	0.08	0.3	3.95
Atmospheric input			
Rain	1.0	3.65	
Dry deposition	0.5	1.8	5.45
Total	27.4	100.00	100.00

Source: Millero, F. J., *Chemical Oceanography,* 2nd ed., CRC Press, Boca Raton, FL, 1996, 308. With permission.

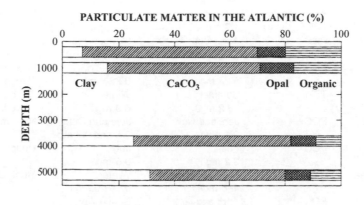

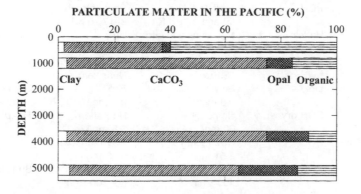

Figure 2.7–3 Particulate matter in Atlantic and Pacific Oceans. (From Millero, F. J., *Chemical Oceanography*, 2nd ed., CRC Press, Boca Raton, FL, 1996, 353. With permission.)

Table 2.7–5 Seasonal Maxima and Minima of Particulate Organic Matter in Estuarine and Coastal Marine Waters

Location	Measurement	Maximum (Time conc.)	Minimum (Time conc.)	Remarks
Estuary, Netherlands	Amino acids	May 1.5–1.6 mg C/l	Feb. Near zero	Lower estuary
	Amino acids	July 4.2 mg C/l	Feb. 0.1–0.2 mg C/l	Upper estuary
	POC	Summer 3 mg/l	Sept.–Feb. 1 mg/l	Lower estuary
	POC	No pattern 8–9 mg/1	No pattern Near zero	Upper estuary
	Amino acids	May 1.7 mg C/l	Oct. <0.1 mg C/l	Lower estuary
	POC	Aug.–Oct. 10–20 mg/l	Nov.–July 2–8 mg/l	
Arabian Sea estuary	POC	June and Aug. 5.24 mg/l	Oct.–Jan. 0.28 mg/l	Max. during monsoons
Estuary, Brazil	TSM	Summer 78.2 mg/l	Winter 6.9 mg/l	TSM about 60% organic
	POC	Jan. and Nov. 2.1 mg/l	July and Oct. 0.6 and 0.3 mg/l	Near mouth
	POC	Feb. 3.2 mg/l	July 0.5 mg/l	Upper estuary
Dutch Wadden Sea	POC	May 4–6 mg/l	Winter 0.5–1.0 mg/l	Tidal inlet
English Channel	POC	April and July 386 and 320 μg/l	Sept.–May 120–144 μg/l	
	PON	April and July 50 and 37 μg N/l	Sept.–March 8–17 μg N/l	
Galveston Bay, TX	Protein	May–June and Aug. 1.7 mg/l	Jan. 0.2 mg/l	0.6 mg/l mid-Aug.
	Carbohydrate	Sept. 530 μg/l	Dec. and Aug. Near zero	Weak seasonal pattern
	Lipid	Sept. 0.8 mg/l	Feb. 0.1 mg/l	Similar to CHO pattern
Gulf of Naples	PON dry wt.	No pattern 30 mg/l	No pattern 3.5 mg/l	*Posidonia* beds
Coastal Arabian Sea	POC	Nov. (30/m) 2.51 mg/l	May (surface) 0.52 mg/l	
Puget Sound	POC	July 400 μg/l	Sept.–March 50–100 μg/l	Surface
N. Dawes Inlet, AK	POC	July 1.22 mg/l	Nov. 0.24 mg/l	
Funka Bay, Japan	POC	July 11.5 g C/m^2	Early Oct. 3.5 g C/m^2	Integrated upper 50 m
	Protein	Feb. 3 g C/m^2	Early Oct. 1.5 g C/m^2	Integrated upper 50 m
Tripoli Harbor	POC	Aug. 1.3 mg/l	Nov.–April 0.9–1.1 mg/l	Surface
Strait of Georgia, B.C.	POC	Summer 400 μg/l	Nov. 170 μg/l	Seasonal means
Coastal S. California	POC	Feb. and March 351 and 235 μg/l	Sept. 191 μg/l	Coastal stations

Note: POC, particulate organic carbon; TSM, total suspended material; PON, particulate organic nitrogen.

Source: Wotton, R. S., Ed., *The Biology of Particles in Aquatic Systems,* CRC Press, Boca Raton, FL, 1990, 97. With permission.

Table 2.7–6   Particulate Organic Carbon Budget for Barataria Bay, Louisiana

Source	C/m²/year (g)	Sink	C/m²/year (g)
Export from salt marshes	297	Consumption in estuary	432
Benthic algae production	244	Export to Gulf of Mexico	318
Phytoplankton production	209		
Total	750		750

Source: Wolff, W. J., in Ecology of Marine Benthos, Coull, B. C., Ed., University of South Carolina Press, Columbia, 1977, 267. With permission.

Table 2.7–7   Seasonal Maxima and Minima of Particulate Organic Matter (POC) in Rivers

Location	Measurement	Maximum (Time, conc.)	Minimum (Time, conc.)	Remarks
Ganges R., Bangladesh	Amino acids	July 2395 µg/l	March 24 µg/l	
	Carbohydrate	July 1672 µg/l	March 46 µg/l	
Indus R., Pakistan	POC	Aug. (1981–82) Up to 16 mg/l	Feb. or April 0.3 mg/l	June and Sept. peaks (1983)
	Amino acids	Aug. or Sept. 659–2009 µg/l	Oct. or April 127–277 µg/l	Irregular sampling
	Carbohydrate	Aug. or Sept. 412–1105 µg/l	Feb./April/May 58–122 µg/l	Irregular sampling
Yangtze R., China	POC	July 17 mg/l	Jan.–Feb. 3 mg/l	Averages
Amazon R., near mouth	POC	Feb.–March 8.2 mg/l	May–June 1–2 mg/l	
Amazon R., upstream	POC	Feb.–March 15–20 mg/l	May–June 3.7 mg/ l	
Guatemalan rivers	POC	June 7 mg/l	Nov.–April Undetermined	Tributaries to 15 mg/l
Orinoco R., Venezuela	POC	No pattern 2.5 mg/l	No pattern 1 mg/l	
Gambia R., West Africa	POC	Early Sept. 2 mg/l	Oct.–Nov. 0.3 mg/l	No relation to discharge
Columbia River	POC	Summer 860 µg/l	Winter 2 µg/l	Highly variable
St. Lawrence River	POC (means)	May and Nov. 0.67 and 0.69 µg/l/l	Feb. 0.24 mg/l	May–terrigenous, Nov.–in situ
Erriff River, Ireland	POC	No pattern 4 mg/l	No pattern 0.25 mg/l	Runoff-related peaks
Bundorragha River, Ireland	POC	No pattern 0.9 mg/l	No pattern 0.3 mg/l	Steeper basin than Erriff
Loire R., France	POC	May–Aug. 6 mg/l	Nov. and March 2–3 mg/l	Increased at low water
Chalk Stream, England	POM	Fall–Winter 10–12 mg/l	Summer 1–2 mg/l	Dry weight

POC = Particulate organic carbon.

Source: Wotton, R. S., Ed., The Biology of Particles in Aquatic Systems, CRC Press, Boca Raton, FL, 1990, 88. With permission.

Table 2.7–8 The Dissolved Organic Constituents of Seawater (specific dissolved organic compounds identified in seawater)

I. Carbohydrates

Name of Compound and Chemical Formula	Concentration	Ref.	Locality
Pentoses $C_5H_{10}O_5$	0–8 mg/l	7, 8	Gulf of Mexico
Hexoses	0.5 µg/l	12	Pacific, off California
	14–36 µg/l	12	Pacific, off California
Rhamnosides $C_6H_{12}O_5$ Rhamnosides	0.1–0.4 mg/l	21	Pacific Ocean coast U.S.A.
Dehydroascorbic acid COCOCOCHCH(OH)CH$_2$OH	0.1 mg/l	29	Gulf of Mexico inshore water

II. Proteins and Their Derivatives

Name of Compound and Chemical Formula	(a) µg/l	(b) µg/l	(c) µg/l	(d) µg/l	Ref.	Locality
Peptides C:N ratio = 13.8:1					14	Gulf of Mexico
Polypeptides and polycondensates of:						
Glutamic acid COOH(CH$_2$)$_2$CH(NH$_2$)COOH		8–13	8–13	0.1–1.8	(a) 24 (by ion-exchange)	Gulf of Mexico
Lysine NH$_2$(CH$_2$)$_4$CH(NH$_2$)COOH	<1	?	Trace–3	0.1–0.9	(b) 27 (by paper chromatography)	Gulf of Mexico
Glycine NH$_2$CH$_2$COOH		—	Trace–3	1.2–3.7	(c) 27 (by ion-exchange)	Gulf of Mexico
Aspartic acid COOHCH$_2$CH(NH$_2$)COOH		3–8	Trace–3	0.1–1.0	(d) 12	Pacific, off California

Compound					Reference	Location
Serine $CH_2OHCH(NH_2)COOH$		?	Trace–3	1.8–5.6	(a) 27 (by ion-exchange)	Gulf of Mexico
Alanine $CH_3CH(NH_2)COOH$		3–8	Trace–3	0.7–3.1	(b) 27 (by paper chromatography)	Gulf of Mexico
Leucine $(CH_3)_2CHCH_2CH(NH_2)COOH$	0.5–1	8–13	Trace–3	0.9–3.8	(c) 27	Gulf of Mexico
Valine $(CH_3)_2CHCH(NH_2)COOH$		Trace–3	Trace–3	0.1–1.7	(d) 12 (by ion-exchange)	Pacific, off California
Cystine $[SCH_2CH(NH_2)COOH]_2$		Trace–3	—	0.0–3.8		
Isoleucine $CH_3CH_2CH(CH_3)CH(NH_2)COOH$		8–13	Trace–3	—		
Leucine $(CH_3)_2CHCH_2CH(NH_2)COOH$		—	—	0.9–3.8		
Ornithine $NH_2(CH_2)_3CH(NH_2)COOH$		—	Trace–3	0.2–2.4		
Methionine sulphoxide $CH_3S(:O)CH_2CH_2CH(NH_2)COOH$	<0.5	—	—	—		
Threonine $CH_3CHOHCH(NH_2)COOH$		—	3–8	0.3–1.3		
Tyrosine $HOC_6H_4CH_2CH(NH_2)COOH$		—	Trace–3	Trace–0.5		
Phenylalanine $C_6H_5CH_2CH(NH_2)COOH$		—	—	0.1–0.9		
Histidine $C_3H_3N_2CH_2CH(NH_2)COOH$		?	Trace–3	Trace–2.4		
Arginine $NH_2C(:NH)NH(CH_2)_3CH(NH_2)COOH$		?	Trace–3	0.1–0.6		
Proline C_4H_8NCOOH		?	—	0.3–1.4		
Methionine $CH_3SCH_2CH_2CH(NH_2)COOH$		—	Trace–3	Trace–0.4		
Tryptophan $C_8H_6NCH_2CH(NH_2)COOH$		—	Trace–3	—		
Glucosamine $C_6H_{13}NO_5$		—	Trace–3	—		

(continued)

Table 2.7–8 The Dissolved Organic Constituents of Seawater (specific dissolved organic compounds identified in seawater) (continued)

Name of Compound and Chemical Formula	Concentration		Ref.	Locality
	(e)	(f) μg/l	(e) 23	Norwegian coastal water
			(f) 12	Pacific, off California
Free amino acids				
Cystine [SCH$_2$CH(NH$_2$)COOH]$_2$	det.			
Lysine NH$_2$(CH$_2$)$_4$CH(NH$_2$)COOH	det.	0.2–3.1		
Histidine C$_3$H$_3$N$_2$CH$_2$CH(NH$_2$)COOH	det.	0.5–1.7		
Arginine NH$_2$C(:NH)NH(CH$_2$)$_3$CH(NH$_2$)COOH	det.	0.0		
Serine CH$_2$OHCH(NH$_2$)COOH	det.	2.3–28.4		
Aspartic acid COOHCH$_2$CH(NH$_2$)COOH	det.	Trace–9.6		
Glycine NH$_2$CH$_2$COOH	det.	Trace–37.6		
Hydroxyproline C$_4$H$_7$N(OH)COOH	det.	Trace–2.8		
Glutamic acid COOH(CH$_2$)$_2$CH(NH$_2$)COOH	det.	1.4–6.8		
Threonine CH$_3$CHOHCH(NH$_2$)COOH	det.	2.8–11.8		
a-Alanine CH$_3$CH(NH$_2$)COOH	det.			
Proline C$_4$H$_8$NCOOH	det.	0.0		
Tyrosine HOC$_6$H$_4$CH$_2$CH(NH$_2$)COOH	det.	Trace–5.0		
Tryptophan C$_8$H$_6$NCH$_2$CH(NH$_2$)COOH	det.	—		
Methionine CH$_3$SCH$_2$CH$_2$CH(NH$_2$)COOH	det.	—		
Valine (CH$_3$)$_2$CHCH(NH$_2$)COOH	det.	0.3–2.7		
Phenylalanine C$_6$H$_5$CH$_2$(NH$_2$)COOH	det.	Trace–2.4		

Compound						
Isoleucine $CH_3CH_2CH(CH_3)CH(NH_2)COOH$	—					
Leucine $(CH_3)_2CHCH_2CH(NH_2)COOH$	det.	0.5–5.5			3,4	Pacific coast near La Jolla
Free compounds						
Uracil $\overline{NHCONHCOCH:CH}$	det.					
Isoleucine $CH_3CH_2CH(CH_3)CH(NH_2)COOH$	det.					
Methionine $CH_3SCH_2CH_2CH(NH_2)COOH$	det.					
Histidine $C_3H_3N_2CH_2CH(NH_2)COOH$	det.					
Adenine $C_5H_3N_4NH_2$	det.					
Peptone	det.					
Threonine $CH_3CHOHCH(NH_2)COOH$	det.					
Tryptophan $C_8H_6NCH_2CH(NH_2)COOH$	det.					
Glycine NH_2CH_2COOH	det.					
Purine $C_5H_4N_4$	det.					
Urea CH_4ON_2	det.				12	Pacific, off California

III. Aliphatic Carboxylic and Hydroxycarboxylic Acids

Compound	mg/l (0–200m)	mg/l (200–600m)	mg/l (>600m)		
Lauric acid $CH_3(CH_2)_{10}COOH$	0.01–0.32	0.01–0.28	0–0.28	26	Coastal waters of Gulf of Mexico
Myristic acid $CH_3(CH_2)_{12}COOH$	0.01–0.10	0.01–0.05	0–0.07		

(continued)

Table 2.7–8 The Dissolved Organic Constituents of Seawater (specific dissolved organic compounds identified in seawater) (continued)

Name of Compound and Chemical Formula	Concentration		Ref.	Locality
Myristoleic acid $CH_3(CH_2)_3CH:CH(CH_2)_7COOH$	Trace–0.02	0–0.05		
Palmitic acid $CH_3(CH_2)_{14}COOH$	0.01–0.17	0–0.38		
Palmitoleic acid $CH_3(CH_2)_5CH:CH(CH_2)_7COOH$	0.02–0.16	0–0.21		
Stearic acid $CH_3(CH_2)_{16}COOH$	0.04–0.09	0–0.10		
Oleic acid $CH_3(CH_2)_7CH:CH(CH_2)_7COOH$	0.01	0.02	0	
Linoleic acid $CH_3(CH_2)_4CH:CHCH_2CH:CH(CH_2)_7COOH$	0.01	0.01	0	
	mg/l (1000–2500m)			
Fatty acids with:			30	Pacific Ocean coastal water
12 C-atoms	0.0003–0.02			
14 C-atoms	0.0004–0.043			
16 C-atoms	0.0027–0.0209			
16 C-atoms + 1 double bond	0.0003–0.003			
18 C-atoms	0.0037–0.0222			
18 C-atoms + 1 double bond	0.0083			
18 C-atoms + 2 double bonds	0.0000–0.0029			
20 C-atoms	Trace–0.0081			
22 C-atoms	Trace–0.0014			
	mg/l			
Acetic acid CH_3COOH	<1.0		20	Pacific Ocean
Lactic acid $CH_3CH(OH)COOH$				

Compound	Value	Units	Ref.	Location
Glycolic acid $HOCH_2COOH$				
Malic acid $HOOCH(OH)CH_2COOH$	0.28		10	Atlantic coastal water
Citric acid $HOOCCH_2C(OH)(COOH)CH_2COOH$	0.14			
Carotenoids and brownish-waxy or fatty matter	2.5		16	North Sea
			31	English Channel

IV. Biologically Active Compounds (see also Reference 25)

Compound	Value	Units	Ref.	Location
Organic Fe compound(s)	3.4–1.6	µg/l	13	Deep-sea water
Vitamin B_{12} (Cobalamin) $C_{63}H_{88}O_{14}N_{14}PCo$	0.2	µg/l (summer)	28	Long Island Sound
Vitamin B_{12}	2.0	µg/l (winter)	9	Oceanic surface water
Vitamin B_{12}	0.2–5.0	µg/l	11	North Pacific Ocean
Vitamin B_{12}	0–2.6	µg/l	19	Sargasso Sea 0–05 m.
Vitamin B_{12}	0–0.03	µg/l	22	Surface water, possibly from land drainage
Thiamine (Vitamin B_1) $C_{12}H_{17}ON_4SCl_2$	0–20	µg/l	9	
Plant hormones (auxins)	3.41	µg/l	5	North Sea near Scotland

V. Humic Acids

Compound	Ref.	Location
"Gelbstoffe" (Yellow substances) melanoidin-like	17,18	Coastal waters
	15	
	1,2	

(continued)

Table 2.7–8   The Dissolved Organic Constituents of Seawater (specific dissolved organic compounds identified in seawater) (continued)

Name of Compound and Chemical Formula	Concentration	Ref.	Locality
VI. Phenolic Compounds			
p-Hydroxybenzoic acid HOC$_6$H$_5$COOH	1–3 µg/l	12	Pacific, off California
Vanillic acid CH$_3$(HO)C$_6$H$_3$COOH	1–3 µg/l		
Syringic acid (CH$_3$O)$_2$(HO)C$_6$H$_2$COOH	1–3 µg/l		
VII. Hydrocarbons			
Pristane: (2, 6, 10, 14-tetramethylpentadecane)	Trace	6	Cape Cod Bay

Note: — = not detected; ? = possibly present; det. = detected.

Source: Duursma, E. K., in *Chemical Oceanography*, Vol. 1, Riley, J. P. and Skirrow, G., Eds., Academic Press, London, 1965, 450. With permission.

REFERENCES

1. Armstrong, F. A. J. and Boalch, G. T., *Nature* (London), 192, 858, 1961.
2. Armstrong, F. A. J. and Boalch, G. T., *J. Mar. Biol. Assoc. U.K.*, 41, 591, 1961.
3. Belser, W. L., *Proc. Natl. Acad. Sci. Wash.*, 45, 1533, 1959.
4. Belser, W. L., in *The Sea*, Hill, M. N., Ed., Vol. II, Wiley-Interscience, New York, 1963, 220–231.
5. Bentley, J. A., *J. Mar. Biol. Assoc. U.K.*, 39, 433, 1960.
6. Blumer, M., Mullin, M. M., and Thomas, D. W., *Science*, 140, 974, 1963.
7. Collier, A., *Spec. Sci. Rep. U.S. Fish Wildl.*, 178, 7, 1956.
8. Collier, A., Ray, S. M., and Magnitzky, A. W., *Science*, 111, 151, 1950.
9. Cowey, C. B., *J. Mar. Biol. Assoc. U.K.*, 35, 609, 1956.
10. Creach, P., *C. R. Acad. Sci. (Paris)*, 240, 2551, 1955.
11. Daisley, K. W. and Fisher, L. R., *J. Mar. Biol. Assoc. U.K.*, 37, 683, 1958.
12. Degens, E. T., Reuter, J. H., and Shaw, K. N. F., *Geochim. Cosmochim. Acta*, 28, 45, 1964.
13. Harvey, H. W., *J. Mar. Biol. Assoc. U.K.*, 13, 953, 1925.

14. Jeffrey, L. M. and Hood, D. W., *J. Mar. Res.*, 17, 247, 1958.
15. Jerlov, N. G., *Göteb. Vetensk Samh. Handl.*, F6. B.6. (14), 1955.
16. Johnston, R., *J. Mar. Biol. Assoc. U.K.*, 34, 185, 1955.
17. Kalle, K., *Dtsch. Hydrogr. Z.*, 2, 117, 1949.
18. Kalle, K., *Kiel. Meeresforsch.*, 18, 128, 1962.
19. Kashiwada, K., Kakimoto, D., Morita, T., Kanazawa, A., and Kawagoe, K., *Bull. Jpn. Soc. Sci. Fish.*, 22, 637, 1957.
20. Koyama, T. and Thompson, T. G., *Preprints International Oceanographic Congress, 1959*, American Association for Advancement of Science, Washington, D.C., 1959, 925.
21. Lewis, G. J. and Rakestraw, N. W., *J. Mar. Res.*, 14, 253, 1955.
22. Menzel, D. W. and Spaeth, J. P., *Limnol. Oceanogr.*, 7, 151, 1962.
23. Palmork, K. H., *Acta Chem. Scand.*, 17, 1456, 1963.
24. Park, K., Williams, W. T., Prescott, J. M., and Hood, D. W., *Science*, 138, 531, 1962.
25. Provasoli, L., in *The Sea*, Hill, M. N., Ed., Vol. II, Wiley-Interscience, New York, 1963, 165—219.
26. Slowey, J. F., Jeffrey, L. M., and Hood, D. W., *Geochim. Cosmochim. Acta*, 26, 607, 1962.
27. Tatsumoto, M., Williams, W. T., Prescott, J. M., and Hood, D. W., *J. Mar. Res.*, 19, 89, 1961.
28. Vishniac, H. S. and Riley, G. A., *Limnol. Oceanogr.*, 6, 36, 1961.
29. Wangersky, P. J., *Science*, 115, 685, 1952.
30. Williams, P. M., *Nature (London)*, 189, 219, 1961.
31. Wilson, D. P. and Armstrong, F. A. J., *J. Mar. Biol. Assoc. U.K.*, 31, 335, 1952.

Table 2.7–9   Concentration Ranges (μg/l) of the Major Identified Groups of Dissolved Organic Substances in Natural Waters

	Rain	Groundwater	River	Lake	Sea
Volatile fatty acids	10	40	100	100	40
Nonvolatile fatty acids	5–17	5–50	50–500	50–200	5–50
Amino acids	—	20–350	50–1000	30–6000	20–250
Carbohydrates	—	65–125	100–2000	100–3000	100–1000
Aldehydes	—	—	~0.1	—	0.01–0.1

Source: Wotton, R. S., Ed., *The Biology of Particles in Aquatic Systems,* CRC Press, Boca Raton, FL, 1990, 119. With permission.

Table 2.7–10   Seasonal Maxima and Minima of Dissolved Organic Matter in Estuaries and Coastal Marine Waters

Location	Measurement	Maximum (Time, conc.)	Minimum (Time, conc.)	Remarks
Estuary, Netherlands	DOC	Summer 4–5 mg/l	Fall 2 mg/l	Near mouth
	AA	March 0.8 mg C/l	Sept. 0.1 mg C/l	Near mouth
	DOC	No pattern 13 mg/l	No pattern 8 mg/l	
	DOC	June and July 2 mg/l	March and Nov. 0.3–0.5 mg/l	Autochthonous
	AA	July 0.6–0.8 mg C/l	Feb.–March 0.1 mg C/l	Highest near mouth
Nile estuary	DOM (O$_2$ demand)	August 9.05 mg O/l	Dec. 2.37 O/l	Temp.-dependent changes
Maine estuary	DOC	Summer 5–7 mg/l	Winter 1–2 mg/l	Up to 12.6 mg/l
	DFAA	No pattern 300 nM/l	No pattern 100 nM/l	Some summer low values
English Channel	DOC	March, July–Aug. and Oct. 1.8–2.2 mg/l	March–April and June 0.8–1.2 mg/l	Lagged Chl *a*
	DON	Late summer 7 μmol N/l	Winter Undetectable	Summarized Station E1
	DFAA	Winter 3–4 μmol N/l	Summer 1 μmol N/l	
	DOC	June and Sept. 0.88 and 0.96 mg/l	March 0.56 mg/l	
	DON	April and Aug. 5–6 μg-at N/l	Oct. 2.5 μg-at N/l	
Southern North Sea	DOC	Spring 1.7 mg/l	Winter 0.4 mg/l	Summer-up, autumn-down
Dutch Wadden Sea	DOC	May–June 4–5 mg/l	Winter 1–2 mg/l	Tidal inlet
Maine coast	DFAA	May 88 μg C/l	Dec. 10 μg C/l	July and Oct. peaks
Louisiana coast	DOC	Oct. 3 mg/l	Jan. 1.5 mg/l	<Full year
Gulf of Naples	DOC	No pattern 31.4 mg/l	No pattern 0.4 mg/l	*Posidonia* seagrass bed
Strait of Georgia, B.C.	DOC	Summer 3 mg/l	Nov. 1 mg/l	

Table 2.7–10   Seasonal Maxima and Minima of Dissolved Organic Matter in Estuaries and Coastal Marine Waters (continued)

Location	Measurement	Maximum (Time, conc.)	Minimum (Time, conc.)	Remarks
Coastal pond, Cape Cod, MA		April 2.9 mg/l	Feb. 1.3 mg/l	Inverse to chlorophyll
Menai Strait, England	DOC	Autumn 3–4 mg/l	Winter 1 mg/l	Spring-fall increase
Irish Sea	DFAA	May and Sept. 31 and 25 μm/l	Feb. and Aug. 5–10 μg/l	Late bloom maximum
	Total AA	Jan. and July 120 and 111 μg/l	Feb. 2 μg/l	Little pattern

Note: DOC, dissolved organic carbon; DOM, dissolved organic material; DFAA, dissolved free amino acid; DON, dissolved organic nitrogen; AA, amino acid; C, carbohydrate.

Source: Wotton, R. S., Ed., *The Biology of Particles in Aquatic Systems,* CRC Press, Boca Raton, FL, 1990, 95. With permission.

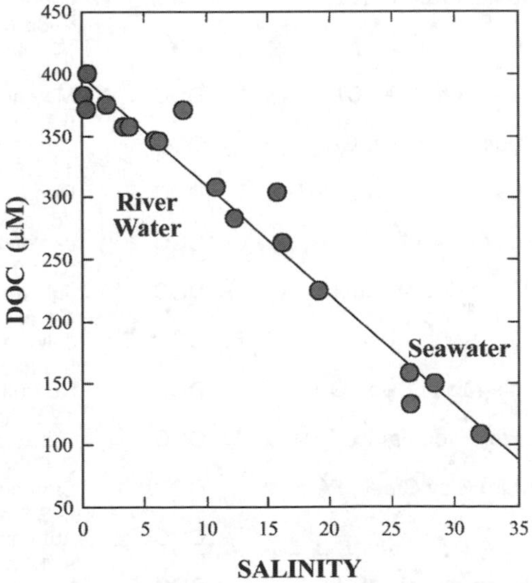

Figure 2.7–4   Dissolved organic carbon (DOC) as a function of salinity in a typical estuary. (From Millero, F. J., *Chemical Oceanography,* 2nd ed., CRC Press, Boca Raton, FL, 1996, 344. With permission.)

Table 2.7–11   Seasonal Maxima and Minima of Dissolved Organic Matter in Rivers

Location	Measurement	Maximum (Time, conc.)	Minimum (Time, conc.)	Remarks
Ganges River, Bangladesh	DOC	July 9.3 mg/l	June 1.3 mg/l	Max near crest
	C	June 1120 μg/l	Oct. 141 μg/l	
	AA	July 616 μg/l	March 150 μg/l	
Brahmaputra River	DOC	July 6.5 mg/l	Rest of year 1.3–2.6 mg/l	Rising water maximum
	AA	Aug. 262 μg/l	March 79 μg/l	
	C	July 985 μg/l	Oct. 155 μg/l	
Indus River	DOC	Aug.–Sept. 22 mg/l	Low-flow period 1.2 mg/l min	Maximum range
Amazon River	DOC	May–June 6.5 mg/l	Feb.–March 4.2 mg/l	
Orinoco River, Venezuela	DOC	May and Dec. 5 and 4 mg/l	June–Nov. 2–3 mg/l	
Gambia River, West Africa	DOC	Sept. 3.7 mg/l	Dec. 1.3 mg/l	
Columbia River	TOC	Spring–Summer 3.2 mg/l	Late fall 1.8 mg/l	~89% dissolved
Tigris River, Iraq	DOM	April 13.7–21.5 mg O/l	Oct. 0.3–1.6 mg O/l	As O_2 demand
Caroni River, Venezuela	DOC	Aug. and Jan. 8 mg/l	Nov. 4 mg/l	Humic rich
Ems River, Netherlands	DOC	Feb. 12 mg/l	Aug.–Sept. 4–5 mg/l	Increase in fall-winter
Guatemalan rivers	DOM	June–July 4–36 mg/l	June or Oct. 3–5 mg/l	Peak discharge July–Aug.
Alaskan rivers	DOC	Variable 4–6 mg/l	Aug.–Sept. 1–2 mg/l	
Shetucket River, CT	DOC	May and Sept. 6.2–10 mg/l	Jan.–April 2–4 mg/l	Max. 26.4 mg/l in runoff
Ogeechee River, GA	DOC	Jan.–May 12–15 mg/l	Aug.–Dec. 6–8 mg/l	To 17 mg/l July storm
Westerwoldse, Aa, Neth.	DOC	Dec. 48 mg/l	July 13 mg/l	Winter pollution
Black Creek, CA	DOC	Jan.–May 31–38 mg/l	June–Dec. 14–28 mg/l	To 42 mg/l in storm
N. Carolina stream	DOC	July and Oct. 1.1 and 1.3 mg/l	April 0.4 mg/l	Undisturbed watershed
	DOC	July and Oct. 0.5 and 0.6 mg/l	April–May 0.3 mg/l	Clear-cut watershed
Little Miami River, OH	DOC	No pattern 12.5 mg/l	No pattern 2.5 mg/l	Pollution related
Loire River, France	DOC	Jan.–April and Dec. 5–6 mg/l	May 2.5 mg/l	
White Clay Creek, PA	DOC	Summer-fall 2–2.5 mg/l	Winter 1–1.5 mg/l	Higher DOC downstream
	DOC	Autumn 9–12 mg/l	Mid-winter 2–4 mg/l	To 18 mg/l fall peak
Hubbard Brook, NH	DOC	No pattern 2 mg/l	No pattern <0.1 mg/l	Max. and min. in fall
Moorland Stream, U.K.	DOM	Aug. Up to 30 mg/l	Feb. 0–3 mg/l	Peak flow in Aug.

Note: DOC, dissolved organic carbon; DOM, dissolved organic material; TOC, total organic carbon; AA, amino acid; C, carbohydrate.

Source: Wotton, R. S., Ed., *The Biology of Particles in Aquatic Systems,* CRC Press, Boca Raton, FL, 1990, 85. With permission.

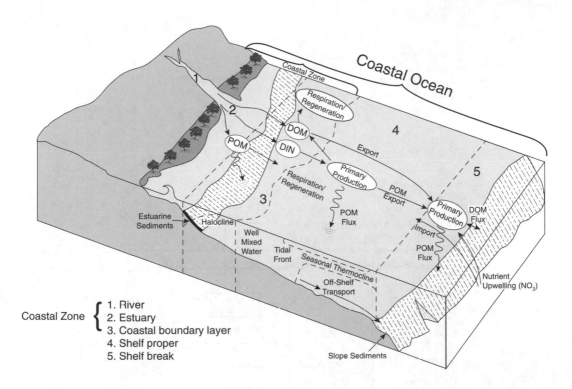

Coastal Zone
{
1. River
2. Estuary
3. Coastal boundary layer
4. Shelf proper
5. Shelf break

Figure 2.7–5   Key biogeochemical fluxes linking land and sea and pelagic and benthic processes, including those of dissolved and particulate organic matter. (From Alongi, D. M., *Coastal Ecosystem Processes*, CRC Press, Boca Raton, FL, 1998, 184. With permission.)

Table 2.7–12   Global Mass Balance of Organic Carbon Fluxes (10^{15} g C/year) on the Continental Shelf and Continental Slope

Continental Shelf	
Primary production	6.9
	⇓
Lost as pelagic respiration	3.7
Deposited to sediments	2.2 ⇒ 2.0 respired, 0.2 buried
Available for export	1.0
	⇓
Continental Slope	
Import from shelf	1.0
Lost as pelagic respiration	0.8
Deposited to sediments	0.2 ⇒ 0.18 respired, 0.02 buried

Source: From Alongi, D. M., *Coastal Ecosystem Processes*, CRC Press, Boca Raton, FL, 1998, 312. With permission.

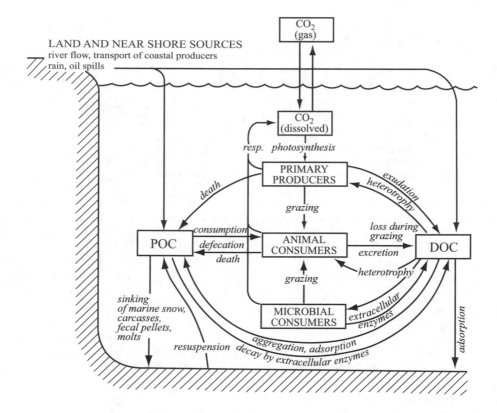

Figure 2.7–6 The transfer of carbon in aerobic marine environments. The boxes are pools while the arrows represent processes. The inorganic parts of the cycle have been simplified. "Marine snow" refers to organic aggregates and debris not shown in the diagram. There is probably some release of DOC from sediments into overlying waters. Viral lysis of primary producer cells could be an additional mechanism that releases DOC. (From Valiela, I., *Marine Ecological Processes,* 2nd ed., Springer-Verlag, New York, 1995, 389. With permission.)

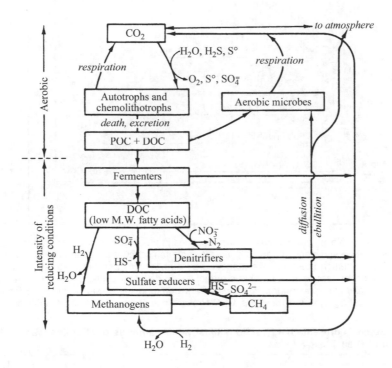

Figure 2.7–7 Carbon transformations in the transition from aerobic to anaerobic situations. The gradient from aerobic to anaerobic can be thought of as representing a sediment profile, with increased reduction of different microbial processes deeper in the sediment. Boxes represent pools or operators that carry out processes; arrows are processes that can be biochemical transformations or physical transport. Elements other than carbon are shown, where relevant, to indicate the couplings to other nutrient cycles. Some arrows indicate oxidizing and some reducing pathways. (From Valiela, I., *Marine Ecological Processes,* 2nd ed., Springer-Verlag, New York, 1995, 416. With permission.)

2.8 DECOMPOSITION OF ORGANIC MATTER

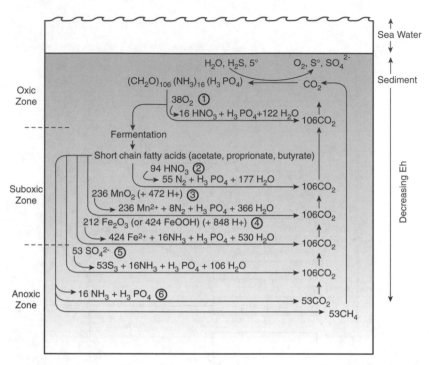

Figure 2.8–1 Model depicting the major oxidation reactions in the decomposition of organic matter with depth in marine sediments. Organic matter is represented by the formula $(CH_2O)_{106}$ $(NH_3)_{16}$ (H_3PO_4). The oxidation reactions are identified by number: (1) aerobic respiration; (2) denitrification; (3) manganese reduction; (4) iron reduction; (5) sulfate reduction; and (6) methanogenesis. Note that nitrogen, phosphorus, and water are common metabolic products. (From Alongi, D. M., *Coastal Ecosystem Processes,* CRC Press, Boca Raton, FL, 1998, 4. With permission.)

Table 2.8–1   Some Representative Reactions Illustrating Pathways of Microbial Metabolism and Their Energy Yields[a]

		Energy Yield (kcal)
Aerobic respiration	$C_6H_{12}O_6$ (glucose) $+ 6O_2 = 4CO_2 + 4H_2O$	686
Fermentation	$C_6H_{12}O_6 = 2CH_3CHOCOOH$ (lactic acid)	58
	$C_6H_{12}O_6 = 2CH_2CH_2OH$ (ethanol) $+ 2CO_2$	57
Nitrate reduction and denitrification	$C_6H_{12}O_6 + 24/6\ NO_3^- + 24/5\ H^+ = 6CO_2 + 12/5\ N_2 + 42/5\ H_2O$	649
Sulfate reduction	$CH_3CHOHCOO^-$ (lactate) $+ 1/2\ SO_4 + 3/2\ H^+ = CH_3COO^-$ (acetate) $+ CO_2 + H_2O + 1/2\ HS^-$	8.9
	$CH_3COO^- + SO_4^- = 2CO_2 + 2H_2O + HS^-$	9.7
Methanogenesis	$H_2 + 1/4\ CO_2 = 1/4\ CH_4 + 1/2\ H_2O$	8.3
	$CH_3COO + 4H_2 = 2CH_4 + 2H_2O$	39
	$CH_3COO = CH_4 + CO_2$ [b]	6.6
Methane oxidation	$CH_4 + SO_4^- + 2H^+ = CO_2 + 2H_2O + HS^-$	3.1
	$CH_4 + 2O_2 = CO_2 + 2H_2O$	193.5
Sulfide oxidation	$HS^- + 2O_2 = SO_4^- + H^+$	190.4
	$HS^- + 8/5\ NO_3^- + 3/5\ H^+ = SO_4^- + 4/5\ N_2 + 4/5\ H_2O$	177.9

[a] Energy yields vary depending on the conditions, so different measurements may be found in different references. The values reported here are representative.
[b] This reaction is sometimes considered fermentation.

Source: Valiela, I., *Marine Ecological Processes,* Springer-Verlag, New York, 1995, 414. With permission.

Table 2.8–2 Contribution of Various Aerobic and Anaerobic Processes to Carbon Mineralization in a Spectrum of Marine and Freshwater Sediments

Site	Rate of Carbon Oxidation (mmol C/m²/day)	Respiratory Mode (1% of C oxidation)		
		O_2 Resp.	SO_4^{2-} Resp.	CH_4 Prod.
Marine				
Limfjorden	36	47	53	—
Sippewissett Salt Marsh	458	10	90	(0)
Sapelo I. Salt Marsh	200	20	70	(10)
Sippewissett Salt Marsh	180	50	50	(0)
Sulfate-Depleted Marine				
Cape Lookout Bight	100	(0)	68	32
Lacustrine				
Blelham Tarn	—	42	2	25
Wintergreen Lake–A	3.1[a]	—	13[b]	87[b]
Wintergreen Lake–B	14.4[a]	—	13[b]	87[b]
Wintergreen	108[c]	—	30	71
Lake Vechten	23	—		70
Lawrence Lake	3[a]	—	30–81[c]	19–70[c]
Experimental Lakes Area	9.5–12	—	16–20[c]	72–82[c]

[a] Assuming 2 cm active depth in sediment.
[b] Percentage of total anoxic mineralization only.
[c] 0 to 15 cm.

Source: Wotton, R. S., Ed., *The Biology of Particles in Aquatic Systems,* CRC Press, Boca Raton, FL, 1990, 133. With permission.

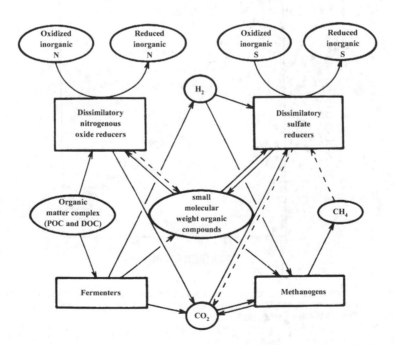

Figure 2.8–2 Conceptual model of anaerobic microbial processes in saltmarsh sediments. Solid lines are confirmed fluxes. Dashed lines are possible fluxes. (From Wiebe, W. J. et al., in *The Ecology of a Salt Marsh,* Pomeroy, L. R. and Wiegert, R. G., Eds., Ecological Studies, Vol. 38, Springer-Verlag, New York, 1981, 137. With permission.)

2.9 OXYGEN

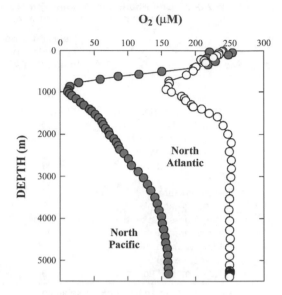

Figure 2.9–1   Profiles of oxygen in the North Atlantic and Pacific Oceans. (From Millero, F. J., *Chemical Oceanography,* 2nd ed., CRC Press, Boca Raton, FL, 1996, 221. With permission.)

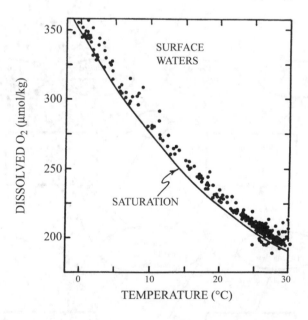

Figure 2.9–2   Comparison of the measured and calculated dissolved oxygen in surface seawaters as a function of temperature. (From Millero, F. J., *Chemical Oceanography,* 2nd ed., CRC Press, Boca Raton, FL, 1996, 221. With permission.)

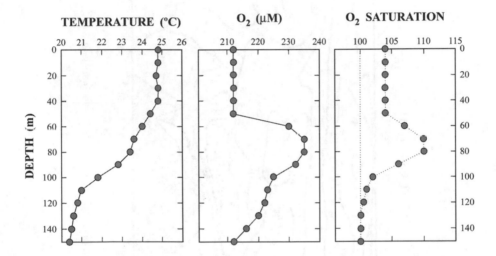

Figure 2.9–3 Profiles of temperature and oxygen in Pacific waters showing increases due to photosynthesis. (From Millero, F. J., *Chemical Oceanography,* 2nd ed., CRC Press, Boca Raton, FL, 1996, 222. With permission.)

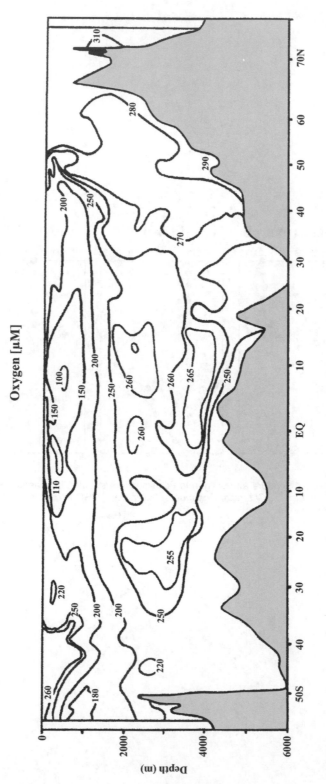

Figure 2.9-4 A section of oxygen in the Atlantic Ocean. (From Millero, F. J., *Chemical Oceanography*, 2nd ed., CRC Press, Boca Raton, FL, 1996, 223. With permission.)

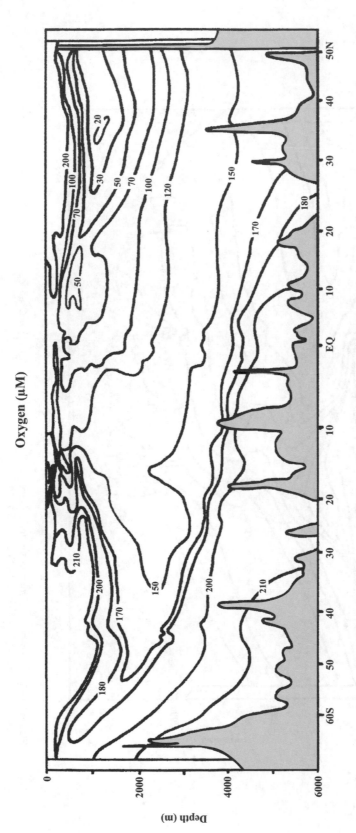

Figure 2.9–5   A section of oxygen in the Pacific Ocean. (From Millero, F. J., *Chemical Oceanography*, 2nd ed., CRC Press, Boca Raton, FL, 1996, 223. With permission.)

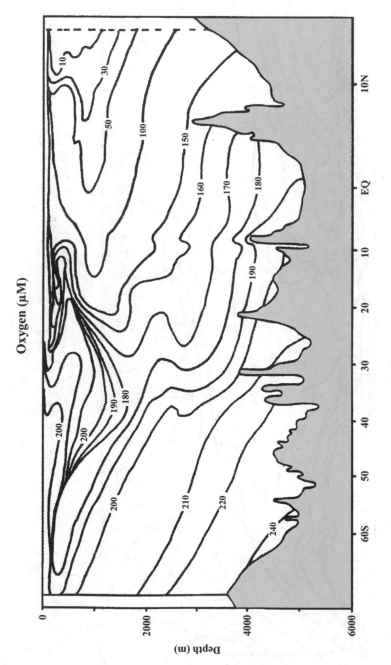

Figure 2.9–6 A section of oxygen in the Indian Ocean. (From Millero, F. J., *Chemical Oceanography*, 2nd ed., CRC Press, Boca Raton, FL, 1996, 224. With permission.)

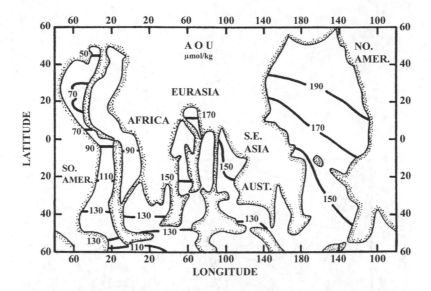

Figure 2.9–7 Apparent oxygen utilization in deep waters of the world oceans. (From Millero, F. J., *Chemical Oceanography,* 2nd ed., CRC Press, Boca Raton, FL, 1996, 226. With permission.)

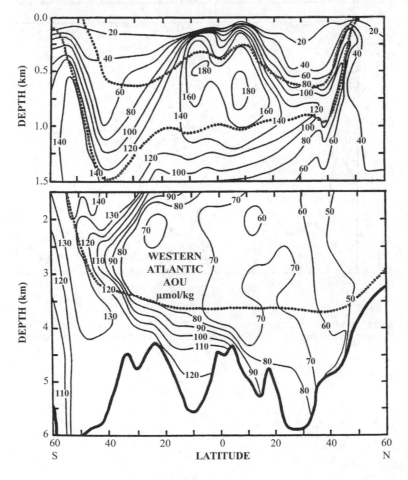

Figure 2.9–8 A section of apparent oxygen utilization of Atlantic Ocean waters. (From Millero, F. J., *Chemical Oceanography,* 2nd ed., CRC Press, Boca Raton, FL, 1996, 227. With permission.)

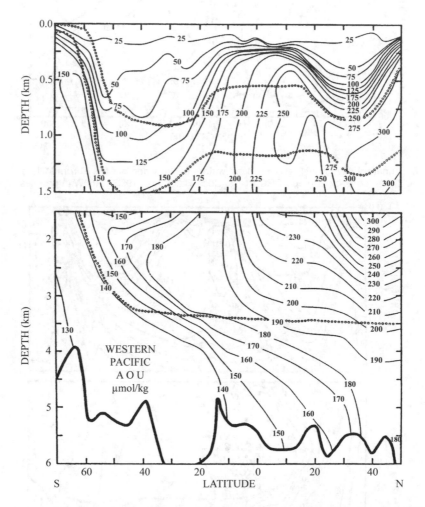

Figure 2.9–9   A section of apparent oxygen utilization of Pacific Ocean waters. (From Millero, F. J., *Chemical Oceanography,* 2nd ed., CRC Press, Boca Raton, FL, 1996, 228. With permission.)

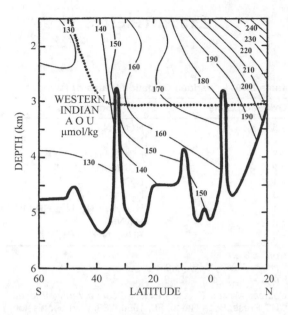

Figure 2.9–10   A section of apparent oxygen utilization of Indian Ocean waters. (From Millero, F. J., *Chemical Oceanography*, 2nd ed., CRC Press, Boca Raton, FL, 1996, 229. With permission.)

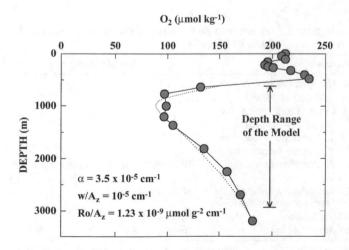

Figure 2.9–11   A comparison of the measured and calculated concentrations of oxygen using a one-dimensional mixing model. (From Millero, F. J., *Chemical Oceanography*, 2nd ed., CRC Press, Boca Raton, FL, 1996, 229. With permission.)

2.10 NUTRIENTS

Table 2.10–1   Various Oxidation States of Nitrogen

Oxidation State	Compound
+5	NO_3^- , N_2O_5
+4	NO_2
+3	HONO,[a] NO_2^- , N_2O_3
+2	HONNOH,[b] $HO_2N_2^-$, $N_2O_2^{2-}$
+1	N_2O
0	N_2
−1	H_2NOH, HN_3, N_3^- , NH_2OH
−2	H_2NNH_2
−3	RNH_4, NH_3^c , NH_4^{+c}

[a] $pK = 3.35$.
[b] $pK_1 = 7.05$, $pK_2 = 11.0$.
[c] $pK_B = 4.75$, $pK_A = 9.48$.

Source: Millero, F. J., *Chemical Oceanography*, 2nd ed., CRC Press, Boca Raton, FL, 1996, 289. With permission.

Table 2.10–2   Transformation of Nitrogen Forms in the Nitrogen Cycle of Estuaries

Process	Transformation
A. Nitrogen fixation (oxygen sensitive)	$N_2 \rightarrow NH_3$
B. Dissimilatory reduction (oxygen sensitive)	
1. Respiratory reduction (nitrate respiration)	$NO_3^- \rightarrow NO_2^-$
2. Denitrification	NO_3^- , NO_2^- , $N_2O \rightarrow$ gaseous products (N_2O, N_2, NH_3)
C. Assimilatory reduction (ammonia sensitive)	NO_3^-, etc. $\rightarrow NH_3$
D. Nitrification	$NH_3 \rightarrow NO_3^-$
E. Ammonification	"R-NH$_2$" $\rightarrow NH_3$

Source: Webb, K. L., in *Estuaries and Nutrients,* Neilson, B. J. and Cronin, L. E., Eds., Humana Press, Clifton, NJ, 1981, 25. With permission.

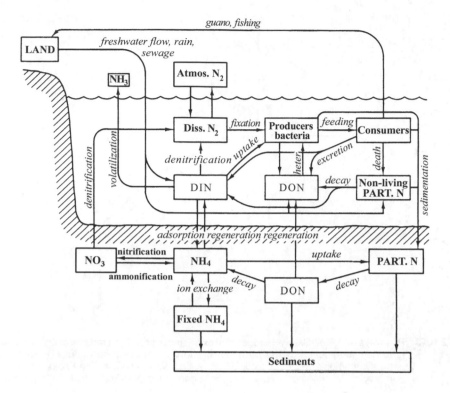

Figure 2.10–1 Simplified scheme of major transformations and transport of nitrogen in marine environments. (From Valiela, I., *Marine Ecological Processes,* 2nd ed., Springer-Verlag, New York, 1995, 438. With permission.)

Figure 2.10–2 Model of the major pools and transformations of the nitrogen cycle. The processes are numbered as follows: 1 = ammonification; 2 = nitrification (ammonium oxidation); 3 = nitrification (nitrite oxidation); 4 = denitrification and dissimilatory nitrate reduction; 5 = nitrogen fixation; 6 = assimilatory nitrate reduction; 7 = assimilatory nitrite reduction; 8 = immobilization and assimilation. (From Alongi, D. M., *Coastal Ecosystem Processes,* CRC Press, Boca Raton, FL, 1998, 6. With permission.)

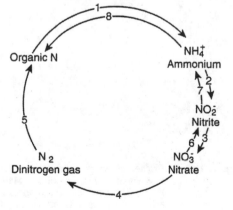

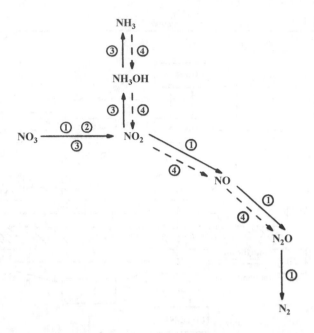

Figure 2.10–3 Pathways of dissimilatory nitrogenous oxide reduction. (1) Denitrification; (2) dissimilatory reduction (terminates at NO_2); (3) dissimilatory ammonia production; (4) "nitrification" N_2O pathway: ammonia to nitrous oxide. (From Wiebe, W. J. et al., in *The Ecology of a Salt Marsh,* Pomeroy, L. R. and Wiegert, R. G., Eds., Springer-Verlag, New York, 1981, 137. With permission.)

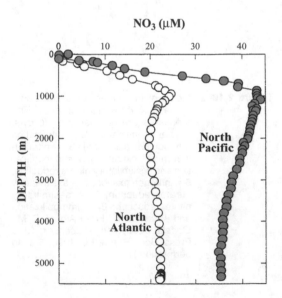

Figure 2.10–4 Profiles of nitrate in the North Atlantic and Pacific Oceans. (From Millero, F. J., *Chemical Oceanography,* 2nd ed., CRC Press, Boca Raton, FL, 1996, 294. With permission.)

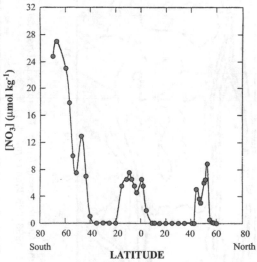

Figure 2.10–5 The concentration of nitrate in the surface waters of the North Pacific, Equatorial Pacific, and Antarctic Oceans. (From Millero, F. J., *Chemical Oceanography,* 2nd ed., CRC Press, Boca Raton, FL, 1996, 323. With permission.)

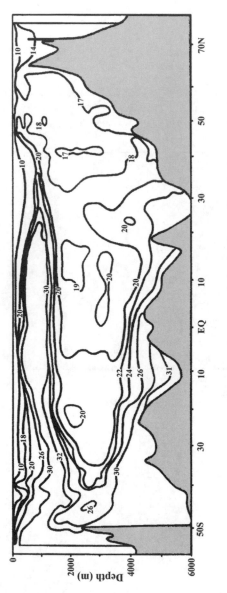

Figure 2.10–6 Section of nitrate in the Atlantic Ocean. (From Millero, F. J., *Chemical Oceanography*, 2nd ed., CRC Press, Boca Raton, FL, 1996, 295. With permission.)

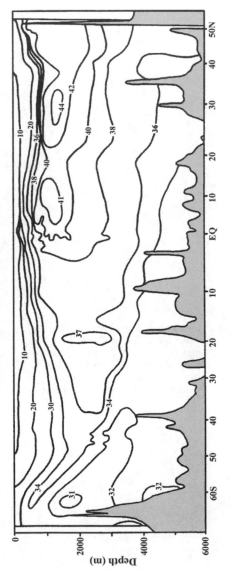

Figure 2.10–7 Section of nitrate in the Pacific Ocean. (From Millero, F. J., *Chemical Oceanography*, 2nd ed., CRC Press, Boca Raton, FL, 1996, 295. With permission.)

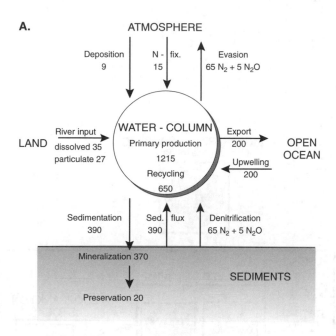

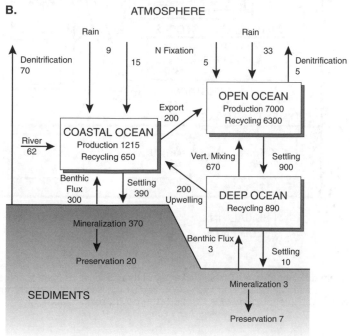

Figure 2.10–8   Preliminary global budget of nitrogen in the coastal zone (A) and for the entire ocean (B). Fluxes are 10^{12} g N yr^{-1}. (From Alongi, D. M., *Coastal Ecosystem Processes,* CRC Press, Boca Raton, FL, 1998, 315. With permission.)

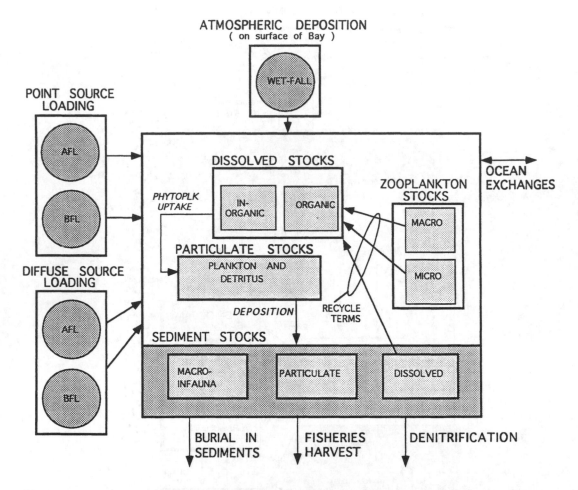

Figure 2.10–9   A schematic diagram of the Chesapeake Bay nutrient budget. Nutrient sources, storages, recycle
pathways, internal losses, and exchanges across the seaward boundary are indicated. (From
Boynton, W. R., et al., *Estuaries,* 18, 285, 1995. With permission.)

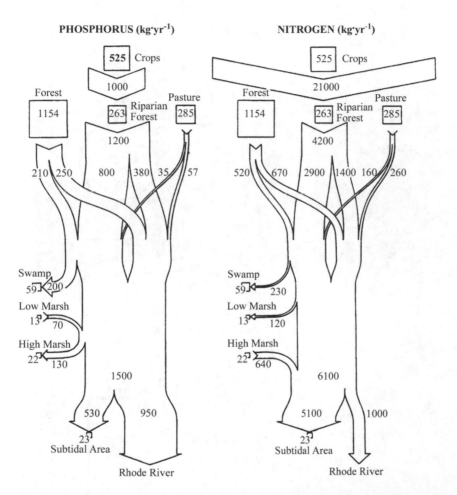

Figure 2.10–10 Net annual nitrogen and phosphorus fluxes (kg/year) through the component ecosystems of the Rhode River landscape. Arrows indicate watershed discharges from upland systems, net uptake by freshwater swamp, and net tidal exchanges by the marshes and subtidal area. Widths of arrows are proportional to the fluxes given by the numbers in arrows. Numbers in boxes give the area of each ecosystem type (ha). (From Correll, D. L. et al., *Estuaries*, 15, 431, 1992. With permission.)

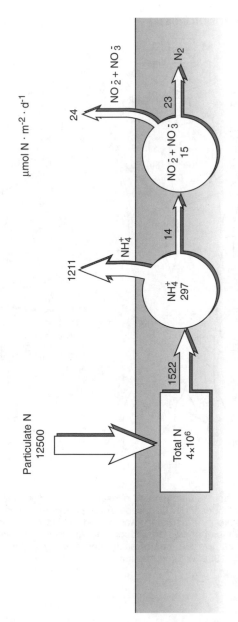

Figure 2.10–11 Preliminary nitrogen budget (mmol N/M²/day) for mid-shelf sediments of the central Great Barrier Reef lagoon. (From Alongi, D. M., *Coastal Ecosystem Processes*, CRC Press, Boca Raton, FL, 1998, 296. With permission.)

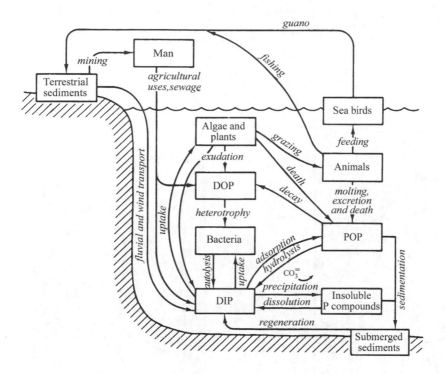

Figure 2.10–12 The phosphorus cycle. Boxes indicate pools of phosphorus, arrows show processes, including transport or transformations, (From Valiela, I., *Marine Ecological Processes,* Springer-Verlag, New York, 1996, 428. With permission.)

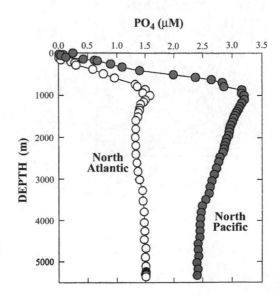

Figure 2.10–13 Profiles of phosphate in the North Atlantic and Pacific Oceans. (From Millero, F. J., *Chemical Oceanography,* 2nd ed., CRC Press, Boca Raton, FL, 1996, 289. With permission.)

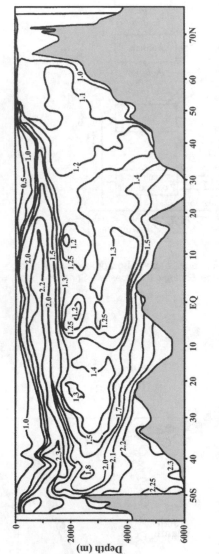

Figure 2.10–14 Section of phosphate in the Atlantic Ocean. (From Millero, F. J., *Chemical Oceanography*, 2nd ed., CRC Press, Boca Raton, FL, 1996, 291. With permission.)

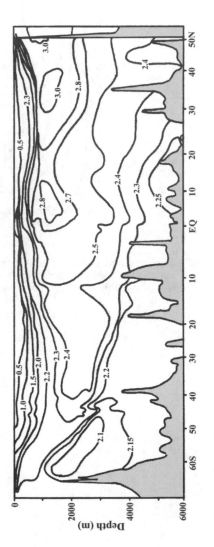

Figure 2.10–15 Section of phosphate in the Pacific Ocean. (From Millero, F. J., *Chemical Oceanography*, 2nd ed., CRC Press, Boca Raton, FL, 1996, 291. With permission.)

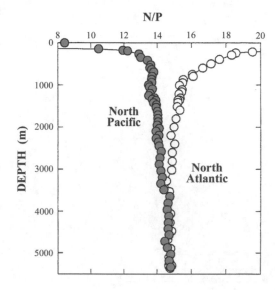

Figure 2.10–16 The nitrogen to phosphate molar ratios in the North Atlantic and Pacific Oceans. (From Millero, F. J., *Chemical Oceanography,* 2nd ed., CRC Press, Boca Raton, FL, 1996, 298. With permission.)

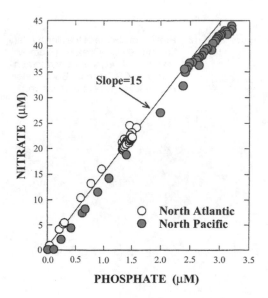

Figure 2.10–17 Correlation of the concentration of nitrogen to phosphate in the North Atlantic and Pacific Oceans. (From Millero, F. J., *Chemical Oceanography,* 2nd ed., CRC Press, Boca Raton, FL, 1996, 298. With permission.)

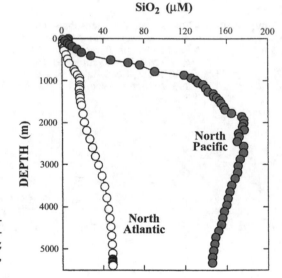

Figure 2.10–18   Profiles of silicate in the North Atlantic and Pacific Oceans. (From Millero, F. J., *Chemical Oceanography*, 2nd ed., CRC Press, Boca Raton, FL, 1996, 301. With permission.)

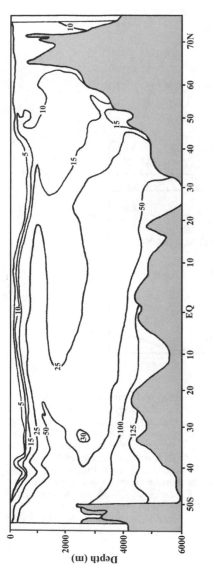

Figure 2.10–19 Section of silicate in the Atlantic Ocean. (From Millero, F. J., *Chemical Oceanography*, 2nd ed., CRC Press, Boca Raton, FL, 1996, 302. With permission.)

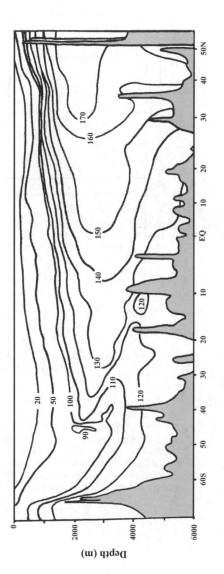

Figure 2.10–20 Section of silicate in the Pacific ocean. (From Millero, F. J., *Chemical Oceanography*, 2nd ed., CRC Press, Boca Raton, FL, 1996, 302. With permission.)

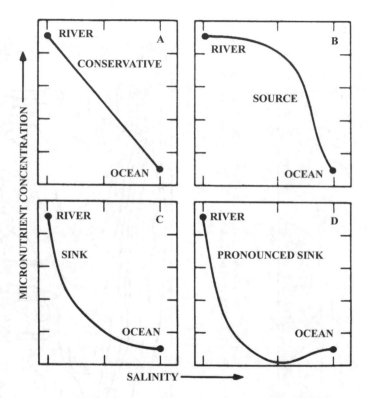

Figure 2.10–21 Idealized micronutrient-salinity relations showing concentration and mixing of nutrient-rich river water with nutrient-poor seawater. (A) Expected concentration-salinity distribution of a substance behaving in a conservative manner (for example, chloride) in an estuary; (B) expected concentration-salinity distribution of a substance for which the estuary is a source (for example, particulate carbon); (C) expected concentration-salinity distribution of a substance for which the estuary is a sink (for example, phosphorus); and (D) expected concentration-salinity distribution of a substance for which the estuary is a pronounced sink, that is, where the concentration of the substance in the estuary is lower than in the river and the ocean (for example, Si). (From Biggs, R. B. and Cronin, L. E., in *Estuaries and Nutrients,* Neilson, B. J. and Cronin, L. E., Eds., Humana Press, Clifton, NJ, 1981, 3. With permission.)

Table 2.10-3   Dissolved and Particulate Nutrient Concentrations in Tropical Aquatic Ecosystems

	NH₄	NO₂-NO₃	DON	PON (μmol/l)	PO₄	DOP	POP	Si
Oceanic (near surface)								
Gulf of Mexico	0.0–0.7	0.0–0.1		0.1–0.6	0.0–0.5			0–3
Subtropical North Pacific		0.0–0.1		0.2–0.3	0.1–0.1			2–3
Equatorial Pacific		0.0–2.0	6.2–13.8					
Equatorial Atlantic	0.0–0.1	0.0–0.1		0.0–0.3	0.0–0.4		0.0–0.1	0–2
South Pacific (Moorea)								
Coastal (near surface)								
Campeche Bank	0.0–2.7	0.0–0.3		0.1–5.1	0.0–0.4		0.0–0.1	0–3
Gulf of Papua/Torres Strait	0.0–3.2	0.0–3.1	0.9–13.7		0.3–0.6	0.0–0.2		1–30
Central GBR (18–20S)	0.0–0.5	0.0–0.5	2.4–14.8	1.0–3.8	0.0–0.3	0.0–0.8	0.0–0.2	0–2
Barbados	0.5–2.7	0.4–5.1			0.1–0.2			
Upwelling								
Peru	0.0–3+	0.0–20+		3.0–14+	0.0–2.5+			0–25
Arabian Sea		0.0–20+			0.0–2.0+			0–16
Estuary (near surface)								
Cochin Backwater (India)		0.0–20.3		0.0–150.0				
Missionary Bay (Australia)	0.1–0.4	2.0–7.0		0.1–0.4	0.2–0.6			
Coral reefs, oceanic and atolls								
Canton Atoll	0.1–1.3	0.0–2.4	1.7–2.3		0.0–0.5			2–3
Enewetak Atoll	0.2–0.3	0.1–0.3	1–23.0		0.0–0.2	0.0–0.2		
Tonga Lagoon	0.1–0.7	0.1–1.0		3.0–10.0	0.1–0.9			17–91
Gilbert Islands	0.3–0.5	0.0–2.6	3.8–5.6		0.0–0.4			
Takapoto Atoll	0.0–0.1	0.0–0.2			0.0–0.1			0–0
Moorea		0.0–0.1			0.0–0.5			0–2
Coral reefs, shelf and fringing								
Kaneohe Bay	0.4–2.4	0.1–2.6	3.4–7.5		0.2–1.0	0.0–0.4		
Jamaica	0.1–3.8		3–5.0		0.0–0.7			
Lizard Island (GBR)	0.1–0.2	0.2–1.0			0.2–0.4			1–2
Davies and Old Reef lagoons (GBR)	0.0–0.2	0.0–0.4			0.0–0.2			0–1
Abrolhos Islands	0.1–11.0	0.8–5.2			0.2–2.9	0.0–4.9		1–7

(continued)

Table 2.10–3   Dissolved and Particulate Nutrient Concentrations in Tropical Aquatic Ecosystems (continued)

	NH$_4$	NO$_2$–NO$_3$	DON	PON (µmol/l)	PO$_4$	DOP	POP	Si
Rivers								
Amazon		0.0–8.5			0.0–0.2			0–128
Ganges	0.7–31.4	35.7–770.0			1.9–652.0			21–2869
Maroni (S. America)		5.0–9.0			0.0–0.2			0–191
Niger	0.0–1.0	0.0–7.6						0–250
Bermejo (S. America)	0.8–35.7	5.5–52.1			0.8–3.3	0.1–0.8	1.5–230.0	
Orinoco	0.0–0.4				0.0–0.2			0–91
Zaire		6.7–7.0			0.1–0.8			165–342
Streams								
Savannah Rivers (Uganda)	<0.4–3.9	0.0–61.0			1.1–30.3			87–603
Wet forest streams (Uganda)	<0.4–9.7	0.0–171.0			0.1–8.2			118–825
Lesser Antilles	<2.8–62.8	<8.1–179.0			0.1–1.5			42–533
Lakes								
Lake Waigani (PNG)	0.0–4.4	0.0–21.6	0.0–723.0		0.0–42.6	0.0–22.0		
Lake Victoria (Uganda)	0.0–28.6	0.0–0.4			0.0–0.3			
Lake Nabugado (Kenya)	0.0–<7				0.0–21.0			
Lake Kyoga (Kenya)	0.0–14.0	14.0			0.0–16.8			0–326
Lake Chilwa (Malawi)		2180.0–5114.0			0.0–229.0			
Lagartijo Reservoir		0.0–4.3			0.0–0.3			0–478
Lakes Robertson/McIlwaine (Zimbabwe)	0.4–8.8	1.3–28.2			0.0–11.8			
Lake Calado (Brazil)	0.0–1.4	0.0–3.2			0.1–0.3			
Lake La Plata (Puerto Rico)	0.0–3.5	0.0–53.0	0.0–46.0		1.5–6.8			
Swamps and Wetlands								
Kawaga Swamp (Uganda)	0.0–8.6	0.0–0.3			0.0–0.9			
North Swamp (Kenya)	0.0–3.2	0.0–4.6	0.0–63.0	0.0–871.0		0.0–1.9		
Tasek Bara (Malaysia)	0.0–54.7	0.7–20.7	4.0–109.0		0.0–3.4	1.1–25.2	0.0–30.6	0–52
Papyrus swamps (Uganda)	0.0–<0.4	0.0–30.3			0.2–1.3			34–558

Note: 0.0 = nondetectable.

2.11 CARBON

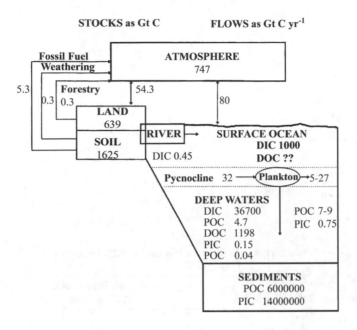

Figure 2.11–1 The global carbon cycle. (From Millero, F. J., *Chemical Oceanography,* 2nd ed., CRC Press, Boca Raton, FL, 1996, 238. With permission.)

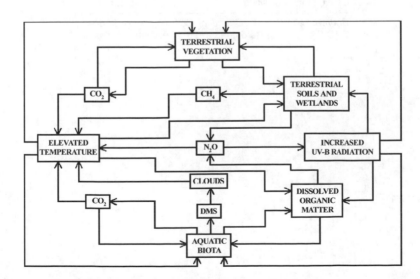

Figure 2.11–2 Conceptual interlinking of the carbon cycle with the global climate system. The top half represents terrestrial and wetland ecosystems, while the bottom half represents aquatic ecosystems. (From Bukata, R. P. et al., *Optical Properties and Remote Sensing of Inland and Coastal Waters,* CRC Press, Boca Raton, FL, 1995, 296. With permission.)

Table 2.11–1   Budget for the Global CO_2 System
              (1980–1989)

System	Average Perturbation (10^{15} g C/yr)
Sources	
Fossil fuel combustion	5.4 ± 0.5
Deforestation	1.6 ± 1.0
Total	7.0 ± 1.2
Sinks	
Atmosphere	3.4 ± 0.2
Oceans (models)	2.0 ± 0.8
Total	5.4 ± 0.8
Unaccounted-for sinks	1.6 ± 1.4

Source: Millero, F. J., *Chemical Oceanography,* 2nd ed., CRC Press, Boca Raton, FL, 1996, 239. With permission.

Table 2.11-2 Atmospheric Concentration of Carbon Dioxide, 1958–1996

Year	Jan.	Feb.	Mar.	April	May	June	July	Aug.	Sept.	Oct.	Nov.	Dec.	Annual
1958	N/A	N/A	315.56	317.29	317.34	N/A	315.69	314.78	313.05	N/A	313.18	314.50	N/A
1959	315.42	316.31	316.50	317.56	318.13	318.00	316.39	314.65	313.68	313.18	314.66	315.43	315.83
1960	316.27	316.81	317.42	318.87	319.87	319.43	318.01	315.74	314.00	313.68	314.84	316.03	316.75
1961	316.73	317.54	318.38	319.31	320.42	319.61	318.42	316.63	314.84	315.16	315.94	316.85	317.49
1962	317.78	318.40	319.53	320.42	320.85	320.45	319.45	317.25	316.11	315.27	316.53	317.53	318.30
1963	318.58	318.93	319.70	321.22	322.08	321.31	319.58	317.61	316.05	315.83	316.91	318.20	318.83
1964	319.41	N/A	N/A	N/A	322.05	321.73	320.27	318.53	316.54	316.72	317.53	318.55	N/A
1965	319.27	320.28	320.73	321.97	322.00	321.71	321.05	318.71	317.66	317.14	318.70	319.25	319.87
1966	320.46	321.43	322.23	323.54	323.91	323.59	322.24	320.20	318.48	317.94	319.63	320.87	321.21
1967	322.17	322.34	322.88	324.25	324.83	323.93	322.38	320.76	319.10	319.24	320.56	321.80	322.02
1968	322.40	322.99	323.73	324.86	325.40	325.20	323.98	321.95	320.18	320.09	321.16	322.74	322.89
1969	323.83	324.27	325.47	326.50	327.21	326.54	325.72	323.50	322.22	321.62	322.69	323.95	324.46
1970	324.89	325.82	326.77	327.97	327.90	327.50	326.18	324.53	322.93	322.90	323.85	324.96	325.52
1971	326.01	326.51	327.02	327.62	328.76	328.40	327.20	325.27	323.20	323.40	324.63	325.85	326.16
1972	326.60	327.47	327.58	329.56	329.90	328.92	327.88	326.16	324.68	325.04	326.34	327.39	327.29
1973	328.37	329.40	330.14	331.33	332.31	331.90	330.70	329.15	327.35	327.02	327.99	328.48	329.51
1974	329.18	330.55	331.32	332.48	332.92	332.08	331.01	329.23	327.27	327.21	328.29	329.41	330.08
1975	330.23	331.25	331.87	333.14	333.80	333.43	331.73	329.90	328.40	328.17	329.32	330.59	330.99
1976	331.58	332.39	333.33	334.41	334.71	334.17	332.89	330.77	329.14	328.78	330.14	331.52	331.98
1977	332.75	333.24	334.53	335.90	336.57	336.10	334.76	332.59	331.42	330.98	332.24	333.68	333.73
1978	334.80	335.22	336.47	337.59	337.84	337.72	336.37	334.51	332.60	332.38	333.75	334.78	335.34
1979	336.05	336.59	337.79	338.71	339.30	339.12	337.56	335.92	333.75	333.70	335.12	336.56	336.68
1980	337.84	338.19	339.91	340.60	341.29	341.00	339.39	337.43	335.72	335.84	336.93	338.04	338.52
1981	339.06	340.30	341.21	342.33	342.74	342.08	340.32	338.26	336.52	336.68	338.19	339.44	339.76
1982	340.57	341.44	342.53	343.39	343.96	343.18	341.88	339.65	337.81	337.69	339.09	340.32	340.96
1983	341.20	342.35	342.93	344.77	345.58	345.14	343.81	342.21	339.69	339.82	340.98	342.82	342.61
1984	343.52	344.33	345.11	346.89	347.25	346.62	345.22	343.11	340.90	341.18	342.80	344.04	344.25

(continued)

Table 2.11–2 Atmospheric Concentration of Carbon Dioxide, 1958–1996 (continued)

Year	Jan.	Feb.	Mar.	April	May	June	July	Aug.	Sept.	Oct.	Nov.	Dec.	Annual
1985	344.79	345.82	347.25	348.17	348.74	348.07	346.38	344.52	342.92	342.62	344.06	345.38	345.73
1986	346.11	346.78	347.68	349.37	350.03	349.37	347.76	345.73	344.68	343.99	345.48	346.72	346.97
1987	347.84	348.29	349.24	350.80	351.66	351.08	349.33	347.92	346.27	346.18	347.64	348.78	348.75
1988	350.25	351.54	352.05	353.41	354.04	353.63	352.22	350.27	348.55	348.72	349.91	351.18	351.31
1989	352.60	352.92	353.53	355.26	355.52	354.97	353.75	351.52	349.64	349.83	351.14	352.37	352.75
1990	353.50	354.55	355.23	356.04	357.00	356.07	354.67	352.76	350.82	351.04	352.70	354.07	354.04
1991	354.59	355.63	357.03	358.48	359.22	358.12	356.06	353.92	352.05	352.11	353.64	354.89	355.48
1992	355.88	356.63	357.72	359.07	359.58	359.17	356.94	354.92	352.94	353.23	354.09	355.33	356.29
1993	356.63	357.10	358.32	359.41	360.23	359.55	357.53	355.48	353.67	353.95	355.30	356.78	356.99
1994	358.34	358.89	359.95	361.25	361.67	360.94	359.55	357.49	355.84	356.00	357.59	359.05	358.88
1995	359.98	361.03	361.66	363.48	363.82	363.30	361.93	359.49	358.08	357.77	359.57	360.69	360.90
1996	362.03	363.22	363.99	364.66	365.36	364.91	363.58	361.41	359.37	359.50	360.65	362.21	362.57

Note: The data in this table were taken at Mauna Loa Observatory in Hawaii and represent averages adjusted to the 15th of each month. The concentration of CO_2 is given in parts per million by volume. Data from other measurement sites may be found in the references.

Source: Lide, D. R. and Frederikse, H. P. R., Eds., *CRC Handbook of Chemistry and Physics*, 79th ed., CRC Press, Boca Raton, FL, 14–25. With permission.

REFERENCES

1. Keeling, C. D. and Whorf, T. P., Carbon Dioxide Information Analysis Center WWW site, available at http://cdiac.esd.ornl.gov/trends_html/trends/co2, August 1997.
2. Keeling, C. D. and Whorf, T. P., in *Trends '93: A Compendium of Data on Global Change*, Boden, T. A., Kaiser, D. P., Sepanski, R. J., and Stoss, F. W., Eds., ORNL/CDIAC-65, Oak Ridge National Laboratory, Oak Ridge, TN, 1994, 16.

CO$_2$ Concentration in Parts per Million (Annual Average)

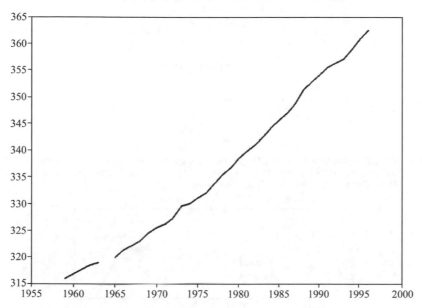

Figure 2.11–3 The concentration of atmospheric carbon dioxide from 1958 to 1996. (From Lide, D. R. and Frederikse, H. P. R., Eds., *CRC Handbook of Chemistry and Physics,* 79th ed., CRC Press, Boca Raton, FL, 14–26. With permission.)

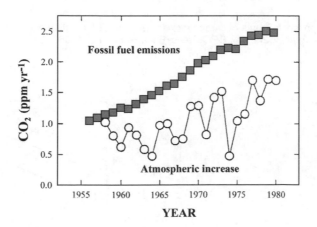

Figure 2.11–4 Annual input of fossil fuel CO$_2$ to the atmosphere compared with the measured CO$_2$ in the atmosphere (1955–1980). (From Millero, F. J., *Chemical Oceanography,* 2nd ed., CRC Press, Boca Raton, FL, 1996, 239. With permission.)

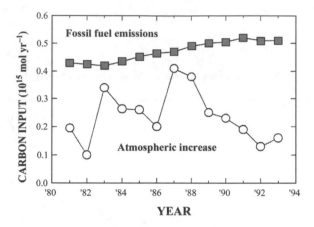

Figure 2.11–5 Annual input of fossil fuel CO_2 to the atmosphere compared with the measured CO_2 in the atmosphere (1980–1994). (From Millero, F. J., *Chemical Oceanography,* 2nd ed., CRC Press, Boca Raton, FL, 1996, 240. With permission.)

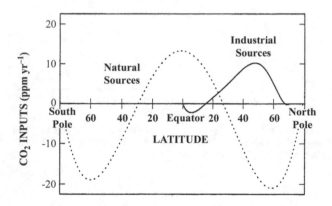

Figure 2.11–6 Anthropogenic inputs of carbon dioxide as a function of latitude. (From Millero, F. J., *Chemical Oceanography,* 2nd ed., CRC Press, Boca Raton, FL, 1996, 238. With permission.)

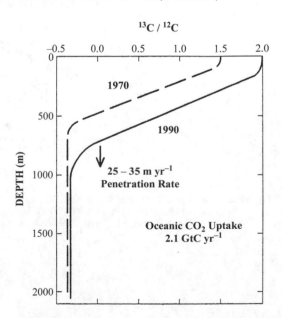

Figure 2.11–7 Changes in the ratio of $^{13}C/^{12}C$ as a function of depth in the ocean. (From Millero, F. J., *Chemical Oceanography,* 2nd ed., CRC Press, Boca Raton, FL, 1996, 240. With permission.)

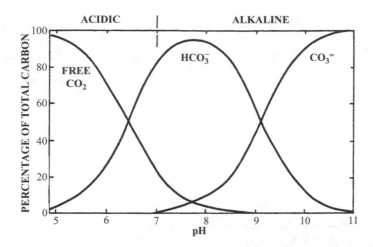

Figure 2.11–8 Percentage of dissolved carbon dioxide (CO_2), bicarbonate (HCO_3^-), and carbonate ($CO_3^=$) as a function of pH. (From Beer, T., *Environmental Oceanography,* 2nd ed., CRC Press, Boca Raton, FL, 1997, 127. With permission.)

Table 2.11–3 Dissociation Constants for Carbonate Calculations in Seawater ($S = 35$)

Temp., °C	pK_0	pK_1	pK_2	pK_B	pK_W	pK_{cal}	pK_{arg}
0	1.202	6.101	9.376	8.906	14.30	6.37	6.16
5	1.283	6.046	9.277	8.837	14.06	6.36	6.16
10	1.358	5.993	9.182	8.771	13.83	6.36	6.17
15	1.426	5.943	9.090	8.708	13.62	6.36	6.17
20	1.489	5.894	9.001	8.647	13.41	6.36	6.18
25	1.547	5.847	8.915	8.588	13.21	6.37	6.19
30	1.599	5.802	8.833	8.530	13.02	6.37	6.20
35	1.647	5.758	8.752	8.473	12.84	6.38	6.21
40	1.689	5.716	8.675	8.416	12.67	6.38	6.23

Source: Millero, F. J., *Chemical Oceanography,* 2nd ed., CRC Press, Boca Raton, FL, 1996, 250. With permission.

Table 2.11–4 Effect of Pressure on the Solubility of Calcium Carbonate in Seawater[a]

Mineral	t°C	$(K'_{sp})_{500}/(K'_{sp})_1$	$(K'_{sp})_{1000}/(K'_{sp})_1$
Aragonite	2	2.11 ± 0.06	4.23 ± 0.27
	22	1.80 ± 0.01	3.16 ± 0.02
Calcite	2	2.18	4.79
	22	1.88	3.56

Note: $K'_{sp} = [Ca^{2+}][CO_3^{2-}]$

[a] The subscripts 1, 500, and 1000 refer to the pressure (atm).

Source: Hawley, J. and Pytkowicz, R. M., *Geochim. Cosmochim. Acta,* 33, 1557, 1969. With permission from Pergamon Press, Ltd., Headington Hill Hall, Oxford.

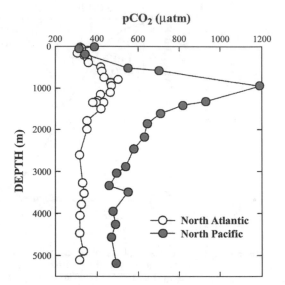

Figure 2.11–9 Depth profile for the partial pressure of carbon dioxide in the Atlantic and Pacific Oceans. (From Millero, F. J., *Chemical Oceanography,* 2nd ed., CRC Press, Boca Raton, FL, 1996, 264. With permission.)

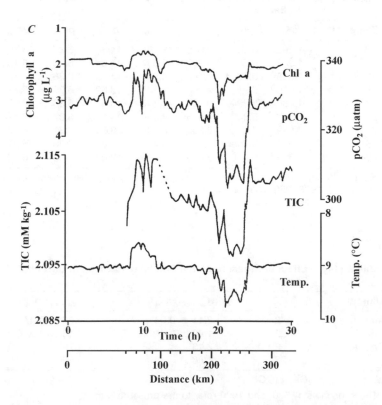

Figure 2.11–10 The variations of pCO_2, TCO_2, chlorophyll, and temperature in the North Atlantic. (From Millero, F. J., *Chemical Oceanography,* 2nd ed., CRC Press, Boca Raton, FL, 1996, 264. With permission.)

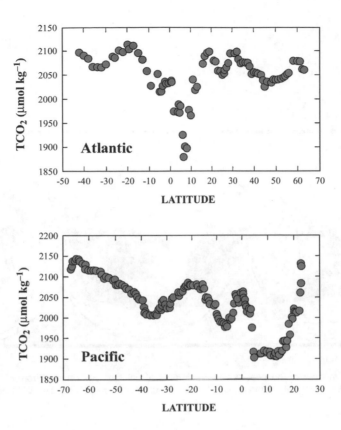

Figure 2.11–11 Surface values of the total carbon dioxide (TCO_2) in the Atlantic and Pacific Oceans. (From Millero, F. J., *Chemical Oceanography,* 2nd ed., CRC Press, Boca Raton, FL, 1996, 270. With permission.)

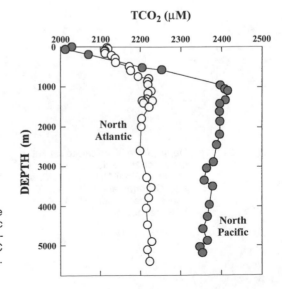

Figure 2.11–12 Depth profile of total carbon dioxide (TCO_2) in the Atlantic and Pacific Oceans. (From Millero, F. J., *Chemical Oceanography,* 2nd ed., CRC Press, Boca Raton, FL, 1996, 271. With permission.)

Figure 2.11–13   The area of the oceans with sediments of calcium carbonate. (From Millero, F. J., *Chemical Oceanography,* 2nd ed., CRC Press, Boca Raton, FL, 1996, 274. With permission.)

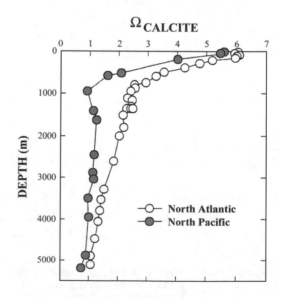

Figure 2.11–14   Depth profile of the calcite saturation state for the Atlantic and Pacific Oceans. (From Millero, F. J., *Chemical Oceanography,* 2nd ed., CRC Press, Boca Raton, FL, 1996, 274. With permission.)

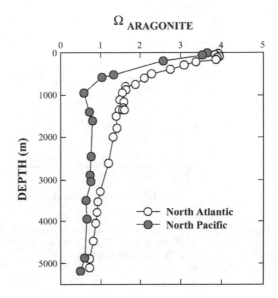

Figure 2.11–15 Depth profile of the aragonite saturation state for the Atlantic and Pacific Oceans. (From Millero, F. J., *Chemical Oceanography,* 2nd ed., CRC Press, Boca Raton, FL, 1996, 275. With permission.)

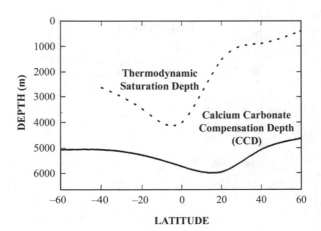

Figure 2.11–16 Comparisons of the thermodynamic saturation state and calcium carbonate compensation (CCD) depths in the Atlantic Ocean. (From Millero, F. J., *Chemical Oceanography,* 2nd ed., CRC Press, Boca Raton, FL, 1996, 275. With permission.)

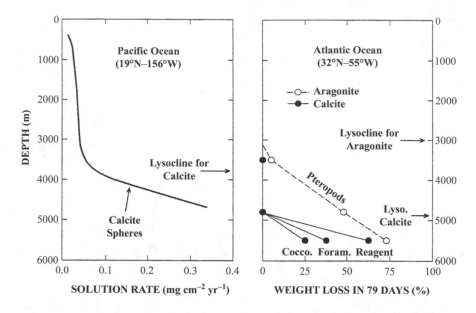

Figure 2.11–17 Depth profiles for the rates of solution of calcium carbonate in the Atlantic and Pacific Oceans. (From Millero, F. J., *Chemical Oceanography,* 2nd ed., CRC Press, Boca Raton, FL, 1996, 276. With permission.)

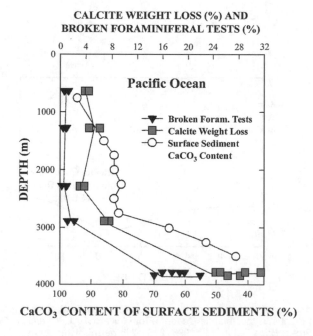

Figure 2.11–18 Depth profiles of the calcite loss and surface sediment concentration of $CaCO_3$ in the Pacific Ocean. (From Millero, F. J., *Chemical Oceanography,* 2nd ed., CRC Press, Boca Raton, FL, 1996, 276. With permission.)

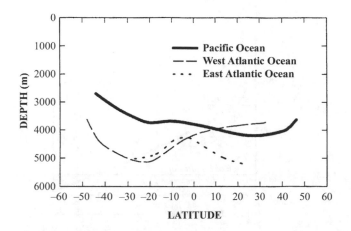

Figure 2.11–19 Depths of the lysocline in various oceans. (From Millero, F. J., *Chemical Oceanography*, 2nd ed., CRC Press, Boca Raton, FL, 1996, 277. With permission.)

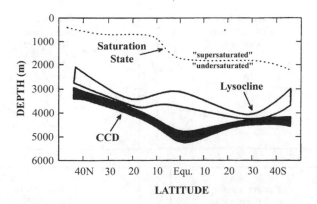

Figure 2.11–20 Comparisons of the saturation level, the carbonate compensation depth (CCD), and the lysocline at various latitudes. (From Millero, F. J., *Chemical Oceanography*, 2nd ed., CRC Press, Boca Raton, FL, 1996, 277. With permission.)

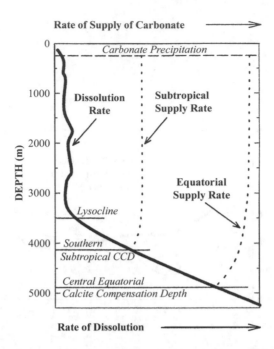

Figure 2.11–21   Depth profile of the dissolution rate and the rate of supply of carbonate to the oceans. (From Millero, F. J., *Chemical Oceanography,* 2nd ed., CRC Press, Boca Raton, FL, 1996, 278. With permission.)

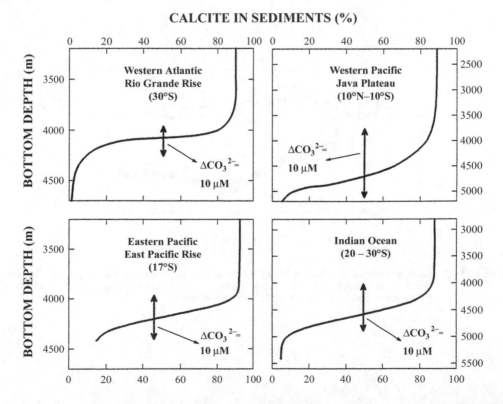

Figure 2.11–22   Calcium carbonate concentrations in the sediments of various oceans. (From Millero, F. J., *Chemical Oceanography,* 2nd ed., CRC Press, Boca Raton, FL, 1996, 278. With permission.)

CHAPTER 3

Physical Oceanography

I. SUBJECT AREAS

This branch of marine science entails the study of four major subject areas: (1) the physical properties of seawater, especially those parameters that influence its density (i.e., temperature and salinity); (2) special properties of scawater (e.g., light, sound, and pressure); (3) energy transfer between the ocean and atmosphere; and (4) water movements in the ocean (e.g., waves, tides, and currents) and the forcing mechanisms responsible for these movements.[1] The physical properties of seawater and the circulation of the oceans are closely coupled to atmospheric conditions. For example, processes operating at the air-sea interface, such as solar heating, evaporation, and precipitation, regulate temperature and salinity characteristics of the surface waters in the ocean. Atmospheric pressure gradients and resultant wind stresses generate waves and wind-driven currents that affect the upper few hundred meters of the water column. Global wind patterns are responsible for the major current systems of the world's oceans. The well-known effects of El Niño demonstrate the potentially catastrophic outcome of oceanic and atmospheric interactions on world climate.

Heat exchange between the atmosphere and oceans initiates the heat transfer process from the equator to the poles and, consequently, the global heat budget. The strong coupling between the atmosphere and oceans is evident from heat budgets that have been formulated for specific regions.[2,3] Wind-driven currents play a critical role in the heat transport process, as well as in the mixing of surface waters. Additional research must be conducted, however, on heat fluxes through the sea surface, as well as on heat exchanges by advection or eddy diffusion, to understand more thoroughly the distribution of temperature in the oceans.

II. PROPERTIES OF SEAWATER

A major part of physical oceanography is the characterization of seawater properties. The distribution of temperature, salinity, and density throughout the ocean basins provides valuable information on water masses, currents, mixing processes, and other physical components of the marine realm. Fundamental knowledge on the physics of the oceans derives from detailed measurements of seawater properties worldwide.

A. Temperature

The surface temperature of the ocean decreases poleward because of declining insolation with increasing latitude. As a result, surface temperatures in the upper 5 m of the water column range from nearly $-2°C$ in the polar oceans to $>27°C$ in tropical waters. Broader bands of high temperature waters (>25°C) occur on the western margins of the tropical Atlantic and Pacific Oceans because of large circulating current systems (gyres). Seasonal temperature variations peak in mid-latitude regions (~40°N and 40°S).

Temperature-vs.-depth profiles reveal three distinct zones in the ocean below the uppermost surface waters (0 to 5 m): (1) the upper zone; (2) the thermocline; and (3) the deep zone. The upper zone from ~5 to 200 m depth is a well-mixed layer typified by nearly isothermal conditions attributable to surface winds. Seasonal temperature changes are essentially restricted to these waters.[1] However, at depths between 200 and 1000 m, water temperature decreases rapidly, marking the position of the thermocline. In low and mid-latitudes, the thermocline is a permanent hydrographic feature, while at high latitudes, the thermocline may only develop seasonally. In addition, a seasonal thermocline often occurs at a depth of ~50 to 100 m in mid-latitude oceanic waters. Low stable temperatures (averaging <4°C) are found below the permanent thermocline.[4] Temperature profiles in low-latitude waters decline from ~20°C in the surface mixed layer to ~2 to 5°C in the deep zone. In the mid-latitudes, temperatures drop from ~10 to 15°C in the surface mixed-layer to <5°C in the deeper waters. In high-latitude (polar) regions, temperatures are consistently low (<4°C) throughout the water column.[1] A cold, dicothermal layer (less than −1.5°C) may form here at depths of 50 to 100 m between warmer surface and deep waters.

Throughout the world's oceans, waters in the deep zone are uniformly dense and cold. Temperatures slowly decrease below the thermocline to a depth of ~3000 to 4000 m. At depths below 3000 m in deep-sea trenches, however, temperatures slowly increase again as a consequence of rising pressure. Because of the vast extent of the deep zone, ~75% of all oceanic water has a temperature between 0 and 4°C.

B. Salinity

In the open ocean, surface salinity ranges from 33 to 37‰. Observed values are a function of latitude and the opposing effects of evaporation and precipitation. Minimum salinity values occur near the equator (<35%) and at 60°N and 60°S latitudes (<34‰) as a result of elevated precipitation, as well as at high-latitude (polar) regions (<33‰) as a result of the melting of ice. Salinity is also locally depressed in some coastal areas that receive large volumes of river runoff. Salinity maxima are found in subtropical waters of the open ocean at ~30°N and 30°S latitudes (>35.5‰) and in some restricted basins, such as the eastern Mediterranean (39‰) and the Red Sea (41‰), where high rates of evaporation predominate.

Salinity-depth profiles for the Atlantic and Pacific Oceans reveal peak salinities at the surface declining to minimum levels at 600 to 1000 m because of the northward movement of sub-Antarctic intermediate water. In the Atlantic Ocean, a sharp salinity gradient (i.e., halocline) is evident between 40°N and 40°S latitudes, separating higher salinity surface waters (35‰) from lower salinity waters below.[4] At depths of 1000 to 2000 m depth, salinity increases slightly. Salinity is relatively uniform in deep ocean waters worldwide, ranging from 34.6 to 34.9‰ at depths below 4000 m.[6]

C. Density

Temperature, salinity, and pressure control the density of seawater. Hydrostatic pressure is a function of depth, increasing about 1 atmosphere (1 bar, 10^5 pascal) every 10 m. Density increases with increasing salinity and pressure and decreasing temperature. In the open ocean, density values range from ~1021 kg/m^3 at the surface to ~1070 kg/m^3 at a depth of 10,000 m. The density distribution of the ocean can be related to large-scale circulation patterns through the application of a geostrophic relationship.[6] Salinity, temperature, and density measurements are used to identify specific water masses in the oceans. Other water properties that may be useful in identifying water masses are dissolved oxygen content, dissolved gases other than oxygen, nutrients (e.g., nitrate, phosphate, and silicate ions), and optical characteristics. However, these properties are nonconservative and, therefore, must be applied with caution. The temperature–salinity diagram (T–S diagram) has proved to be an effective method of characterizing oceanic water masses. By plotting temperature vs. salinity from near the sea surface to the seafloor and inspecting the resultant distribution curves, water masses can be clearly distinguished. Bivariate histograms constructed for specific regions are also of value in characterizing ocean waters.

III. OPEN OCEAN CIRCULATION

Ocean circulation is divided into two components: (1) wind-driven (surface) currents and (2) thermohaline (deep) circulation. Winds drive most surface currents and only affect the upper few hundred meters of the water column, representing ~10% of the volume of the oceans. Wind action, therefore, primarily generates horizontal circulation, although there may be consequent vertical water motion (e.g., equatorial upwelling or downwelling). Thermohaline circulation, in contrast, principally gives rise to vertical water movements. However, horizontal flow is also detected in these deeper zones. The density structure of the ocean determines thermohaline circulation and controls vertical mixing in the deep sea.

A. Wind-Driven Circulation

1. Ocean Gyres

The major surface currents in the oceans arise from global wind patterns. For example, the trade winds produce the northward-flowing western boundary currents, and the westerlies, the eastern boundary currents. It is these long-term winds (e.g., prevailing westerlies and trade winds), together with the Coriolis force, that generate a series of gyres or large circulating current systems in all ocean basins, centered at approximately 30°N and 30°S.[7] The gyres circulate clockwise in the North Atlantic and North Pacific Oceans and counterclockwise in the South Atlantic, South Pacific, and Indian Oceans. The rotation of the earth displaces the gyres toward the western boundary of the oceans, creating stronger currents along this perimeter and effectively separating coastal ocean waters from the open ocean. The displacement of ocean gyres toward the west resulting from the rotation of the earth causes a steeper slope of the sea surface than toward the east. The steepness of the slope controls the current strength.[8] Thus, western boundary currents (Gulf Stream, Kiroshio, Brazil, and East Australian currents) are more intense, deeper, and narrower than eastern boundary currents (California, Humboldt, Canary, and Benguela currents), which are characteristically slow, wide, shallow, and diffuse. As a consequence, the western boundary currents transport significant volumes of warm water northward, thereby transferring heat toward the poles. Surface currents flowing along the western coasts of continents toward the equator are deflected away from the landmasses due to the Coriolis force. The overall effect of this force is to deflect water currents to the right in the Northern Hemisphere and to the left in the Southern Hemisphere.

2. Meanders, Eddies, and Rings

Meanders, eddies, and rings are common features of western boundary currents. The Gulf Stream, a strong geostrophic current with transports up to 150 Sv, does not follow a straight path along its course. North of Cape Hatteras, for example, large wavelike meanders develop in the system and may form self-contained loops or whirl-like eddies called rings. More than a dozen times a year, meander cutoffs of the Gulf Stream create these sizable swirling eddies: north of the Gulf Stream the rings have warm-water centers (warm-core rings) and south of the Gulf Stream, cold-water centers (cold-core rings). The anticyclonic, warm-core rings rotate clockwise, and the cyclonic, cold-core rings circulate counterclockwise at velocities up to 150 cm/s.[9] The swirl velocities peak near the sea surface and gradually decline with depth. Measuring ~100 to 300 km in diameter, the rings move with the surrounding waters at speeds of 5 to 10 km/day. Some may persist for months or years, but eventually coalesce with the Gulf Stream. Prior to being reabsorbed in the Gulf Stream, the rings can be tracked for many kilometers across the ocean via satellite infrared imaging, shipboard measurements, surface drifters, and subsurface floats. Rings similar to those of the Gulf Stream also occur along the Brazil Current in the South Atlantic Ocean and the Kiroshio Current in the North Pacific Ocean.

Although ocean eddies are associated with a wide range of length and timescales, the most energetic types have length, time, and velocity scales of 20 to 200 km, 7 to 70 days, and 0.04 to 0.4 m/s, respectively. Mesoscale eddies, time-dependent current and temperature patterns with horizontal dimensions of 100 to 500 km and periods of one to several months, are ubiquitous phenomena in the world's oceans. However, eddy fields are most dense in the North Atlantic Ocean and North Pacific Ocean. Mesoscale ocean currents serve as a primary source of these eddies, being formed by instability mechanisms. Having rotary currents with velocities of ~0.1 m/s, mesoscale eddies travel a few kilometers per day.[10] They are strongest near the most intense mesoscale currents.[11-15] The coupling between intense mesoscale eddies and energetic ocean currents at mid-latitudes has been of major interest to physical oceanographers for several decades. Mesoscale eddies and the role they play in the dynamics of the oceans remain a focus of many physical oceanographic studies.[1,4,16,17]

3. Equatorial Currents

In the tropics, major surface currents consist of the North Equatorial Current driven by the northeast trade winds and the South Equatorial Current driven by the southeast trade winds. These two westward-flowing currents, which are part of the main anticyclonic current gyres of the Northern and Southern Hemispheres, join to become the Antilles Current off the West Indies in the Atlantic Ocean. Flowing mainly in the well-mixed, warm surface layer, the North and South Equatorial Currents attain mean velocities of 0.3 to 1.0 m/s. Current speeds decrease substantially below the thermocline. The Equatorial Counter Current with speeds up to ~0.6 m/s flows eastward between the North and South Equatorial Currents. In the equatorial Pacific, an undercurrent called the Cromwell Current, is found ~300 km from the equatorial countercurrent at depths of ~200 m in the western Pacific and ~50 m in the eastern Pacific. The Cromwell Current flows with velocities up to 1.7 m/s, and volume transports of ~40 Sv. It is ~300 km wide and 200 m deep.

A unique feature of the tropical Pacific Ocean is the El Niño Southern Oscillation (ENSO), an episodic (2 to 7 years) warm current that develops off the coasts of Ecuador and Peru. An ENSO event takes place on average every 4 years and actually involves large-scale warming of the entire tropical Pacific. It alternates over a period of about 3 years with the La Niña, an opposite cold phase.[18] ENSO originates from the strong coupling, and sometimes chaotic interplay, of the ocean and atmosphere, and its global climate effects are often devastating, as evidenced by natural disasters, economic crashes, and ecological breakdowns (droughts, floods, and decimation of Peruvian anchoveta fisheries).[19] According to Neelin and Latif (p. 32),[20] "ENSO arises as a coupled cycle in which anomalies in SST [sea surface temperature] in the Pacific cause the trade winds to strengthen or slacken and, in turn, drive changes in ocean circulation that produce anomalous SSTs. Ocean-atmosphere feedback can amplify perturbations in either the equatorial SST or the Walker Circulation—the thermodynamic circulation of air parallel to the equator."

A strong equatorial gradient typifies the long-term average SST in the tropical Pacific, with temperatures being cooler in the east and warmer in the west. During an intense El Niño event (e.g., in 1982–1983 and 1997–1998) the southeast trade winds collapse, being replaced by westerly winds driving warm surface waters to flow eastward along the equator and then south along the west coast of South America.[8] In response to the eastward expansion of the area of warm surface water during El Niño, the convective zone of the western tropical Pacific also drifts eastward causing the eastern tropical Pacific to experience a decrease in surface pressure, an increase in rainfall, and a relaxation of the trade winds. Surface pressure increases and rainfall decreases to the west. A thermocline becomes depressed in the east and elevated in the west. Other effects that are apparent include an intensified eastward North Equatorial Countercurrent, a weakened westward South Equatorial Current that is replaced by an eastward equatorial jet in the west, and a very weak equatorial upwelling.[19] Gradually, normal atmospheric pressure gradients and trade winds are reestablished.

In the case of La Niña, intense trade winds persist for months. Warm surface waters contract toward the western tropical Pacific, where low surface pressure and heavy rains are confined. SSTs in the central and eastern Pacific decrease to unusually low levels, such as during the unusually cold conditions during December 1998 and January 1999 when SSTs along the equatorial Pacific declined by at least 1°C below average. Eastward tropical surface currents slacken and western currents strengthen, as does equatorial upwelling. La Niña likewise influences global climate; for example, a significant drought over North America during 1988 was ascribed to a strong La Niña event in the tropical Pacific.[19] Changes in SSTs during both La Niña and El Niño appear to affect the location of atmospheric convective zones that can have dramatic ramifications on world climate. Future research must address the ocean–atmosphere interaction responsible for the development of complex La Niña and El Niño events. This information is necessary for formulating predictive models critical to forecasting potentially devastating climatic conditions.

4. Antarctic Circumpolar Current

Prevailing westerly winds near the Antarctic continent drive the Antarctic Circumpolar Current, a continuously eastward-flowing system. Density differences also play a role in the genesis of this current, which is sometimes called the West Wind Drift. The volume transport of the Antarctic Circumpolar Current is the largest in all the oceans, amounting to $\sim 1.1 \times 10^8$ cm³/s. It is a massive current extending from below the sea surface to the ocean bottom at a depth of ~4000 m. Antarctic Circumpolar water has a mean salinity of ~34.7‰. Its temperature rises from ~0 to 0.5°C at the seafloor to a maximum of ~1.5 to 2.5°C at a depth of ~300 to 600 m.[6] The Antarctic Polar Front Zone, often called the Antarctic Convergence, marks the northern margin of the Antarctic Circumpolar Current. This zone separates Subantarctic and Antarctic surface waters.[1]

5. Convergences and Divergences

Wind stress across the ocean surface is an integral factor in the development of convergences and divergences, which form at the boundary between major surface currents. Along a line or zone of convergence, denser water sinks below lighter water originating from opposing sides. Equatorial, tropical, subtropical, and polar convergence lines provide examples. Aside from these large-scale convergences, small-scale convergences occur elsewhere in the oceans. Divergences, in turn, develop when wind stress leads to horizontal flow of surface waters away from a common center or zone. In this case, subsurface waters rise to replace the diverging horizontal flow. The upwelling of more nutrient-rich, cold subsurface waters is often associated with divergences in the world's oceans (see below).

6. Ekman Transport, Upwelling, and Downwelling

As wind-driven currents flow below the sea surface, Coriolis force causes each successively deeper layer of water to be deflected slightly to the right of the layer above it in the Northern Hemisphere and to the left of the layer above it in the Southern Hemisphere. A depth is eventually reached (i.e., the Ekman depth) where the water flows in a direction opposite to that of the uppermost layer. Current velocity also decreases with increasing depth. When represented by vectors, the resultant change of current direction and velocity with depth follows a spiral pattern (i.e., the Ekman spiral).

Effects of the spiraling current extend to depths of ~100 to 200 m. Net water transport (i.e., bulk water movement) over the entire wind-driven spiral, known as the Ekman transport, is at a 90° angle to the generating wind.

Ekman transport can greatly affect surface currents in coastal areas, generating both upwelling (rising) and downwelling (sinking) of ocean waters. At the eastern boundary of oceans, winds blowing parallel to the coast produce Ekman transport that moves surface waters away from adjacent landmasses. This outflow is compensated by the upwelling of nutrient-rich subsurface waters from depths of ~100 to 200 m. Vertical circulation associated with coastal upwelling creates some of the most biologically productive waters on earth. In contrast, at the western boundary of oceans, Ekman transport drives seawater toward the coast. This process results in the downwelling of surface waters often depleted by biological demands, yielding conditions of lower biological production.

Upwelling and downwelling also occur in the open ocean as noted previously. For example, the convergence of ocean gyre surface currents at about 30°N and 30°S latitudes due to Coriolis deflection causes downwelling. At lower latitudes, Coriolis deflection of the North and South Equatorial Currents to the right (north) in the Northern Hemisphere and to the left (south) in the Southern Hemisphere induces divergent surface flow along the equator, fostering upwelling of subsurface waters.

7. Langmuir Circulation

In some areas of the ocean, winds blowing across the sea surface at speeds >3.5 m/s are responsible for the formation of Langmuir cells characterized by corkscrewlike water motions, in helical vortices alternately right and left handed. Langmuir circulation involves the development of a series of convective cells with their long axes aligned parallel to the direction of the generating wind. Each cell measures ~10 to 50 m wide, ~100 to 1000 m long, and ~5 to 6 m deep.[4] Because adjacent cells rotate in opposite directions, alternating zones of convergence (followed by downwelling) and divergence (followed by upwelling) are evident. Debris, foam, oil slicks, and other material typically accumulate along the lines of convergence. Langmuir circulation appears to arise from instability in well-mixed surface waters and the short-term response of these waters to wind drag.

B. Surface Water Circulation

1. Atlantic Ocean

In the North Atlantic Ocean, a clockwise gyre begins off the west coast of Africa with the westward flow of the North Equatorial Current to the West Indies, where it combines with the South Equatorial Current to become the Antilles Current. Waters derived from the Gulf of Mexico and flowing between Florida and Cuba join with the Antilles Current to become the Florida Current. The Gulf Stream is the northern extension of the Florida Current. It mixes with cold, low-salinity water from the Labrador Current moving south from the Labrador Sea and, farther to the northeast beyond the Grand Banks of Newfoundland, flows into the North Atlantic Current. This latter current diverges into a northward component (i.e., the Norwegian Current) that continues on toward the Arctic and a southward component that becomes the Canary Current off the northwest coast of Africa, thereby completing the gyre.

In the South Atlantic, the southern component of the South Equatorial Current (i.e., the Brazil Current) flows south along the South American coastline. The Falkland Current flowing northward from the Drake Passage separates the Brazil Current from the South American coast at ~25°S. At the subtropical convergence, the Brazil Current turns east and becomes part of the Circumpolar Current. Off the southwest coast of Africa, this current turns northward and flows along the west coast of the continent as the Benguela Current. It bends to the west as the equator is approached, with a northward-flowing component becoming the South Equatorial Current.

2. Pacific Ocean

The surface water circulation in the Pacific Ocean, similar to that in the Atlantic Ocean, consists of two gyres: (1) a clockwise gyre in the North Pacific and (2) a counterclockwise gyre in the South Pacific. In the North Pacific, the North Equatorial Current flows westward near the equator, driven by the trade winds at velocities of 25 to 30 cm/s. Closer to the equator, the North Equatorial Counter Current flows eastward at speeds of 35 to 60 cm/s. The Asian continent forms a barrier to the North Equatorial Current, which turns northward and becomes the Kuroshio Current, a strong western boundary current flowing parallel to the coast. Its northward extension is the Oyashio Current. Farther to the east, the Oyashio Current becomes the California Current, which flows southward along the North American coastline, transporting cold water from the North Pacific toward the equator.

In the South Pacific, the westward-flowing South Equatorial Current turns to the south off Australia and becomes the East Austrialian Current. At the subtropical convergence, waters flow eastward as the Circumpolar Current and then northward as the Peru Current along the South American coastline.

3. Indian Ocean

Asian landmasses restrict the flow of surface currents in the Indian Ocean north of the equator to the Bay of Bengal and Arabian Sea. The Northeast Trade Winds drive the North Equatorial Current eastward to the east coast of Africa between November and March each year. The remaining months of the year, southeast monsoon winds replace the Northeast Trade Winds, generating an eastward-flowing Monsoon Current and causing the South Equatorial Current to turn north and supply the Somali Current, which continues north along the east African coast. The Equatorial Counter Current disappears during this monsoon period.

A counterclockwise gyre exists year-round in the Indian Ocean. The South Equatorial Current flows west and then turns south to supply the Agulhas Current along the south coast of Africa. A part of this current turns west around the south coast of Africa and joins the Benguela Current in the South Atlantic. The Circumpolar Current transports waters eastward along the Subtropical Convergence at 40°S. The West Australian Current flows northward along the west coast of Australia to the South Equatorial Current, thus completing the gyre.

4. Southern Ocean

Waters of the Southern Ocean are continuous with those of the Atlantic, Pacific, and Indian Oceans to the north and bounded by the Antarctic continent to the south. The Antarctic Polar Front and Subtropical Convergence divide the surface waters of the Southern Ocean into two zones: the Antarctic zone and Subantarctic zone. The Antarctic zone is the region extending from the Antarctic continent to the Antarctic Polar Front. The Subantarctic zone, in turn, stretches from the Antarctic Polar Front, located at ~50°S in the Atlantic and Indian Oceans and ~60°S in the Pacific Ocean, to the Subtropical Convergence observed at ~40°S in all oceans.

The eastward-flowing Antarctic Circumpolar Current and westward-flowing East Wind Drift dominate surface water circulation in the Southern Ocean. These two gyres originate from different wind patterns. Westerly winds drive the surface flow of the Antarctic Circumpolar Current, and prevailing easterly winds near the coast of Antarctica generate the westward surface flow of the East Wind Drift. The Coriolis force produces a northward component to the surface current flow of the Antarctic Circumpolar Current, leading to the formation of convergences within the Antarctic Polar Frontal Zone.

5. Arctic Sea

The Lomonosov Ridge divides the Arctic Sea into two basins, the eastern Eurasian Basin and the western Canadian Basin. Surface water flow follows a generally clockwise circulation pattern from the Canadian Basin to the Eurasian Basin at velocities of ~1 to 5 cm/s.[6] Waters from the North Atlantic enter the Norwegian Sea via the northward-flowing Norwegian Current, which is a continuation of the North Atlantic Current flowing at speeds up to 30 cm/s.[5] The East Greenland current transports Arctic water and ice, as well as some water from the Norwegian Current, south along the east coast of Greenland, around the southern tip of Greenland, and into the Labrador Sea. The West Greenland Current flowing north along the west coast of Greenland results from this surface flow. The southward-flowing Baffin Land Current continues south as the Labrador Current, transporting cold waters (<0°C) down the west side of the Labrador Sea into the North Atlantic. Some of the Labrador Current water originates from Hudson Bay.

C. Thermohaline Circulation

Density differences between water masses caused by variations in temperature and salinity drive vertical and deep water circulation in the oceans. These subsurface water movements, referred to as thermohaline circulation, affect ~90‰ (by volume) of the world's oceans. Subsurface water masses are subdivided into three broad categories: central, intermediate, as well as deep and bottom water types found at depths of ~0 to 1 km, 1 to 2 km, and >2 km, respectively.

1. Atlantic Ocean

Atlantic deep and bottom waters derive from several different sources. Antarctic Bottom Water (AABW), originating in the Weddell and Ross Seas, has temperature and salinity values of ~0.5°C and 34.7‰, respectively. Bottom topography influences the northward flow of this water mass. For example, the Mid-Atlantic Ridge separates warmer waters (minimum 2.4°C) in the eastern basin from colder waters (minimum 0.4°C) in the western basin. The Walfish Ridge with a sill depth of ~3500 m acts as a barrier to bottom water flow in the eastern basin.[5] AABW extends across the equator far into the North Atlantic to about 40°N.

Antarctic Deep Water (AADW), characterized by temperatures of ~4°C and salinities of ~35‰, forms farther north off Antarctica than AABW. It flows northward at the surface to the Atlantic Convergence, and then sinks beneath less dense subpolar water. At depth, the AADW is found between denser AABW below and less dense North Atlantic Bottom Water (NABW) (2.5 to 3°C, 34.9‰) and North Atlantic Deep Water (NADW) (3 to 4°C, 34.9‰) above. The Norwegian and Labrador Seas represent the primary sources of NABW and NADW. These water masses comprise the largest volume of deep waters extending from the North Atlantic to the Southern Ocean.

Antarctic Intermediate Water (AAIW) and Arctic Intermediate Water (AIW) overlie NADW. The Mediterranean Intermediate Water (MIW), a less extensive water mass at mid-depths (2 to 3 km) of the North Atlantic, is generated in the Mediterranean Sea. The low-salinity AAIW (2.5 to 5°C, 33.9 to 34.5‰) has its origin in the Antarctic Polar Frontal Zone. This water mass sinks down to a depth of ~1 km, and flows southward to ~20°N where it encounters the leading edge of the southward-flowing, higher-salinity AIW (34.9‰), which forms in the subarctic. The dense, warm MIW (~6 to 12°C, 35.5 to 36.5‰) flows westward into the North Atlantic from the Mediterranean Sea.

North Atlantic Central Water (NACW) (~5 to 20°C, 35.0 to 36.5‰) and South Atlantic Central Water (SACW) (5 to 20°C, 34.5 to 36.0‰) occur above AAIW and AIW between 40°N and 40°S latitudes. They reach greatest depths (~500 to 900 m) in mid-latitudes and become shallower near the equator (~300 m) and at high latitudes. These water masses appear to form by sinking on the equator-side of the subtropical convergences.[6]

2. Pacific Ocean

Pacific Subarctic Water (PSW) and Oceanic or Common Water (CoW) constitute the deepest water masses in the Pacific Ocean. PSW is characterized by relatively low temperature and salinity values (~2 to 4°C, 33.5 to 34.5‰). It originates at the Subarctic Convergence in the western Pacific, covers the entire North Pacific seafloor, and extends across the equator to nearly 10°S latitude. The Arctic Sea is not a significant source of bottom and deep water for the North Pacific because the shallow Bering Strait with the Aleutian Ridge serves as a partial barrier to flow from the Bering Sea. In addition, the density of waters in the Bering Sea is not high enough to displace bottom waters.

CoW is the largest oceanic water mass, being generated by the mixing of other water masses (i.e., mainly AABW and NADW), in the South Atlantic with subsequent injection by the Antarctic Circumpolar Current into the Pacific Ocean. It occupies all areas in the South Pacific below ~2 km and is identified by temperature and salinity measurements of 1.5°C and 34.7‰, respectively. Above this water mass, AAIW (2.2°C, 33.8‰) occurs at depths ranging from ~500 to 800 m. This low-salinity water mass develops at the Antarctic Polar Front. Similarly, in the North Pacific, the North Pacific Intermediate Water (NPIW) (4°C, 34.0 to 34.5‰) overlies PSW at approximately the same depths as the AAIW. The mixing processes responsible for NPIW formation are poorly understood.

Central waters are found above the intermediate water masses in both the North and South Pacific. North Pacific Central Water (NPCW) (8 to 20°C, 34.0 to 34.8‰) and South Pacific Central Water (SPCW) (20 to 20°C, 34.5 to 36.0‰) extend from ~50°N to 40°S latitudes. Both are relatively warm, low-density water masses observed above a depth of ~800 m.

3. Indian Ocean

In contrast to the Atlantic and Pacific Oceans, the Indian Ocean does not receive deep and bottom waters from the Arctic region. As noted previously, the Asian continent limits the northern extent of the Indian Ocean to only 25°N. Hence, the Atlantic/Antarctic plays a particularly important role in the formation of water masses found throughout the Indian basin.

CoW is by far the most extensive water mass in the Indian Ocean, occupying all depths below ~1000 m, although a tongue of warm, saline Red Sea Intermediate Water (RSIW) (23°C, 40.0‰) flows into CoW in the northern and western Indian Ocean near the equator at a depth of ~3000 m. As in the Pacific Ocean, CoW in the Indian Ocean consists of a mixture of AABW and NADW. The lower CoW in the Indian Ocean exhibits properties similar to AADW, and the upper CoW, properties similar to NADW.[4] AAIW forms at the Antarctic Polar Front and flows northward over the CoW to ~15°W. It can be traced to depths of ~1200 m.

Central waters in the Indian Ocean are divided into Indian Equatorial Central Water (IECW) and South Indian Central Water (SICW). IECW, with relatively uniform salinity (34.9 to 35.5‰), occurs from 25°N to the equator at depths down to 1000 m. SICW, with a wider range of salinity (34.5 to 36.0‰), is observed at somewhat shallower depths (~500 m) from the equator to ~40°S.

4. Arctic Sea

Vertical profiles of the Arctic Sea reveal three distinct water masses: (1) surface or Arctic Water from 0 to 200 m; (2) Atlantic Water from 200 to 900 m; and (3) bottom water from 900 m to the seafloor. Salinity largely determines density in Arctic Sea waters. In the surface water mass, which extends down to ~50 m depth, salinity ranges from ~28.0 to 33.5‰, and temperature (controlled by melting or freezing) from −1.8 to −1.5°C. Subsurface water between ~25 and 100 m is isothermal and characterized by a strong halocline. Temperature increases substantially below 100 m depth.

Atlantic Water between 200 and 900 m results from the mixing of subsurface Arctic Water and Atlantic Water. Its temperature ranges from ~0 to 3°C, and its salinity from 34.8 to 35.1‰. This water mass flows counterclockwise, which is opposite to the (clockwise) circulation of the Arctic Water mass above it.

The largest volume (~60%) of water in the Arctic Sea consists of bottom water (900 m to the seafloor) originating from the Norwegian Sea. The minimum temperature, recorded at 2500 m in the Eurasian Basin, amounts to −0.4°C. Salinity is quite uniform, ranging from 34.9 to 34.99‰. It tends to rise slightly with increasing depth.

IV. ESTUARINE AND COASTAL OCEAN CIRCULATION

A. Estuaries

An estuary is defined as a semienclosed coastal body of water that is either permanently or periodically open to the sea and within which seawater is measurably diluted with seawater derived from land drainage. Water circulation in estuaries consists of tidal and nontidal components. The tidal component results from extrinsic forces, that is, tide-producing forces of the sun and moon, which are responsible for the periodic vertical displacement of estuarine waters and the associated horizontal movement of tidal currents.[21] The nontidal component originates from factors independent of astronomical forces, specifically water movements induced by the interaction of fresh and saline waters of different densities.[22]

The intensity of tidal currents relative to river flow, in combination with the geometric configuration of the estuary, largely determines circulation patterns. While tides and tidal currents provide the ultimate driving force for much of the turbulence and mixing in most stratified and well-mixed estuaries, other mechanisms (e.g., surface wind stress and other meteorological forcing) can be important modifying factors, especially in shallow coastal bays.[23,24] Marked changes in estuarine circulation can be induced by nontidal forcing from the nearshore ocean (e.g., coastal Ekman convergence or divergence), storm surges, and variability in the neap–spring tidal cycle.

Seawater (density = 1.025 g/cm) is more dense than fresh water (density = 1.0 g/cm³); consequently, these two fluids tend to form separate water masses in estuaries, with fresh water overlying seawater. Although water temperature influences the density value of both fresh water and seawater, it has a much smaller effect than the concentration of dissolved salts on density levels. In a density-stratified system lacking currents, the mixing between freshwater and seawater masses is ascribed to diffusive and advective processes. Diffusion in estuarine water is defined as a flux of salt, and advection as a flux of salt and a flux of water.[25] Vertical advection, the upward-breaking of internal waves at the interface between the freshwater and seawater layers, is the primary mixing agent in the absence of currents, causing a diffuse boundary layer between the two water masses and a gradual increase in salinity in the freshwater layer downestuary.[26] Although the upward flux of salt by diffusion does not produce the mass flux of water and salt as advection, it plays an important role in the mixing of waters along the vertical, lateral, and longitudinal axes of estuaries.

The mixing of estuarine waters is facilitated by current action. The interaction of tidal currents, wind stress, internal friction, and bottom friction can reduce or eliminate density stratification of the water column. The turbulence produced by both internal shear and bottom friction is a critical factor for estuarine mixing. In well-mixed systems, turbulence generated by bottom friction predominates, whereas in highly stratified estuaries, turbulence produced by internal shear is paramount.[27] Both internal shear and bottom friction are important in the mixing of partially stratified systems.[28]

Four types of estuaries are recognized based on circulation:

1. Type A, salt wedge estuaries, which are highly stratified and characterized by river inflow that completely dominates the circulation system;
2. Type B, partially mixed estuaries, which are moderately stratified and characterized by less dominating river inflow;
3. Type C, vertically homogeneous estuaries, which have a lateral salinity gradient and are characterized by greater tidal action relative to river inflow;
4. Type D, sectionally homogeneous estuaries, which are both laterally and vertically homogeneous and characterized by tidal flow that greatly exceeds river discharges.[29,30]

Studies of spatial variability in circulation have concentrated on flow in the longitudinal sense, along the estuary, and its changes in the vertical plane, with increasing depth. Little or no consideration is often given to the variation of flow across the estuary in a lateral direction. In many cases, lateral differences in velocity or density fields produce transverse effects that can be large compared with vertical variations.[31] Secondary flows may arise where circulation is not evenly distributed across the estuary because of the lack of consistency in the cross-sectional form of the basin in a longitudinal direction, the effect of the Coriolis force, and differences in lateral density fields.[32]

In many stratified and well-mixed estuaries, longitudinal density gradient circulation and mixing are common. The longitudinal surface slope, which acts in a downestuary direction, and the longitudinal, density gradient force, which acts in an upestuary direction, drive this type of circulation.[33] According to Officer,[33] when river discharge is low, the surface slope force dominates in the upper part of the water column, with a net flow downestuary. The density gradient force, in turn, dominates in the lower part of the water column, with a net flow upestuary. Lateral effects, surface wind stress, meteorological forcing, and the basin geometry may obscure this circulation pattern, at least for certain periods of time, particularly in larger estuaries. Wind stress interacting with tidal and river flows often governs the degree of turbulent mixing and controls the vertical salinity and velocity structure of estuaries.[34-38]

In summary, the magnitude of river discharge relative to tidal flow is the principal factor determining the type of water circulation observed in estuaries. The river discharge acting against tidal motion and interacting with wind stress, internal friction, and bottom friction controls the degree of turbulent mixing and consequently the vertical salinity and velocity structure.[39] The characteristics of this system vary as the volume of river discharge, the range of the tide, and the geomorphology of the estuarine basin change.[40] Water flow in the estuary is strongly three dimensional because of the generally irregular structure of the basin, spatial density currents, and the effect of the Coriolis force. These factors all contribute to the complexity of estuarine circulation patterns.

B. Coastal Ocean

1. Currents

There are a number of factors that affect coastal water circulation. Particularly noteworthy are the strength of tidal currents, river runoff, meteorological conditions, shoreline configuration, as well as water depth and topography of the continental shelf. Meteorological forcing plays a prominent role in coastal ocean circulation.

The proximity of landmasses greatly influences coastal physical oceanography. Aside from being a barrier to flow, the landmasses are a source of freshwater input via river discharges and groundwater influx. Hence, the salinity of nearshore ocean waters often varies considerably, with surface waters showing reduced levels compared with the open ocean, particularly near the mouths of large rivers. River-generated haloclines result in greater stability of the water column, an increase in the sharpness of the thermocline, and an attenuation of vertical mixing. These conditions promote the development of higher temperatures in the surface layer during summer. Another effect of the landmasses is to limit current directions such that horizontal flows trend parallel to the coast.[1] This pattern of flow can hinder the transfer of fresh water across the system, leading to greater residence times. Winds, tides, and river discharges largely drive coastal currents. As winds and runoff increase, coastal currents intensify.[8] Tidal currents may facilitate vertical mixing and a breakdown of water column stratification.[6]

As discussed above, upwelling can dominate coastal circulation along the eastern boundaries of ocean basins. These regions have received considerable attention not only because of the physical significance of the upwelling process, but also because of their importance in generating high biological production. Strong winds parallel to the coastline produce the most intense upwelling effects; here, deep cold, nutrient-rich waters transported to the surface support some of the most productive fisheries on earth.

Far offshore, basin boundary currents of the open ocean form a seaward border to coastal circulation. The direction of these major boundary currents is frequently opposite to the current flow of coastal waters. Geostrophic currents on the continental shelf may be highly variable.[8]

2. Fronts

Circulation and mixing along the continental shelf and in estuaries change abruptly at frontal sites.[41,42] Oceanic and estuarine fronts are defined as boundaries between horizontally juxtaposed water masses of dissimilar properties.[43] Four types of fronts are germane to coastal and estuarine waters: (1) shelf break fronts; (2) upwelling fronts; (3) shallow sea fronts; and (4) river plume fronts.

Salt wedge estuaries are typified by well-developed fronts marking areas of convergent flows with strong vertical motions and sharp gradients (approaching discontinuity) in physical parameters, especially velocity and density fields.[44] Along an estuarine front, isohalines rise steeply to the surface, and sharp horizontal gradients in salinity are apparent.[45] Surface water sinks at the frontal interface and, because of convergence, a line of foam or floating organic and detrital material generally collects at the surface. Changes in color and turbidity of the water often delineate the site of the front as it advances downestuary over a layer of water.

Consistently high river discharges in narrow estuaries promote the development of fronts.[46] However, episodic, heavy fluvial inflow also may enhance their formation.[47] In addition, tidal dynamics must be considered in estuarine frontal development, together with basin bathymetry.[48] The buoyancy that issues from estuaries and river mouths contributes to density stratification of coastal waters.[49]

3. Waves

Surface waves exert considerable influence on physical processes in nearshore oceanic and estuarine environments. Trapped waves (e.g., Kelvin waves, Rossby waves, edge waves, and seiches) likewise are important elements of shallow water coastal environments.[1] Invariably, meteorological forcing plays an integral role in wave generation in these waters.

a. Kelvin and Rossby Waves

Although catastrophic events such as earthquakes and volcanic eruptions occasionally generate major undulations of the sea surface (e.g., tsunamis), wind stress produces most surface waves in the ocean. Edge waves, seiches (i.e., standing waves), internal waves, as well as tides, also are encountered in estuarine and oceanic waters. In regard to ocean circulation, Kelvin waves and planetary or Rossby waves are of paramount importance.

Kelvin waves, which can travel as surface (barotropic) waves or as baroclinic waves, represent perturbations of the sea surface or of the thermocline that propagate parallel to and close to the coast. As water piles up against a coastal boundary, an offshore horizontal pressure gradient force develops. The coast may act as a waveguide to constrain the way in which the water can move in response to the forces acting on it. The Coriolis force directed toward the coast and a horizontal pressure gradient force that is coupled to the slope of the sea surface and opposed to the Coriolis force are necessary conditions for the propagation of Kelvin waves.[10]

Rhines[50] states, "A Rossby wave is not a surface wave of the familiar sort, but a great undulation of the whole ocean mass that carries signals from one shore to another over weeks, months, and years." The need for the potential vorticity to be conserved accounts for their existence in the ocean.[10] They propagate zonally, along the equator, as well as along other lines of latitude.

b. Edge Waves

These waves travel parallel to the coastline, being produced by storms moving up the coast or even by breaking waves piling water on beaches. The amplitude of edge waves generally ranges from a few centimeters to as much as a meter. Peak amplitudes occur near the shoreline and rapidly decline seaward. Their velocity is given by

$$C^2 = g/K(\sin t) \tag{3.1}$$

where C is the celerity, g is gravity, K is the wave number, and t is the slope of the bottom, which typically ranges from 10^{-2} to 10^{-4}. As specified by Knauss,[1] "Their characteristic phase speed is that of a deep water wave modified by the slope of the sea bottom."

c. Seiches

Unlike surface waves with crests and troughs that travel from one point on the sea surface to another, seiches are stationary waves that do not move horizontally. They oscillate pendulum fashion around a fixed point (node) on the water surface. Maximum vertical movement takes place at points on the water surface termed *antinodes*. Hence, seiches are essentially "trapped" as water moves beneath them.[4]

Seiches develop in response to atmospheric, seismic, and tidal forces, and they have periods that are variable and strongly related to the basin geometry. The period of seiches, for example, ranges from a few minutes up to a day. In some fjord-type estuaries (e.g., Oslo Fjord), tidal currents interact with a sill to form internal seiches.[51] These seiches usually have ranges and periods substantially greater than those on the surface.

Seiches sometimes act synergistically with progressive waves in estuaries to produce large water movements. In spite of the small magnitude of the tide-generating forces, resonance can generate large tidal movements.[21] Resonance is likely to take place when the natural period of oscillation of an estuary, which depends on its dimensions, is similar to the period of one of the tide-generating forces.

d. Internal Waves

Internal waves occur beneath the sea surface between water layers of different density.[52] These gravity waves propagate along a pycnocline associated either with a halocline or thermocline, and are formed by multiple mechanisms, including traveling atmospheric pressure fields, interactions of currents with bottom topography, tidal oscillations, or second-order interactions between surface waves of different frequencies.[53,54] The shear of rapidly flowing surface layers also induces internal waves in stratified estuaries. Internal waves travel more slowly than surface waves, but can attain much greater heights. They mix waters below the surface and may be important in the movement of sediment.

e. Tides

Changes in the position of the sun and moon relative to points on the earth's surface cause variations in gravitational forces that produce tides.[1] The periodic rise and fall of the tide and the associated tidal currents that are generated have a dramatic effect on circulation in the marine realm, especially in shallow coastal oceanic and estuarine waters.[4,6,8] Tidal currents typically range from 0.5 to 1 m/s, but are generally stronger and more variable in coastal regions. Ocean tides are essentially waves of very large wavelength equal to about one half of the earth's circumference ($>20,000$ km) which increase in height as coastal waters are approached. Shallow depths, bathymetric irregularities, and changes in coastal geomorphology alter the landward progression and shape of the waves.

In estuaries, the tide may have the characteristics of a progressive, standing, or mixed wave. Larger estuaries and estuaries where the opening to the ocean is restricted and the entering wave attenuated by bottom friction and reflections from boundary irregularities typically exhibit progressive tidal waves.[55] Standing wave characteristics are evident in estuaries without a restricted connection to the ocean, where wave attenuation is small, and the amplitude and period of the entering wave equal that of the reflected wave. In these cases, the tide rises and falls at the same time throughout the embayment and current flow peaks at mid-flood and mid-ebb. A mixed tide characterized by a progressive wave in one area and a standing wave in another is apparent in some elongated estuaries. The standing wave tends to form near the entrance and also near the head of tide, with both progressive and standing wave characteristics observed between these areas.

Based on the number and pattern of high and low tides per tidal day (24 h and 50 min), four general types of tides are recognized in estuaries (i.e., semidiurnal, diurnal, mixed, and double tides). Most coastal regions have two high and two low tides per tidal day. Estuaries along the Atlantic Coast of the United States are typified by semidiurnal tides, and those along the Pacific Coast have mixed tides. In some parts of the Gulf of Mexico, estuaries with diurnal tides predominate.

f. Surface Waves

Surface waves are classified by size, typically ranging from small capillary waves or ripples to large waves having periods of up to 5 s.[8] When wind velocity exceeds 4 km/h across the ocean, waves develop because of friction on the sea surface.[56] Bowden[57] defines wind stress on the sea surface as follows:

$$\tau_S = C_D \, \rho_a \, W^2 \tag{3.2}$$

where W is the wind speed measured at a standard height, ρ_a is the density of the air, and C_D is a drag coefficient dependent on the height at which the wind is measured, the roughness of the water surface, and the stability of the air in the first few meters above the surface.

Waves increase in size with increasing wind velocity, length of time the wind blows in one direction, and the distance the wind blows over the water surface in a constant direction (i.e., fetch). Although the waveform advances laterally with little net movement of water, individual water particles oscillate in approximately circular paths, moving forward on the crest of the wave, then vertically, and finally backward under the trough. The diameter of the circular orbits decreases with increasing water depth, being nearly equal to the wave height at the surface and declining to near zero at a depth of one half the wavelength. Below this depth, there is little water movement with wave passage.

In deep water, the waveform is symmetrical and can be described by its length (the horizontal distance between adjacent crests), height (the vertical distance between a wave crest and adjacent trough), period (the time interval between successive waves), and velocity (wavelength/wave period). As shallow water is approached, water particles near the seafloor do not transcribe circular orbits, but move in a horizontal, back-and-forth motion along the bottom. Sediment is slowly transported in the direction of wave propagation as water particles move backward and forward along the seafloor.[22] When the water depth becomes less than one half the wavelength, the interference of the bottom with the motion of water particles affects wave characteristics, reducing the wave velocity and length and increasing the wave height. The seafloor strongly influences the motion of water particles at depths less than 1/20 of the wavelength, although at the surface, the orbits of water particles may be only slightly altered into the form of an ellipse.[10] Continued shoaling of water makes the wave unstable; the crest increases in steepness, can no longer support itself, and it breaks forward along the shore where rapid water movements release energy, transport sediments, and generate currents. Waves break when the wave steepness (the ratio of wave height to wavelength) exceeds 1:7,[10] and the ratio of water depth to wave height is about 4:3.[22] Several types of breaking waves are recognized: spilling, plunging, collapsing, and surging. The type of breaker at a given locality depends on the types of waves approaching the shoreline and the complexity and slope of the seafloor.

The velocity of shallow water waves can be calculated by the following expression:

$$V \simeq \sqrt{gd} \qquad (3.3)$$

where g is the acceleration due to gravity and d is the depth to the seafloor. In comparison, the velocity of deep water waves may be determined by the equations:

$$V = \frac{gT}{2\pi} \qquad (3.4)$$

$$V = \sqrt{\frac{gL}{2\pi}} \qquad (3.5)$$

where T is the wave period, L is the wavelength, and g is the acceleration due to gravity.

Waves approaching a shoreline may be refracted or diffracted. As wavefronts enter shallow water at an oblique angle, they bend such that the waves strike nearly parallel to the shore. The portion of a wavefront initially advancing into shallow water "feels" bottom first causing the circular orbits of water particles to become elliptical. This process of wave retardation from bottom effects slows down the segment of the wavefront closest to the shore, enabling the wavefront to change the direction of its approach to shore. Waves, although refracted as the shore is approached, usually break at a slight angle to the shore; this action produces longshore currents parallel to shore and an accompanying longshore drift (zigzag movement) of sediment. Water that accumulates on the shore from breaking waves returns rapidly offshore in rip currents that cut narrow, intermittent channels away from the shore. Longshore currents and longshore drift are more pronounced in coastal ocean waters than in estuaries, but have been documented in both environments. Rip currents are less obvious is estuaries, being important only in those systems having a wide mouth.

The refraction of waves on an irregular shoreline concentrates energy on headlands and disperses it in adjacent bays. Because water shoals near a headland, approaching waves refract and converge on the projecting landmass, eroding it more quickly than the neighboring shoreline. As waves enter a bay, they are refracted in shallow water along the perimeter, whereas in deep water, they travel with greater velocity. The divergence of wave energy in bays reduces erosion and increases sedimentation. Through time, therefore, shoreline irregularities become smoother as a result of this nonuniform distribution of wave energy.

If waves impinge perpendicularly on an obstacle, for example, a barrier beach, they may be reflected and their energy transferred to waves traveling in another direction.[10] Waves that pass the end of an obstacle, such as the terminus of a spit, may be diffracted into sheltered regions of an estuary. Subsequent to passage through the mouth of an estuary, waves generally decrease in velocity as a result of bottom effects and the opposing river flow.

Wave heights and wavelengths in estuaries are usually less than those in nearshore oceanic areas because of the reduced fetch of the wind. Wavelengths commonly range from 15 to 25 m, and wave periods are usually less than 5 s. In addition, waves tend to be more irregular in estuaries than in the ocean. For an ideal, triangular-shaped estuary in which the base opens to the ocean, the other two sides are long and bordered by land, and the apex forms the head, the fetch of the wind is very short from most directions. Only along the longitudinal axis is the fetch sufficiently great to allow the formation of large waves. Waves entering estuaries from the nearshore ocean will be diffracted as they travel through the mouth, dissipating much of their energy before the main body of the estuary is reached. Diffraction of ocean waves becomes accentuated in lagoon-type, bar-built estuaries having a narrow, shallow mouth. Although most estuaries are calm compared with the open ocean, wave action can be significant in systems with a wide, deep mouth where ocean waves meet with less interference during passage in an upestuary direction.

g. Tsunamis

Tsunamis refer to potentially destructive gravitational sea waves caused by earthquakes, volcanic eruptions, and mass movements in and around ocean basins. Also known as seismic sea waves, tsunamis have relatively small amplitudes (up to a few meters), but long wavelengths (150 to 250 km) and periods (10 to 60 min).[58] Their velocity can be determined by the celerity equation:

$$V = gH \tag{3.6}$$

where V is the wave velocity, g is the acceleration of gravity (9.81 m/s^2), and H is the ocean depth. In the open ocean, tsunamis may travel for thousands of kilometers at velocities of 700 to 900 km/h. They may be only 3 to 5 m in height. In coastal waters, the velocities decrease substantially, while the wave heights increase up to 30 to 70 m. Because of their great velocity and wave heights, tsunamis are extremely destructive in coastal environments, and thousands of coastal inhabitants have been killed during individual tsunamis events.[59,60] More tsunamis occur in the Pacific Ocean (more than two per year) than in the Atlantic and Indian Oceans because of greater earthquake and volcanic activity in the Pacific basin.

REFERENCES

1. Knauss, J. A., *Introduction to Physical Oceanography*, 2nd ed., Prentice-Hall, Upper Saddle River, NJ, 1997.
2. Etter, P. C., Lamb, P. J., and Portis, D. H., Heat and freshwater budgets of the Caribbean Sea with revised estimates for the central American seas, *J. Phys. Oceanogr.*, 17, 1232, 1987.
3. Gallimore, R. G. and Houghton, D. D., Approximation of ocean heat storage by ocean-atmosphere energy exchange: implications for seasonal cycle mixed-layer ocean formulations, *J. Phys. Oceanogr.*, 17, 1214, 1987.
4. Pinet, P. R., *Invitation to Oceanography*, Jones and Bartlett Publishers, Sudbury, MA, 1998.
5. Millero, F. J., *Chemical Oceanography*, 2nd ed., CRC Press, Boca Raton, FL, 1996.
6. Pickard, G. L. and Emery, W. J., *Descriptive Physical Oceanography: An Introduction*, 4th ed., Pergamon Press, Oxford, 1982.
7. Levinton, J. S., *Marine Ecology*, Prentice-Hall, Englewood Cliffs, NJ, 1982.
8. Gross, M. G., *Oceanography: A View of the Earth*, 5th ed., Prentice-Hall, Englewood Cliffs, NJ, 1990.
9. Richardson, P. L., Tracking ocean eddies, in *Oceanography: Contemporary Readings in Ocean Sciences*, 3rd ed., Pirie, R. G., Ed., Oxford University Press, New York, 1996, 88.
10. Brown, J., Colling, A., Park, D., Phillips, J., Rothery, D., and Wright, J., *Ocean Circulation*, Pergamon Press, Oxford, 1989.
11. Emery, W. J., On the geographical variability of the upper level mean and eddy fields in the North Atlantic and North Pacific, *J. Phys. Oceanogr.*, 13, 269, 1983.
12. Krauss, W. and Kase, R. H., Mean circulation and eddy kinetic energy in the eastern North Atlantic, *J. Geophys. Res.*, 89, 3407, 1984.
13. Schmitz, W. J., Jr., Holland, W. R., and Price, J. F., Mid-latitude mesoscale variability, *Res. Geophys. Space Phys.*, 21, 1109, 1983.
14. Schmitz, W. J., Jr., Abyssal eddy kinetic energy in the North Atlantic, *J. Mar. Res.*, 42, 509, 1984.
15. Kraus, W., Fahrbach, E., Aitsam, A., Elken, J., and Koske, P., The North Atlantic Current and its associated eddy field southeast of Flemish Cap, *Deep-Sea Res.*, 34, 1163, 1987.
16. Robinson, A. R., Overview and summary of eddy science, in *Eddies in Marine Science*, Robinson, A. R., Ed., Springer-Verlag, New York, 1983, 3.
17. Krauss, W. and Boning, C. W., Langrangian properties of eddy fields in the northern North Atlantic as deduced from satellite tracked buoys, *J. Mar. Res.*, 45, 259, 1987.

18. Philander, G., El Niño and La Niña, in *Oceanography: Contemporary Readings in Ocean Sciences,* 3rd ed., Pirie, R. G., Ed., Oxford University Press, New York, 1996, 72.

19. Sallenger, A. H., Jr., Krabill, W., Brock, J., Swift, R., Jansen, M., Manizade, S., Richmond, B., Hampton, M., and Eslinger, D., Airborne laser study quantifies El Niño-induced coastal change, *Eos,* 80, 89, 1999.

20. Neelin, J. D. and Latif, M., El Niño dynamics, *Phys. Today,* 51, 32, 1998.

21. Glen, N. C., Tidal measurement, in *Estuarine Hydrography and Sedimentation,* Dyer, K. R., Ed., Cambridge University Press, Cambridge, 1979, 19.

22. Perkins, E. J., *The Biology of Estuaries and Coastal Waters,* Academic Press, London, 1974.

23. National Academy of Sciences, National Research Council, Variability of circulation and mixing in estuaries, in *Fundamental Research on Estuaries: The Importance of an Interdisciplinary Approach,* National Academy Press, Washington, D.C., 1983, 15.

24. Kennish, M. J. and Lutz, R. A., Eds., *Ecology of Barnegat Bay, New Jersey,* Springer-Verlag, New York, 1984.

25. Biggs, R. B. and Cronin, L. E., Special characteristics of estuaries, in *Estuaries and Nutrients,* Neilson, B. J. and Cronin, L. E., Eds., Humana Press, Clifton, NJ, 1981, 3.

26. Biggs, R. B., Estuaries, in *The Encyclopedia of Beaches and Coastal Environments,* Schwartz, M. L., Ed., Hutchinson Ross, Stroudsburg, PA, 1982, 393.

27. Abraham, G., On internally generated estuarine turbulence, in *Proc. 2nd I.A.H.R. Int. Symp. Stratified Flows,* Trondheim, 344, 1980.

28. Dyer, K. R., Mixing caused by lateral internal seiching within a partially mixed estuary, *Estuarine Coastal Shelf Sci.,* 15, 443, 1982.

29. Pritchard, D. W., Observations of circulation in coastal plain estuaries, in *Estuaries,* Lauff, G. H., Ed., Publ. 83, American Association for the Advancement of Science, Washington, D.C., 1967, 37.

30. Pritchard, D. W., Estuarine circulation patterns, *Proc. Am. Soc. Civ. Eng.,* 81, 1, 1955.

31. Dyer, K. R., Lateral circulation effects in estuaries, in *Estuaries, Geophysics and the Environment,* National Academy of Science, Washington, D.C., 1977, 22.

32. Dyer, K. R., Estuaries and estuarine sedimentation, in *Estuarine Hydrography and Sedimentation,* Dyer, K. R., Ed., Cambridge University Press, Cambridge, 1979, 1.

33. Officer, C. B., Physics of estuarine circulation, in *Estuaries and Enclosed Seas,* Ketchum, B. H., Ed., Elsevier, Amsterdam, 1983, 15.

34. Elliott, A. J., Observations of the meteorologically induced circulation in the Potomac estuary, *Estuarine Coastal Mar. Sci.,* 6, 285, 1978.

35. Smith, T. J. and Takhar, H. S., A mathematical model for partially mixed estuaries using the turbulence energy equation, *Estuarine Coastal Shelf Sci.,* 13, 27, 1981.

36. Blumberg, A. E. and Goodrich, D. M., Modeling of wind-induced destratification in Chesapeake Bay, *Estuaries,* 13, 236, 1990.

37. Lee, J. M., Wiseman, W. J., Jr., and Kelly, F. J., Barotropic, subtidal exchange between Calcasieu Lake and the Gulf of Mexico, *Estuaries,* 13, 258, 1990.

38. Paraso, M. C. and Valle-Levinson, A., Meteorological influences on sea level and water temperature in the lower Chesapeake Bay: 1992, *Estuaries,* 19, 548, 1996.

39. Smith, T. J. and Takhar, H. S., A mathematical model for partially mixed estuaries using the turbulence energy equation, *Estuarine Coastal Shelf Sci.,* 13, 27, 1981.

40. Dyer, K. R., Localized mixing of low salinity patches in a partially mixed estuary (Southampton water, England), in *Estuarine Comparisons,* Kennedy, V. S., Ed., Academic Press, New York, 1982, 21.

41. O'Donnell, J., Surface fronts in estuaries: a review, *Estuaries,* 16, 12, 1993.

42. Gelfenbaum, G. and Stumpf, R. P., Observations of currents and density structure across a buoyant plume front, *Estuaries,* 16, 40, 1993.

43. Bowman, M. J., Introduction and historical perspective, in *Oceanic Fronts in Coastal Processes,* Bowman, M. J. and Esaias, W. E., Eds., Springer-Verlag, New York, 1978, 2.

44. Huzzey, L. M., The dynamics of a bathymetrically arrested estuarine front, *Estuarine Coastal Shelf Sci.,* 15, 537, 1982.

45. Simpson, J. H. and James, I. D., Coastal and estuarine fronts, in *Baroclinic Processes on Continental Shelves, Coastal and Estuarine Science,* Vol. 3, Mooers, C. N. K., Ed., American Geophysical Union, Washington, D.C., 1986, 63.

46. Ingram, R. G., Characteristics of the Great Whale River plume, *J. Geophys. Res.,* 86, 2017, 1981.
47. Wolanski, E. and Collis, P., Aspects of aquatic ecology of the Hawksbury River. I. Hydrodynamical processes, *Aust. J. Mar. Freshwater Res.,* 27, 565, 1976.
48. Sarabun, C. C., Jr., Observations of a Chesapeake Bay tidal front, *Estuaries,* 16, 68, 1993.
49. Wright, L. D., *Morphodynamics of Inner Continental Shelves,* CRC Press, Boca Raton, FL, 1995.
50. Rhines, P. B., Physical oceanography, *Oceanus,* 36, 78, 1992.
51. Stigebrandt, A., Vertical diffusion driven by internal waves in a sill fjord, *J. Phys. Oceanogr.,* 5, 468, 1976.
52. Baines, P. G., Internal tides, internal waves, and nearinertial motions, in *Baroclinic Processes on Continental Shelves,* Mooers, C. N. K., Ed., American Geophysical Union, Washington, D.C., 1986, 19.
53. Wright, L. D., Internal waves, in *The Encyclopedia of Beaches and Coastal Environments,* Schwartz, M. L., Ed., Hutchinson Ross, Stroudsburg, PA, 1982, 492.
54. Haury, L. R., Wiebe, P. H., Orr, M. H., and Briscoe, M. G., Tidally generated high-frequency internal wave packets and their effects on plankton in Massachusetts Bay, *J. Mar. Res.,* 41, 65, 1983.
55. Ketchum, B. H., Estuarine characteristics, in *Estuaries and Enclosed Seas,* Ketchum, B. H., Ed., Elsevier, Amsterdam, 1983, 1.
56. Lounsbury, J. F. and Ogden, L., *Earth Science,* 2nd ed., Harper & Row, New York, 1973.
57. Bowden, K. F., Physical factors: salinity, temperature, circulation, and mixing processes, in *Chemistry and Biogeochemistry of Estuaries,* Olausson, E. and Cato, I., Eds., John Wiley & Sons, Chichester, England, 1980, 37.
58. Costa, J. E. and Baker, V. R., *Surficial Geology: Building with the Earth,* John Wiley & Sons, New York, 1981.
59. Camfield, F. E., Tsunami effects on coastal structures, in *Coastal Hazards: Perception, Susceptibility, and Mitigation,* Finkl, C. W., Jr., Ed., The Coastal Education and Research Foundation, Fort Lauderdale, FL, 1994, 177.
60. Schubert, C., Tsunamis in Venezuela: some observations on their occurrence, in *Coastal Hazards: Perception, Susceptibility, and Mitigation,* Finkl, C. W., Jr., Ed., The Coastal Education and Research Foundation, Fort Lauderdale, FL, 1994, 189.

3.1 DIRECT AND REMOTE SENSING (OCEANOGRAPHIC APPLICATIONS)

Table 3.1–1   Instruments Commonly Used in Oceanographic Investigations

Characteristic	Instrument
Physical Characteristics	
Motion	Drogue, current meter
Temperature	Bathythermograph, thermistor
Salinity	Conductivity meter
Clarity	Turbidity meter (nephelometer), Secchi disk
Depth	Plumb line, echo sounder, pressure sensor
Waves	Buoyed accelerometer (waverider)
Tide	Tide gauge
Chemical Characteristics	
Chemicals, nutrients	Sampling bottle, pump, and tube
Dissolved oxygen (D.O.)	D.O. probe
Particulates	Settleables and floatables collectors
Bottom composition	Grab sampler, corer
Biological Characteristics	
Particulates	Plankton net, mussel buoy
Fauna	Television, baited camera, trawl
Microbes	Sterile bag sample
Meteorological Characteristics	
Radiation	Pyranometer
Wind	Anemometer
Precipitation	Rain gauge, pluviograph
Evaporation	Evaporation pan

Source: Beer, T., *Environmental Oceanography,* 2nd ed., CRC Press, Boca Raton, FL, 1997, 262. With permission.

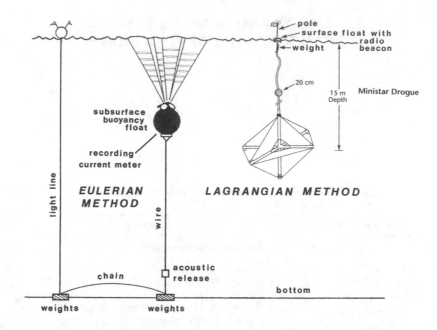

Figure 3.1–1 Subsurface currents may be measured by Eulerian methods which moor a recording current meter at the appropriate depth, or by Lagrangian methods such as a drogued buoy. (From Beer, T., *Environmental Oceanography,* 2nd ed., CRC Press, Boca Raton, FL, 1997, 263. With permission.)

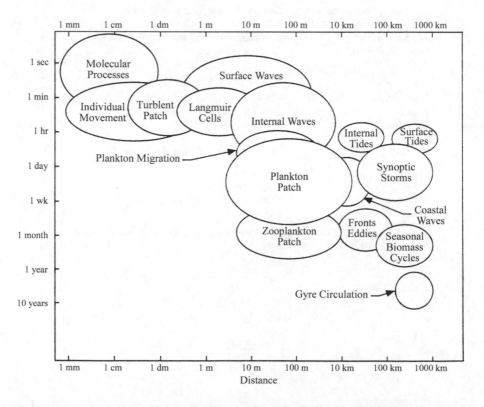

Figure 3.1–2 Time and space scales characteristic of physical and biological processes in the oceans. (From Millero, F. J., *Chemical Oceanography,* 2nd ed., CRC Press, Boca Raton, FL, 1996, 319. With permission.)

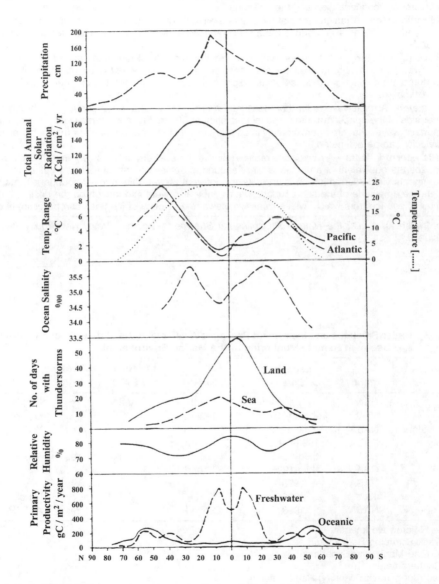

Figure 3.1–3 Latitudinal changes in various physical and biophysical parameters. (From Connell, D. W. and Hawker, D. W., Eds., *Pollution in Tropical Aquatic Systems,* CRC Press, Boca Raton, FL, 1992. With permission.)

Table 3.1–2   Satellite-Borne Remote-Sensing Techniques Used to Measure the Topography of the
Sea Surface and Other Physical Parameters

Altimeter—A pencil-beam microwave radar that measures the distance between the spacecraft and the earth. Measurements yield the topography and roughness of the sea surface from which the surface current and average wave height can be estimated.

Color scanner—A radiometer that measures the intensity of radiation emitted from the sea in visible and near-infrared bands in a broad swath beneath the spacecraft. Measurements yield ocean color, from which chlorophyll concentration and the location of sediment-laden waters can be estimated.

Infrared radiometer—A radiometer that measures the intensity of radiation emitted from the sea in the infrared band in a broad swath beneath the spacecraft. Measurements yield estimates of sea-surface temperature.

Microwave radiometer—A radiometer that measures the intensity of radiation emitted from the sea in the microwave band in a broad swath beneath the spacecraft. Measurements yield microwave brightness temperatures, from which wind speed, water vapor, rain rate, sea-surface temperature, and ice cover can be estimated.

Scatterometer—A microwave radar that measures the roughness of the sea surface in a broad swath on either side of the spacecraft with a spatial resolution of 50 km. Measurements yield the amplitude of short surface waves that are approximately in equilibrium with the local wind and from which the surface wind velocity can be estimated.

Synthetic aperture radar—A microwave radar similar to the scatterometer except that it electronically synthesizes the equivalent of an antenna large enough to achieve a spatial resolution of 25 m. Measurements yield information on features (swell, internal waves, rain, current boundaries, etc.) that modulate the amplitude of the short surface waves; they also yield information on the position and character of sea ice from which, with successive views, the velocity of ice floes can be estimated.

Source: Walsh, J. J., *On the Nature of Continental Shelves,* Academic Press, New York, 1989. With permission.

Table 3.1–3   Radar Satellites and Instrument Packages Available in the Mid-1990s and Beyond Which
Are Useful in Remote-Monitoring of the Marine Environment

Satellite	Agency	Launch Date	Instrument Packages			
			SAR	SCATT	RAlt	Other
ERS-1	ESA	July 1991	AMI[a]	AMI	Yes	ATSR[b]
JERS-1	JSA	February 1992	L-band			Oc col.
Topex-Poseidon	NASA	June 1992			Yes	Yes
SIR-C	NASA	1994–1995	C, L, X-band			
ERS-2	ESA	Mid-1995	AMI	AMI	Yes	ATSR-2, GOME[c]
Radarsat	CSA	Mid-1995	C-band			
Geosat	NASA	1995			Yes	
ADEOS	JSA	1996		NSCAT[d]		Oc col.
POEM	NASA	1998	C-band		Yes	AATSR,[e] Oc col.

[a] Advanced Microwave Imager.
[b] Along-Track Scanning Radiometer.
[c] Global Ozone Monitoring Experiment.
[d] NASA Scatterometer.
[e] Advanced Along-Track Scanning Radiometer.

Source: Dobson, F. W., in *Oceanographic Applications of Remote Sensing,* Ikeda, M. and Dobson, F. W., Eds., CRC Press, Boca Raton, FL, 1995, 292. With permission.

Table 3.1–4 Overview of the Remote Sensing Satellite (ERS-1) SAR Validation Experiment for Mesoscale Ocean Variability

Principal Investigator	Study Region	Objectives
Nilsson et al.	East Australia	Mapping the East Australian Current with SAR
Gower	West coast of Canada	Surface feature mapping with SAR
Johannessen, J. A. et al.	Norwegian Coast	Coastal ocean studies with ERS-1 SAR during NORCSEX
Scoon and Robinson	English Channel	SAR imaging of dynamic features in the English Channel
Keyte et al.	Iceland-Faeroe frontal region	Comparison of ERS-1 SAR and NOAA AVHRR images of the Iceland-Faeroe front
Shemdin et al.	Gulf of Alaska	SAR ocean imaging in the Gulf of Alaska
Tilley and Beal	Gulf Stream	ERS-1 and ALMAZ SAR ocean wave imaging over the Gulf Stream and Grand Banks
Alpers and La Violette	Gibraltar and Messina Straits	Study of internal waves generated in the Strait of Gibraltar and Messina by using SAR
Font et al.	Western Mediterranean	Comparison of ERS-1 SAR images to *in situ* oceanographic data
Liu and Peng	Gulf of Alaska	Waves and mesoscale features in the marginal ice zone
Johannessen, O.M. et al.	Greenland and Barents Seas	ERS-1 SAR ice and ocean signature validation during SIZEX

Note: Detailed reports on these studies including highlights are found in Reference 9 of original source.

Source: Johannessen, J. A. et al., in *Oceanographic Applications of Remote Sensing,* Ikeda, M. and Dobson, F. W., Eds., CRC Press, Boca Raton, FL, 1995, 29. With permission.

Table 3.1–5 Mesoscale Ocean Current Variability: Relationship between Types of Ocean Current and Ice Edge Features Imaged by Synthetic Aperture Radar (SAR) and the Dominating Environmental Conditions That Lead to the Image Expressions

	SAR Image Expressions				
	Natural Film	Wave-Current Interaction		Wind Stress	Sea Ice
Type		Shear/Convergence	Convergence		
Eddies	X	X		X	X
Fronts					
Nonthermal	X	X	X		
Thermal	X	X		X	X
Internal waves	X		X		X

Note: For sea-ice applications the ice acts as a tracer (i.e., spiral of ice reflecting eddies and regular spaced ice bands in connection with internal waves).

Source: Johannessen, J. A. et al., in *Oceanographic Applications of Remote Sensing,* Ikeda, M. and Dobson, F. W., Eds., CRC Press, Boca Raton, FL, 1995, 30. With permission.

Table 3.1–6   A Summary of Some Key Information on Present and Past Satellite Missions Equipped with Reflector Array (RA) Sensor Systems

	GEOS-3	Seasat	Geosat	ERS-1	TOPEX/ Poseidon
Operational	1975–1978	1978	1985–1989 ERM[a] 1987–1989	1991–	1992–
Agency	NASA	NASA	US Navy	ESA	NASA CNES
Data distribution	NASA JPL	NASA JPL	NOAA NGS	ESA ESRIN	JPL/DAAC CLS/AVISO
Frequency (GHz)	13.9	13.5	13.5	13.8	T: 13.6 and 5.30 P: 13.65
Auxiliary sensors used in the processing of geophysical parameters	TRANET tracking network Laser RA	TRANET LRA Multichannel microwave radiometer (SMMR)	TRANET	PRARE tracking system failed LRA Along-track scanning radiometer (ATSR)	LRA DORIS tracking system Dual-frequency altimeter for ionospheric correction TOPEX microwave radiometer (TMR) GPS demonstration receiver
Inclination	115[a]	108[a]	108[a]	98.5°	66°
Orbital height (km)	843	800	800	780	1336
Repeat cycle of orbit (days)	Various	3	17 d during ERM	3, 35, or 168	10
					T ≈ 90% P ≈ 10%

[a] ERM: Exact repeat mission, an unclassified 17-day repeat cycle.

Source: Pettersson, L. H. et al., in *Oceanographic Applications of Remote Sensing,* Ikeda, M. and Dobson, F. W., Eds., CRC Press, Boca Raton, FL, 1995, 428. With permission.

Table 3.1–7   Typical Amplitudes and Spatial Scales of the Various Components of the Altimeter Height Measurements

Height Component	Vertical Range	Typical Spatial Scale
Ocean dynamic topography		
Mesoscale eddies	Tens of cm	Tens to hundreds of km
Continental boundary	Eastern: ≈ 20 cm	≈ 500 km
Currents	Western: ≈ 100 cm	≈ 100 km
Ocean gyres	≈ 100 cm	>1000 km
Equatorial currents	≈ 20 cm	>500 km
Sea-state effect (EM-bias)	≈ 10 cm	>20 km
Ocean tide	Up to several m	Tens of km
Inverse barometric effect	Up to 50 cm	Tens of km
Geoid	−100 m to +80 m	Spatial variation on all scales

Source: Pettersson, L. H. et al., in *Oceanographic Applications of Remote Sensing,* Ikeda, M. and Dobson, F. W., Eds., CRC Press, Boca Raton, FL, 1995, 429. With permission.

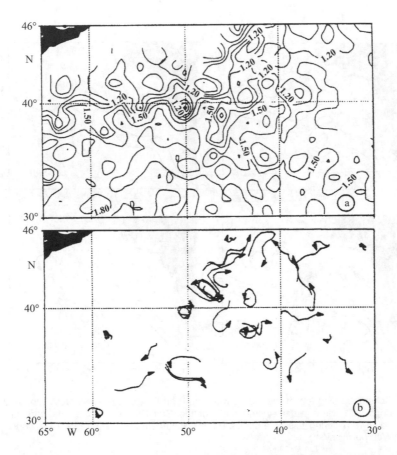

Figure 3.1–4   (a) Objectively analyzed sea surface height anomalies in the North Atlantic from Geosat altimeter data for the 17-day period from May 24 to June 9, 1987 superimposed upon the climatological mean dynamic topography with contour intervals of 15 cm; and (b) trajectories of drifting buoys for the same period. (From Willebrand, W. J. et al., *J. Geophys. Res.*, 95, 3007, 1990. With permission.)

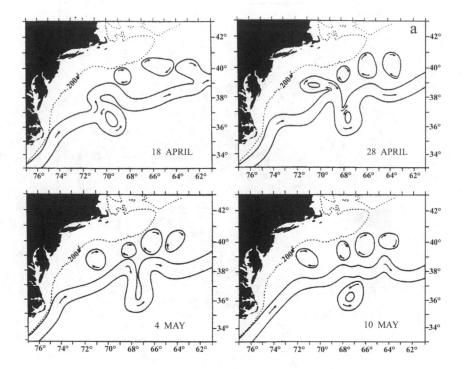

Figure 3.1–5   Schematic diagram of the Gulf Stream path and rings based on NOAA infrared images. Two rings with warm water existed north of the Gulf Stream, and a cold-core ring was formed southward during this time series. (From Richardson, P. L., *J. Phys. Oceanogr.*, 10, 90, 1980. With permission.)

3.2 LIGHT

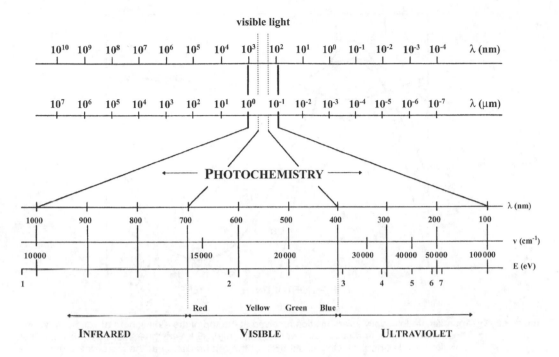

Figure 3.2–1   Electromagnetic radiation. (From Millero, F. J., *Chemical Oceanography,* 2nd ed., CRC Press, Boca Raton, FL, 1996, 363. With permission.)

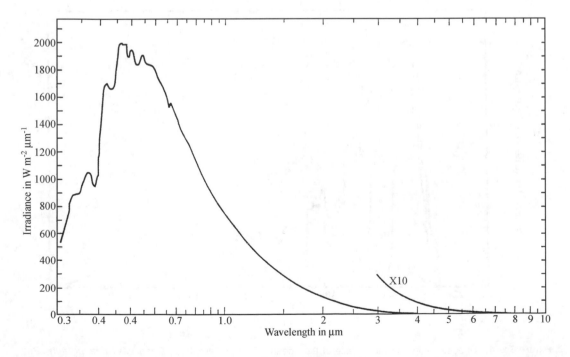

Figure 3.2–2   Solar spectral irradiance at the top of the atmosphere in the range 0.3 to 10 μm. (From Lide, D. R. and Frederikse, H. P. R., Eds., *CRC Handbook of Chemistry and Physics,* 79th ed., CRC Press, Boca Raton, FL, 1998, 14–15. With permission.)

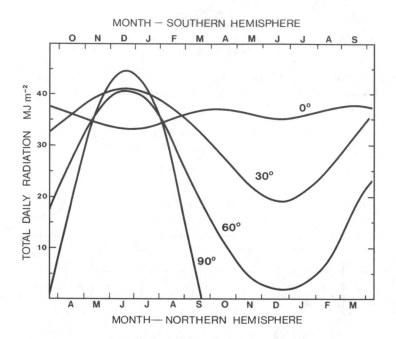

Figure 3.2–3   Approximate total daily solar radiation received at the top of the atmosphere at various latitudes. These curves assume a radiant solar energy flux of 1.35 kW/m² (the solar constant) and then allow for latitude and time. Only about 55% of the total radiation at the top of the atmosphere reaches the earth's surface. The exact percentage depends on cloudiness and atmospheric constituents. (From Beer, T., *Environmental Oceanography,* 2nd ed., CRC Press, Boca Raton, FL, 1997, 273. With permission.)

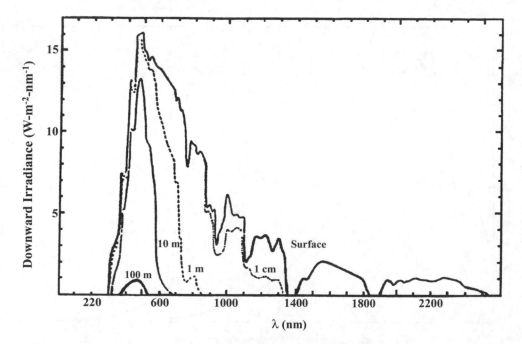

Figure 3.2–4   Spectra of light in the surface of the ocean. (From Millero, F. J., *Chemical Oceanography,* 2nd ed., CRC Press, Boca Raton, FL, 1996, 366. With permission.)

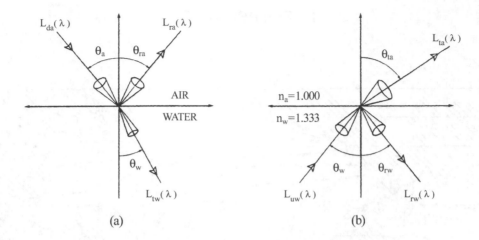

Figure 3.2–5 (a) Interaction of above-water downwelling incident radiance with the air–water interface. (b) Interaction of subsurface upwelling radiance with the air–water interface. (From Bukata, R. P. et. al., *Optical Properties and Remote Sensing of Inland and Coastal Waters,* CRC Press, Boca Raton, FL, 1995, 67. With permission.)

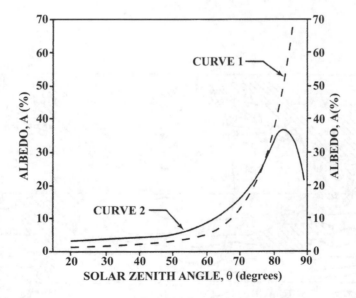

Figure 3.2–6 Effect of surface roughening (including foam and whitecaps) on water surface albedo. (From Bukata, R. P. et. al., *Optical Properties and Remote Sensing of Inland and Coastal Waters,* CRC Press, Boca Raton, FL, 1995, 61. With permission.)

3.3 TEMPERATURE

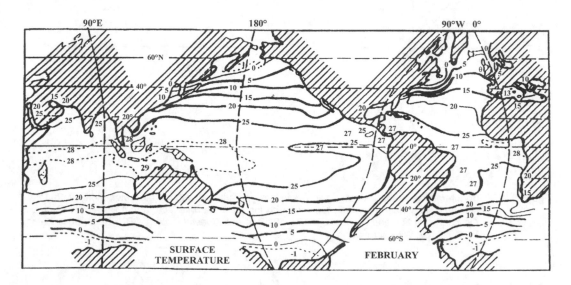

Figure 3.3–1 Temperature of ocean surface waters (February). (From Millero, F. J., *Chemical Oceanography,* 2nd ed., CRC Press, Boca Raton, FL, 1996, 8. With permission.)

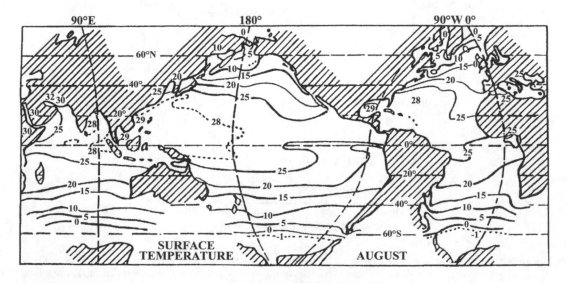

Figure 3.3–2 Temperature of ocean surface waters (August). (From Millero, F. J., *Chemical Oceanography,* 2nd ed., CRC Press, Boca Raton, FL, 1996, 9. With permission.)

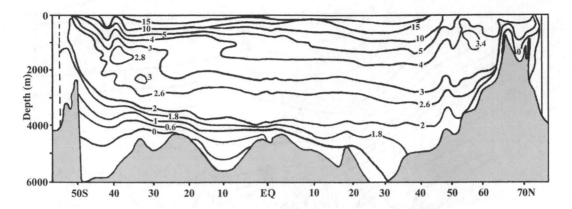

Figure 3.3–3   North–south section of potential temperature in the Atlantic Ocean. (From Millero, F. J., *Chemical Oceanography,* 2nd ed., CRC Press, Boca Raton, FL, 1996, 13. With permission.)

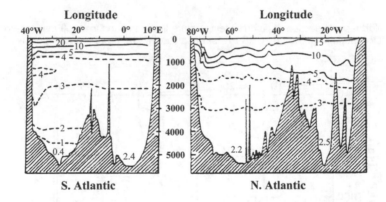

Figure 3.3–4   East–west section of temperature in the Atlantic Ocean. (From Millero, F. J., *Chemical Oceanography,* 2nd ed., CRC Press, Boca Raton, FL, 1996, 29. With permission.)

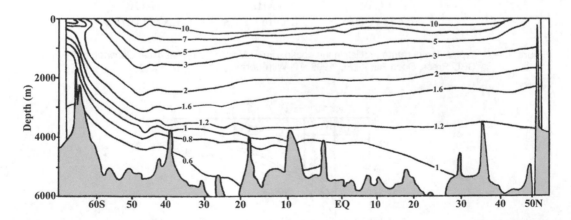

Figure 3.3–5   North–south section of potential temperature in the Pacific Ocean. (From Millero, F. J., *Chemical Oceanography,* 2nd, ed., CRC Press, Boca Raton, FL, 1996, 14. With permission.)

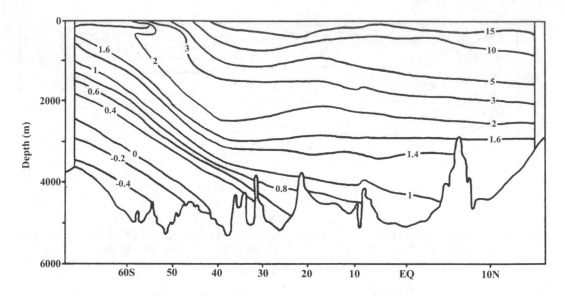

Figure 3.3–6 North–south section of potential temperature in the Indian Ocean. (From Millero, F. J., *Chemical Oceanography,* 2nd ed., CRC Press, Boca Raton, FL, 1996, 15. With permission.)

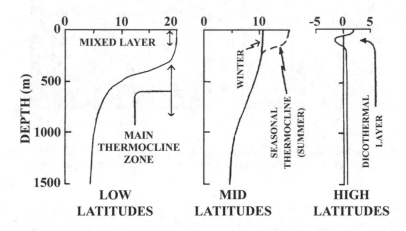

Figure 3.3–7 Typical temperature profiles in the ocean. (From Millero, F. J., *Chemical Oceanography,* 2nd ed., CRC Press, Boca Raton, FL, 1996, 10. With permission.)

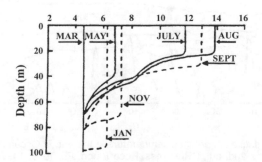

Figure 3.3–8 Growth and decay of the thermocline in the ocean. (From Millero, F. J., *Chemical Oceanography,* 2nd ed., CRC Press, Boca Raton, FL, 1996, 10. With permission.)

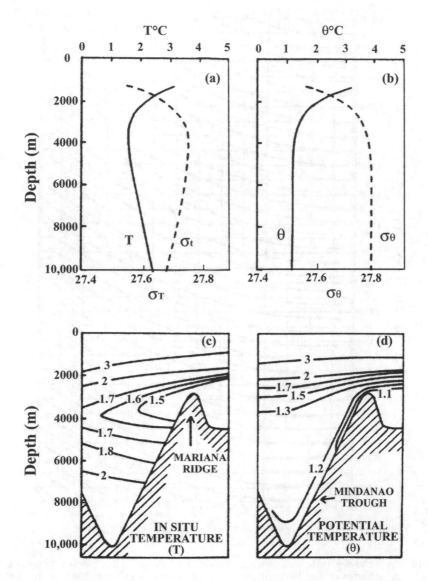

Figure 3.3–9 *In situ* and potential temperature in the Mindanao trench. (From Millero, F. J., *Chemical Ocean-ography,* 2nd ed., CRC Press, Boca Raton, FL, 1996, 11. With permission.)

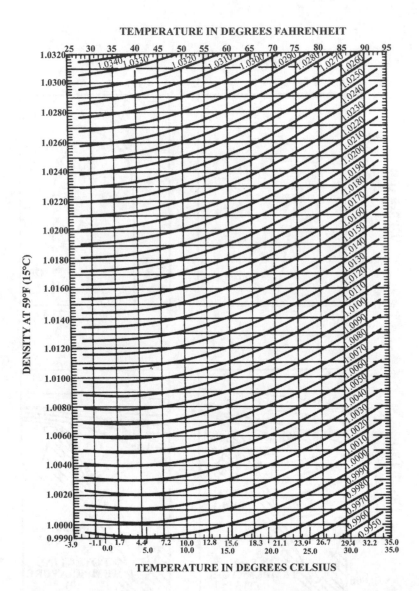

Figure 3.3–10 Seawater density at various temperatures. The purpose of this graph is to provide the density of seawater at any temperature apt to be encountered when the density at the standard temperature of 59°F (15°C) is known. To convert a density at 59°F (15°C) to density at another temperature, enter the graph horizontally from the left with the known density and downward from the top or upward from the bottom with the desired temperature; the position of the point of intersection with respect to the curves gives the density at the desired temperature. Interpolate between curves when necessary. For example, by this method, water having a density of 1.0162 at 59°F is found to have a density of 1.0124 at 85°F. The densities are referred to the density of fresh water at 4°C (39.2°F) as unity. (From National Ocean Survey, National Oceanic and Atmospheric Administration, Rockville, MD.)

3.4 SALINITY

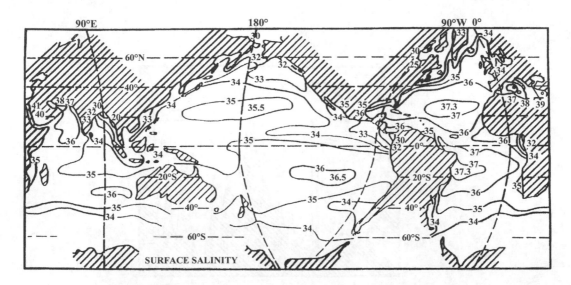

Figure 3.4–1   Salinity of ocean surface waters. (From Millero, F. J., *Chemical Oceanography,* 2nd ed., CRC Press, Boca Raton, FL, 1996, 16. With permission.)

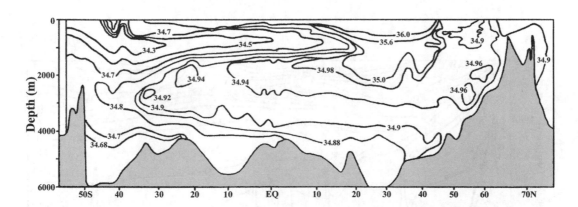

Figure 3.4–2   North–south section of salinity in the Atlantic Ocean. (From Millero, F. J., *Chemical Oceanography,* 2nd ed., CRC Press, Boca Raton, FL, 1996, 18. With permission.)

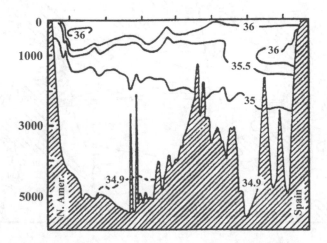

Figure 3.4–3   East–west section of salinity in the Atlantic Ocean. (From Millero, F. J., *Chemical Oceanography*, 2nd ed., CRC Press, Boca Raton, FL, 1996, 29. With permission.)

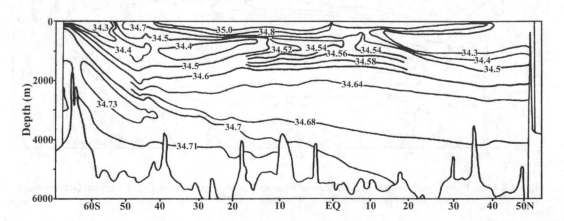

Figure 3.4–4   North–south section of salinity in the Pacific Ocean. (From Millero, F. J., *Chemical Oceanography*, 2nd ed., CRC Press, Boca Raton, FL, 1996, 19. With permission.)

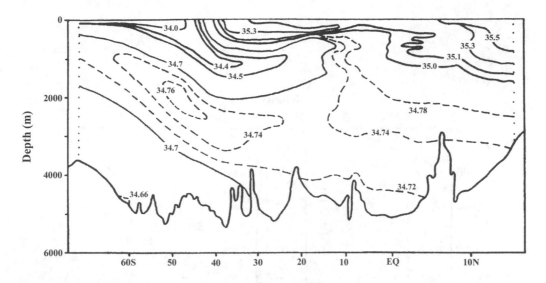

Figure 3.4–5 North–south section of salinity in the Indian Ocean. (From Millero, F. J., *Chemical Oceanography,* 2nd ed., CRC Press, Boca Raton, FL, 1996, 20. With permission.)

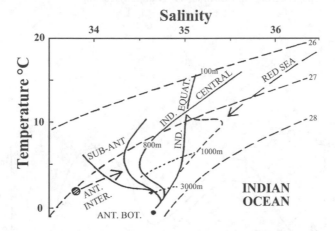

Figure 3.4–6 T–S diagram for the Indian Ocean. (From Millero, F. J., *Chemical Oceanography,* 2nd ed., CRC Press, Boca Raton, FL, 1996, 37. With permission.)

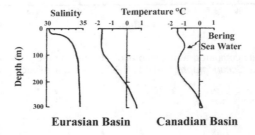

Figure 3.4–7 Typical temperature and salinity profiles for the Arctic Sea. (From Millero, F. J., *Chemical Oceanography,* 2nd ed., CRC Press, Boca Raton, FL, 1996, 39. With permission.)

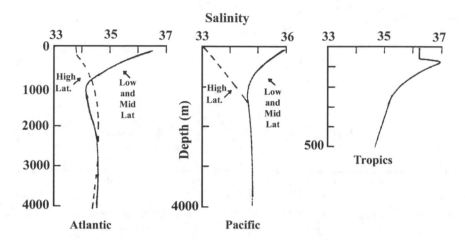

Figure 3.4–8 Typical salinity profiles in the oceans. (From Millero, F. J., *Chemical Oceanography,* 2nd ed., CRC Press, Boca Raton, FL, 1996, 17. With permission.)

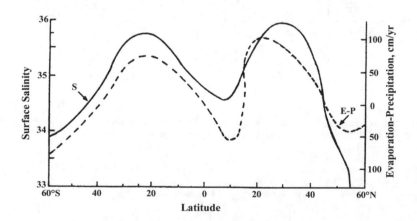

Figure 3.4–9 Surface salinity compared with evaporation (E) minus precipitation (P) (cm/year). (From Millero, F. J., *Chemical Oceanography,* 2nd ed., CRC Press, Boca Raton, FL, 1996, 17. With permission.)

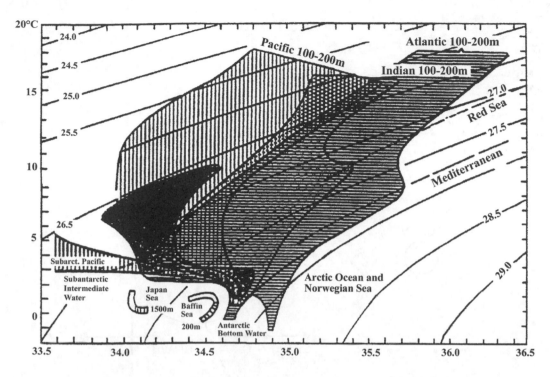

Figure 3.4 10 Temperature–salinity (T–S) diagram for waters of the oceans. (From Millero, F. J., *Chemical Oceanography,* 2nd ed., CRC Press, Boca Raton, FL, 1996, 23. With permission.)

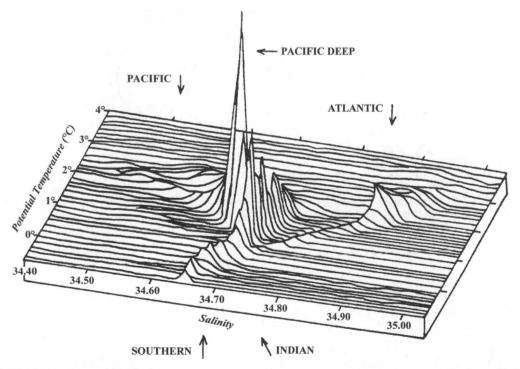

Figure 3.4–11 Three-dimensional distribution of ocean waters as a function of potential temperature and salinity. (From Millero, F. J., *Chemical Oceanography,* 2nd ed., CRC Press, Boca Raton, FL, 1996, 23. With permission.)

3.5 TIDES

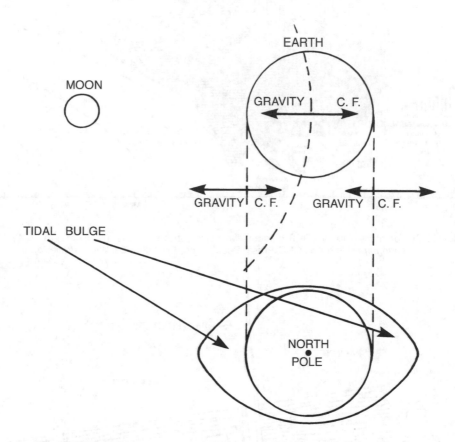

Figure 3.5–1 Schematic representation of the balance between the gravitational and rotational forces (marked C.F. for centrifugal force) at the earth's center and their imbalance at the surface, leading to tidal bulges toward and away from the moon. (From Beer, T., *Environmental Oceanography,* 2nd ed., CRC Press, Boca Raton, FL, 1997, 88. With permission.)

Table 3.5–1   Average Tidal Constituents for the World

Constituent	Name	Relative Magnitude	Period (hours)
		Semidiurnal	
M_2	Main lunar S.D.	1.0	12.42
S_2	Main solar S.D.	0.466	12.00
N_2	Lunar elliptic S.D.	0.192	12.66
K_2	Luni-solar S.D.	0.127	11.97
		Diurnal	
K_1	Soli-lunar diurnal	0.584	23.93
O_1	Main lunar diurnal	0.415	25.82
P_1	Main Solar diurnal	0.193	24.07
Q_1	Lunar elliptic diurnal	0.080	26.87
		Long Period	
M_f	Lunar fortnightly	0.172	327.8
M_m	Lunar monthly	0.091	661.3
S_{sa}	Solar semiannual	0.080	4383.3

Source: Wright, L. D., *Morphodynamics of Inner Continental Shelves,* CRC Press, Boca Raton, FL, 1995, 65. With permission.

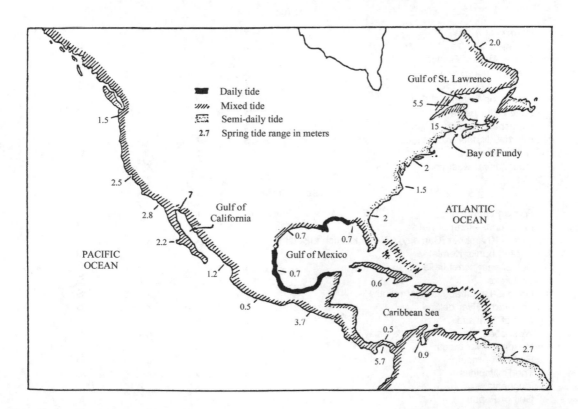

Figure 3.5–2   Types of tides and ranges of spring tides (in meters) on North American coastlines. (From U.S. Naval Oceanographic Office, Oceanographic Atlas of the North Atlantic Ocean, Section 1: Tides and Currents, H. O. Publ. 700, Washington, D.C., 1968. With permission of NAVOCEANO.)

Table 3.5–2   Characteristics of the Most Important Intertidal Areas Tidal Range

	Tidal Range			
	Micro	Meso	Macro	Tidal Regime
Europe				
Iceland	x	x	x	x
White Sea		x	x	x
Wadden Sea[a]	x	x		S
Ooster, Wester Schelde		x		S
The Wash[b]			x	S
Dyfi estuary	x			S
Baie du Mont-St. Michel			x	S
Bassin d'Arcachon			x	S
Tejo River mouth, Sado estuary		x		S
Rias northern Spain		x		S
Thames estuary (Essex coast)		x	x	S
Norfolk flats			x	S
Dee estuary			x	S
Bristol Channel/Seven estuary			x	S
Solway Firth			x	S
France, Atlantic coast			x	S
Baltic (wind flats)	x			M
Mediterranean, Caspian Sea (wind flats)	x			M,S
Africa				
Northwest Africa	x			S
Senegal, Gambia	x			S
Guinea-Bissau-Guinea		x		S
Sierra Leone-Benin	x	x		S
Niger delta	x	x		S
Cameroon-Equatorial Guinea		x		S
Zaire River estuary	x			S
Langebaan lagoon	x			S
Kwazulu-Natal-Mozambique[a]	x	x	x	S
Tanzania-Kenya		x		S
Madagascar west coast			x	S
South Asia				
Persian Gulf	x	x		M
Indus delta-Rann of Kutch		x		M
Gulf of Khambhat (Cambay), Gulf of Kachchh (Kutch)		x	x	M
(Mud banks Kerala)				M
Deltas east coast India	x			S
Sri Lanka	x			S
Ganges-Brahmaputra river mouth[b]			x	S
Irrawaddy River delta			x	S
Gulf of Thailand	x	x		D
West coast Burma, Thailand, Malaysia		x		S
Mekong River mouth[b]		x		M
Red River mouth[b]			x	D
The Philippines	x	x		M
Indonesia[a]	x	x		M,D,S
Gulf of Papua		x		M
Northeast Asia				
North Siberia	x			S(M)
East Siberia (wind flats)	x			S
Gulf of Anadyr	x			S

Tabel 3.5–2   Characteristics of the Most Important Intertidal Areas Tidal Range (continued)

| | Tidal Range | | | |
	Micro	Meso	Macro	Tidal Regime
Sea of Okhotsk			x	M
West Korea[b]			x	S
Bohai Bay/Huang He delta	x	x		S
China coast[a]			x	S
West Taiwan[a]			x	S

Australia

	Micro	Meso	Macro	Tidal Regime
South and East Australia	x			M,S
West Australia	x			M
Northwest/North Australia[a]			x	S
North Australia		x	x	S,M
Gulf of Carpentaria	x	x		M
Broad Sound			x	M
New Zealand	x	x		S
Pacific Islands	x			S

North America

	Micro	Meso	Macro	Tidal Regime
Mackenzie River delta	x			S
Yukon River delta	x			M
Kuskokwim River delta	x			M
Bristol Bay		x		M
Cook Inlet			x	M
Queen Charlotte Islands			x	M,S
Fraser River delta		x		M
Yaquina Bay, San Francisco Bay		x		M
Colorado River delta			x	M
Lagoons Mexico/Texas (wind flats)	x			D,M
Louisiana/Mississippi delta[a]	x			D,M
Southwest Florida–Belize	x			M
Northwest Florida	x			S
Andros, Caicos, Grand Cayman	x			S
Georgia-Maine		x		S
Bay of Fundy			x	S
St. Lawrence River estuary			x	M
Hudson Bay-Labrador-Baffin Island	x	x	x	S

Central/South America

	Micro	Meso	Macro	Tidal Regime
Colombia coast[b]			x	M
Gulf of Guayaquil			x	M
San Sebastian Bay			x	S
San Antonio Bay			x	S
Bahia Blanca		x		S
Mar Chiquita	x			S
South American east coast	x	x	x	S,M
Amazon-Orinoco delta[b]		x	(x)	S

[a] Partially along an open coast.
[b] Along an open coast.
Micro = microtidal; meso = mesotidal; macro = macrotidal; tidal regime: S = semidiurnal; M = mixed; D = diurnal.

Source: Eisma, D., *Intertidal Deposits: River Mouths, Tidal Flats, and Coastal Lagoons,* CRC Press, Boca Raton, FL, 1998, 3. With permission.

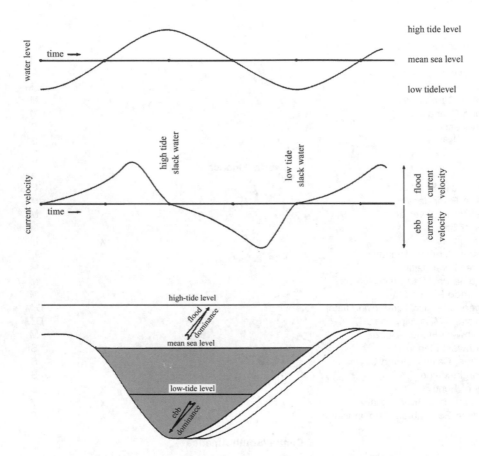

Figure 3.5–3 The asymmetrical character of the tide in most inshore tidal environments leads to a dominance of the ebb current at deeper levels, i.e., in tidal channels, and a dominance of the flood current at higher levels, i.e., on the intertidal flats. As a result, sub- to intertidal sedimentary successions often show a current reversal (ebb → flood) in the vertical direction. (From Eisma, D., *Intertidal Deposits: River Mouths, Tidal Flats, and Coastal Lagoons,* CRC Press, Boca Raton, FL, 1998, 356. With permission.)

3.6 WIND

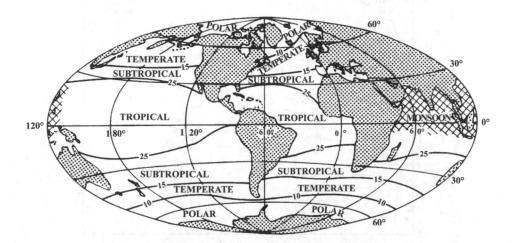

Figure 3.6–1 Climatic zonation of the open oceans. Zone boundaries tend to follow latitudes and line up with climatic belts on land. Temperature, seasonality, and water budget (evaporation–precipitation balance) are the most important descriptors. Temperate and polar can be separated by another zone: subpolar. Approximate temperatures of surface waters in °C shown at the boundaries. (From Seibold, E. and Berger, W. H., *The Sea Floor: An Introduction to Marine Geology,* 2nd ed., Springer-Verlag, Berlin, 1993, 187. With permission.)

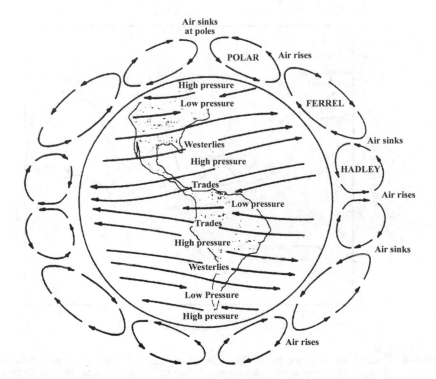

Figure 3.6–2 The major wind systems of the earth. (From Millero, F. J., *Chemical Oceanography,* 2nd ed., CRC Press, Boca Raton, FL, 1996, 24. With permission.)

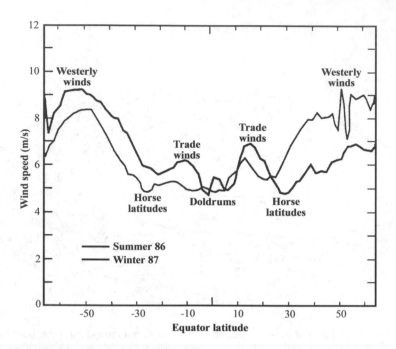

Figure 3.6–3 Average winds for the North Atlantic Ocean for summer 1986 and winter 1987. (From Dobson, E. B., in *Oceanographic Applications of Remote Sensing,* Ikeda, M. and Dobson, F. W., Eds., CRC Press, Boca Raton, FL, 1995, 200. With permission.)

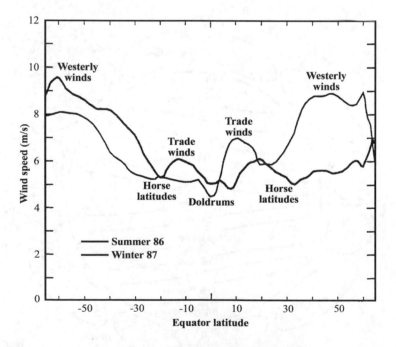

Figure 3.6–4 Average winds for the North Pacific Ocean for summer 1986 and winter 1987. (From Dobson, E. B., in *Oceanographic Applications of Remote Sensing,* Ikeda, M. and Dobson, F. W., Eds., CRC Press, Boca Raton, FL, 1995, 200. With permission.)

Table 3.6–1 Beaufort Scale of Wind Force

Beaufort Number	Wind Speed (m s⁻¹)	Wind Description	Wind State	Sea Description	Sea State	Wave Height (m)
0	<0.5	Calm	Still	Mirror-like	Calm seas	0
1	0.5–1.5	Light air	Smoke moves, wind vanes still	Wavelet-scales		
2	2–3	Light breeze	Wind felt on face, leaves rustle	Short waves, none break	Smooth seas	0–0.1
3	3.5–5	Gentle breeze	Light flags extended	Foam has glassy appearance, not yet white	Slight seas	0.1–0.5
4	5.5–8	Moderate breeze	Dust and papers moved	Longer waves with white areas	Moderate seas	0.5–1.25
5	8.5–10.5	Fresh breeze	Small trees sway	Long pronounced waves with white foam crests	Rough seas	1.25–2.5
6	11–13.5	Strong breeze	Large branches move	Large waves, white foam crests all over	Very rough seas	2.5–4
7	14–16.5	Moderate gale	Whole trees move	Wind blows foam in streaks	Very rough seas	4–6
8	17–20	Gale	Twigs broken off trees	Higher waves	Very rough seas	4–6
9	20.5–23.5	Strong gale	Some houses damaged	Dense foam streaks	High seas	4–6
10	24–27.5	Whole gale	Trees uprooted	High waves with long overhanging crests	Very high seas	6–9
11	28–33	Violent storm	Extensive damage	Ships in sight hidden in wave troughs		9–14
12	>33	Hurricane	Devastation	Air–sea boundary indistinguishable		>14

Source: Beer, T., *Environmental Oceanography*, 2nd ed., CRC Press, Boca Raton, FL, 1997, 211. With permission.

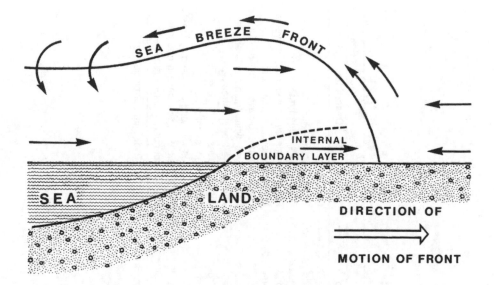

Figure 3.6–5   Vertical structure of a sea breeze front showing a marked change in wind direction and an abrupt drop in temperature across the front. (From Beer, T., *Environmental Oceanography,* 2nd ed., CRC Press, Boca Raton, FL, 1997, 212. With permission.)

3.7 WAVES AND THEIR PROPERTIES

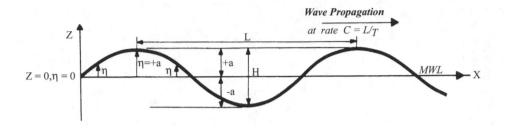

Figure 3.7–1 Definition sketch for the dimensions of a surface gravity wave. (From Wright, L. D., *Morpho-dynamics of Inner Continental Shelves,* CRC Press, Boca Raton, FL, 1995, 55. With permission.)

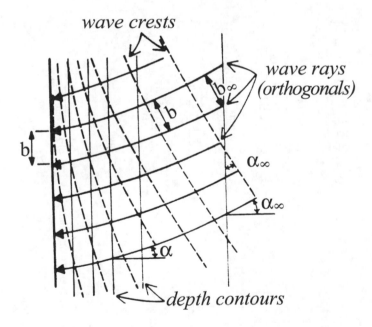

Figure 3.7–2 Definition sketch for refraction of shoaling gravity waves. (From Wright, L. D., *Morphodynamics of Inner Continental Shelves,* CRC Press, Boca Raton, FL, 1995, 60. With permission.)

Table 3.7–1 Properties of Linear Gravity Waves

Property	General Expression	Deep Water (h/L > 0.5; short waves)	Shallow Water (h/L < 0.05; long waves)
Velocity potential, ϕ	$\phi = a\dfrac{g}{\omega}\dfrac{\cosh k(z+h)}{\cosh kh}\sin(kx-\omega t)$	$\phi = a\dfrac{g}{\omega}e^{kz}\sin(kx-\omega t)$	$\phi = a\dfrac{g}{\omega}\sin(kx-\omega t)$
Phase speed, C	$C \equiv \dfrac{\omega}{k} = \dfrac{L}{T} = \dfrac{g}{\omega}\tanh kh$	$C_\infty = \dfrac{g}{\omega} = g\dfrac{T}{2\pi}$	$C = (gh)^{\frac{1}{2}}$
Wavelength, L	$L = g\dfrac{T^2}{2\pi}\tanh kh$	$L_\infty = g\dfrac{T^2}{2\pi}$	$L = T(gh)^{\frac{1}{2}}$
Group velocity, C_g	$C_g = Cn = C\dfrac{1}{2}\left[1+\dfrac{2kh}{\sinh 2kh}\right]$	$C_g^\infty = \dfrac{1}{2}C_\infty$	$C_g = C = (gh)^{\frac{1}{2}}$
Pressure, P_w	$P_w = \rho ga\dfrac{\cosh k(z+h)}{\cosh kh}\cos(kx-\omega t)$	$p_w = \rho gae^{kz}\cos(kx-\omega t)$	$p_w = \rho g\eta$
Horizontal orbital velocity, $\tilde{u}_o$	$\tilde{u}_o = a\omega\dfrac{\cosh k(z+h)}{\sinh kh}\cos(kx-\omega t)$	$\tilde{u}_o = a\omega e^{\omega z}\sin(kx-\omega t)$	$\tilde{u}_o = \dfrac{Ca}{h}\cos(kx-\omega t)$
Vertical orbital velocity, $\tilde{w}_o$	$\tilde{w}_o = a\omega\dfrac{\sinh k(z+h)}{\sinh kh}\sin(kx-\omega t)$	$\tilde{w}_o = a\omega e^{\omega z}\cos(kx-\omega t)$	$\tilde{w}_o = 0$
Energy density, E	$E = \dfrac{1}{2}\rho ga^2$	$E = \dfrac{1}{2}\rho ga^2$	$E = \dfrac{1}{2}\rho ga^2$
Energy flux, E_f	$E_f = EC_g$	$E_f = EC_g$	$E_f = EC_g$
Radiation stress, S_{xx} (x flux of x momentum)	$S_{xx} = E\left[\dfrac{1}{2}+\dfrac{2kh}{\sinh 2kh}\right]$	$S_{xx} = \dfrac{1}{2}E$	$S_{xx} = \dfrac{3}{2}E$
Radiation stress, S_{yy} (y flux of y momentum)	$S_{yy} = E\left[\dfrac{kh}{\sinh 2kh}\right]$	$S_{yy} = 0$	$S_{yy} = \dfrac{1}{2}E$
Radiation stress, S_{xy} (x flux of y momentum)	$S_{xy} = \dfrac{1}{2}E\left[1+\dfrac{2kh}{\sinh 2kh}\right]\sin\alpha\cos\alpha$	$S_{xy} = \dfrac{1}{2}E\sin\alpha\cos\alpha$	$S_{xy} = E\sin\alpha\cos\alpha$

Source: Wright, L. D., *Morphodynamics of Inner Continental Shelves*, CRC Press, Boca Raton, FL, 1995, 56. With permission.

Table 3.7–2 Typical Properties of Ideal Waves[a]

λ Wavelength (m)	k Wave Number (m⁻¹)	T Period (s)	ω Angular Frequency (s⁻¹)	C Celerity (m/s)	Wave Type in 10 m of Water
1.00E + 03	6.28E − 03	1.01E + 02	6.21E − 02	9.89E + 00	Shallow water gravity
1.00E + 02	6.28E − 02	1.07E + 01	5.85E − 01	9.32E + 00	Intermediate gravity
1.00E + 01	6.28E − 01	2.53E + 00	2.48E + 00	3.95E + 00	Deep water gravity
1.00E + 00	6.28E + 00	8.01E − 00	7.84E + 00	1.25E + 00	Deep water gravity waves
1.00E − 01	6.28E + 01	2.50E − 01	2.52E + 01	4.01E − 01	Deep water gravity waves
1.00E − 02	6.28E + 02	4.02E − 02	1.56E + 02	2.49E − 01	Deep water capillary waves
1.00E − 03	6.28E + 03	1.47E − 03	4.29E + 03	6.83E − 01	Deep water capillary waves
1.00E − 04	6.28E + 04	4.64E − 05	1.35E + 05	2.16E + 00	Deep water capillary waves

[a] The notation for numbers in this table follows computer syntax: 100 is 1.00E2, where the E2 denotes 10 to the power 2, or 10–. One thousandth is 1.00E−3.

Source: Beer, T., *Environmental Oceanography,* 2nd ed., CRC Press, Boca Raton, FL, 1997, 63. With permission.

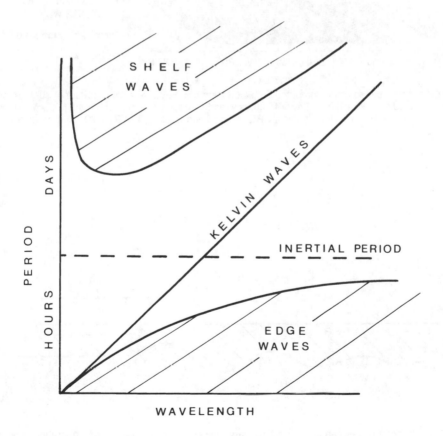

Figure 3.7–3   Trapped wave nomenclature in terms of wave period and wavelength. A diagram relating these two quantities (or frequency and wave number) is called a dispersion diagram. (From Beer, T., *Environmental Oceanography,* 2nd ed., CRC Press, Boca Raton, FL, 1997, 80. With permission.)

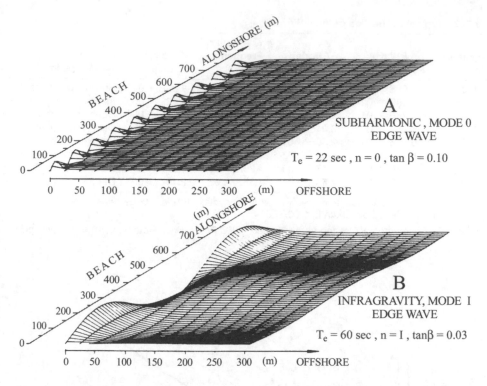

Figure 3.7–4   Three-dimensional computer-simulated view of nearshore edge waves: (A) subharmonic mode 0 edge wave; (B) infragravity-frequency mode 1 edge wave. (From Wright, L. D., *Morphodynamics of Inner Continental Shelves,* CRC Press, Boca Raton, FL, 1995, 72. With permission.)

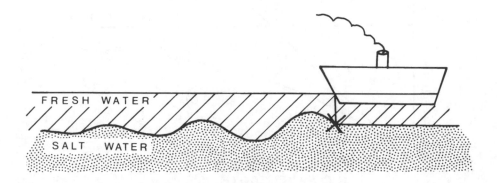

Figure 3.7–5   Internal waves can exist at a boundary between different layers of fluid. In this case, the ship's propeller is generating internal waves rather than moving the ship forward. (From Beer, T., *Environmental Oceanography,* 2nd ed., CRC Press, Boca Raton, FL, 1997, 81. With permission.)

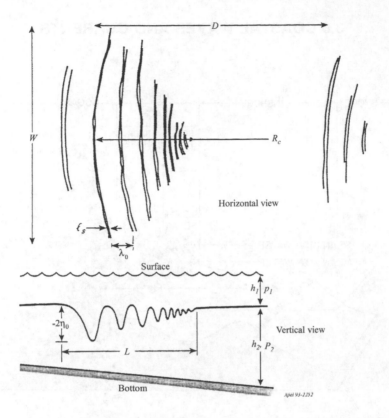

Figure 3.7–6 Schematic internal soliton packets on the continental shelf, showing characteristic variations in the amplitude, wavelength, and crest length of internal waves. (From Apel, J. R., in *Oceanographic Applications of Remote Sensing,* Ikeda, M. and Dobson, F. W., Eds., CRC Press, Boca Raton, FL, 1995, 60. With permission.)

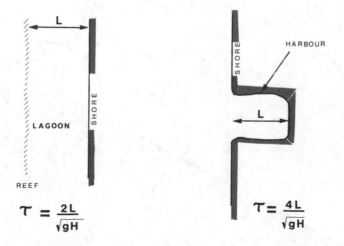

Figure 3.7–7 Seiching and harbor resonance. The formulae relate the seiching or resonant period to the distance, *L,* as shown, and to the depth of water, *H.* (From Beer, T., *Environmental Oceanography,* 2nd ed., CRC Press, Boca Raton, FL, 1997, 77. With permission.)

3.8 COASTAL WAVES AND CURRENTS

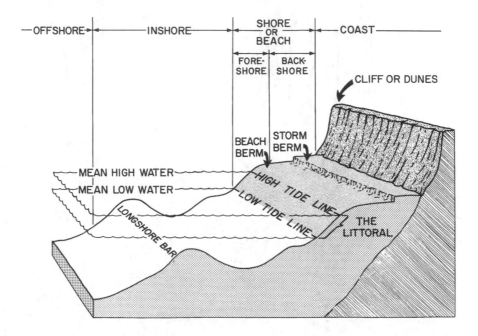

Figure 3.8–1 Beach profile illustrating terminology. (From Beer, T., *Environmental Oceanography,* 2nd ed., CRC Press, Boca Raton, FL, 1997, 30. With permission.)

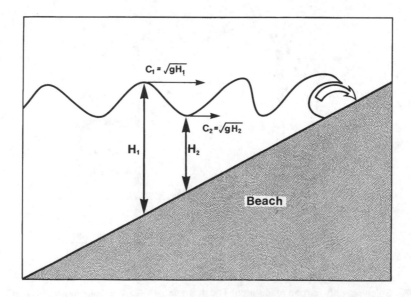

Figure 3.8–2 Breakers occur when the height of the wave is significantly affected by the bottom so that the wave speed, $\sqrt{gH}$, differs from crest to trough. The faster moving crest eventually overtakes the trough and topples to form a breaker. (From Beer, T., *Environmental Oceanography,* 2nd ed., CRC Press, Boca Raton, FL, 1997, 33. With permission.)

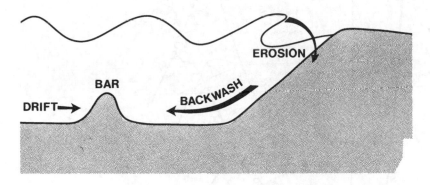

Figure 3.8–3   During storm activity, large waves erode the sand dunes until a bar forms out at sea which protects
the beach from further erosion. During calm periods the sand in the bar is slowly redeposited
on the beach. (From Beer, T., *Environmental Oceanography,* 2nd ed., CRC Press, Boca Raton,
FL, 1997, 36. With permission.)

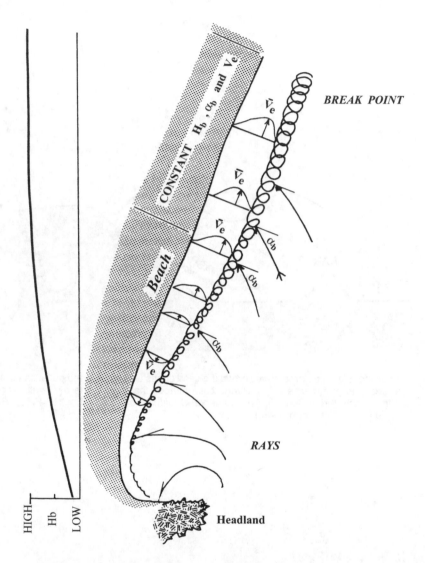

Figure 3.8–4   Longshore currents generated by a combination of oblique wave breaking and alongshore gradients in breaker height related to the presence of a headland. (From Wright, L. D., *Morphodynamics of Inner Continental Shelves,* CRC Press, Boca Raton, FL, 1995, 80. With permission.)

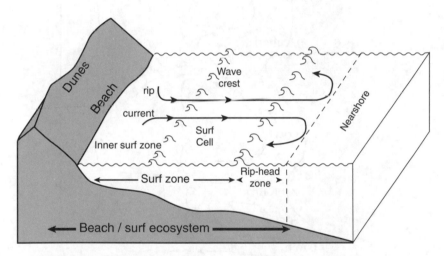

Figure 3.8–5 Idealized view of the structure and maintenance of rip currents and surf-cell circulation within an intermediate-type sandy beach/surf-zone ecosystem. (From Alongi, D. M., *Coastal Ecosystem Processes,* CRC Press, Boca Raton, FL, 1998, 31. With permission.)

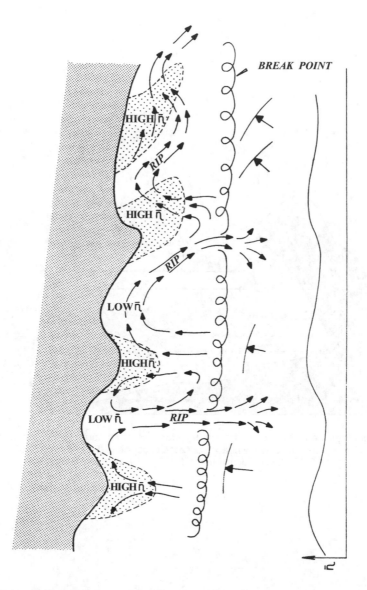

Figure 3.8–6 Combined rip circulation and longshore currents over complex surf zone topography. (From Wright, L. D., *Morphodynamics of Inner Continental Shelves,* CRC Press, Boca Raton, FL, 1995, 81. With permission.)

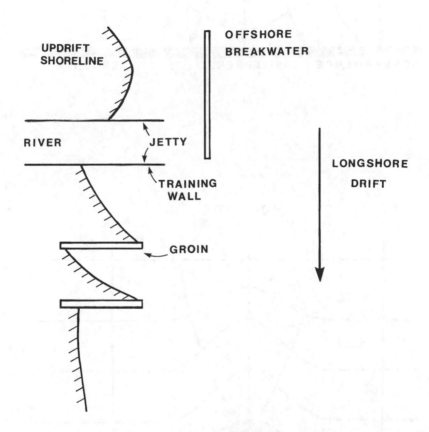

Figure 3.8–7 A sketch of the effects of a jetty, breakwater, groin combination. (From Beer, T., *Environmental Oceanography,* 2nd ed., CRC Press, Boca Raton, FL, 1997, 45. With permission.)

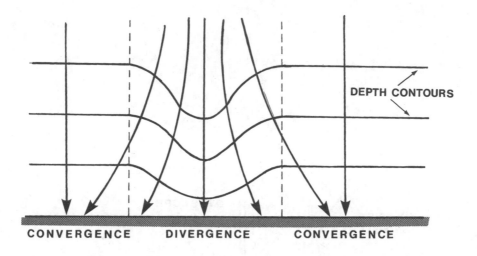

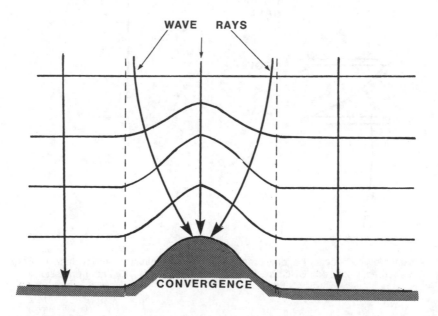

Figure 3.8–8 An undersea canyon refracts waves away, and a headland refracts waves toward itself. Zones of convergence have substantial wave activity. Divergence zones are calm. (From Beer, T., *Environmental Oceanography,* 2nd ed., CRC press, Boca Raton, FL, 1997, 32. With permission.)

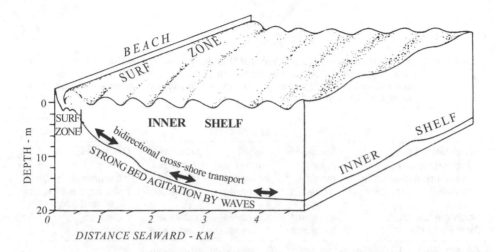

Figure 3.8–9 The inner continental shelf: a region where waves frequently agitate the seabed. (From Wright, L. D., *Morphodynamics of Inner Continental Shelves,* CRC Press, Boca Raton, FL, 1995, 3. With permission.)

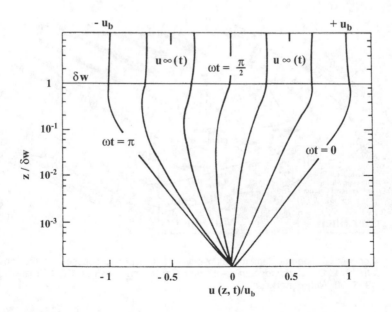

Figure 3.8–10 Oscillatory boundary layer flows under waves. (From Wright, L. D., *Morphodynamics of Inner Continental Shelves,* CRC Press, Boca Raton, FL, 1995, 109. With permission.)

Table 3.8–1   Summary of the Relative Effects of Unsteady Flow, Wave–Current Interactions, Distributed Bed Roughness, Sediment Transport, and Stratification on Boundary Layer Properties (as Compared to Steady, Neutrally Stratified Flow over a Flat Fixed Bed)

	z_o	ξ_{xz} ($z < 10$ cm)	C_{100}	$\bar{\tau}$
Accelerating flow	Decreases	Decreases	Decreases	Decreases
Decelerating flow	Increases	Increases	Increases	Increases
Wave–current interaction	Increases	Increases	Increases	Increases
Distributed bed roughness	Increases	Increases	Increases	Local increases; space-averaged decreases
Sediment transport	Increases	Increases	Increases	Increases
Stratification	Increases	Decreases	Decreases	Decreases

Source: Wright, L. D., *Rev. Aquat. Sci.,* 1, 75, 1989. With permission.

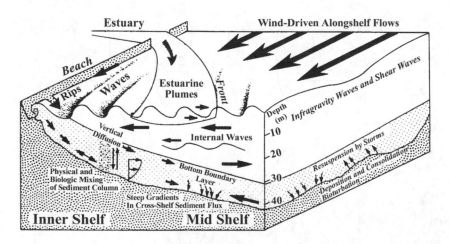

Figure 3.8–11   Conceptual diagram demonstrating physical transport processes on the inner shelf. (From Wright, L. D., *Morphodynamics of Inner Continental Shelves,* CRC Press, Boca Raton, FL, 1995, 50. With permission.)

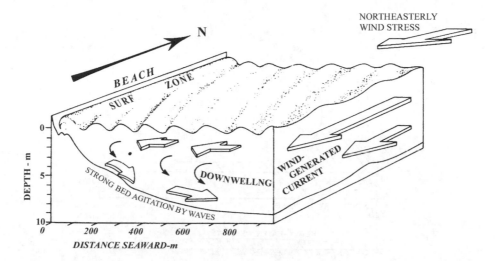

Figure 3.8–12 Conceptual diagram depicting coastal jetlike flows generated by oblique winds. Strong along-shelf flows in this example (which is based on "northeasters" in the Middle Atlantic Bight) are accompanied by high waves and by shoreward transport at the surface and seaward transport near the bed. (From Wright, L. D., *Morphodynamics of Inner Continental Shelves,* CRC Press, Boca Raton, FL, 1995, 53. With permission.)

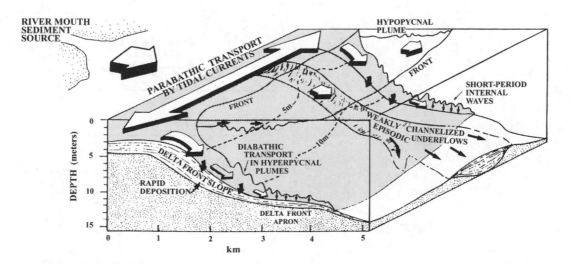

Figure 3.8–13 Conceptual diagram illustrating along-shelf and across-shelf transports by a combination of tidal currents and hyperpycnal flows. (From Wright, L. D., *Morphodynamics of Inner Continental Shelves,* CRC Press, Boca Raton, FL, 1995, 95. With permission.)

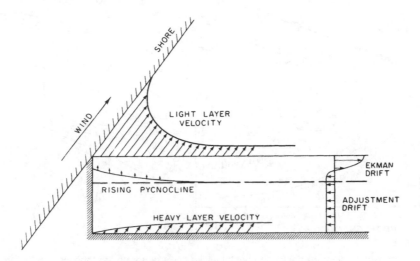

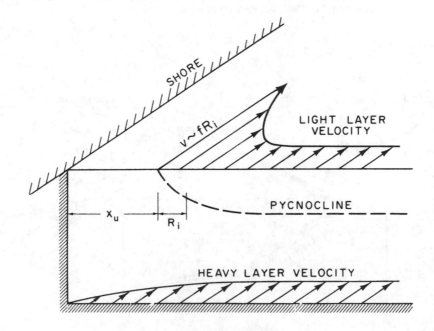

Figure 3.8–14   Evolution of an upwelling front and offshore coastal jet in the Northern Hemisphere. (From Csanady, G. T., *Eos, Trans. Am. Geophys. Union,* 62, 9, 1981. With permission.)

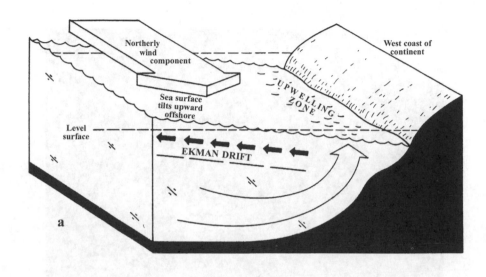

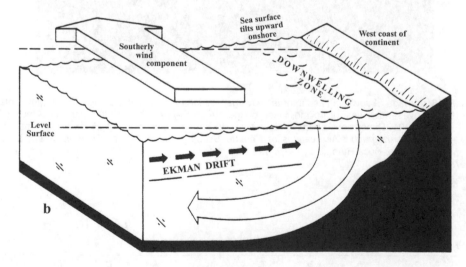

Figure 3.8–15   Upwelling and downwelling Ekman effects produced by surface winds along coastlines. (a) Ekman transport driven by a surface wind that has a northerly component is in the offshore direction. These are the necessary conditions for upwelling. Notice the upward tilt of sea level in the offshore direction. (b) Ekman transport is onshore, moving warm water against the coastline and creating a tilt in the sea surface near the shore. Strong winds will actually force enough water against the shore that the warm water sinks somewhat. (From Neshyba, S., *Oceanography: Perspectives on a Fluid Earth,* John Wiley & Sons, New York, 1987, 179. Reproduced by permission of copyright owner, John Wiley & Sons, Inc.)

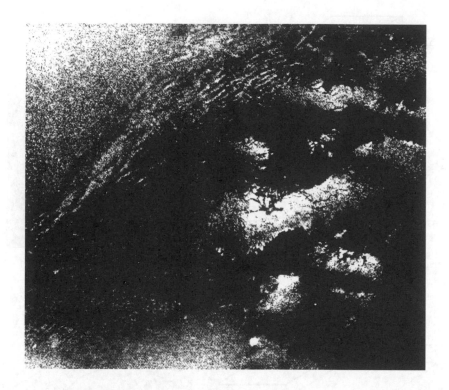

Figure 3.8–16   Image of upwelling and internal wave generation occurring at the edge of the continental shelf west-southwest of Ireland, taken by the 10-cm SAR on the Russian satellite Almaz-1 in July 1991. Dimensions approximately 40×40 km^2, resolution about 25 m. (From Apel, J. R., in *Oceanographic Applications of Remote Sensing,* Ikeda, M. and Dobson, F. W., Eds., CRC Press, Boca Raton, FL, 1995, 63.)

Table 3.8–2   Typical Scales for Continental Shelf Internal Waves

L (km)	$2\eta_0$ (m)	h_1 (m)	h_2 (m)	l_s (m)	λ_0 (m)	W (km)	D (km)	R_c (km)	$\Delta\rho/\rho$
1–5	0–15	20–50	100	100	50–500	0–30	15	25–∞	0.001

Source: Apel, J. R., in *Oceanographic Applications of Remote Sensing, Ikeda,* M. and Dobson, F. W., Eds., CRC Press, Boca Raton, FL, 1995, 63.

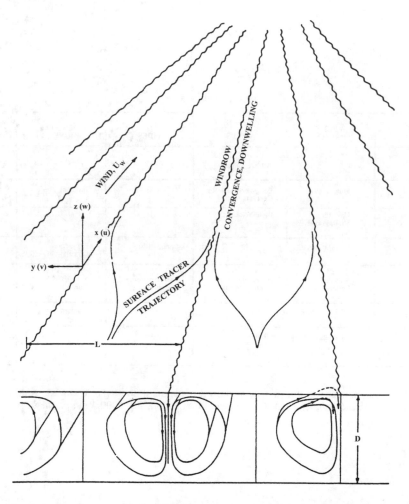

Figure 3.8–17 Schematic diagram of typical Langmuir circulation cells. (From Leibovich, S., *Annu. Rev. Fluid Mech.*, 15, 391, 1983. With permission.)

SHELF CURRENTS

TIME SCALES OF VARIATION	WIND DRIVEN	WAVES	TIDES	DENSITY DRIVEN	DEEP OCEAN
Secs		Orbital Velocity			
Hours	Surges	Longshore and Rip Currents	Tidal Currents	Internal Tides	
Days weeks	Upwelling Shelf Waves	Drift Velocity			
Months	Circulation		Sprint-Neaps	Estuarine Circulation	Rings Eddies
Seasonal	Seasonal Patterns	Hurricanes Winter Storms	Low frequency variability	Seasonal Run-off	Ocean Currents
Annual	Winter Ice			Insolation	
Long Term	Extreme Storm Conditions				Oceanic Circulation

Figure 3.8–18   Contributions to the velocity field on a continental shelf. (From Davies, A. M., Ed., *Modeling Marine Systems,* Vol. 1, CRC Press, Boca Raton, FL, 1990. With permission.)

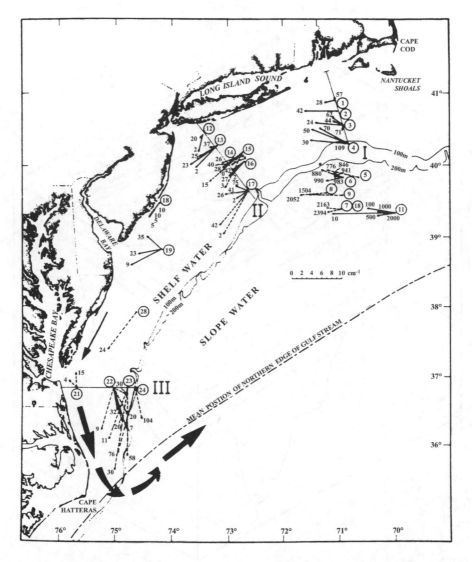

Figure 3.8–19 Regional coastal circulation in the Middle Atlantic Bight. A net southerly drift is deflected seaward off Cape Hatteras and is entrained into the Gulf Stream. (From Wright, L. D., *Morphodynamics of Inner Continental Shelves,* CRC Press, Boca Raton, FL, 1995, 83. With permission.)

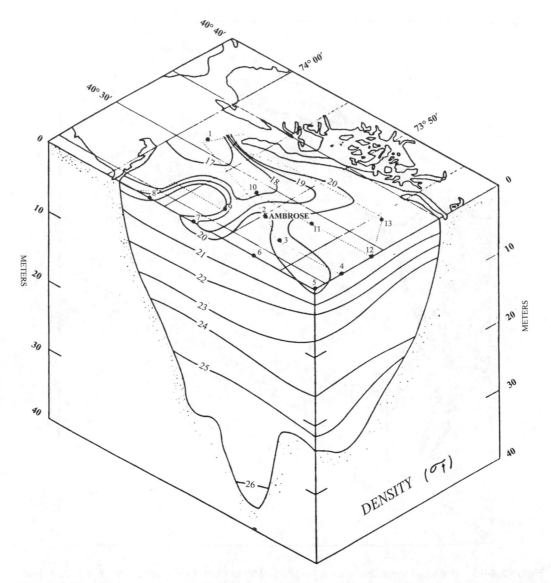

Figure 3.8–20   Isometric diagram of density (σ_r) in the New York Bight Apex, August 13, 1976, showing the Hudson Estuary plume with its associated fronts. (From Kjerfve, B., Ed., *Hydrodynamics of Estuaries,* Vol. 1, *Estuarine Physics,* CRC Press, Boca Raton, FL, 1988. With permission.)

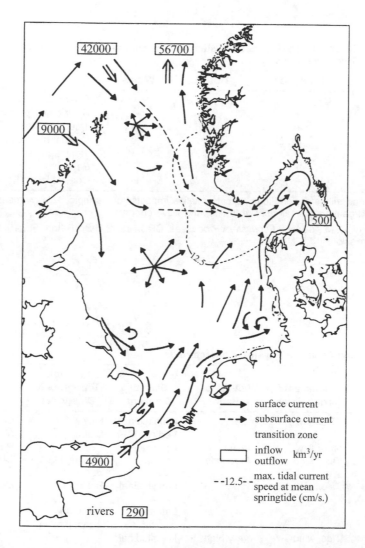

Figure 3.8–21 The general circulation in the North Sea. Flow directions in the central and northern part of the North Sea are variable (indicated by a rosette of arrows) because of variable wind directions, but are mainly directed toward the east, as the winds are predominantly westerly. (From Eisma, E. and Kalf, J., *J. Geol. Soc.* (London), 144, 161, 1987. With permission.)

3.9 CIRCULATION IN ESTUARIES

Table 3.9–1   Classification of Coastal Systems Based on Relative Importance of River Flow, Tides, and Waves

Type	River Flow	Tide	Waves	Description
I	+	−	−	River delta
II	+	−	+	River delta (plus barriers)
III	+	+	−	Tidal river delta
IV	0	+	−	Coastal plain estuary
V	−	+	+	Tidal lagoon
VI	−	+	−	Bay
VII	−	−	+	Coastal lagoon

Note: Plus and minus designations indicate relative impacts; for example, − + + means that river discharge is very small relative to tidal and wave energy.

Source: Alongi, D. M., *Coastal Ecosystem Processes,* CRC Press, Boca Raton, FL, 1998, 187. With permission.

Table 3.9–2   General Estuarine Physical Characteristics

Estuarine Type	Dominant Mixing Force	Width/Depth Ratio	Salinity Gradient	Probable Topographic Categories	Example
A, Highly stratified	River flow	Low	Longitudinal Vertical	Fluvial Deltaic	Columbia River Mississippi River
B, Moderately stratified	River flow Tide, wind	Low Moderate	Longitudinal Vertical Lateral	Fjord Coastal plain Bar built	Puget Sound Chesapeake Bay Albemarle Sound
C, Vertically homogeneous	Tide, wind	High	Longitudinal Lateral	Bar built (Lagoon) Coastal plain	Baffin Bay Delaware Bay
D, Vertically and laterally homogeneous	Tide, wind	Very high	Longitudinal	?	?

Source: Biggs, R. B., et al., *Rev. Aquat. Sci.,* 1, 189, 1989. With permission.

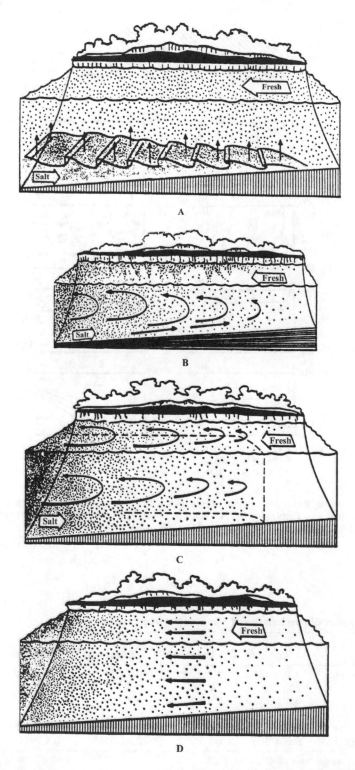

Figure 3.9–1 Classification of estuaries based on circulation. (A) Highly stratified, salt wedge estuary, type A; (B) partially mixed, moderately stratified estuary, type B; (C) completely mixed, vertically homogeneous estuary with lateral salinity gradient, type C; (D) completely mixed, sectionally homogeneous estuary with longitudinal salinity gradient, type D. (From Pritchard, D. W., *Proc. Am. Soc. Civil Eng.,* 81, 1, 1955. With permission.)

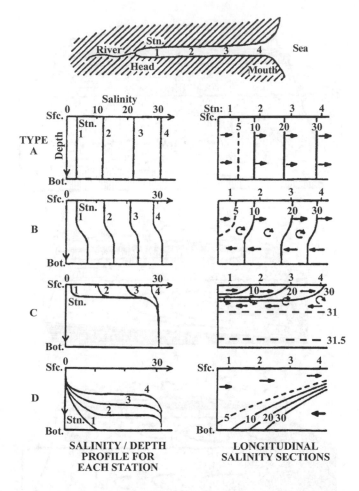

Figure 3.9–2   The salinity depth profiles and sections for types of estuaries (A = vertically mixed, B = slightly stratified, C = highly stratified, and D = salt wedge). (From Millero, F. J., *Chemical Oceanography*, 2nd ed., CRC Press, Boca Raton, FL, 1996, 42. With permission.)

Table 3.9–3   Physical Characteristics of Five Contrasting Estuaries

Characteristic	Amazon	St. Lawrence	Savannah	Scheldt	Tamar
Catchment area, km²	7.1×10^6	1.0×10^6	2.5×10^4	2.1×10^4	9×10^2
Mean river flow, m³ s⁻¹	200,000	14,160	300	100	19
Tidal range,[a] m	4.3/3.3	2.6/1.4	2.7/1.9	~4.0	4.7/2.2
Flushing time, days	[b]	~30	~12	60–90	7–14
Stratification	Salt wedge	Partly mixed	Partly mixed	Well mixed	Partly mixed
Sediment discharge, t a⁻¹	0.5×10^9	5.1×10^6	2.0×10^6	0.1×10^6	0.1×10^5

[a] Spring/neap tidal range near mouth.
[b] Not appropriate for salt wedge estuaries.

Source: Millward, G. E. and Turner, A., in *Trace Elements in Natural Waters,* Salbu, B. and Steinnes, E., Eds., CRC Press, Boca Raton, FL, 1995, 224. With permission.

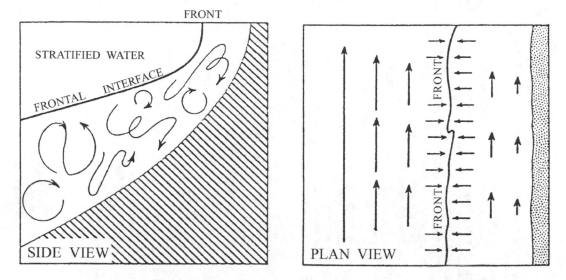

Figure 3.9–3 Schematic diagram of an estuarine frontal zone. The wiggly lines represent random turbulent motions and the straight lines, tidal currents. (From Kjerfve, B., Ed., *Hydrodynamics of Estuaries,* Vol. 1, *Estuarine Physics,* CRC Press, Boca Raton, FL, 1988. With permission.)

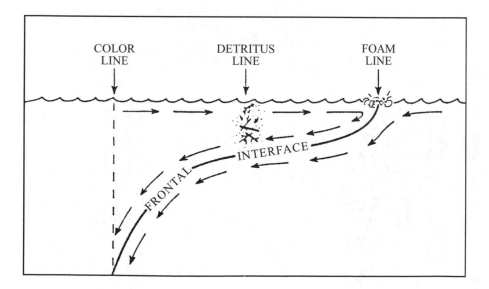

Figure 3.9–4 Schematic diagram of a shallow front. (From Kjerfve, B., Ed., *Hydrodynamics of Estuaries,* Vol. 1, *Estuarine Physics,* CRC Press, Boca Raton, FL, 1988. With permission.)

Table 3.9–4 Characteristics of Estuarine Circulation and Water Quality for Different Environments of the Coastal Biophysical Regions of the United States

Biophysical Region	Smooth Shoreline	Indented Shoreline	Marshy Shoreline	Unrestricted River Entrance	Embayment, Coastal Drainage Only	Embayment, Continuous Upland River Flow	Fjord
North Atlantic	Deep nearshore, oceanic water, longshore currents, some suspended sand and clay	Deep nearshore, oceanic water, erratic tidal currents, eddies, and tidal pools	Strong currents in many small channels through marsh, some turbidity, high oxygen	Highly stratified, some turbidity, high oxygen, temperatures warmer in summer, colder in winter than ocean	Little turbidity, water of oceanic character, strong tidal currents through inlets	Little turbidity, high oxygen, may be stratified, upper layer fresh, with temperatures warmer in summer colder in winter than ocean	
Middle Atlantic	Oceanic water, longshore currents, suspended mud, clay, silt	Generally shallow, suspended mud and sand, oceanic water	Moderate currents in well-defined channels, high dissolved organic material, little turbidity, high oxygen	Moderate stratification, suspended mud and silt, high oxygen, strong currents	Generally shallow, small tides, clear water with lowered salinity, high oxygen	Variable stratification, suspended mud and silt, high oxygen, small amounts of organic material	
Chesapeake	Longshore tidal currents, highly variable salinities, small amounts of organic material	Moderate tidal currents, highly variable salinities, some turbidity	Poorly defined channels, small currents, dissolved organic material, moderate fluctuations of oxygen	Moderate stratification, suspended mud and silt, high oxygen, strong currents	Generally shallow, small tides, clear water with lowered salinity, high oxygen	Variable stratification, suspended mud and silt, high oxygen, small amounts of organic material	
South Atlantic	Primarily tidal and wave-induced currents, oceanic water with mud, clay, and silt	Moderate tidal currents, highly variable salinities, some turbidity	Small currents, high color, high dissolved organics, highly variable oxygen, high temperatures	Strong stratification, high suspended mud and clay, strong currents, dissolved organics, moderate oxygen	Some color, small currents, generally shallow, high dissolved organics, highly fluctuating oxygen	Slight and variable stratification, river water cooler than ocean, slight color, some oxygen fluctuation, moderate to high suspended sediment	

Region						
Caribbean	Clear oceanic water, gentle currents, warm temperature throughout the year	Clear oceanic water, gentle currents, eddies, warmer than ocean	High dissolved organics, high color, suspended mud, very small currents, hot	Slightly turbid, strong currents, river cooler than ocean water	Very small currents, generally shallow, warm, clear oceanic water	Slightly turbid, eddying currents, slight stratification, high oxygen	Stagnant below sill depth, very little oxygen, high salinity, hydrogen sulfide
Gulf of Mexico	Clear, generally warm oceanic water, longshore currents	Very small currents, oceanic water with slight turbidity, warmer than ocean	High dissolved organics, high color, very small currents, slightly to moderately turbid, high temperature	Slightly turbid, strong currents, river cooler than ocean water	Very small currents except in inlet, shallow, warm, slight turbidity from sand and silt, highly fluctuating oxygen	Slight and variable stratification, river water cooler than ocean, some oxygen fluctuation	
Southwest Pacific	Strong wave action, cool oceanic water, some silt and clay turbidity	Moderate suspended solids, erratic currents, high oxygen, cool	High suspended solids, erratic tidal currents, warmer than ocean and rivers	Strong stratification, offshore bar formation, cool, high oxygen	Some suspended silt, erratic currents, cool, high oxygen	Moderate to strong stratification, high suspended silt, strong currents, high oxygen, cool	
Northwest Pacific	Strong wave action, cold oceanic water, some silt and clay turbidity	Moderate suspended solids, erratic currents, high oxygen, cold	High suspended solids, erratic tidal currents, warmer than ocean and rivers	Strong stratification, offshore bar formation, cold, high oxygen	Some suspended silt, erratic currents, cold, high oxygen	Moderate to strong stratification, high suspended silt, strong currents, high oxygen, cold	
Alaska	Very cold oceanic water, usually ice, salinities slightly depressed	Very cold oceanic water, overlain by some freshwater, high oxygen	Very cold water, variable salinities, much fine silt, debris from freezing	Strong currents, high suspended solids, frequently glacial in origin, very cold	Very cold organic water, much ice, surface layer of freshwater, high oxygen	High turbidity with glacial debris, seasonal freeze-ups, strong currents during runoffs	
Pacific Islands	Clear, warm oceanic water, strong wave action	Clear oceanic water, gentle currents, eddies, warmer than ocean	High dissolved organics, color, high suspended mud, very small currents, hot	Slightly turbid, strong currents, river cooler than ocean water	Very small currents, generally shallow, warm, clear oceanic water	Slightly turbid, eddying currents, slight stratification, high oxygen	

Source: U.S. Department of Interior, The National Estuarine Pollution Study, Vol. 4, Federal Water Pollution Control Administration, Washington, D.C., 1969.

Table 3.9–5   Large Rivers of the World Ranked by Length, Mean
               Annual Discharge at Mouth, and Average Suspended
               Load

Length (km)

Nile	Africa	6650
Amazon	South America	6437
Mississippi/Missouri	North America	6020
Yangtze	Asia	5980
Yenisey	Asia	5540
Ob	Asia	5410
Hwang-Ho	Asia	4845
Zaire	Africa	4700
Amur	Asia	4444
Lena	Asia	4400

Mean Annual Discharge at Mouth (1000 m³/s)

Amazon	South America	212.5
Zaire	Africa	39.7
Yangtze	Asia	21.8
Brahmaputra	Asia	19.8
Ganges	Asia	18.7
Yenisey	Asia	17.4
Mississippi/Missouri	North America	17.3
Orinoco	South America	17.0
Lena	Asia	15.5
Paraná	South America	14.9

Average Annual Suspended Load

	Total $t \times 10^6$	Load/Discharge
Ganges	1625.6	86.9
Brahmaputra	812.9	41.0
Yangtze	560.8	25.7
Indus	488.7	87.3
Amazon	406.4	1.9
Mississippi	349.5	20
Nile	123.8	44.2

Source: Kjerfve, B., Ed., *Hydrodynamics of Estuaries,* Vol. 1, *Estuarine Physics,* CRC Press, Boca Raton, FL, 1988. With permission.

Table 3.9–6 Estimates of Water (km³/year) and Sediment (10⁶ metric ton/year) Discharge Rates from Large Tropical and Some Subtropical Rivers

River	Country	Area	Water Discharge	Sediment Discharge
Amazon	Brazil	6.1	5700	1200
Zaire	Zaire	3.8	1292	43
Nile	Egypt	3.0	30	120
Zambesi	Mozambique	1.4	546	48
Niger	Nigeria	1.2	192	40
Tigris–Euphrates	Iraq	1.05	47	55
Orinoco	Venezuela	0.99	1080	150
Ganges	Bangladesh	0.98	1280	520
Indus	Pakistan	0.97	237	250
Mekong	Vietnam	0.79	466	160
Brahmaputra	Bangladesh	0.61	690	540
Irrawaddy	Burma	0.43	428	260
Limpopo	Mozambique	0.41	5	33
Volta	Ghana	0.40	37	18
Godavari	India	0.31	84	170
Senegal	Senegal	0.27	13	2
Krishna	India	0.25	35	64
Magdalena	Colombia	0.24	237	220
Rufiji	Tanzania	0.18	9	17
Chao Phya	Thailand	0.16	30	11
Mahandi	India	0.14	72	60
Hungho	Vietnam	0.12	123	130
Narmada	India	0.089	45	125
Fly	New Guinea	0.076	110	115
Moulouya	Morocco	0.051	2	7
Sebou	Morocco	0.040	5	26
Tana	Kenya	0.032	5	32
Purari	New Guinea	0.031	78	105
Cheliff	Algeria	0.022	2	3
Meddjerdah	Algeria	0.021	2	13
Damodar	India	0.020	10	28
Solo	Indonesia	0.016	8	19
Porong	Indonesia	0.012	6	20
Total estimated discharge in tropics			12,906	4604
Percentage of estimated world total			32%	17%

Note: Rivers are listed by decreasing drainage basin area ($\times 10^6$ km²).

Source: Alongi, D. M., *Coastal Ecosystem Processes,* CRC Press, Boca Raton, FL, 1998, 240. With permission.

Table 3.9–7   Methods of Measuring River Flows, Inflows, and Tributary Flows

Method	Principle of Operation Used to Define Flow Rate, Q	Measurement Uncertainty	Flow Range or Criteria (m^3/s)	Min. Depth or Diameter (m)
	Open-Channel Flows			
Velocity area	$Q = AU$. Velocity measured with Price or pygmy meters, pitot tubes, acoustic meter, electromagnetic meters, propeller meters, deflection vanes, or floats; float and vane measurements and point measurements require correction to obtain the average velocity	2.2% using Price meters, 10–25% using floats	0.085–2.43 m/s using Price meters	0.18
Moving boat	Adaptation of the velocity-area method; current meter moved across the stream at a fixed depth, and velocity is recorded for each subsection used to compute mean flow as $$Q = C_{mb} \Sigma (A_i V_v \sin\alpha)$$	Probably on the order of 3–7%	Larger flows accessible by boat	1.5–1.8
Slope-area	Flow related to estimates or measurements of slope, S, or head loss for pipes, channel geometry, and some form of the friction coefficient such as the Manning n (friction factor for pipes): $$Q = (A/n)R^{2/3}S^{1/2}$$	10–20% for pipe lengths of 61–305 m; 20–50% for less accurate estimates of R, A, S, and n	Used for high Q and preliminary estimates	Depth and width should be as uniform as possible over channel reaches
Tracers	Salt, dyes, and radioisotopes are used to measure flow from the dilution of a tracer: $$Q = q_t (C_t - C_D)/(C_D - C_o)$$ where C_t, C_D, and C_o = tracer concentration being injected and measured upstream and downstream of the injection or from the tracer travel time between two stations, Δt: $Q = (L/\Delta t)A$ where length between stations, L, and area, A, must be measured	2–3% in well-defined small channels in treatment plants	Useful for shallow depth where current meters cannot be used; not useful for large wide flows with limited lateral mixing	Limited by lateral mixing
Width contraction	Flow related to head loss through a bridge opening or channel constriction: $$Q = C_c A_d [2g (\Delta h + \alpha(U_1^2/2g) + h_f)]^{1/2}$$ where α = velocity head coefficient, $\Delta h = h_1 - h_3$, and h_f = head loss through the constriction	Order of 10% or more	High Q or highly constricted with measurable $\Delta h \simeq 0.01$–0.08	Small to moderate where constriction is likely

Flumes	Usually forces the flow into critical or supercritical flow in the throat by width contractions, free fall, or steepening of the flume bottom so that discharge can be empirically determined as a function of depth and flume characteristics; the approach channel must be straight and free of waves, eddies, and surging to provide a uniform velocity distribution across the throat; has less head loss and solids deposition but is more expensive to construct than weirs; most common device used in sewage plants	3–6% or better with calibration	Higher Q than similar size weir	—
Parshall	Scaled from standard dimensions of 1 in. to 50 ft (0.025–15 m) throat widths $$Q = C_p W h_A^N$$	3–6% (7% when submerged)	0.0001–93.9; see range for each size	0.03–0.09; see each size
Trapezoidal supercritical flow	Supercritical depth is maintained in three scaled sizes (throat widths = 1, 3, and 8 ft or 0.3, 0.91, and 2.44 m) $$Q = (A_o)^{1/2}(T_o)^{1/2}g^{1/2}$$	~5% (Similar Venturi flumes are error prone; 1–10% if head loss is small)	0.02–56 (see range for each size; accurate at lower flows; up to 736 m³/s for 37-m wide flumes)	0.03
Cutthroat	More sensitive to lower flow; less sensitive to submergence (<80%); less siltation; fits existing channels and ditches Flat bottom passes solids better than a Parshall flume and is better adapted for existing channels; accurate at higher degrees of submergence; for free flow: $$Q = KW^{1.025} h_A^n \quad \text{(English units)}$$	—	—	0.061
H, HS, and HL types	Simply constructed and installed agricultural runoff flumes that are reasonably accurate for a wide range of flows; flumes can be easily attached to the end of a pipe; approach flow must be subcritical and uniform; the flume slope must be ≤1%	1% for 30% submergence; 3% for 50% submergence	Max. flows HS:0.0023–0.023; H:0.010–0.89; and HL:0.586–3.31 m³/s	0.061
San Dimas	Developed for sediment-laden flows, having a 3% bottom slope to create supercritical flow; not rated for submerged flow $$Q = 6.35 W^{1.04} h^{1.5-n''} \quad \text{(English units)}$$	Not sensitive or accurate at low Q	0.0042–5.7	0.061
Weirs	Control of flow produces relationship between discharge and head; inexpensive and accurate but requires at least 0.15 m (0.5 ft) head loss and must be maintained to clean weir plate and remove solids	Accurate under proper conditions	Most useful for smaller Q where 0.15 m (0.5 ft) head loss can be afforded	0.061 + P

(continued)

Table 3.9-7   Methods of Measuring River Flows, Inflows, and Tributary Flows (continued)

Method	Principle of Operation Used to Define Flow Rate, Q	Measurement Uncertainty	Flow Range or Criteria (m³/s)	Min. Depth or Diameter (m)
Thin plate	Sharp crest exists when $h/L_c \geq 15$ and air is free to circulate under nappe (jet of water over weir); L_c = weir thickness	2–3% when calibrated and maintained; 5–10% when silted in	0.007–1.68; best when $0.061\ m \leq h \leq h/2b$	$h \geq 0.061$ m for nappe to spring free of weir
Rectangular	$Q = (2/3)C_1(2g)^{1/2}bh^{1/5}$ where b = horizontal length of crest; neglects correction of $-0.001–0.004$ m	3–4%	Best when $Q = 0.0081$ to 23 m³/s	0.03 m or 15 L_c and $P \geq 0.1$ m
V-Notch	More sensitive to low flows: $Q = (8/15)C_2(2g)^{1/2}[\tan(\theta/2)]h_c^{5/2}$ Tailwater level lower than the vertex of the notch; common angles are 22.5, 45, 60, 90, and 120°	3–6%	$Q = 0.0003–0.39$ but usually limited to 0.028	$0.049 \leq h \leq 0.061$
	$Q = C_3(2/3)(2g)^{1/2}[b + (4/5)h\tan(\theta/2)]h^{3/2}$	—	—	—
Trapezoidal Cipolletti	Trapezoidal weir having same equation for discharge as rectangular weir and is designed with side slopes of 1 horizontal to 4 vertical lengths to avoid the need for contraction adjustments to C_1	Less accurate than rectangular or V-notch	$Q = 0.0085$ to 272	$h = 0.061 - 0.61 - 0.5b$
Compound weir	V-notch weir that changes side slope to a wider angle to handle higher flows at higher h; the Q vs. h relationship is ambiguous as the sides change slope; for a 1 ft (0.3 m) deep 90° cut into rectangular notches of widths 2, 4, and 6 ft (0.61, 1.2, and 1.8 m): $Q = 3.9h_{vn}^{1.72} - 1.5 + 3.3bh_m^{1.5}$	Has not been fully investigated	—	—
Other types	Sutro or proportional and approximate linear relationship between Q and h; other types include approximate exponential and poebing weirs	Has not been fully investigated	—	—

Type	Equation/Description	Comments	Conditions	
Broad crested	$Q = C_B b (h + h_v)^{3/2}$; $h_t = h + h_v$ (Weir is broad enough to consider to the fluid above at hydrostatic pressure; notches for low flow modify equation for Q; typical streamwise cross sections are rectangular, triangular, trapezoidal, and rounded $$C_B = C_b C_v K_n g^{1/2}$$ where K_n is a constant depending on notch shape.)	More uncertainty for lower h; difficult to predict without in-place calibration	$0.08 \leq h_t/L \leq 0.50$ (No effect of submergence until $h_t/h > 0.65$ to 0.85 depending on weir profile)	0.061
Rectangular notch	Flat crest: $Q = C_b C_v (2/3)(2g/3)^{1/2} bh^{2/3}$ Trenton type with 1:1 slopes upstream and downstream of the flat crest: $Q = 3.5bh^{1.65}$ (English units) The Crump triangular weir with a 50% slope on the upstream face and 20% slope on the downstream face: $$Q = 1.96 b h_t^{3/2}$$	—	$0.08 < h_t/L \leq 0.33$. 0.33. $\leq h_t/L \leq 1.5$ to 1.8 is a short-crested weir that has not been fully investigated; $1.5 \leq h_t/L \leq 3$ is usually unstable. $h_t/L \geq 3$ similar to sharp-crested weirs	0.061
Triangular notch	$Q = C_b C_v (16/25)(2g/5)^{1/2}[\tan(\theta/2)]h^{5/2}$ where $Q = 2.5bh^{1.65}$ for flat vee	—	Usually ≥ 28 (1000 cfs)	0.061
Truncated triangular notch	$Q = C_b C_v T (2/3)(2g/3)^{1/2}(h - h_b)^{3/2}$	—	$h_t \geq 1.25 h_b$	—
Parabolic notch	$Q = C_b C_v (3/g/4)^{1/2} h_t^2$	—	—	—
Trapezoidal notch	$Q = C_b (Th_c + mh_c^2)[2g(h_t - h_c)]^{1/2}$ where T = top width of flow over weir Most frequently used gauging control in the U.S.; requires calibration for the full range of flows; above $h = 0.21$ m: $$Q = 8.5(h - 0.2)^{3.3}$$ (English units)	—	—	—
Columbus type control	Consists of an upward convex notch below $h = 0.21$ m and slopes 1:5 for 0.611 m on either side of the notch and 1:10 to the remaining distance to the banks	—	—	—

(continued)

Table 3.9–7   Methods of Measuring River Flows, Inflows, and Tributary Flows (continued)

Method	Principle of Operation Used to Define Flow Rate, Q	Measurement Uncertainty	Flow Range or Criteria (m³/s)	Min. Depth or Diameter (m)
Submerged orifice	Flow is related to the head difference across the orifice: $$Q = 0.61\,(1 + 0.15 r_o) A_d [2g(\Delta h)]^{1/2}$$ for a contracted or suppressed rectangular orifice	Properly operated meter gates: 2%, but as much as 18% noted	0.003–1.43	0.38
Acoustic meter	The difference in travel time, of an acoustic pulse across the flow and back (t_{AB} and t_{BA}) over a path length, L_A, diagonal to the flow is related to average velocity along the path: $$Q = A C_A L_A\,(1/t_{AB} - 1/t_{BA})/(2\,\cos\alpha)$$ where α = angle of path with flow	1–7% for parallel flow, $\leq$14% in poorly developed flow	Used in large rivers; no practical upper limit for Q	Depends on W, density gradients, and allowable error
Electromagnetic coils	Experimental method involving coils buried in the bed or suspended in the flow. See pipe method $$Q = C_c A E_m / W H$$	—	Generally used in shallow flows	No practical limits
Rating navigation locks, dam crests, and gates	Q determined by calibration in model studies or by in-place measurements; generally locks behave like submerged orifices: $$Q = C \Delta h$$ and dams act like weirs: $$Q = C_b\, b h_t^{3/2}$$	5–30%	—	—
Superelevation in bends	The difference in water surface elevation across a bend is related to Q	—	Requires high vel. for measurable Δh	No constraint

Partially Full Pipe, Culvert, and Sewer Line Flow

Method	Description	Accuracy	Applicability	Limits
Velocity-area	See pipe and open-channel methods: $Q = UA$; U related to point velocities in at least four ways: (1) $U = 0.9u_{max}$ (2) $U = u_{0.4}$ (3) $U = (u_{0.2} + u_{0.4} + u_{0.8})/3$ (4) U = average of three vertical profiles at the quarter points across the pipe plus four measurements at both walls 1/8 the distance across the flow	All methods are expected to be accurate to at least 10%	Should be applicable to all flows; use method (1) or (2) if the flow is rapidly varying	Use method (1) if depth ≤ 5 cm (2 in.)
Tracers	See open-channel method	5%	Requires good mixing	—
Volumetric	Flow from a pipe or channel is diverted to a bucket, tank, sump, or pond of known dimensions and the increase in weight or volume is timed to measure the average flow rate; orifice buckets may also be useful	1% or better depending on how well dimensions are known	Useful for smaller flows, but unique conditions may make application to any flow possible	No practical minimum for flows encountered
Slope-area	The Manning equation is applied to uniform pipe reaches; see channel method and culvert method; the effects of manholes (changing channel shape, slope, or direction) must be avoided or the increased friction losses taken into account; design slopes should be verified	10–20% or 20–50% for less precise estimates of S and h	Requires a 200–1000 d uniform approach flow	
Culvert equations	Q computed from continuity and energy equations	2–8%	Not readily available	0.15
Palmer-Bowlus flume	Constriction inserted into existing partially full pipes where relationship exists between Q and h: $$Q = h^{3/2}\left[\frac{w^{(26+h)}}{8(L_w + h)}\right]^{3} g^{1/2}$$ Does not have fully standardized design; more accurate but less resolution than Parshall flume; fits pipe $d = 0.10$ to 1.1 m and larger; portable; for full pipe flow: $$Q = 8.335(\Delta h/d)^{0.512}d^{5/2} \text{ (English units)}$$	3–5% depending on care in construction; 10% for low Q	0.0010 to 0.51 for $d =$ 0.15 to 0.76 m; submergence ≤ 85%; requires uniform approach for 25 times diameter with slope ≤ 2% and 0.4 ≤ depth ≤ 0.9 d	0.61h$_{min}$ = 0.061

(continued)

Table 3.9–7 Methods of Measuring River Flows, Inflows, and Tributary Flows (continued)

Method	Principle of Operation Used to Define Flow Rate, Q	Measurement Uncertainty	Flow Range or Criteria (m³/s)	Min. Depth or Diameter (m)
USGS sewer flowmeter	A U-shaped fiberglass constriction of standard dimensions is inserted in a pipe to form a Venturi flume where $Q = f(\Delta h, h)$. Full: $$Q = 5.74d^{5/2} (\Delta h/d)^{0.52}$$ Transition between channel and pipe flow: $$Q = d^{5/2} [2.6 \pm (|0.590 - h_2/d|/0.164)^{1/2}]$$ Channel flow: (1) Supercritical: $$Q = 5.58d^{5/2} (h_1/d)^{1.58}$$ (2) Subcritical—culvert slope <0.020: (a) $h_1/d \geq 0.30$: $Q = 2.85d^{5/2} (h_1/d - 0.191)^{1.76}$ (b) $h_1/d < 0.30$: $Q = 1.15d^{5/2} (h_1/d - 0.177)^{1.38}$ (3) Subcritical—culvert slope $\geq$0.020: $$Q = 1.70ad^{5/2} (h_1/d)^{2.71} \quad a = 2.15 + (9.49)(10)^{11} (S - 0.008)^{6.76}$$	Depends on field calibration	Not available	0.61 and 1.52
Wenzel flume	(all equations in English units) Symmetrical or asymmetrical constrictions contoured to the side of the pipe result in unique Q vs. h for partially full and full flows; open bottom does not trap solids $$Q = C_w [2gA_2\Delta h/[1 - (A_2/A_1)^2]]^{1/2}$$ $$C_w = (\{1 + K_e + [A_2^2 A_1^2/(A_1^2 - A_2^2)] [(f_1 L_1/4R_1 A_1^2 + f_3 L_3/4R_3 A_3^2, 3/2)]\})^{-1/2}$$	At least 5%; 25% at low flow	As much as 30:1 variation in flows; not valid for steep $S \geq 0.020$	0.20 (8 in.)
Trajectory methods	The extent of a jet leaving the end of a pipe, and pipe flow geometry are related to discharge	Generally not accurate enough for U.S. EPA NPDES inspections	—	—

Method	Description	Accuracy	Range	Size (m)
Purdue	For level pipe flowing full or partially full where the water surface elevation below the top of the pipe at the outlet (if $a/d < 0.8$), or at 6 12, or 18 in. (0.15, 0.30, or 0.46 m), the distance from the top of the pipe down to the water surface has been experimentally related to flow	Q will be underestimated if the pipe slopes downward	$Q = 0.00032$ to 0.10	$d = 0.051$ to 0.15
California pipe	For a level pipe of length $6d$ or more, flowing only partially full, discharging freely into air, and having a negligible approach, U: $$Q = 8.69(1 - a/d)^{1.88} d^{2.48}$$ (a and d measured in ft)	—	Confirmed for $d = 0.076$ to 0.25 m and $a/d \geq 0.5$, but probably useful for larger d	0.076
Vertical pipes	In gal/min, with d and height of jet, H', in. and $H' > 1.4d$: $$Q = 5.01 d^{1.99}(H')^{0.53}$$ For $H' < 0.37d$: $$Q = 6.17 d^{1.25}(H')^{1.35}$$		$d = 0.051$ to 0.30 and $H' = 0.013$ to 1.52	0.051
Parabolic discharge nozzle	Attached to the end of a pipe with a free outfall; Q is related to h^2 by laboratory calibration; nozzle length is $\simeq 4d$	5%, 1% at Q_{max}	0.0000–0.85	0.15, 0.20, 0.25, 0.3, 0.41, 0.51, 0.61, 0.76, 0.91
Kennison discharge nozzle	Attached to the end of a pipe with a free outfall; Q is related to h by laboratory calibration; nozzle length is $\simeq 2d$	5%, 1% at Q_{max}	≤ 0.85	Same as parabolic nozzle
Acoustic meter	See pipe method			
Electromagnetic meter	See open-channel and pipe methods; generally used to measure one or more point velocities for the velocity-area method	2% under optimum conditions	$U = -1.67$ to 6.1 m/s (-5.5–20 fps)	Insertable in most pipes
Pressurized Pipe Flow				
Differential head meters	Flow constrictions produce pressure losses related to flow, Q	Better results when calibrated	Min. $\Delta h = 25.4$ mm (1 in.) for water and 51 mm (2 in.) for sewage	—
Venturi throat	$$Q = \frac{C_d A_d (2 g \Delta h)^{1/2}}{(1 - r^4)^{1/2}}$$	0.5–3% depending on calibration	Well tested for diameters up to 0.81 (32 in.)	0.051
Nozzle	Same as Venturi throat	1–1.5%	5:1 range for Q unless calibrated	0.051
Orifice	Same as Venturi throat	0.5–4.4%		0.038

(continued)

Table 3.9-7 Methods of Measuring River Flows, Inflows, and Tributary Flows (continued)

Method	Principle of Operation Used to Define Flow Rate, Q	Measurement Uncertainty	Flow Range or Criteria (m³/s)	Min. Depth or Diameter (m)
Centrifugal meter	Flow rate related to pressure difference between the inside and outside of a pipe bend: $$Q = C_d A_d (2g\Delta h)^{1/2}$$ For uncalibrated 90° bends with moderate or higher Reynolds no. $C_d = [r_b/(2d\ t)]^{1/2}$	<10% if calibrated; predicted C_d error $\simeq$ 10%	May be installed in any pipe for which Δh in measurable (1 in H_2O)	0.038
Pipe friction meter	Flow rate related to friction loss in fully developed pipe flow: $$Q = C_d (\Delta h)^{1/2}$$	Unknown	Requires high flow rates to yield measurable Δh	Any pipe
Velocity area	where $C_d = A_d d^{1/2}/fL$ and f is the friction factor $Q = A_d U$, A_d determined from inside diameter; U determined from point velocity measurements by pitot tubes, small propeller meters, electromagnetic probes, etc.; velocities best measured at 0.026, 0.082, 0.146, 0.226, 0.342, 0.6658, 0.774, 0.854, 0.918, and 0.974d and then averaged for round pipes; relationships between U and a single point velocity can be derived; see channel method as well	0.5% for pitot tube meas. 5% for some propeller meters	$d \leq 1.5\,m$ and $U_{max} = 1.5$ to 6 m/s	Depends on pitot tube or velocimeter diameter; $\simeq$ 0.10 is the limit for a 3.8 cm diameter pitot tube to make precise measurements
Tracers	Salt or flourescent dyes are used to determine flow rate from the dilution of a known amount of tracer mass at a downstream location where the tracer is fully mixed or from the time of travel between two locations at which the tracer is well mixed; the latter method requires that the distance between station and flow cross-sectional area be measured; see channel method	3%	Requires good mixing	Should be applicable to any size pipe
Gibson	Pressure rise following valve closure is related to U: $$Q = (Ag/L)(\text{area ABCA})$$	Believed to be very accurate	Requires at least 25d approach length	Unknown
Mechanical meters Displacement	— Flow displaces piston or disk; oscillations are counted as a function time; calibration required	1% new, more with age	0.001–0.28 Depends on manufacturer's specifications	Varies Generally used for smaller pipes

Method	Principle	Accuracy	Range	Size limit
Inferential	Flow rotates a turbine whose rate of rotation is related to flow rate by calibration; requires an approach length of at least 20–30d to develop the velocity profile; vanes control spiraling from bends	Needs constant checks: 2–5%	0.15–5.2 m/s (0.5 to 17 ft/s) but inaccurate below 0.3–0.46 m/s (1–1.5 ft/s)	0.61–2.3 (2–7.5 ft)
Variable area	Rotometer consists of vertical tapered tube with a metal float that rises as flow increases	1% at Q_{max}	Small flows	Depends on manufacturer
Acoustic meter	See channel-flow method for time of travel: $$Q = AC_AL_A(1/t_{AB} - 1/t_{BA})/2 \cos\alpha$$ Reflected Doppler meters detect a frequency shift from a signal reflected back from a point where sound beams are crossed in the flow; the shift is related to the point velocity which must be related to the average velocity by calibration to calculate the flow rate; time of flight acoustic meters determine a turbulent signature of the flow at a station and correlates that with a downstream turbulent signature to determine time of travel over a known distance through a known pipe volume: $$Q = \pi d^2L/4\Delta t$$	2–5% and higher at bends	Time of travel and Doppler meters require approach lengths of at least 10–20d of straight pipe or must be calibrated to determine C_A; time of flight meters require a limited approach of a few diameters	No practical limit
Electromagnetic meter	all types can be attached to outside of pipes Inducted voltage, E_m, caused by water flow perpendicular to magnetic field, proportional to velocity, U: $$Q = AE_m/HdC_e$$ Meter must be in contact with fluid; regular checking of the electrodes is necessary to avoid fouling; magnetic coils can be embedded in new pipe or in a ring insert, or a probe can be inserted to measure point velocities or profiles to be used in the velocity-area method	Larger of 1–2% or 0.002 m/s	Depends on manufacturer, at least up to 16.7 (590 ft³/s)	At least 0.10–1.52 m diameter (4–60 in.)
Calibrated pumps, turbines, valves, and gates	Empirical relationships between Q and power, or valve or gate opening for appropriate heads are defined with laboratory models or by calibration in place	Depends on in-place calibration; laboratory calibration ≈10% or more for accurate in-place calibration	—	—

(continued)

Table 3.9–7   Methods of Measuring River Flows, Inflows, and Tributary Flows (continued)

DEFINITION OF SYMBOLS USED IN THE TABLE

A = cross-sectional area of channels and pipes; subscript i refers to the area of individual subsections; subscript c refers to area of a width constriction or area of a throat of a flume

U = cross-sectional average velocity

C_{mb} = coefficient relating point velocity measurements to the mean velocity in a cross section; for large rivers $C_{mb} = 0.87$ to 0.92 and 0.9 is typically used

n = Manning's roughness coefficient describing friction loss in channels

R = hydraulic radius (A/wetted perimeter); subscripts refer to section numbers such as 1 and 3 for the Wenzel flume

C_c = discharge coefficient for a width constriction

g = gravitational acceleration

C_p = discharge coefficient for Parshall flumes

N = exponent in the Parshall flume equation

T_c = top width of the water surface in the trapezoidal flume at the point where critical flow occurs

K = free flow discharge coefficient for cutthroat flume; the standard 1.5-, 3-, 4.5-, and 9-ft (0.46-, 0.91-, 1.37-, and 2.74-m)-long flumes varying from 6.1 to 3.5 (English units)

n' = free flow discharge exponent for a cutthroat flume; the standard 1.5-, 3-, 4.5-, and 9-ft (0.46-, 0.91-, 1.37, and 2.74-m)-long flumes varying from 2.15 to 1.56 (English units)

n'' = discharge exponent for the San Dimas flume = $0.179W^{0.32}$ (English units)

h = water surface height of the approaching flow above the weir crest or bottom of the notch; for precise estimates 0.001 m is added to h to account for the effects of surface tension and viscosity; h is measured at least three to four times h_{max} upstream of the weir face for sharp-crested rectangular and V-notch weirs; two to three times h for Cipolletti weirs; and two to three times ht for broad-crested weir; generally the minimum distance upstream should be $4h_{max}$

P = height of lip of weir above stream bottom

C_1 = discharge coefficient for rectangular thin-plate weir; $f(h/P, b/B, E)$ for free falling, free discharge that springs free of the plate without clinging (thickness of the weir is ≤ 1/15 of the depth of flow over the weir)

C_s = $[1 − (h/D)m]^{0.385}$ where $m = 1.44$ for a fully contracted weir and 1.50 for a suppressed weir

C_2 = discharge coefficient for a V-notch weir ≃0.58 for a fully contracted weir flow where $h/P ≤ 0.4$, $h/B ≤ 0.2$, $0.049 < h ≤ 0.381$ m, $P ≥ 0.46$ m, $B ≥ 0.91$ m; values cannot be predicted for partially contracted weir flow without calibration

h_e = $h + K_h$; K_h varies from 0.0025 to 0.0085 depending on the angle of the notch of the weir

C_3 = $0.63C_v$; C_v = velocity head correction factor

C_b = discharge coefficient for broad-crested weir; for flat-crested weirs, C_b depends on the rounding or slope of the upstream face, h/L, and $h_i/(h_i + P)$; for a horizontal upstream face with a sharp corner, the basic discharge coefficient is $C_b = 0.848$ if $0.08 \leq h/L \leq 0.33$ and $h'(h + P) \leq 0.35$; for $h/L > 0.33$ and $h'(h + P) > 0.35$; the value of C_b is corrected for $h/L \leq 1.5$ if $h'(h + P) \leq 0.35$; a final correction is made for the approach velocity head varying from 1.00 to 1.2 as a function of $C_b A_b/A_a$, where A_b is the area of flow over the control structure and A_a is the area of flow in the approach section where h is measured; for rounding of the upstream corner: $C_b = [1 - 2x(L - r_b)/B][1 - x(L - r_b)/h]^{3/2}$ where x is a parameter that accounts for boundary-layer effects and r_b is the radius of rounding of the upstream corner; for field installation of well-finished concrete, $x \simeq 0.005$; where clean water flows over precise cut blocks, $x \simeq 0.003$. C_b varies between 0.85 and 1.00 for triangular weirs having $h/L = 0.08$ to 0.7

h_v = $\alpha U^2/2g$ for the approach flow

C_v = discharge coefficient correcting for the effect of an approach velocity for weirs and orifices

r_o = ratio of suppressed portion of the perimeter of a submerged orifice to the entire perimeter

C_A = discharge coefficient for acoustic gauging stations relating cross-sectional average velocity to average velocity along the acoustic path; $C_A = f$ (water depth and tidal condition)

C_e = discharge coefficient for electromagnetic meters determined in the laboratory for portable meters attached to the outside of a pipe; for open-channel coils, $C_c = [1 + (W\sigma_b/2h\sigma_w)]/\beta$ where σ_b and σ_w conductivities of the bed and river water, respectively, and β is a correction factor for the end effects of the magnetic field and for incomplete coverage of the cross-sectional area by the field in case of limited coils

H = magnetic field intensity in Tesla

E_m = electromotive force in volts generated by the movement of a conduction fluid such as slightly contaminated water through a magnetic field

A_d = area of Venturi throat, orifice, or flow nozzle opening or pipe diameter for bend meters

d = pipe diameter

Δ_h = pressure drop or hydraulic head difference measured across a Venturi throat or flume, orifice, flow nozzle, bend meter, or along a straight length of pipe; also the difference in water surface elevations across a submerged orifice or width contraction

h = water depth measurement in partially full pipes where subscripts 1 and 2 refer to locations upstream of and locations in the throat of a flume

S = slope of a pipe or channel

C_w = discharge coefficient for the Wenzel flume that accounts for energy loss and velocity head corrections; for full pipe flow, C_w is given above; K_c is an entrance loss coefficient of approximately 0.2; f is the Darcy-Weisbach friction factor; L_j is the length; valid when backwater conditions do not exist; for partially full flow, C_w seems to be $\simeq 1$

C_d = discharge coefficient for Venturi throat, orifice, and flow nozzle or bend meter; function of Reynolds number, Re, for Venturi throat when pressure taps are properly located in the straight pipe section of the throat and 0.5 and 0.25 pipe diameters upstream of the meter; $C_d = 0.96$ to 0.99 as a function of Re $\leq 10^6$ for flow nozzle when pressure taps are located one pipe diameter upstream of the beginning of the nozzle and before the end of the nozzle, for precise measurements, C_d is determined by calibration for bend meters and for straight pipe sections between pressure taps; the best results are obtained in prediction C_d when the flowmeter is well maintained and correctly installed; indirect flow measurements are used to check the meter rating and to determine if debris lodges in the meter; proper installation places the meter downstream of at least 10 diameters of straightened flow and avoids solids deposition; vanes are used to straighten flows if necessary

r = radio of throat diameter, d_2, to pipe diameter, d

r_b = radius of the center line of a pipe bend

f = friction factor for pipes

L = length of straight pipe between pressure tapes for a pipe friction meter

Source: McCutcheon, S. C., *Water Quality Modeling,* Vol. 1, *Transport and Surface Exchange in Rivers,* CRC Press, Boca Raton, FL, 1989, 277. With permission.

3.10 OCEAN CIRCULATION

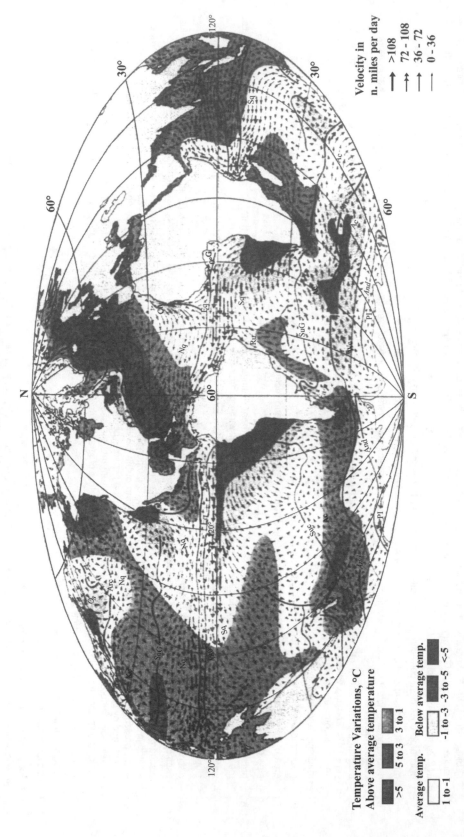

Figure 3.10–1 Surface circulation and temperature variations in the oceans. (From Millero, F. J., *Chemical Oceanography*, 2nd ed., CRC Press, Boca Raton, FL, 1996, 25. With permission.)

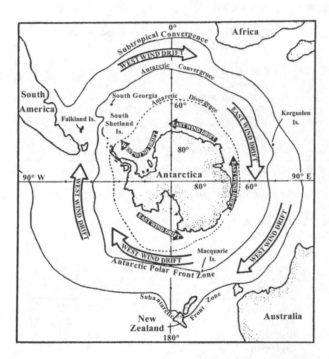

Figure 3.10–2　Circulation of the Southern Ocean, which is bounded by the Antarctic continent and the seafloor south of the Subtropical Convergence (Subantarctic Front Zone). The predominant clockwise trajectory of the West Wind Drift (Antarctic Circumpolar Current) extends south of the Antarctic Convergence (Antarctic Polar Zone). South of the West Wind Drift is the counterclockwise East Wind Drift and the Antarctic Divergence between them. (From Berkman, P. A., *Rev. Aquat. Sci.,* 6, 295, 1992. With permission.)

Table 3.10–1 Gyres and Surface Currents of the Major Oceans

Pacific Ocean	Atlantic Ocean	Indian Ocean	Antarctic Circulation
North Pacific Gyre	North Atlantic Gyre	Indian Ocean Gyre	Antarctic Circupolar Gyre
North Pacific Current	North Atlantic Current	South Equatorial Current	West Wind Drift
California Current[a]	Canary Current[a]	Agulhas Current[b]	Other Major Currents
North Equatorial Current	North Equatorial Current	West Wind Drift	East Wind Drift
Kuroshio (Japan) Current[b]	Gulf Stream[b]	West Australian Current[a]	
South Pacific Gyre	South Atlantic Gyre	Other Major Currents	
South Equatorial Current	South Equatorial Current	Equatorial Countercurrent	
East Australian Current[b]	Brazil Current[b]	North Equatorial Current	
West Wind Drift	West Wind Drift	Leeuwin Current	
Peru (Humboldt) Current[a]	Benguela Current[a]		
Other Major Currents	Other Major Currents		
Equatorial Countercurrent	Equatorial Countercurrent		
Alaskan Current	Florida Current		
Oyashio Current	East Greenland Current		
	Labrador Current		
	Falkland Current		

[a] Denotes as eastern boundary current of a gyre, which is relatively *slow, wide,* and *shallow* (and is also a *cold water* current).

[b] Denotes a western boundary current of a gyre, which is relatively *fast, narrow,* and *deep* (and is also a *warm water* current).

Source: Thurman, H. V. and Trujillo, A. P., *Essentials of Oceanography,* 6th ed., Prentice-Hall, Upper Saddle River, NJ, 1999, 211. With permission.

Table 3.10–2   Flows in Major Oceanic Currents

Current	Flow[a] ($m^3 s^{-1}$)
Gulf Stream	100×10^6
Agulhas	60×10^6
Kuroshio	65×10^6
California	12×10^6
West Australian	10×10^6
East Australian	20×10^6
Antarctic circumpolar	120×10^6
Peru	18×10^6
Benguela	15×10^6
Equatorial undercurrent	40×10^6
Brazil	10×10^6
Equatorial countercurrent	25×10^6
Pacific North Equatorial	30×10^6
Pacific South Equatorial	10×10^6
Flinders	15×10^6

[a] Both hydrologists and oceanographers often use nonstandard terms when referring to flows. Hydrologists call 1 m^3 s^1 a *cumec* and oceanographers call 10^6 m^3 s^1 a *Sverdrup.*

Source: Beer, T., *Environmental Oceanography,* 2nd ed., CRC Press, Boca Raton, FL, 1997, 145. With permission.

Figure 3.10–3 Map of Brazil showing the Guiana, South Equatorial, and Brazil Currents. (From Schaeffer-Novelli, Y. et al., *Estuaries,* 13, 204, 1990. With permission.)

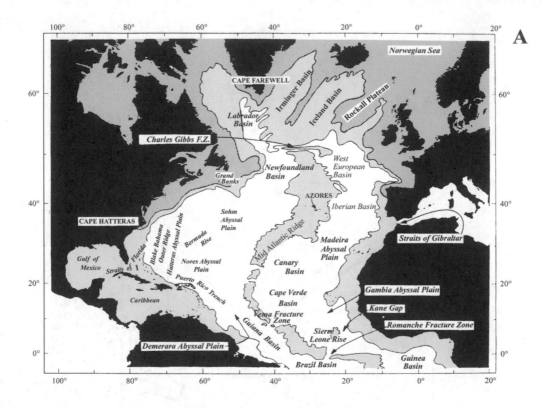

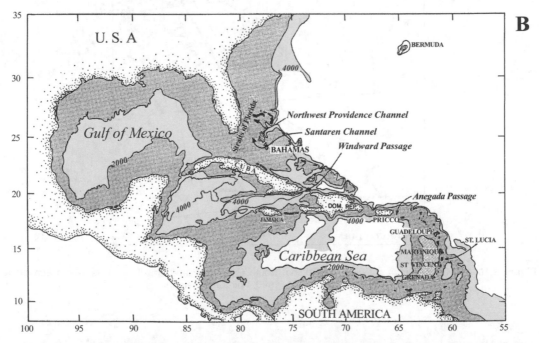

Figure 3.10–4   Maps of the North Atlantic: (A) and (B) summary topographic features and significant place names; (C) a schematic two-layer transport streamline field adapted from Stommel (1957, 1965), (U) denotes upper layer and (L) denotes lower. (From Schmitz, W. J., Jr. and McCartney, M. S., *Rev. Geophys.*, 31, 29, 1993. With permission.)

C

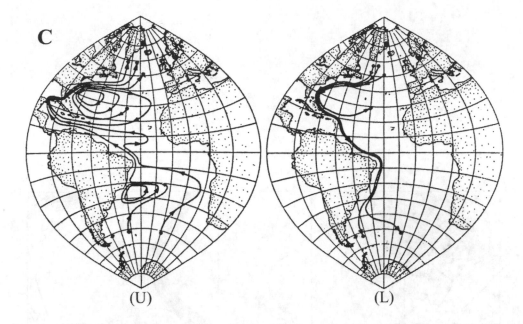

(U)   (L)

Figure 3.10–4   (continued)

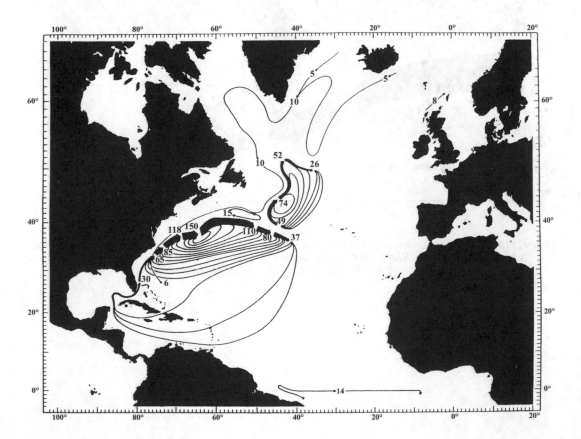

Figure 3.10–5   The North Atlantic circulation total transport, in Sverdrups (10^6 m³/s). (From Schmitz, W. J., Jr. and McCartney, M. S., *Rev. Geophys.,* 31, 29, 1993. With permission.)

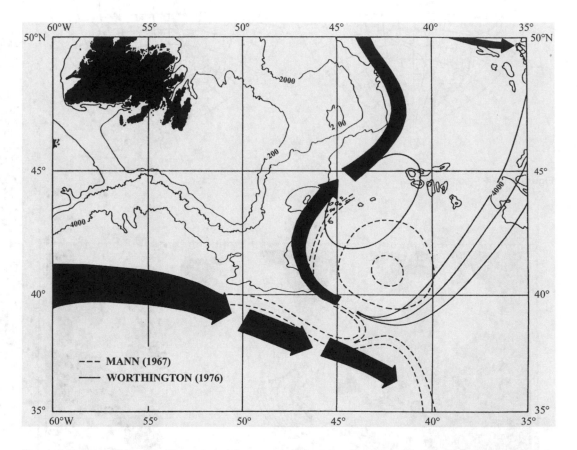

Figure 3.10–6 A schematic illustration of flow patterns near the Grand Banks of Newfoundland: solid lines adapted from Worthington (1976), dashed lines from Mann (1967). (From Schmitz, W. J., Jr. and McCartney, M. S., *Rev. Geophys.,* 31, 29, 1993. With permission.)

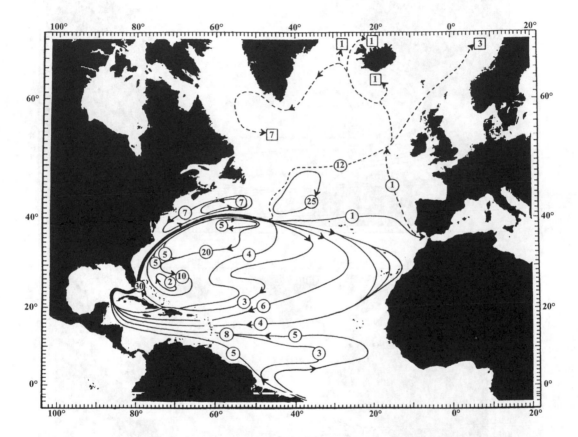

Figure 3.10–7 Schematic illustration of transport (values enclosed in circles are in Sverdrups) for the North Atlantic at temperatures above 7°C (nominal). Numbers in squares (in Sverdrups) denote sinking. (From Schmitz, W. J., Jr. and McCartney, M. S., *Rev. Geophys.,* 31, 29, 1993. With permission.)

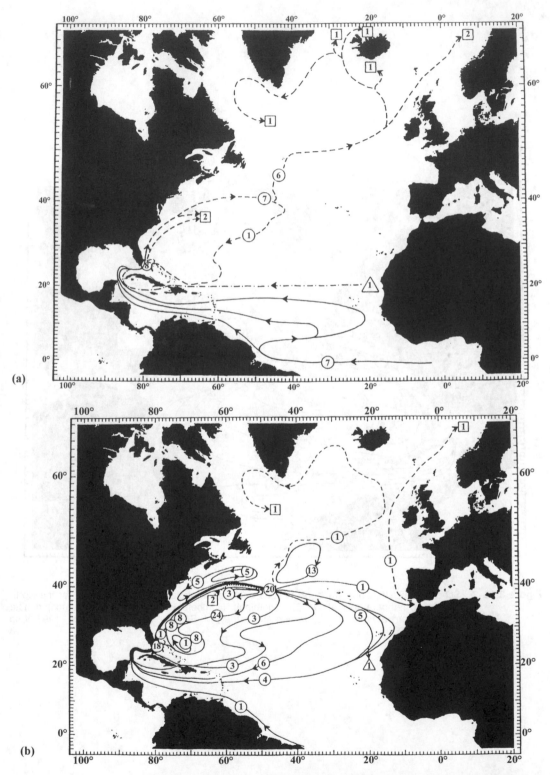

Figure 3.10–8   The composite circulation in Figure 3.10–7 is split into three layers: (a) near surface, (b) 12–24°C, (c) 7–12°C, (d) 7–12°C, including the circulation of Mediterranean water and subpolar mode water. Transports in circles are in Sverdrups. Squares represent sinking; triangles denote upwelling. (From Schmitz, W. J., Jr. and McCartney, M. S., *Rev. Geophys.,* 31, 29, 1993. With permission.)

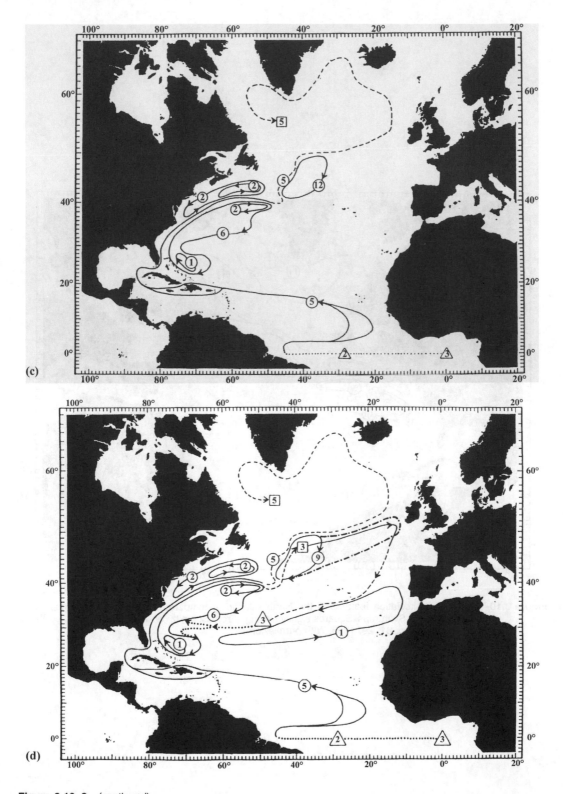

Figure 3.10–8   (continued)

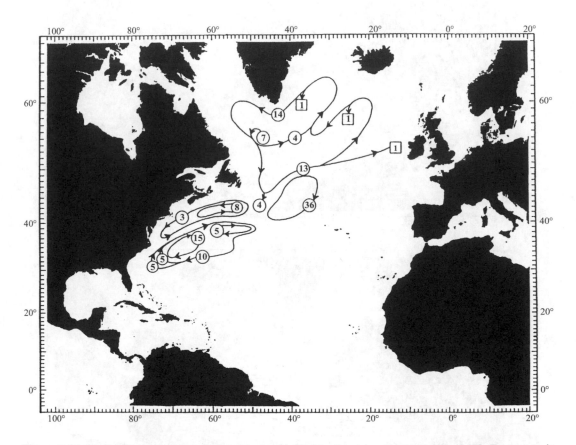

Figure 3.10–9 Selected circulation features in the North Atlantic at temperatures below 4°C. Transports in circles are in Sverdrups. Squares signify sinking. (From Schmitz, W. J., Jr. and McCartney, W. S., *Rev. Geophys.,* 31, 29, 1993. With permission.)

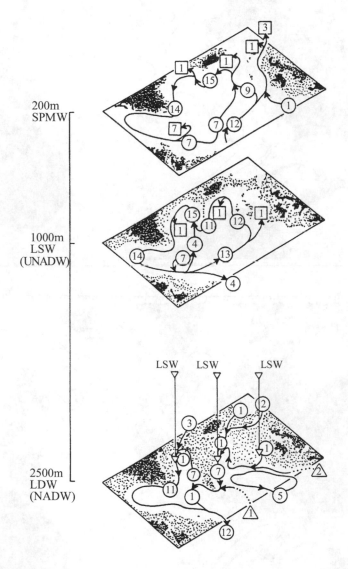

Figure 3.10–10 Circulation patterns for the northern North Atlantic. Transports in circles are in Sverdrups. Squares represent sinking; triangles denote upwelling. (From Schmitz, W. J., Jr. and McCartney, M. S., *Rev. Geophys.*, 31, 29, 1993. With permission.)

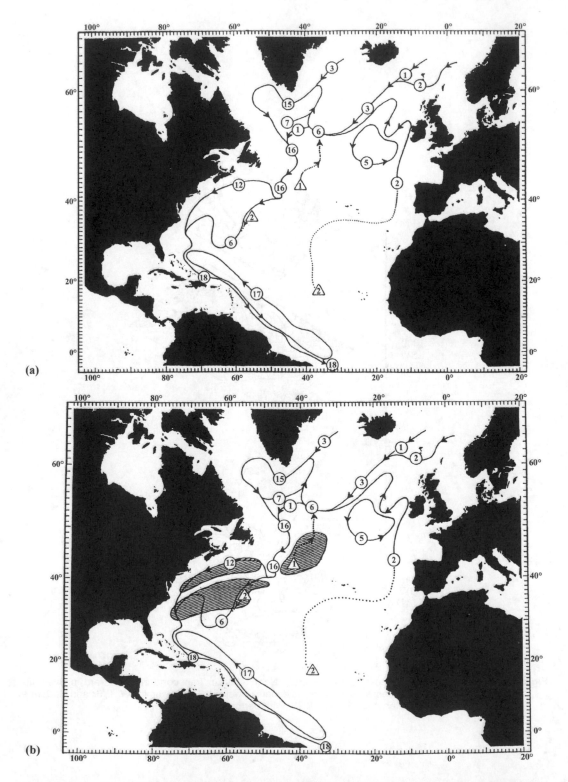

Figure 3.10–11 Circulation patterns for thermohaline forced deep water (1.8–4°C) in the North Atlantic. (a) Based on McCartney (1992). (b) Deep gyres in Figure 3.10–9 added as hatched areas. Transports in circles are in Sverdrups. Squares denote sinking; triangles represent upwelling. (From Schmitz, W. J., Jr. and McCartney, M. S., *Rev. Geophys.,* 31, 29, 1993. With permission.)

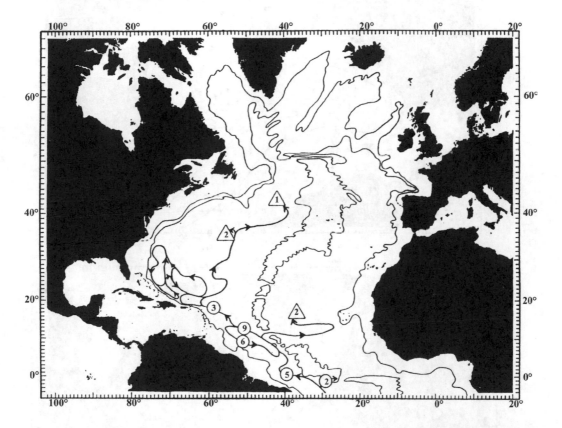

Figure 3.10–12 Circulation patterns for bottom water (1.3–1.8°C). Transports in circles are in Sverdrups (10^6 m³/s). Triangles denote upwelling. (From Schmitz, W. J., Jr., and McCartney, M. S., *Rev. Geophys.*, 31, 29, 1993. With permission.)

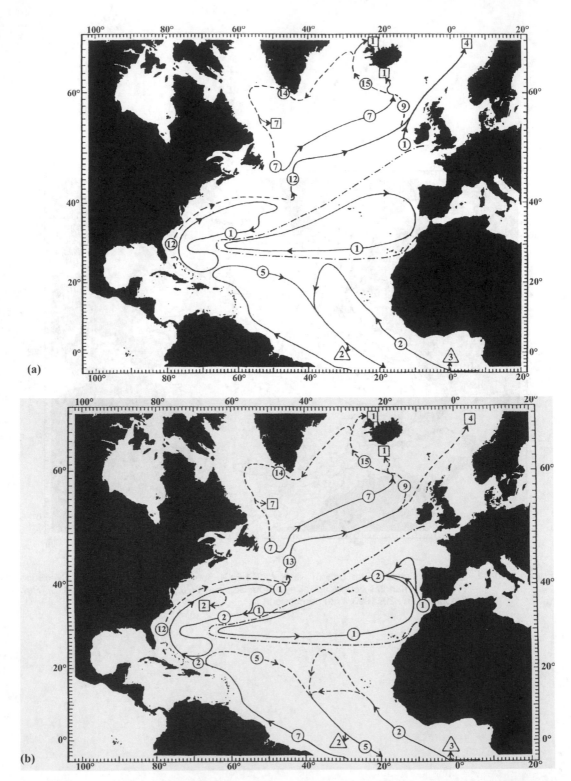

Figure 3.10–13 Circulation patterns for intermediate water (4–7°C). Transports in circles are in Sverdrups (10⁶ m³/s). (a) The simplest version. (b) With additional and modified features. Squares denote sinking; triangles signify upwelling. (From Schmitz, W. J., Jr. and McCartney, M. S., *Rev. Geophys.*, 31, 29, 1993. With permission.)

COLD CORE RING

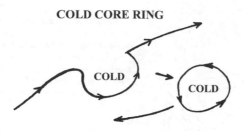

WARM CORE RING

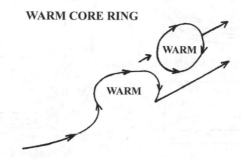

Figure 3.10–14   The formation of Gulf Stream rings. (From Millero, F. J., *Chemical Oceanography,* 2nd ed., CRC Press, Boca Raton, FL, 1996, 27. With permission.)

Structure of Rings

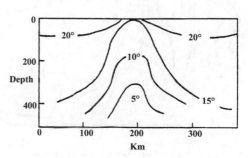

Figure 3.10–15   The temperature and structure of a Gulf Stream ring. (From Millero, F. J., *Chemical Oceanography,* 2nd ed., CRC Press, Boca Raton, FL, 1996, 27. With permission.)

Higher	NO_3, PO_4, SiO_2, Chlorophyll
Lower	T, S

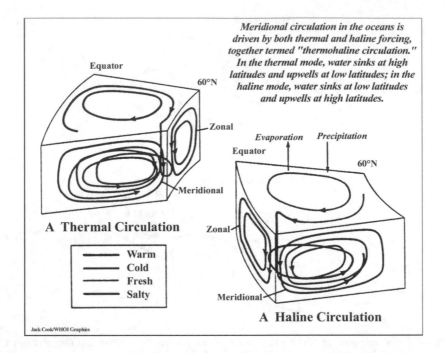

Figure 3.10–16   Meridional circulation in the oceans is driven by both thermal and haline forcing, together termed *thermohaline circulation.* In the thermal mode, water sinks at high latitudes and upwells at low latitudes; in the haline mode, water sinks at low latitudes and upwells at high latitudes. (From Huang, R. X., Freshwater driving forces: a new look at an old theory. *Oceanus,* 36, 37, 1992. Courtesy of *Oceanus* Magazine/Woods Hole Oceanographic Institution.)

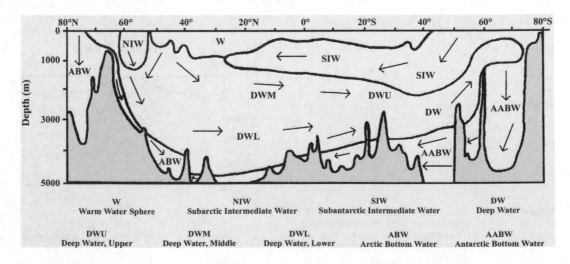

Figure 3.10–17   Characteristic water masses in the Atlantic Ocean. (From Millero, F. J., *Chemical Oceanography,* 2nd ed., CRC Press, Boca Raton, FL, 1996, 52. With permission.)

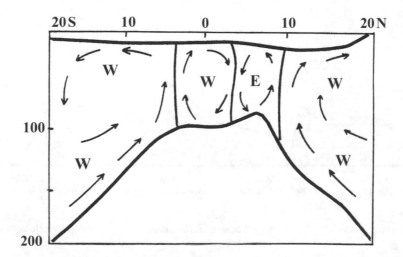

Figure 3.10–18 Vertical circulation in the equatorial Atlantic. (From Millero, F. J., *Chemical Oceanography*, 2nd ed., CRC Press, Boca Raton, FL, 1996, 31. With permission.)

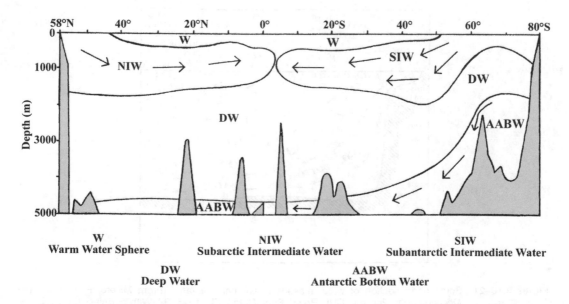

Figure 3.10–19 Characteristic water masses in the Pacific Ocean. (From Millero, F. J., *Chemical Oceanography*, 2nd ed., CRC Press, Boca Raton, FL, 1996, 53. With permission.)

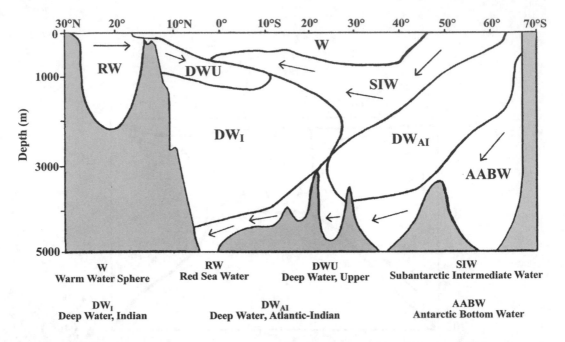

Figure 3.10–20 Characteristic water masses in the Indian Ocean. (From Millero, F. J., *Chemical Oceanography,* 2nd ed., CRC Press, Boca Raton, FL, 1996, 54. With permission.)

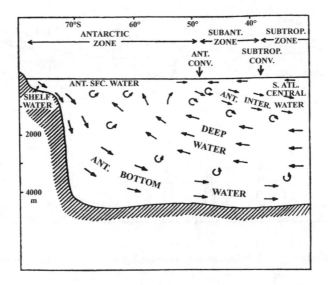

Figure 3.10–21 South–north section of water masses in southern oceans. (From Millero, F. J., *Chemical Oceanography,* 2nd ed., CRC Press, Boca Raton, FL, 1996, 33. With permission.)

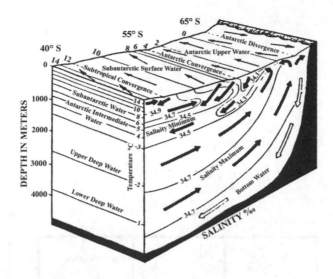

Figure 3.10–22 Vertical profile of the major water masses and thermohaline characteristics in the Southern Ocean including the flow of Antarctic bottom water (AABW), North Atlantic deep water (NADW), and Antarctic intermediate water (AAIW), which are predominant features of the oceans of the world. (From Berkman, P. A., *Rev. Aquat. Sci.,* 6, 295, 1992. With permission.)

Figure 3.10–23 Classification of Pacific Ocean water masses. SAW, Subarctic water; WNPCW, Western North Pacific Central water; ENPCW, Eastern North Pacific Central water; PEW, Pacific Equatorial water; WSPCW, Western South Pacific Central water; ESPCW, Eastern South Pacific Central water; SAAW, Subantarctic water; AACPW, Antarctic Circumpolar water. (From Neshyba, S., *Oceanography: Perspectives on a Fluid Earth,* John Wiley & Sons, New York, 1987. With permission.)

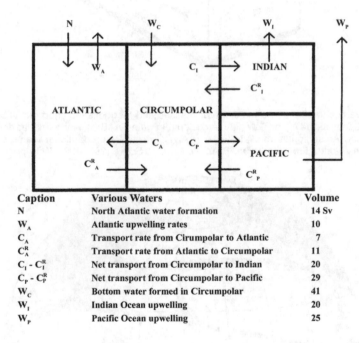

Caption	Various Waters	Volume
N	North Atlantic water formation	14 Sv
W_A	Atlantic upwelling rates	10
C_A	Transport rate from Cirumpolar to Atlantic	7
C_A^R	Transport rate from Atlantic to Circumpolar	11
$C_I - C_I^R$	Net transport from Circumpolar to Indian	20
$C_P - C_P^R$	Net transport from Circumpolar to Pacific	29
W_C	Bottom water formed in Circumpolar	41
W_I	Indian Ocean upwelling	20
W_P	Pacific Ocean upwelling	25

Figure 3.10–24 Box model for the deep-water circulation of the world ocean waters. (From Millero, F. J., *Chemical Oceanography,* 2nd ed., CRC Press, Boca Raton, FL, 1996, 56. With permission.)

CHAPTER **4**

Marine Geology

I. PLATE TECTONICS THEORY

Global plate tectonics, the unifying theory that revolutionized the field of marine geology during the 1960s, provides the conceptual framework for understanding the origin and evolution of the ocean basins and continental margins. According to this compelling theory, the earth's outer shell consists of a mosaic of broad, rigid lithospheric plates up to 250 km thick under the continents and about 100 km thick under the oceans. These plates overlie a partially molten, plastic asthenosphere that facilitates their movement. Although the lithospheric plates behave as rigid bodies, the boundaries of adjacent plates are dynamic, interacting in three ways: (1) by diverging or spreading, as at mid-ocean ridges; (2) by converging, as at deep-sea trenches; and (3) by slipping sideways past each other, as at transform faults. The lithospheric plates are in constant motion in response to tensional forces at divergent boundaries, compressional forces at convergent boundaries, and shear forces at transform faults. As a result, tectonic activity concentrates along the plate boundary zones, where the frequency of earthquake and volcanic activity far exceeds that in intraplate areas.

Lithosphere is generated at centers of divergence (i.e., mid-ocean ridges), which represent constructive plate margins. In contrast, it is resorbed or destroyed at convergent plate boundaries, specifically subduction zones marked by deep-sea trenches. At these destructive plate margins, one lithospheric plate overrides another, with the subducted (overridden) plate sinking into the mantle. This subduction process ensures that the earth maintains constancy in size.

There are three types of lithospheric plate convergence: (1) ocean-to-ocean plates (e.g., Mariana Arc or Tonga Arc); (2) continent-to-ocean plates (e.g., Peru–Chile Trench and adjacent Andean Cordillera); and (3) continent-to-continent plates (e.g., Indian and China plate collisions and resultant Himalayan uplift).[1] Because of the relatively low density of continental crust, the convergence of continental plates does not culminate in lithospheric subduction but suturing (fusing) of the landmasses.

Aside from constructive and destructive plate margins where lithosphere is created and destroyed, respectively, conservative plate margins exist at transform faults. Here, lithospheric plates normally do not undergo spreading or subduction and hence are neither generated nor consumed. They move parallel to each other to accommodate their relative motions. However, leaky transform faults do occur which exhibit a spreading component. Transform faults and associated fracture-zone extensions (i.e, inactive arms of transform faults) form topographic irregularities across the seafloor manifested by linear clefts and bordering scarps which offset the axis of mid-ocean ridges. The clefts may be more than 1 km deeper than the adjacent ridge crests, but they become shallower at increasing distances from the ridge axis. The rate of movement of opposing plates across transform faults is proportional to the spreading rates observed at neighboring ridge axis sites.

The plate tectonics paradigm, which embraces continental drift and seafloor spreading, explains the major topographic features of the ocean basins. For example, the 75,000-km-long, globally encircling mid-ocean ridge system arises from volcanic eruptions at diverging plate boundaries in the deep sea. The volcanic ridges reach highest elevations along the ridge crests, and as the newly formed lithosphere cools, subsides, and accelerates away from the ridge axis, the volcanic terrain becomes less rugged. At greater distances from the ridge axis, abyssal hills thinly veneered with marine sediment become the dominant topographic feature and are gradually supplanted landward by broad abyssal plains covered by thick deposits of sand, silt, and clay (turbidites) primarily transported by turbidity currents from the outer continental shelf and continental slope. Thick turbidites also underlie the continental rise, a gently sloping topographic feature that extends from the abyssal plain to the continental slope.

The deepest areas of the ocean basins are trenches located near continental margins at sites of lithospheric plate subduction. These long and narrow troughs, which range in depth from ~7 to 11 km at their axes, result from the downward thrusting of lithospheric plates. They are bordered by volcanic island arcs or a continental margin magmatic belt. A volcanic island arc (e.g., Aleutians) occurs when the overriding lithospheric plate is oceanic. A continental-margin magmatic belt (e.g., Andes mountains) occurs when the overriding plate is a continent.[2] Extensive continental shelves develop along passive plate margins, and grade into relatively steeply inclined continental slopes. Steep-walled submarine canyons cut by turbidity currents deeply incise outer shelf and slope sediments at periodic points along the continental margins, commonly offshore of large rivers.

II. SEAFLOOR TOPOGRAPHIC FEATURES

A. Mid-Ocean Ridges

The central and most striking morphological feature of the ocean floor is the mid-ocean ridge (MOR), a continuous volcanic mountain range extending through all the ocean basins. It is an immense physiographic feature covering about one third of the ocean floor (>1000 km) or ~25% of the earth's surface.[3] The MOR exhibits a broadly convex profile with highest elevations at the axis and gradually declining slopes on either flank. The ridge crests lie at depths of ~2.5 to 3.0 km worldwide; they rise ~1 to 3 km above the adjoining ocean floor. Axial areas are either sediment-free or only thinly veneered with sediment, and consequently the local relief is very rugged. Greater sediment accumulation away from the ridge axis tends to smooth the topography on the flanks.

Most MORs occur in the central part of the ocean basins. An exception is the MOR in the North Pacific. Here, the ridge system occupies the eastern part of the ocean basin as the East Pacific Rise, which trends northeastward into the Gulf of California, and the Galapagos Rift, which trends eastward from the East Pacific Rise near the equator. Farther northward, the Gorda Ridge, Juan de Fuca Ridge, and Explorer Ridge all lie within 500 km of the North American coastline.

Topographic profiles of MORs in the Atlantic Ocean differ markedly from those in the Pacific Ocean because of lower magma supply and seafloor spreading rates. The Mid-Atlantic Ridge is characterized by a well-developed axial rift valley (~25 to 30 km wide and ~1 to 2 km deep) flanked by very rugged terrain. In contrast, the East Pacific Rise lacks a rift valley, but is typified by an axial summit caldera or axial summit graben along its crest and broader, less rugged flanks. It varies from ~2 to 4 km in height and 2 to 4 km in width along its length. Gradients away from the ridge crest for the East Pacific Rise are ~1 in 500 compared with ~1 in 100 for the Mid-Atlantic Ridge.[4] MOR profiles in the Indian Ocean are similar to those in the Atlantic Ocean, having a well-defined rift valley set within very rugged relief.

Differences in the aforementioned topographic profiles are ascribable to variations in seafloor spreading rates. The rate of formation of new ocean crust varies considerably at superfast-, fast-, intermediate-, and slow-spreading centers. At superfast-spreading centers, for example, seafloor genesis exceeds 15 cm/year. Somewhat lower rates of spreading (9 to 15 cm/year) occur at fast-spreading centers. The East Pacific Rise at 13°N latitude (11 to 12 cm/year) and 20°S latitude (16 to 18 cm/year) provides examples.[5] Intermediate rates of seafloor spreading range from 5 to 9 cm/year,[6] for example, the Galapagos Rift has spreading rates of 6 to 7 cm/year.[7] Low rates of seafloor spreading (~2 to 4 cm/year) have been documented along the Mid-Atlantic Ridge.[8-10]

New ocean crust forms at MORs through an interplay of magmatic construction, hydrothermal convection, and tectonic extension.[11-14] These spreading centers are sites of active basaltic volcanism, shallow-focus earthquakes, and high rates of heat flow. Ocean crust formation requires the occurrence, at least intermittently, of a molten reservoir, or shallow crustal chamber, beneath the MOR axis.[15,16] The magma chamber is envisioned as the focal point in the ocean crust where igneous differentiation occurs, and fractionation processes construct a vertically heterogeneous crustal column.[16-23] As such, it directly controls the igneous structure and composition of the ocean crust.[17] In addition to being a source of lava erupted onto the seafloor, magma chambers yield the heat that drives hydrothermal circulation along the ridge axis.

Magma chambers vary in size, shape, and longevity. In some areas they appear to be only ~2 to 4 km wide and less than 1 km thick, whereas in other regions they seem to be even smaller, essentially thin (tens to hundreds of meters high), narrow (<1 to 2 km wide) melt lenses overlying a zone of crystal mush in the mid-crust.[16] Seismic tomography of fast-spreading centers (e.g., East Pacific Rise) presents an image of a small, steady-state melt storage region surrounded by a wider reservoir of very hot rock.[22-25] At slow-spreading centers (e.g., Mid-Atlantic Ridge), in contrast, magma supply rates are much lower, and no evidence exists for the presence of a steady-state axial magma chamber in the crust or upper mantle.[26-28] Here, melt replenishment and storage probably take place episodically.

Through adiabatic decompression associated with lithospheric rifting, partial melting of peridotite at depths of ~20 to 80 km below the seafloor, and buoyant diapirism of the molten asthenosphere, mantle-derived melt accumulates and undergoes magmatic differentiation in magma chambers prior to its emplacement in the crust and eruption onto the seafloor. Gabbros form in the chambers through slow cooling and crystal fractionation of the melt. Lateral and vertical injection of dikes from the molten reservoir into existing crust generates sheeted diabase dikes. Volcanic eruptions spill pillow and sheet flow across the seafloor, which is subsequently covered by sediments. A vertical profile of newly formed ocean crust, therefore, consists of three principal layers: (1) layer 1, the sedimentary layer, which increases in thickness away from the ridge axis; (2) layer 2, the basement, comprising interspersed basalt pillows and sheet flows grading downward into a complex of vertical dikes, all of which may be heavily fractured; and (3) layer 3, the oceanic layer, composed of gabbros commonly altered by metamorphism. The Mohorovicic discontinuity marks the base of the oceanic layer.

The thickness of the sedimentary layer typically varies from 0 m at the spreading axis to more than 1 km in areas far removed from the axis. It depends on the source of the sediments, biological productivity of overlying waters, and the age of the crust. The basement layer averages ~1.6 km in thickness, and the oceanic layer ~4 to 5 km in thickness. The original basalt topography, therefore, becomes buried under a thick apron of sediments as the lithosphere ages and gradually moves away from the ridge axis. As the lithosphere accelerates away from the ridge axis, it cools, increases in density, and sinks at a rate of ~2 km over the first 35 million years.[3]

The fundamental unit of crustal accretion of the global ridge system, because of buoyant diapirism, is the ridge segment. Evidence exists that discrete magma chambers are centered beneath individual ridge segments. The spacing, migration, and longevity of ridge segments along the axis control the architecture of oceanic crust generated at MOR systems.[17] Two-dimensional geophysical profiling, high-resolution mapping (e.g., SEABEAM), and petrological observations reveal a hierarchy of ridge segmentation patterns at a scale finer than that predicted by considerations of gravitational instability, which can subdivide the global ridge system into discrete accretionary units separated by various types of ridge axis discontinuities (e.g., transform faults, overlapping spreading centers, oblique shear zones, intervolcano gaps, and deviations from axial linearity).[29-31] Langmuir et al.[32] advocated that axial ridge discontinuities, regardless of their scale, represent boundaries between segments with separate magma sources and, consequently, are basic divisions of the magmatic, crustal-generation process. In effect, these discontinuities partition diapir-induced segments into units that behave independently of one another. Thus, they are deemed to be much more than just surface crustal features along diverging spreading boundaries between lithospheric plates.

The largest (first-order) discontinuities, transform faults, offset opposing ridge segments by more than 30 km. They typically partition the MOR at intervals of ~300 to 900 km along fast-spreading centers, but only ~200 km or less along slow-spreading centers. Their longevity is ~10^7 years.

Quiescent periods of seafloor spreading are punctuated by episodes of renewed rifting and divergence of lithospheric plates at MORs which commence a new cycle of mantle upwelling, peridotite melting, diapir formation, and melt intrusion and extrusion. Older crust derived from the previous cycle of mantle upwelling accelerates away from the MOR axis, ultimately to be covered by a layer of deep-sea sediments. This process is much more rapid along the East Pacific Rise than the Mid-Atlantic and Mid-Indian Ridges.

While magmatic/volcanic processes are directly responsible for the emplacement and accretion of hot rock at divergent plate boundaries, tectonic processes transport the ocean crust away from the site of its genesis. Lithospheric extension of the crust together with thermal contractive forces creates faults and fissures that serve as conduits for the deep penetration of seawater that cools the dike complex and underlying magma chamber. Convective hydrothermal circulation at MORs culminates in the release of heated fluids or hot springs on the seafloor through diffuse, white smoker, and black smoker vents. Hydrothermal convection not only plays an important role in heat transfer at oceanic ridge crests but also in mass transfer, manifested in the metal-rich sulfide deposits concentrated in stockworks and mounds along upwelling zones and on the oceanic crust at and adjacent to discharge sites. In addition to cooling oceanic crust and generating potential ore deposits, hydrothermal vents support lush and exotic animal communities and appear to be a chief factor modulating global ocean chemistry. It is clear, therefore, that processes operating at MORs can influence conditions throughout the world's oceans.

B. Deep Ocean Floor

The physiographic region between the MORs and continental margins is the ocean basin floor. It consists of several subdivisions: (1) abyssal hills; (2) abyssal plains; (3) seamounts; (4) aseismic ridges; and (5) deep-sea trenches. This region comprises nearly one half of the total oceanic area.

1. Abyssal Hills

The topography of the seafloor becomes less rugged and lower moving away from the MORs as the lithosphere cools, subsides, and is buried by sediment. Between the MOR flanks and the level abyssal plains, groups of domes or elongated volcanic hills with slopes (~1° to 15°) thinly veneered by fine sediment rise no more than 1 km above the adjoining seafloor. They range from ~100 m to 100 km in width, and ~5 to 50 km in length.[1,3] Their bases are usually covered by sediment. This varied hilly topography, with lower relief than MOR terrain, comprises the abyssal hill province. The abyssal hills are the most abundant geomorphic structures on earth, characterizing >30% of the ocean floor.[33]

Although common in all the major ocean basins, abyssal hills attain greatest prominence in the Pacific Ocean, where they cover a substantial portion of the seafloor. Macdonald et al.[33] ascertained that Pacific abyssal hills are created on the flanks of the fast-spreading East Pacific Rise as horsts and grabens that lengthen with time. They do not form along the spreading axis because no significant relief is produced here either by volcanic processes or faulting.

2. Abyssal Plains

Farther removed from the MOR province, continued lithospheric subsidence and pelagic and turbidity current deposition result in the burial of the abyssal hills and the creation of the flat-lying abyssal plains. Turbidites, in particular, blanket the original topographic irregularities, generating an extremely flat seafloor with gradients of <0.05°. Sediment thickness here ranges from ~100 m to >1 km.[1,3] Broad aprons of sediment on abyssal plains merge landward with thick deposits on continental rises. An abrupt increase in the slope of the ocean floor marks the foot of the rise region.

3. Seamounts

Ocean floor volcanoes are termed seamounts. These mountainous peaks often exceed 1 km in height, and some (i.e., oceanic islands) rise above sea level. Small seamounts (<2 to 3 km high) outnumber large seamounts (>3 km high) due, in part, to the general subsidence of lithosphere away from spreading centers and local subsidence associated with isostatic adjustment to the seamount load. Seamounts are numerous throughout the ocean basins of the world, occurring as isolated edifices, flat-topped "guyots," clusters, or chains. The Pacific Ocean alone may have as many as 1 million seamounts. Most seamounts form on or near MORs, although some originate in the middle of lithospheric plates.[34] The inner valley floor of the northern Mid-Atlantic Ridge is characterized by well-developed axial volcanic ridges and volcanic cones that are principal sites of seamount construction. For example, more than 450 near-circular volcanic cones have been mapped on the inner valley floor between the Kane and Atlantis fracture zones (24 to 30°N).[35–38] As newly formed lithosphere accelerates away from the ridge axis, seamounts generated near MORs are transported away from the spreading centers.

Some seamounts form in intraplate areas far removed from divergent plate boundaries, indicating that thermal anomalies exist in the surrounding crust. While numerous seamounts are randomly distributed in intraplate areas, others occur in linear chains that are younger than the underlying oceanic crust. One of the best examples is the Hawaiian-Emperor Chain. Volcanoes comprising this island chain progressively increase in age to the northwest in the direction of lithosphere plate movement. Along the Hawaiian Chain, volcanic islands increase in age by 43 Ma over a distance of 3400 km. Each island appears to have originated from hot-spot volcanism coupled to mantle plume development. According to this concept, the volcanic island chain has been generated by the movement of the Pacific lithospheric plate over a stationary hot spot or mantle plume. This process produces an active volcano over the hot spot and a chain of progressively older extinct volcanoes in the direction of seafloor spreading. Other notable examples of volcanic island-chains in the Pacific basin include the Austral Seamount Chain, Marshall-Ellice Island Chain, and Line Islands–Tuamoto Chain. However, evidence supporting the hot-spot model for some island chains, such as the Line Islands–Tuamoto Chain, is less compelling because of anomalous ages recorded on many of the seamounts.

4. Aseismic Ridges

Some of the most conspicuous and perplexing topographic features on the deep ocean floor are aseismic ridges that have no association with earthquake activity. These structural highs, which in some cases extend linearly for thousands of kilometers across the seafloor, reach heights of 2 to 3 km above the adjacent ocean bottom. A number of aseismic ridges occur as isolated features, whereas others are contiguous with MORs or continental landmasses.

Aseismic ridges may represent: (1) isolated continental fragments or microcontinents; (2) uplifted oceanic crust; and (3) linear volcanic features. Among the most prominent ridges are linear and plateau-like constructs found in the Atlantic and Indian Oceans. Some ridges of continental origin, separated during rifting or seafloor spreading, include the Jan Mayen Ridge and Rockall

Plateau in the North Atlantic Ocean, the Agulhas Plateau in the South Atlantic Ocean, and the Mascarene Plateau in the western Indian Ocean. The Rio Grande Rise and Walvis Ridge in the South Atlantic and the Ninetyeast Ridge in the Indian Ocean provide examples of linear aseismic volcanic features. The latter linear construct stretches for more than 4000 km in the eastern Indian basin. This impressive feature appears to have formed near an active spreading center and has subsided with age. The Rio Grande Rise and Walvis Ridge, in turn, constitute twin aseismic ridges on opposing sides of the ocean purportedly produced during the early opening phases of the South Atlantic and subsequently separated by seafloor spreading.[1] Similarly, the Ceará Rise and Sierra Leone Rise located at opposite sides of the Atlantic Ocean at ~7°N are a pair of aseismic ridges generated by volcanic activity.

5. Deep-Sea Trenches

Oceanic trenches originate along seismically active marginal zones at sites of lithospheric subduction. These narrow, steep-sided troughs, which range in width from 30 to 100 km and reach depths of 10 to 11 km, are associated with active volcanic and earthquake activity. They exhibit great continuity; for example, the Peru–Chile Trench, Java Trench, and Aleutian Trench are ~5900, 4500, and 3700 km long, respectively. The deepest trenches occur in the western Pacific in sediment-starved areas off island arcs. Examples are the Mariana Trench (~11 km), Tonga Trench (~10.8 km), Philippine Trench (~10 km), Kermadec Trench (~10 km), and Japan Trench (~9.7 km). Trench slopes generally range from 8° to 15°, but steeper slopes up to 45° have been documented.[3]

The trench walls tend to be highly faulted. The outer trench wall, for example, is subject to normal faulting as a result of the downthrusting oceanic plate that extends into the asthenosphere. Thrust-faulted sediments scraped off the descending oceanic plate often accumulate on the inner trench wall.[4]

Faulting produces steep-sided steps along the trench walls. The outer ridge of the trench, which rises ~200 m to 1000 m above the adjacent ocean floor, separates the trench from the ocean floor. The trench–island arc province only covers ~1.2% of the earth's surface. However, the trenches encircling the Pacific basin are the sites of most earthquakes (>80%) on earth.[1]

As noted previously, deep-sea trenches occur at subduction-type margins where (1) two oceanic plates converge and one plate is subducted beneath the other (i.e., island–arc subduction-zone types) or (2) an oceanic plate is subducted beneath a continent (i.e., continental subduction-zone types). The effect of these processes is the development of a volcanic island arc and often a back-arc basin landward of the trench on the overriding plate as in case 1, and the formation of a volcanic coastal mountain range (e.g., Andes) as in case 2. These destructive plate margins are typified by intense seismicity and volcanism.

According to Hawkins et al. (p. 177),[39] "The arc–trench system encompasses the region which begins at the outer swell of a trench and includes the trench slopes, the fore-arc region, the active arc and, where present, the inactive or remnant volcanic arc and the back-arc basin separating the two volcanic arcs." While island arc–trench systems are generally considered to be zones of convergence, many also exhibit zones of extension of the upper mantle and crust, as evidenced by development of fore-arc subsidence, extensional faulting, and the opening of back-arc basins. This is true of systems in the western Pacific that lack sediment filling, such as the Mariana Trough and Lau Basin. The extent of coupling of lithospheric plates appears to largely control the morphology and tectonic evolution of trench fore-arc provinces. If the subducted and overriding plates are decoupled, arc systems may be dominated by extension rather than compression. Back-arc basins developing in this setting behave effectively as small ocean basins, with Atlantic-type margins, in which mantle-derived magma forms new ocean crust by upwelling along a spreading center. Such systems differ markedly from those at convergent margins near continents.

Fundamental differences exist between the margins of the Pacific, Atlantic, and Indian oceans. The circum-Pacific region is characterized by subducting lithospheric plates, trenches, island arcs, and back-arc, or marginal, basins set between the intraoceanic arcs and continents. The seismically active circum-Pacific region is typified by young volcanic mountain chains, such as the Cascades, Central American, and Andes coastal continental ranges. In contrast, the passive nonconvergent margins of the Atlantic Ocean lack significant volcanic and seismic activity. Where volcanic loops and associated island arcs are found in the Atlantic basin (e.g., the Lesser Antilles and Scotia Arcs), they are considered extensions of the circum-Pacific belt. Similarly, volcanism along margins of the Indian Ocean is relatively slight compared with the circum-Pacific belt, an exception being the Mediterranean-Asian orogenic belt. Since Indian Ocean margins are predominantly nonconvergent, volcanism is limited. The Indonesian Arc, a zone of intense volcanism, represents a spur of the circum-Pacific belt.[1]

C. Continental Margins

Continental margins, which comprise ~20% of the total area of the ocean, occur between the deep ocean basins and continents. There are two basic types of continental margins: Atlantic and Pacific types. Atlantic-type continental margins consist of extensive continental shelf, slope, and rise components. They develop on aseismic, or passive, margins where large volumes of sediment accumulate along steadily subsiding crust subsequent to the rifting of continents and during the genesis of new ocean basins. Pacific-type continental margins, in turn, are typified by a narrow continental shelf and steep slope extending down to a deep trench (Chilean type) or into a shallow marginal basin that leads to an island arc and trench system (island arc type). Sediments that accumulate along these seismic, or active, margins are typically folded, sheared, and altered by volcanic and tectonic activity. They form by the convergence of lithospheric plates and resultant crustal deformation. Hence, they differ dramatically from the thick and relatively uniform stratigraphic sequences of sediment found on passive, Atlantic-type continental margins.

1. Continental Shelf

The gently sloping (~0.5°) submerged edge of a continent extending from the shoreline to an abrupt change in slope (i.e., the shelf break) at a depth of ~100 to 150 m is termed the continental shelf. The width of the continental shelf averages ~75 km, but ranges from a few kilometers along convergent plate margins on the Pacific coasts of North and South America to more than 1000 km in the Arctic Sea.[2,40–42] Terrigenous sediments predominate on the shelf, being underlain by crustal rocks that are continental in character.

Although continental shelves appear to be nearly flat plains, they are not devoid of positive and negative relief features. For example, sand ridges and ridges of structural origin are common on some shelves (e.g., Atlantic shelves of the United States, Argentina, and England). Diapirs, reefs, structurally controlled banks, and terraces produce considerable local relief amounting to tens of meters in various regions. Relict features (e.g., glacial moraines) also are important in many areas, such as in the North Atlantic.[1,43–46]

Negative relief features frequently encountered on continental shelves include linear depressions, marginal basins, and submarine canyons. Relict river channels (e.g., Hudson Channel), reef channels, shelf valleys, and ice-flow gouges are examples of major linear depressions.[47–49] Negative relief features, particularly submarine canyons, can incise deeply into shelf sediments.

2. Continental Slope

Seaward of the shelf break, the seafloor becomes steeply inclined (~4°), marking the upper limit of the continental slope. The water depth over the continental slope increases from ~100 to 150 m at the shelf break to ~1.5 to 3.5 km at the base of the slope. The width of this zone varies between ~20 and 100 km.[4] On passive margins, the continental slope joins at its base with the continental rise. On active margins, such as off Peru and Chile, the slope typically grades to a deep trench rather than a rise.

The continental slope consists of thick accumulations of clay, silt, and sand originating principally from continental sources. These sediments are often cut by deep erosional valleys (i.e., submarine canyons), as well as other types of negative slope features, which serve as conduits for the movement of shelf sediments to the deep ocean floor. Submarine canyons are impressive morphological features with steep V-shaped walls, intersected by tributary channels, and a topographic relief of ~1 to 2 km. Similar to river beds, they follow curved or sinuous paths that empty as channels into fan systems. Some are larger than the Grand Canyon of the Colorado River.

The head of submarine canyons often lie on the continental shelf off large rivers (e.g., Columbia and Hudson Rivers) and valleys on land. Turbidity currents play an integral role in the genesis of the canyons carrying terrigenous sediments downslope and gradually cutting the canyon walls and incising the canyon floor to the base of the continental slope. Sediments moving through the canyons ultimately collect in deep-sea fans, continental rises, abyssal plains, and trenches. Some submarine canyons appear to have been cut by subaerial erosion during glacially mediated lowstands of sea level.

Aside from turbidity currents, other submarine processes such as mudflows and slumping facilitate the movement of sediment downslope. These exogenic processes also scar the surface of the continental slope. In addition, contour currents flowing horizontally along the slope cause slides that redistribute sediment and promote its movement downslope.

3. Continental Rise

Sediments transported seaward to the foot of the continental slope form wedgelike or fanlike deposits up to 2 km thick as the gradient of the seafloor decreases abruptly to ~1°. These thick aprons of sediment, the continental rises, may spread seaward for more than 500 km, where they coalesce with the abyssal plains. Continental rises cover a total of ~14% (~5.0 × 10^7 km^2) of the ocean floor.[50] They are essentially absent on convergent continental margins, and attain greatest areal coverage (~25%) in the Atlantic.[51] Divergent margins in the Indian Ocean also exhibit highly developed rises.

Continental rises are not devoid of surface features. For example, distributary channels cut across their surface at various points, serving as pathways for the flow of turbidity currents to the abyssal plains. These currents flow for hundreds of kilometers, and as their velocities slacken, turbidites are deposited on the seafloor. These turbidites are important in shaping not only the continental margins but also the abyssal plains in the deep ocean floor.

III. SEDIMENTS

A. Deep Ocean Floor

There are five types of deep-sea sediments: terrigenous, biogenic, authigenic, volcanogenic, and cosmogenous sediment. Terrigenous sediment includes fine to coarse-grained sedimentary particles that are land derived. The detrital products of weathering and mechanical disintegration of preexisting continental rocks are delivered to the ocean basins by rivers, winds, and glaciers. Biogenic sediment consists of shells and skeletal remains of marine organisms, particularly plankton. Principal biogenic deposits are calcareous and silicious oozes that cover about one half of the deep ocean floor. Authigenic sediment refers to particles formed by chemical or biochemical reactions at their place of occurrence on the ocean floor. Examples are manganese and phosphate nodules on the deep seafloor, as well as sulfide minerals precipitated at sites of hydrothermal vents. Volcanogenic sediment orginates mainly as ejecta, or pyroclastic fragments (e.g., tephra). It encompasses air-fall volcanic deposits, submarine volcanic material, and marine epiclastics, or reworked volcanic rock fragments. Seafloor sediments near island arcs and at sites of active intraplate volcanism often contain large volumes of volcanic ash, dust, and other ejecta. Cosmogenous sediment constitutes all the fine particles of extraterrestrial origin in deep-sea sediments (e.g., black magnetic spherules). This material comprises a minor fraction of the total sediment pool on the deep ocean floor.

Terrigenous and biogenic particles form the bulk of deep-sea sediments, and the following discussion focuses on them. However, authigenic sediment and volcanogenic sediment are locally significant. In addition, some authigenic material may be of considerable economic importance (e.g., manganese nodules). They will also be examined.

1. Terrigenous Sediment

Land-derived sedimentary particles enter the deep abyss via several processes: (1) river and atmospheric transport followed by slow settling through the water column; (2) bottom transport by gravity flows (e.g., slumping, debris flows, grain flows, and turbidity currents); and (3) near-bottom transport by geostrophic bottom currents.[52] Most terrigenous material in the deep sea is supplied by rivers. This is especially notable in regions where rivers drain large mountain belts, such as the Himalayas, and bulk emplacement processes transport large quantities of sediment to the deep ocean floor.[4]

The largest fraction of coarse sediments (sand and gravel) delivered by influent systems, however, is concentrated in estuaries and along the inner continental shelf. Substantial concentrations of silt occur on the continental slope and rise, although this sediment is also common on the continental shelf and in the deep sea. Hemipelagic sediments comprising a mixture of terrigenous sediment and pelagic biogenic material characterize middle and upper slope sediments in many regions. Clays exist in low-energy environments throughout the marine realm.

Although a significant amount of fine-grained terrigenous material settles out of suspension to the seafloor, most coarse-grained sediments found on the deep ocean floor are the product of gravity transport, including sediment gravity flows (e.g., debris flows, grain flows, fluidized sediment flows, and turbidity currents) in which sediment is transported down the continental slope under the influence of gravity, as well as slides and slumps.[53] Turbidity currents, in particular, account for the thick sediment deposits in deep-sea channels, deep-sea fans, and abyssal plains, as stated previously. Indeed, some slides and slumps may generate turbidity currents.

Coarse sediments on the deep ocean floor are most typically observed in graded turbidite beds near the mouths of submarine canyons. However, ice rafting is a major process responsible for the accumulation of unsorted coarse-grained sediments in polar regions. For example, glacial marine sediments completely surround Antarctica where thick ice shelves generate high densities of icebergs laden with sedimentary material. The inner margins of the Ross and Weddell Seas are especially notable in this regard. In Antarctica, icebergs are the principal agents of sediment transport. Here, sediments are coarser than in the Arctic where icebergs, ice flows, and ice islands transport and deposit much higher concentrations of finer material (i.e., clay and silt).

Pelagic muds comprise the largest fraction of sediments in the deep ocean basins. These deposits consist of terrigenous (deep-sea clays) and biogenic (calcareous and silicious oozes) components. Among the terrigenous sediments, four clay mineral groups predominate: chlorite, illite, kaolinite, and montmorillonite (smectite). Of these four groups, illite and montmorillonite are most abundant. Minor amounts of clay-sized and silt-sized feldspar, quartz, pyroxene, and other minerals are also present. Iron oxides on the sediment particles yield a reddish color from which the term *red clay* is derived.

Pelagic sediments in high-latitude regions are rich in chlorite and illite. Those in low latitudes have larger quantities of kaolinite. The distribution of montmorillonite is closely coupled to areas of oceanic volcanism. Although the clay deposits may have multiple sources, most derive from river and wind transport from continental masses.[3]

2. Biogenous Sediment

Nearly all skeletal remains deposited on the deep ocean floor are those of planktonic organisms. In contrast, the skeletal remains of benthic organisms (e.g., arthropods, corals, mollusks) predominate on the continental shelf. Biogenic oozes in the deep sea consist of 30% or more of pelagic microorganism shells. The remainder comprises clay minerals. The shells of coccolithophores (<30 μm in size), foraminifera (~50 to 500 μm), and pteropods (~0.3 to 10 mm) constitute most of the skeletal remains in calcareous oozes, and those of diatoms (~5 to 50 μm) and radiolaria (~40 to 150 μm), most of the skeletal remains in siliceous oozes. These oozes cover more than 50% of the ocean floor.

a. Calcareous Oozes

There are three types of calcareous oozes in the deep sea: (1) foraminifera (=*Gobigerina*) ooze (≥30% foraminifera tests); (2) nannofossil (=coccolith) ooze (≥30% nannofossils); and (3) pteropod ooze (≥30% pteropod and heteropod shells). Foraminifera and nannofossil oozes are quantitatively most significant. Planktonic foraminifera reach highest numbers in the North Pacific and lowest numbers in the Sargasso Sea. Despite the greater numerical abundance of foraminifera in the Pacific Ocean, sedimentation rates of formaminifera tests peak in the Atlantic Ocean. Total abundances in the Indo-Pacific are about twice as great in the Atlantic. Worldwide concentrations in surface waters generally range from 1 to 10^7 individuals per cubic meter. Large foraminifera dominate in warm fertile regions, and small foraminifera, in cold fertile (phosphate-rich) areas. This pelagic taxonomic group contributes more than any other to calcareous deep-sea sands, especially in parts of the open ocean characterized by strong surface currents. The species diversity of planktonic foraminifera decreases from tropical to polar latitudes.[54]

Coccolithophores are numerically more abundant (often >10,000×) than foraminifera in deep-sea sediments. Often <10 μm in size, coccolithophores secrete skeletal edifices known as coccospheres, which consist of more than 30 discoidal structures termed coccoliths. Calcareous oozes in temperate latitudes generally contain large amounts of this discoidal material.

The geographic distribution of pteropods and heteropods in sediment assemblages corresponds closely with the abundance of living representatives inhabiting productive surface waters.[55] Even though their aragonitic shells are usually larger than other pelagic calcareous organisms, they are highly susceptible to dissolution. As a consequence, pteropod and heteropod skeletal remains are rarely observed in sediment assemblages at great depths in the ocean basins. They attain highest concentrations in shallow tropical waters.

Collectively, the calcareous oozes cover ~1.4 × 10^8 km² of the ocean floor, being deposited at rates of ~1 to 3 cm per 1000 years.[1] The Atlantic Ocean is carbonate rich and the Pacific Ocean carbonate poor. The Indian Ocean exhibits intermediate carbonate concentrations. Because Pacific Ocean waters are saturated with calcium carbonate only in the uppermost few hundred meters, most calcareous deposits exposed on the seafloor in the Pacific undergo continuous dissolution.[56] Therefore, calcium carbonate is less abundant in the Pacific basin than in other basins, where significant volumes of calcium carbonate in seafloor sediments occur at greater depths. For example, only about one third of the Atlantic Ocean seafloor lies below the calcite saturation depth compared with approximately two thirds of the Pacific Ocean seafloor. However, in the deepest parts of all oceans (below ~4.5 to 5 km), virtually no calcium carbonate exists on the seafloor, and the transition zone separating carbonate-rich and carbonate-poor sediments appears to be acute. The depth of this transition zone or boundary, referred to as the calcium carbonate compensation depth (CCD), varies from place to place because of differences in surface productivity and, hence, rate of carbonate supply, as well as the temperature, pressure, and acidity of the water. It averages ~4.5 km, but exceeds 5 km in many parts of the Atlantic Ocean, while rising to shallower depths (~4.2 to 4.5 km) in much of the Pacific Ocean. The CCD represents the horizon where the rate of calcium carbonate sedimentation equals the rate of calcium carbonate dissolution.

As a result of the aforementioned chemical, physical, and biological controls, topographic highs in the Atlantic, Indian, and Pacific Oceans, such as mid-ocean ridges, oceanic rises, and platforms, generally exhibit high concentrations of carbonate, while deep basins are enriched in red clays with little or no carbonate. High productivity of surface waters accounts for elevated carbonate levels in the western North Atlantic basin, in waters off West Africa, and in the equatorial Pacific. Dilution by terrigenous material reduces carbonate percentages in parts of the Indian Ocean basin, whereas dilution by biogenic noncarbonate sediment lowers carbonate percentages in the Southern Ocean. Fluctuating surface productivity and variation in dissolution rates are thought to be largely responsible for conspicuous carbonate cycles in deep-sea sediments.

Aragonite dissolves much more rapidly than calcite with increasing depth. Only a small fraction of the water column is close to equilibrium or supersaturation with respect to aragonite, thus explaining the preferential accumulation of calcite, relative to aragonite, in deep-sea sediments. This also explains, in part, why aragonitic pteropod and heteropod tests are confined to shallow tropical waters (<3 km).

b. Siliceous Oozes

The distribution of siliceous oozes in deep-sea sediments correlates closely with planktonic productivity in surface waters. For example, diatom oozes, which are dominated by diatom frustules (>30% of the sediment), occur in greatest concentrations in high-latitude regions south of the Antarctic convergence and parts of the North Pacific and Bering Sea. Coastal upwelling sites along eastern boundary currents also are favorable locations for biogenic silica accumulation.[57] Most conspicuous, however, is the broad belt (~1000 to 2000 km wide) of diatom ooze surrounding Antarctica that has accumulated from upwelling of nutrient-rich intermediate waters and the resultant high rates of phytoplankton production. The rate of siliceous deposition here amounts to ~0.2 g/cm^{2}ka. Similarly, extensive upwelling in areas of surface water divergence near the equator has stimulated high rates of phytoplankton production dominated by radiolarians. Although the deposition of siliceous skeletal debris here is considerably less than near Antarctica, it is still significant (~0.1 g/cm^{2}ka in some areas).[58]

Three primary factors control the amount of siliceous skeletal remains in ocean floor sediments: (1) the productivity of siliceous organisms in surface waters; (2) the extent of dilution by calcareous, terrigenous, and volcanic particles; and (3) the degree of siliceous shell dissolution. As shown above, production of planktonic siliceous organisms peaks in coastal waters, as well as at major oceanic divergences, where nutrient enrichment (i.e., silica) occurs. The extent of dilution by nonsiliceous material reflects the relative concentrations of siliceous and nonsiliceous particles in the sediments. A distinct negative correlation exists between the amount of silica and carbonate in deep-sea sediments. In addition, areas receiving large quantities of volcanic ash and other nonsiliceous material exhibit much lower percentages of silica than the aforementioned oceanic divergence zones.

There are significant differences in shell dissolution rates of planktonic siliceous organisms. The taxonomic order of increasing resistance to dissolution is as follows: silicoflagellates, diatoms, delicate radiolarians, and robust radiolarians. Most skeletal material of siliceous organisms (>95%) dissolves during descent through the water column, mainly in the upper 1 km. Larger, robust shells are much more likely to survive passage through the water column and burial in surface sediments than small, thin shells. Because of differential preservation, death assemblages of pelagic microorganisms in seafloor sediments may differ substantially from the original life assemblage in surface waters.

3. Pelagic Sediment Distribution

The major ocean basins (i.e., Atlantic, Pacific, and Indian Oceans) exhibit important differences in the relative amounts of pelagic sediments. The Pacific Ocean, for example, has widespread deep-sea clay deposits (i.e., red clays) covering 49% of the seafloor. In contrast, deep-sea clays in the

Atlantic and Indian basins cover only 26 and 25% of the seafloor, respectively. Calcareous oozes are much more extensive in the Atlantic (67%) and Indian (54%) Oceans than in the Pacific (36.1%). Siliceous oozes, however, are more widespread in the Indian Ocean (20.5%) than in the Pacific (15%) and Atlantic (7%) Oceans.[1,3,4]

Globally, more than 85% of ocean floor sediments consist of calcareous oozes (47.5%) and deep-sea clays (38%).[1,55] Bottom sediments containing the highest concentrations of calcareous and siliceous oozes underlie fertile, high-productivity surface waters. Diatom ooze predominates in the northern Pacific Ocean and off Antarctica, and radiolarian ooze, along the equator. Calcareous oozes, which are three to nine times more abundant than siliceous oozes, cover extensive areas of the seafloor above the CCD, often blanketing bathymetric highs (e.g., seamount crests, plateaus, and mid-ocean ridge flanks). Pelagic clays are most abundant in the deepest parts of the ocean basins below the CCD and away from continents and areas of high surface productivity. These clays accumulate at very slow rates (<1 cm per 1000 years) compared with those of terrigenous sediments on the continental margins (>5 cm per 1000 years), and biogenic oozes on the deep ocean floor (1 to 3 cm per 1000 years).[3]

4. Authigenic Sediment

Those deposits that form *in situ* on the deep seafloor are termed authigenic sediments. Among the most conspicuous authigenic deposits include manganese nodules on abyssal plains and other areas of the ocean floor, as well as metalliferous sediments at hydrothermal vents along seafloor spreading centers. Manganese nodules are perhaps the best known authigenic material in the deep sea. These dark spherical masses, which range from ~1 to 10 cm in diameter, form by the precipitation of metal oxides in concentric layers around particulate nuclei (e.g., sand grains, rock particles, biogenic material, etc.) on the seafloor. They are relatively high-grade ores containing ~20% by weight of manganese and iron oxides, and smaller concentrations of other metals (e.g., cobalt, copper, lead, nickel, and zinc). Of particular importance are the relatively high concentrations of copper and nickel, which rival or exceed those mined on land.

Manganese nodules cover extensive areas of the deep ocean floor. For example, up to 20 to 50% of the Pacific basin appears to be blanketed by these deposits.[2] The South Atlantic and Southern Ocean in areas influenced by the Circum-Antarctic Current also harbor large numbers of nodules. The spherical masses accumulate slowly (~1 to 4 mm per million years), growing in concentric layers at the sediment–water interface by rolling on the seabed and exposing surface areas to chemical precipitation.[4] They typically occur in areas of low sedimentation rates (<4 to 8 mm per 1000 years) commonly associated with deep-sea clays and oozes.[1]

Indigenous nonbiogenic deposits originating from hydrothermal vent emissions are locally important on mid-ocean ridge crests and back-arc basins (e.g., Lau and Manus Basins in the western Pacific), as manifested by massive sulfide mounds and chimney structures along active seafloor spreading centers. Metal sulfides precipitate directly from white smoker vents, which release milky fluids with temperatures ranging from 200 to 330°C. They also precipitate from black smoker vents discharging jets of water blackened by sulfide precipitates at temperatures between ~300 and 400°C. White smoker vents precipitate high concentrations of zinc sulfide (e.g., sphalerite and wurtzite). Black smoker vents, in turn, yield large amounts of copper and iron sulfides (e.g., bornite, chalcocite, chalcopyrite, covellite, cubanite isocubanite, marcasite, pyrite, and pyrrhotite).[59,60] Massive sulfide deposits can be immense. For example, the sulfide mound at the TAG site at 26°N on the Mid-Atlantic Ridge rises ~50 m above the adjoining seafloor, and it is ~200 m across.[61–63] Sulfide minerals not only form at the seafloor surface but also in conduits within the subseafloor plumbing system.

Aside from zeolites, barite is also a common authigenic mineral in deep-sea sediments. It may derive from biogenic sources or submarine hydrothermal activity. Barite constitutes up to 10% by weight of the carbonate-free fraction in some deep-sea sediments.[1,64]

5. Volcanogenic Sediment

There are two major components of volcanogenic sediments in the deep sea: (1) pyroclastic debris resulting from aerial and subaerial volcanic explosions; and (2) volcaniclastic material or reworked fragments of volcanic rocks derived from various chemical and mechanical processes. Pyroclastic debris, as defined above, is primary volcanic material. Volcaniclastic material is secondary, having been recycled from primary sources. Although volcanogenic sediments often comprise the bulk of marine deposits near island arcs and other volcanic systems, they also may constitute an important fraction of the sediments a considerable distance from volcanic centers. For example, global ash encircles the earth, and tropospheric ashfalls are detected thousands of kilometers from volcanic centers. The eruption explosivity, duration, and debris volume, as well as global wind patterns, all are important factors in long-distance transport of volcanic ejecta. Apart from atmospheric deposition, volcanogenic sediment may be delivered to the deep-sea via riverine inflow, ice transport, and ocean currents.

Submarine volcanic sediment derives from three principal sources of volcanism: (1) island-arc and ocean-margin volcanism; (2) mid-ocean ridge and oceanic intraplate volcanism; and (3) continental volcanism. Submarine volcanism produces far less marine sediment than subaerial eruptions. Air-fall volcanic material, mainly tephra, consists largely of volcanic glass, or ash. Based on particle size, tephra may be ash (<2 mm), lapilli (2 to 64 mm), or bombs (>64 mm).[1] Tephra aprons generated near island arcs can be several kilometers thick.

6. Cosmogenic Sediment

Extraterrestrial material comprises an insignificant component of deep-sea sediments. Microtektites (i.e., micrometeorites), measuring ~0.030 to 1.0 mm in diameter, accumulate at a rate of ~0.02 μm per 1000 years in the deep sea. They appear to originate from high-velocity meteorite impacts. Prominent depositional sites include areas off Australia, Indonesia, the Philippines, the Ivory Coast of Africa, and the states of Texas and Georgia in the United States.[1]

7. Deep-Sea Sediment Thickness

As alluded to above, sediments gradually increase in thickness from the mid-ocean ridges to the continental rises. The neovolcanic zone of mid-ocean ridge systems are essentially barren of sediment, being floored by newly erupted basaltic lava. Beyond the narrow (1 to 4 km) expanse of the neovolcanic zone, older crust in the zone of crustal fissuring out to ~7 km from the ridge axis is only lightly dusted with pelagic sediment. Thicker deposits rest on the flanks of the ridge crest and in off-axis areas. As the rugged volcanic terrain continues to diverge from the axial region, it slowly cools, contracts, and subsides, thereby accounting for the greater depths of the seafloor away from the ridge axis. Sediments blanket nearly all relief of volcanic ridges by the time the seafloor reaches an age of ~100 Myr and depth of ~5 to 6 km. Sediments may exceed 10 km in thickness on older seafloor where terrigenous sediment input is substantial.[65] However, the sediment column for typical basins in the Atlantic and Pacific are only ~300 to 500 m thick. The total volume of sediment on the deep ocean bottom amounts to ~2.5 × 10^7 km^3.[1]

B. Continental Margins

The continental margins encompass the continental shelf, slope, and rise topographic provinces, occupying ~20% of the ocean floor. The continental slopes and rises underlie only ~13% of the total area of the oceans, but they harbor ~66% of the total marine sediment volume, amounting to $\sim 2 \times 10^8$ km^3 and $\sim 1.5 \times 10^8$ km^3, respectively. While these two topographic provinces receive the bulk of terrigenous sediment from the continents, the continental shelves also are major repositories, with a total sediment volume of $\sim 7.5 \times 10^7$ km^3.[1,3]

1. Continental Shelves

Continental shelves are subdivided into inner, middle, and outer zones. The inner zone extends from the shoreline to ~ 30 m depth. It is characterized by high-energy environments and generally coarse-grained seafloor sediments. Being dynamically, morphologically, and sedimentologically contiguous with beaches and surf zones as well as with tidal inlets, river mouths, and estuaries, inner shelf areas are influenced by an array of transport phenomena, including (1) wind-driven along-shelf and across-shelf (upwelling and downwelling) flows; (2) surface gravity waves; (3) tidal currents; (4) internal waves; (5) infragravity oscillations; (6) buoyant plumes (positive and negative); and (7) wave-driven surf zone processes.[46]

Because of shallow depths, wind stress effects on the inner shelf are transmitted directly to the seabed.[66] For example, strong bed agitation by wind-driven waves causes resuspension of bottom sediments, while currents advect the sediments along and across the shelf where they accumulate in lower-energy environments. The mid-shelf region often serves as a sediment sink, with little material being transported to the outer shelf. Waves primarily drive surf zone processes; however, wind-driven or tidal currents prevail seaward of the surf zone.[67] Geostrophic flows predominate over the outer shelf region.[68] Much sediment movement on the shelf occurs during storms when high-energy conditions impinge on the seabed.[40]

Coarse to fine sands that blanket the inner shelf are modern sediments deposited by intense sediment transport processes operating in the region. Modern fine sediments (clay and silt) mainly accumulate in shelf depressions, off of river mouths, and in coastal embayments. Little fine sediment is actively accumulating on the mid- and outer shelf. Coarse sediments (sand and gravel) overlying extensive areas of these two shelf regions represent relict material deposited during major glacial periods and low stands of sea level. Approximately 60 to 70% of the sediment found on continental shelves worldwide is relict in nature.[3,4,40]

Shelf deposits vary with latitude. Biogenic sediment is most common in tropical and subtropical waters having low terrigenous sediment input. Carbonate shelves or platforms comprised of organism shells exist off southern Florida, the Bahamas, Yucatan Peninsula of Mexico, northern Australia, and elsewhere. These shelves consist largely of sand- or gravel-sized skeletal material. However, in some areas, inorganic precipitation generates considerable amounts of ooliths and pellet-aggregates which can constitute an important component of the shallow water carbonates.

Terrigenous sediments dominate continental shelf deposits in temperate latitudes. For example, off the East Coast of the United States, Cape Hatteras forms a promontory separating a southern shelf region dominated by carbonate sediments and a northern shelf region dominated by terrigenous sediments. In the mid-Atlantic region, fluvial sediment flux and coastal sediment reworking, together with climatic and oceanographic factors, preclude significant carbonate sediment production.

Glacial marine sediment predominates on the continental shelves in polar latitudes. The North Atlantic is heavily laden with glacial moraine deposits. Icebergs and sea ice have delivered large quantities of unsorted sediments to shelf regions in the Arctic and adjacent seas. In addition, ice movement has excavated deep ravines, depressions, and fjords that now receive glacial debris.

Continental shelves typically have thick sediment wedges. This is particularly true of Atlantic-type (passive) margins, which have been gradually subsiding for millions of years and accumulating immense masses of sediment. Here, the total sediment thickness may exceed 10 km, and sediment sequences are nearly continuous. The situation is much different along Pacific-type (active) margins where subducting lithospheric plates produce highly deformed, folded, and sheared sedimentary beds often interlayered with volcanogenic sediment and lavas.[40] In this case, accretionary prisms are generated between trenches and nearby volcanic edifices, with a portion of the accumulating wedge subducted along with the downgoing lithospheric slab.

Continental shelf sedimentation depends on the type, size, distribution, and quantity of sediment supply, energy conditions at the site of deposition, and sea level fluctuations.[40] Sedimentation rates are typically greatest along the inner shelf, as well as in marginal basins, estuaries, and coastal embayments. In these areas, sediments can accumulate at rates of several centimeters per year, especially at locations subjected to heavy riverine influx. By comparison, characteristic deposition rates for continental slopes and deep ocean basins amount to 40 to 200 mm per 1000 years and 1 to 20 mm per 1000 years, respectively.[3]

2. Continental Slopes and Rises

The thickest masses of sediments occur along continental slopes and rises, being derived largely from terrigenous sources and delivered by bulk emplacement processes. Some of these deposits contain admixtures of biogenic material. Continental slopes and rises are most extensively developed on Atlantic-type margins where tectonically quiescent conditions enable thick terrigenous sediment masses to accumulate. In contrast, lithospheric plate collisions in the Pacific basin are not conducive to the formation of such deposits. This is most conspicuous at the Peru–Chile Trench, where a continent-ocean collision has generated a steep continental slope without a rise. Sediments moving from the slope collect in the adjacent trench. Farther to the north along northern California, the slope and rise essentially consist of coalescing deep-sea fans.[3] In the island-arc subduction zone systems of the western Pacific, sediments collecting in deep-sea trenches are subducted along with the downgoing lithospheric plate.

While deep-sea trenches rimming the Pacific margins prevent the movement of lithogenous sediment from continental margins toward the center of the ocean basin, turbidity currents and other bulk emplacement processes (e.g., slumping and debris flows) operate unimpeded along the Atlantic margins, spreading large volumes of terrigenous debris downslope and across the continental rises and onto the abyssal plains. Turbidity currents transport much of this sediment through submarine canyons and deep-sea channels. Turbidites and associated clastic sediment build coalescing submarine fans at the bases of the continental slopes that prograde seaward to form the continental rises. These wedge-shaped aprons of sediment may exceed 2 km in thickness. Similar turbidite deposits cover the continental slopes.

REFERENCES

1. Kennett, J. P., *Marine Geology,* Prentice-Hall, Englewood Cliffs, NJ, 1982.
2. Press, F. and Siever, R., *Earth,* 4th ed., W. H. Freeman, New York, 1986.
3. Seibold, E. and Berger, W. H., *The Sea Floor: An Introduction to Marine Geology,* 2nd ed., Springer-Verlag, Berlin, 1993.
4. Brown, J., Colling, A., Park, D., Phillips, J., Rothery, D., and Wright, J., *The Ocean Basins: Their Structure and Evolution,* Pergamon Press, Oxford, 1989.
5. Francheteau, J. and Ballard, R. D., The East Pacific Rise near 21°N, 13°N, and 20°S: inferences for along-strike variability of axial processes of the mid-ocean ridge, *Earth Planet. Sci. Lett.,* 64, 93, 1983.
6. Macdonald, K. C., Mid-ocean ridges: fine scale tectonic, volcanic, and hydrothermal processes within the plate boundary zone, *Annu. Rev. Earth Planet. Sci.,* 10, 155, 1982.
7. Grassle, J. F., The ecology of deep-sea hydrothermal vent communities, in *Advances in Marine Biology,* Vol. 23, Blaxter, J. H. S. and Southward, A. J., Eds., Academic Press, London, 1986, 301.
8. Carbotte, S., Welch, S. M., and Macdonald, K. C., Spreading rates, rift propagation, and fracture zone offset histories during the past 5 my on the Mid-Atlantic Ridge; 25°–27°30′S and 31°–34°30′S, *Mar. Geophys. Res.,* 13, 51, 1991.

9. Sempere, J.-C., Lin, J., Brown, H. S., Schouten, H., and Purdy, P. M., Segmentation and morphotectonic variations along a slow-spreading center: the Mid-Atlantic Ridge (24°00′N–30°40′N), *Mar. Geophys. Res.*, 15, 153, 1993.

10. Tucholke, B. E., Lin, J., Kleinrock, M. C., Tivey, M. A., Reed, T. B., Goff, J., and Jaroslow, G. E., Segmentation and crustal structure of the western Mid-Atlantic Ridge flank, 25°25′–27°10′N and 0–29 m.y., *J. Geophys. Res.*, 102, 10,203, 1997.

11. Macdonald, K. C. and Fox, P. J., The mid-ocean ridge, *Sci. Am.*, 262, 72, 1990.

12. Macdonald, K. C., Scheirer, D. S., and Carbotte, S. M., Mid-ocean ridges: discontinuities, segments, and giant cracks, *Science,* 253, 986, 1991.

13. Alt, J. C., Subseafloor processes in mid-ocean ridge hydrothermal systems, in *Seafloor Hydrothermal Systems: Physical, Chemical, Biological, and Geological Interactions,* Humphris, S. E., Zierenberg, R. A., Mullineaux, L. S., and Thomson, R. E., Eds., Geophysical Monograph 91, American Geophysical Union, Washington, D.C., 85, 1995.

14. Buck, W. R., Carbotte, S. M., and Mutter, C., Controls on extrusion at mid-ocean ridges, *Geology,* 25, 935, 1997.

15. Detrick, R. S., Buhl, P., Vera, E., Mutter, J., Orcutt, J., Madsen, J., and Brocher, T., Multi-channel seismic imaging of a crustal magma chamber along the East Pacific Rise, *Nature,* 326, 35, 1987.

16. Sinton, J. M. and Detrick, R. S., Mid-ocean ridge magma chambers, *J. Geophys. Res.,* 97, 197, 1992.

17. Detrick, R. S. and Langmuir, C. H., The geometry and dynamics of magma chambers, in *Mid-Ocean Ridge—A Dynamic Global System,* National Academy Press, Washington, D.C., 1988, 123.

18. Nicolas, A., Reuber, I., and Benn, K., A new magma chamber model based on structural studies in the Oman Ophiolite, *Tectonophysics,* 151, 87, 1988.

19. Burnett, M. S., Caress, D. W., and Orcutt, J. A., Tomographic image of the magma chamber at 12°50′N on the East Pacific Rise, *Nature,* 339, 206, 1989.

20. Marsh, B. D., Magma chambers, *Annu. Rev. Earth Planet. Sci.,* 17, 439, 1989.

21. Kent, G. M., Harding, A. J., and Orcutt, J. A., Evidence for a smaller magma chamber beneath the East Pacific Rise at 9°30′N, *Nature,* 344, 650, 1990.

22. Toomey, D. R., Purdy, G. M., Solomon, S., and Wilcox, W., The three dimensional seismic velocity structure of the East Pacific Rise near latitude 9°30′N, *Nature,* 347, 639, 1990.

23. Sinton, J. M., Smaglik, S. M., Mahoney, J. J., and Macdonald, K. C., Magmatic processes at superfast spreading oceanic ridges: glass compositional variations along the East Pacific Rise, 13°–23°S, *J. Geophys. Res.,* 96, 6133, 1991.

24. Burnett, M. S., Caress, D. W., and Orcutt, J. A., Tomographic image of the magma chamber at 12°50′N on the East Pacific Rise, *Nature,* 339, 206, 1989.

25. Toomey, D. R., Tomographic imaging of spreading centers, *Oceanus,* 34, 92, 1991.

26. Purdy, G. M. and Detrick, R. S., Crustal structure of the Mid-Atlantic Ridge at 23°N from seismic refraction studies, *J. Geophys. Res.,* 91, 3739, 1986.

27. Detrick, R. S., Mutter, J. C., Buhl, P., and Kim, I. I., No evidence from multichannel reflection data for a crustal magma chamber in the MARK area of the Mid-Atlantic Ridge, *Nature,* 347, 61, 1990.

28. Cashman, K. V., Magmatic processes in ridge environments, *Ridge Events,* 2, 20, 1991.

29. Fox, P. J., Macdonald, K. C., and Batiza, R., Tectonic cycles and ridge crest segmentation, in *The Mid-Oceanic Ridge: A Dynamic Global System,* National Academy Press, Washington, D.C., 1988, 115.

30. Macdonald, K. C., Fox, P. J., Perram, L. J., Eisen, M. F., Haymon, R. M., Miller, S. P., Carbotte, S. M., Cormier, M.-H., and Shor, A. N., A new view of the mid-ocean ridge from the behaviour of ridge-axis discontinuities, *Nature,* 335, 217, 1988.

31. Sauter, D., Nafziger, J.-M., Whitechurch, H., and Munschy, M., Segmentation and morphotectonic variations of the central Indian Ridge (21°10′S–22°25′S), *J. Geophys. Res.,* 101, 20,233, 1996.

32. Langmuir, C. H., Bender, J. F., and Batiza, R., Petrologic and tectonic segmentation of the East Pacific Rise 5°30′–14°30′N, *Nature,* 322, 422, 1986.

33. Macdonald, K. C., Fox, P. J., Alexander, R. T., Pockalny, R., and Gente, P., Volcanic growth faults and the origin of Pacific abyssal hills, *Nature,* 380, 125, 1996.

34. Smith, D. K., Seamount abundances and size distributions, and their geographic variations, *Rev. Aquat. Sci.,* 5, 197, 1991.

35. Smith, D. K. and Cann, J. R., Hundreds of small volcanoes on the medial valley floor of the Mid-Atlantic Ridge at 24°–30°N, *Nature,* 348, 152, 1990.

36. Smith, D. K. and Cann, J. R., The role of seamount volcanism in crustal construction at the Mid-Atlantic (24°–30°N), *J. Geophys. Res.*, 97, 1645, 1992.

37. Smith, D. K. and Cann, J. R., Building the crust at the Mid-Atlantic Ridge, *Nature,* 365, 707, 1993.

38. Kennish, M. J. and Lutz, R. A., Morphology and distribution of lava flows on mid-ocean ridges: a review, *Earth-Sci. Rev.,* 43, 63, 1998.

39. Hawkins, J. W., Bloomer, S. H., Evans, C. A., and Melchior, J. T., Evolution of intra-oceanic arc-trench systems, in *Geodynamics of Back-arc Regions,* Carlson, R. L. and Kobayashi, K., Eds., *Tectonophysics,* 102, 175, 1984.

40. Pinet, P. R., *Invitation to Oceanography,* Jones and Bartlett Publishers, Boston, 1998.

41. McGregor, B. A., The submerged continental margin, *Am. Sci.,* 72, 275, 1984.

42. Walsh, J. J., *On the Nature of Continental Shelves*, Academic Press, New York, 1989.

43. Swift, D. J. P., Continental shelf sedimentation, in *Geology of Continental Margins,* Burk, C. A. and Drake, C. L., Eds., Springer-Verlag, New York, 1974, 117.

44. Green, M. O., Side-scan sonar mosaic of a sand ridge field: southern Mid-Atlantic Bight, *Geo-Marine Lett.,* 6, 35, 1986.

45. McBride, R. A. and Moslow, T. F., Origin, evolution, and distribution of shoreface sand ridges, Atlantic inner shelf, U.S.A., *Mar. Geol.,* 97, 57, 1991.

46. Wright, L. D., *Morphodynamics of Inner Continental Shelves,* CRC Press, Boca Raton, FL, 1995.

47. Knebel, H. J., Anomalous topography on the continental shelf around Hudson Canyon, *Mar. Geol.,* 33, M67, 1979.

48. Freeland, G. L., Stanley, D. J., Swift, D. J. D., and Lambert, D. N., The Hudson Shelf Valley: its role in shelf sediment transport, in *Sedimentary Dynamics of Continental Shelves,* Nittrouer, C. A., Ed., Elsevier Scientific Publishing Company, Amsterdam, 1981, 399.

49. Ashley, G. M. and Sheridan, R. E., Depositional model for valley fills on a passive continental margin, SEPM Special Publication No. 51, Tulsa, OK, 1994, 285.

50. Emery, K. O., Structure and stratigraphy of divergent continental margins, in *Geology of Continental Margins, Short Course,* American Association of Petroleum Geologists, Houston, TX, 1977, B-1.

51. Emery, K. O., Continental margins—classification and petroleum prospects, *Am. Assoc. Petrol. Geol. Bull.,* 64, 297, 1980.

52. Emiliani, C. and Milliman, J. D., Deep-sea sediments and their geological record, *Earth-Sci. Rev.,* 1, 105, 1966.

53. Middleton, G. V. and Hampton, M. A., Subaqueous sediment transport and deposition by sediment gravity flows, *in Marine Sediment Transport and Environmental Management,* Stanley, D. J. and Swift, D. J. P., Ed., John Wiley & Sons, New York, 1976, 197.

54. Berger, W. H., Ecologic patterns of living planktonic foraminifera, *Deep-Sea Res.,* 16, 1, 1969.

55. Berger, W. H., Biogenous deep-sea sediments: production, preservation, and interpretation, in *Chemical Oceanography,* Vol. 5, 2nd ed., Riley, J. P. and Chester, R., Eds., Academic Press, London, 1976, 265.

56. Morse, J. W. and Mackenzie, F. T., *Geochemistry of Sedimentary Carbonates,* Elsevier, Amsterdam, 1990.

57. Calvert, S. E., Deposition and diagnosis of silica in marine sediments, in *Pelagic Sediments on Land and Under the Sea, Hsü,* K. J. and Jenkyns, H., Eds., Special Publication 1, Blackwell Scientific, Oxford, England, 1974, 273.

58. Falkowski, P. G. and Woodhead, A. D., Eds., *Primary Productivity and Biogeochemical Cycles in the Sea,* Plenum Press, New York, 1992.

59. Haymon, R. M., Hydrothermal processes and products on the Galapagos Rift and East Pacific Rise, in *The Geology of North America,* Vol. N, *The Eastern Pacific Ocean and Hawaii,* Winterer, E. L., Hussong, D. M., and Decker, R. W., Eds., Geological Society of America, Boulder, CO, 1989, 173.

60. Von Damm, K. L., Seafloor hydrothermal activity: black smoker chemistry and chimneys, *Annu. Rev. Earth Planet. Sci.,* 18, 173, 1990.

61. Rona, P. A., Klinkammer, G., Nelsen, T. A., Trefry, J. H., and Elderfield, H., Black smokers, massive sulphides and vent biota at the Mid-Atlantic Ridge, *Nature,* 321, 33, 1986.

62. Thompson, G., Humphris, S. E., Schroeder, B., Sulanowska, M., and Rona, P. A., Active vents and massive sulfides at 26°N (TAG) and 23°N (Snake Pit) on the Mid-Atlantic Ridge, *Can. Min.,* 26, 697, 1988.

63. Rona, P. A., Bogdanov, Y. A., Gurvich, E. G., Rimski-Korsakov, N. A., Sagalevitch, A. M., Hannington, M. D., and Thompson, G., Relict hydrothermal zones in the TAG Hydrothermal Field, Mid-Atlantic Ridge 26°N, 45°W, *J. Geophys. Res.,* 98, 9715, 1993.

64. Cronan, D. S., Authigenic minerals in deep-sea sediments, in *The Sea, Marine Chemistry,* Vol. 5, Goldberg, E. D., Ed., Wiley-Interscience, New York, 1974, 491.

65. Heirtzler, J. R., The evolution of the deep ocean floor, in *The Environment of the Deep Sea,* Ernst, W. G. and Morin, J. G., Eds., Prentice-Hall, Englewood Cliffs, NJ, 1982, 3.

66. Mitchum, G. T. and Clarke, A. J.,The frictional nearshore response to forcing by synoptic scale winds, *J. Phys. Oceanogr.,* 16, 934, 1986.

67. Nittrouer, C. A. and Wright, L. D., Transport of particles across continental shelves, *Rev. Geophys.,* 32, 85, 1994.

68. Huyer, A., Shelf circulation, in *The Sea,* Vol. 9, Part B, *Ocean Engineering Science,* LeMehauté, B. and Hanes, D. M., Eds., John Wiley & Sons, New York, 1990, 423.

4.1 COMPOSITION AND STRUCTURE OF THE EARTH

Table 4.1–1   Astronomical Constants

Defining Constants

Gaussian gravitational constant	$k = 0.01720209895$ m^3 kg^{-1} s^{-2}
Speed of light	$c = 299792458$ m s^{-1}

Primary Constants

Light-time for unit distance (1 AU)	$\tau_A = 499.004782$ s
Equatorial radius of earth	$a_e = 6378140$ m
Equatorial radius of earth (IUGG value)	$a_e = 6378136$ m
Dynamical form-factor for earth	$J_2 = 0.001082626$
Geocentric gravitational constant	$GE = 3.986005 \times 10^{14}$ m^3 s^{-2}
Constant of gravitation	$G = 6.672 \times 10^{-11}$ m^3 kg^{-1}s^{-2}
Ratio of mass of moon to that of earth	$\mu = 0.01230002$
	$1/\mu = 81.300587$
General precession in longitude, per Julian century, at standard epoch J2000	$\rho = 5029''.0966$
Obliquity of the ecliptic at standard epoch J2000	$\varepsilon = 23°26'21''.448$

Derived Constants

Constant of nutation at standard epoch J2000	$N = 9''.2025$
Unit distance (AU $= c\tau_A$)	AU $= 1.49597870 \times 10^{11}$ m
Solar parallax ($\pi_0 = \arcsin(a_e/\text{AU})$)	$\pi_0 = 8''.794148$
Constant of aberration for standard epoch J2000	$\kappa = 20''.49552$
Flattening factor for the earth	$f = 1/298.257 = 0.00335281$
Heliocentric gravitational constant ($GS = A^3 k^2/D^2$)	$GS = 1.32712438 \times 10^{20}$ m^3 s^{-2}
Ratio of mass of sun to that of the earth (S/E) $= (GS)/(GE)$	$S/E = 332946.0$
Ratio of mass of sun to that of earth + moon	$(S/E)/(1 + \mu) = 328900.5$
Mass of the sun ($S = (GS)/G$)	$S = 1.9891 \times 10^{30}$ kg

Ratios of Mass of Sun to Masses of the Planets

Mercury	6023600
Venus	408523.5
Earth + moon	328900.5
Mars	3098710
Jupiter	1047.355
Saturn	3498.5
Uranus	22869
Neptune	19314
Pluto	3000000

Note: The constants in this table are based primarily on the set of constants adopted by the International Astronomical Union (IAU) in 1976. Updates have been made when new data were available. All values are given in SI units; thus masses are expressed in kilograms and distances in meters.

The astronomical unit of time is a time interval of 1 day (1 d) equal to 86400 s. An interval of 36525 days is one Julian century (1 cy).

Source: Lide, D. R. and Frederikse, H. P. R., Eds., *CRC Handbook of Chemistry and Physics,* 79th ed., CRC Press, Boca Raton, FL, 1998. With permission.

REFERENCES

1. Seidelmann, P. K., *Explanatory Supplement to the Astronomical Almanac,* University Science Books, Mill Valley, CA, 1990.
2. Lang, K. R., *Astrophysical Data: Planets and Stars,* Springer-Verlag, New York, 1992.

Table 4.1–2   Properties of the Solar System

This table gives various properties of the planets and characteristics of their orbits in the solar system. Certain properties of the sun and of the earth's moon are also included.

Explanations of the column headings:

- *Den.*: mean density in g/cm^3
- *Radius:* radius at the equator in km
- *Flattening:* degree of oblateness, defined as $(r_e - r_p)/r_e$, where r_e and r_p are the equatorial and polar radii, respectively
- *Potential coefficients:* coefficients in the spherical harmonic representation of the gravitational potential U by the equation

$$U(r, \phi) = (GM/R)[1 - \Sigma J_n(a/r)^n P_n(\sin\phi)]$$

where G is the gravitational constant, r the distance from the center of the planet, a the radius of the planet, M the mass, ϕ the latitude, and P_n the Legendre polynomial of degree n.

- *Gravity:* acceleration due to gravity at the surface
- *Escape velocity:* velocity needed at the surface of the planet to escape the gravitational pull
- *Dist. to sun:* semi-major axis of the elliptical orbit (1 AU = 1.496 × 10^8 km)
- *ε*: eccentricity of the orbit
- *Ecliptic angle*: angle between the planetary orbit and the plane of the earth's orbit around the sun
- *Inclin.*: angle between the equatorial plane and the plane of the planetary orbit
- *Rot. period*: period of rotation of the planet measured in earth days
- *Albedo*: ratio of the light reflected from the planet to the light incident on it
- T_{sur}: mean temperature at the surface
- P_{sur}: pressure of the atmosphere at the surface

The following general information on the solar system is of interest:

Mass of the earth = M_e = 5.9742 × 10^{24} kg
Total mass of planetary system = 2.669 ×10^{27} kg = 447 M_e
Total angular momentum of planetary system = 3.148 × 10^{43} kg m^2/s
Total kinetic energy of the planets = 1.99 × 10^{35} J
Total rotational energy of planets = 0.7 × 10^{35} J

Properties of the sun:

Mass = 1.9891 × 10^{30} kg = 332946.0 M_e
Radius = 6.9599 × 10^8 m
Surface area = 6.087 × 10^{18} m^2
Volume = 1.412 × 10^{27} m^3
Mean density = 1.409 g/cm^3
Gravity at surface = 27398 cm/s^2
Escape velocity at surface = 6.177 × 10^5 m/s
Effective temperature = 5780 K
Total radiant power emitted (luminosity) = 3.86 × 10^{26} W
Surface flux of radiant energy = 6.340 × 10^7 W/m^2
Flux of radiant energy at the earth (Solar Constant) = 1373 W/m^2

REFERENCES

1. Seidelmann, P. K., Ed., *Explanatory Supplement to the Astronomical Almanac,* University Science Books, Mill Valley, CA, 1992.
2. Lang, K. R., *Astrophysical Data: Planets and Stars,* Springer-Verlag, New York, 1992.
3. Allen, C. W., *Astrophysical Quantities,* 3rd ed., Athlone Press, London, 1977.

Table 4.1–2 Properties of the Solar System (continued)

Planet	Mass 10²⁴ kg	Den. g/cm³	Radius km	Flattening	Potential Coefficients 10³ J₂	10⁶ J₃	10⁶ J₄	Gravity cm/s²	Escape Vel. km/s
Mercury	0.33022	5.43	2439.7	0				370	4.25
Venus	4.8690	5.24	6051.9	0	0.027			887	10.4
Earth	5.9742	5.515	6378.140	0.00335364	1.08263	−2.54	−161	980	11.2
(Moon)	0.073483	3.34	1738	0	0.2027			162	2.37
Mars	0.64191	3.94	3397	0.00647630	1.964	36		371	5.02
Jupiter	1898.8	1.33	71492	0.0648744	14.75	−580		2312	59.6
Saturn	568.50	0.70	60268	0.0979624	16.45	−1000		896	35.5
Uranus	86.625	1.30	25559	0.0229273	12			777	21.3
Neptune	102.78	1.76	24764	0.0171	4			1100	23.3
Pluto	0.015	1.1	1151	0				72	1.1

Planet	Dist. to Sun AU	ε	Ecliptic Angle	Inclin.	Rot. Period d	Albedo	No. of Satellites
Mercury	0.38710	0.2056	7.00°	0°	58.6462	0.106	0
Venus	0.72333	0.0068	3.39°	177.3°	−243.01	0.65	0
Earth	1.00000	0.0167		23.45°	0.99726968	0.367	1
(Moon)				6.68°	27.321661	0.12	
Mars	1.52369	0.0933	1.85°	25.19°	1.02595675	0.150	2
Jupiter	5.20283	0.048	1.31°	3.12°	0.41354	0.52	16
Saturn	9.53876	0.056	2.49°	26.73°	0.4375	0.47	18
Uranus	19.19139	0.046	0.77°	97.86°	−0.65	0.51	15
Neptune	30.06107	0.010	1.77°	29.55°	0.768	0.41	8
Pluto	39.52940	0.248	17.15°	118°	−6.3867	0.3	1

Planet	T_sur K	P_sur bar	Atmospheric Composition CO₂	N₂	O₂	H₂O	H₂	He	Ar	Ne	CO
Mercury	440	2 × 10⁻¹⁵					2%	98%			
Venus	730	90	96.4%	3.4%	69 ppm	0.1%			4 ppm		20 ppm
Earth	288	1	0.03%	78.08%	20.95%	0 to 3%			0.93%	18 ppm	1 ppm
Mars	218	0.007	95.32%	2.7%	0.13%	0.03%			1.6%	3 ppm	0.07%
Jupiter	129						86.1%	13.8%			
Saturn	97						92.4%	7.4%			
Uranus	58						89%	11%			
Neptune	56						89%	11%			
Pluto	50	1 × 10⁻⁵									

Source: Lide, D. R. and Frederikse, H. P. R., Eds., *CRC Handbook of Chemistry and Physics*, 79th ed., CRC Press, Boca Raton, FL, 1998, 14–3. With permission.

Table 4.1–3   Interior, Masses, and Dimensions of the Principal Subdivisions of the Earth

	Mass $(10^{25}$ g)	Mean Density (g/cm^3)	Surface Area $(10^6$ km²)	Radius or Thickness (km)	Volume $(10^9$ km³)	Mean Moment of Inertia (spherical symmetry) $(10^{42}$ g cm²)
Core	192	11.0	151	3471	1175	90
Mantle	403	4.5			898	705
Below 1000 km	240	5.1	362	1900	474	333
Above 1000 km	163	3.9	505	970–990	424	372
Crust	2.5	2.8	510		8.9	(7)
Continental	2.0	2.75	242[a]	30	7.3	
Oceanic	0.5	2.9	268[a]	6	1.6	
Oceans and marginal seas	0.14	1.03	361[a]	3.8	1.4[a]	(<1)
Whole earth	597.6	5.52	510	6371	1083	802
Atmosphere	0.00051	0.0013[b]	—	8[c]	—	—

[a] See Reference 2.
[b] Surface value.
[c] Scale height of the "homogeneous" atmosphere.

REFERENCES

1. MacDonald, G. J. F., in Geodetic data, *Handbook of Physical Constants,* Clark, S. P., Jr., Ed., Geological Society of America, Memoir 97, 1966.
2. Poldervaart, A., Chemistry of the earth crust, Geol. Soc. Am. Spec. Pap. No. 62, p. 119, 1955.
3. Schmucker, U., in *Handbook of Geochemistry,* Vol. I, Wedepohl, K. H., Ed., Springer-Verlag, Berlin, 1969, chap. 6.

Source: Wedepohl, K. H., *Handbook of Geochemistry,* Vol. I, Springer-Verlag, Berlin, 1969. With permission.

Table 4.1–4   The Surface Areas of the Earth

	10^6 km²		10^6 km²
Continental shield region	105	Land about	
Region of young folded belts	42	29.2% of total	149
Volcanic islands in deep oceanic and suboceanic region	2		
Shelves and continental slopes region	93	Ocean about	
		70.8% of total	361
Deep oceanic region	268	Total surface	510

Source: Poldervaart, A., Geol. Soc. Am. Spec. Pap. 62, 119, 1955. With permission.

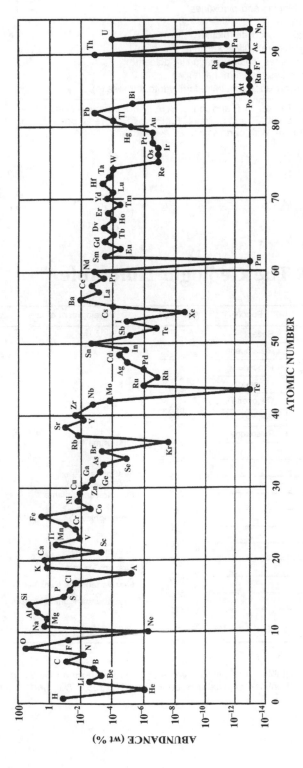

Figure 4.1–1 Crustal abundance of elements 1 to 93. (From Parker, R. L., Data of Geochemistry, 6th ed., U.S. Geological Survey Professional Paper 440-D, U.S. Geological Survey, Washington, D.C., 1967.)

Table 4.1–5   Major Mineral Components of the Earth's Crust

1. Feldspars (plagioclase, orthoclase)
2. Amphiboles and pyroxenes
3. Quartz
4. Micas (muscovite, biotite)
5. Clay minerals (illite, kaolinite, smectite)
6. Carbonates (calcite, aragonite, dolomite)
7. Evaporites (gypsum, halite)
8. Oxyhydroxides (goethite, ferrihydrite, birnessite)

Source: Bricker, O. P. and Jones, B. F., in *Trace Elements in Natural Waters,* Salbu, B. and Steinnes, E., Eds., CRC Press, Boca Raton, FL, 1995, 9. With permission.

The Geologic Timetable

Era	Period	Epoch	Age (Ma)*	Development of Life	Orogenic Events
Cenozoic	Quaternary	Recent			Pacific Coast Ranges form
		Pleistocene	1.6 —		
	Tertiary	Pilocene	5.3 —		
		Miocene	23.7 —		
		Oligocene	36.6 —		
		Eocene	57.8 —		
		Paleocene	66.4 —	First primates	Rocky Mts. form
Mesozoic	Cretaceous		144 —		
	Jurassic		208 —	First birds	Sierra Nevada form
	Triassic		245 —	First mammals First flowering plants	
Paleozoic	Permian		286 —	First reptiles	
	Pennsylvanian		320 —	First insects	Appalachian Mts. form
	Mississippian		360 —	First amphibians	
	Devonian		408 —	First land plants	
	Silurian		438 —	First vertebrates (fish)	
	Ordovician		505 —	First abundant animal fossils	
	Cambrian		570 —		
Precambrian			4500 —	First algae, bacteria	

Figure 4.1–2 The geologic timetable. *Ma = million years ago. (From Thurman, H. V., *Introductory Oceanography,* 8th ed., Prentice-Hall, Upper Saddle River, NJ, 1997. With permission.)

4.2 OCEAN BASINS

Table 4.2–1   Major Features of the Principal Ocean Basins

	Ocean			
	Pacific	Atlantic	Indian	World Ocean
Ocean area (10^6 km²)	180	107	74	361
Land area drained (10^6 km²)	19	69	13	101
Ocean area/drainage area	9.5	1.6	5.7	3.6
Average depth (m)	3940	3310	3840	3730
Area as % of total:				
Shelf and slope	13.1	19.4	9.1	15.3
Continental rise	2.7	8.5	5.7	5.3
Deep ocean floor	42.9	38.1	49.3	41.9
Volcanoes and volcanic ridges[a]	2.5	2.1	5.4	3.1
Ridges[b]	35.9	31.2	30.2	32.7
Trenches	2.9	0.7	0.3	1.7

[a] Volcanic ridges are those related to volcanic island chains that are not part of constructive plate margins, e.g., the Walvis Ridge in the South Atlantic. They do not include island arcs.

[b] These are the ridges that correspond to constructive plate margins, e.g., the Mid-Atlantic Ridge.

Source: Brown, J. et al., *The Ocean Basins: Their Structure and Evolution,* Pergamon Press, Oxford, 1989, 28. With permission.

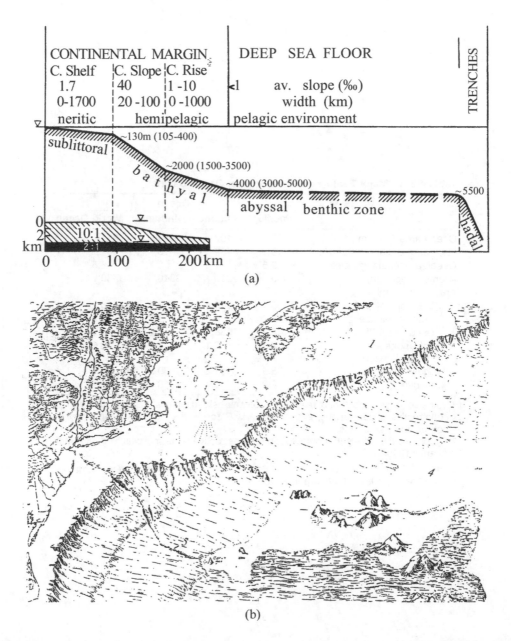

Figure 4.2–1   (a) Schematic diagram illustrating the depth zones of the seafloor. (b) Physiographic diagram of the seafloor off of the northeast coast of the United States. (From Heezen, B. C. et al., Geol. Soc. Am. Spec. Pap. 65, 1, 1959. With permission.)

4.3 CONTINENTAL MARGINS

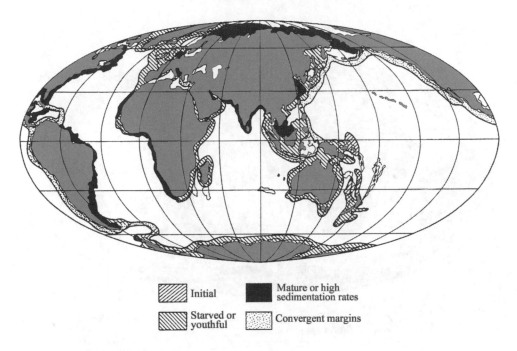

Figure 4.3–1 Global distribution of four types of continental margins. (From Emery, K. O., *Am. Assoc. Petrol. Geol. Bull.,* 64, 297, 1980. With permission.)

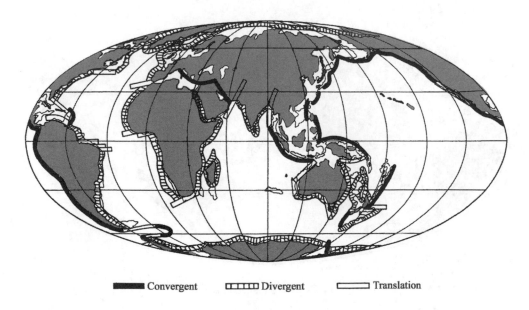

Figure 4.3–2 Global distribution of convergent, divergent, and translation type continental margins. (From Emery, K. O., *Am. Assoc. Petrol. Geol. Bull.,* 64, 297, 1980. With permission.)

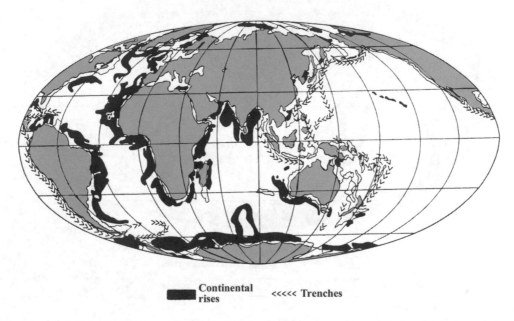

■■ Continental   <<<<< Trenches
       rises

Figure 4.3–3   Global distribution of continental rises and ocean trenches. (From Emery, K. O., *Am. Assoc. Petrol. Geol. Bull.,* 64, 297,1980. With permission.)

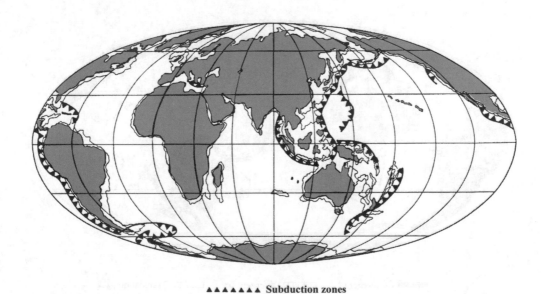

▲▲▲▲▲▲▲ Subduction zones

Figure 4.3–4   Global distribution of subduction zones. (From Kennett, J. P., *Marine Geology,* Prentice-Hall, Englewood Cliffs, NJ, 1982, 354. With permission.)

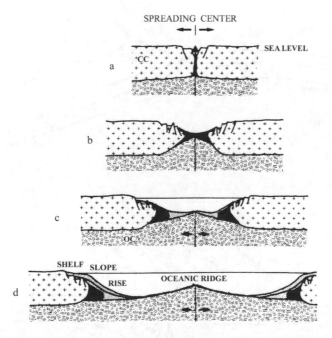

Figure 4.3–5 Evolution of Atlantic-type (=passive) continental margins. (a) Uplift of mantle material causes extension of the continental crust resulting in graben formation. (b) The continental crust thins, subsides, and splits apart, and coarse terrigenous sediments and volcanogenic deposits accumulate. (c) Rifting is followed by drifting and subsidence of the continental margins; mantle material forms new ocean crust. (d) The ocean basin widens due to seafloor spreading, and sediments cover the older parts of the seafloor. (From Seibold, E. and Berger, W. H., *The Sea Floor: An Introduction to Marine Geology,* 2nd ed., Springer-Verlag, Berlin, 1993, 45. With permission.)

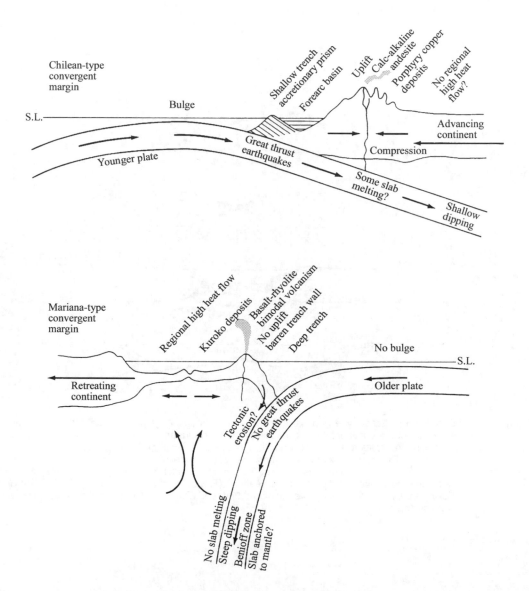

Figure 4.3–6 Schematic diagram illustrating Pacific-type (= active) continental margins. (From Uyeda, S. and Kanamori, H., *J. Geophys. Res.,* 84, 1049, 1979. With permission.)

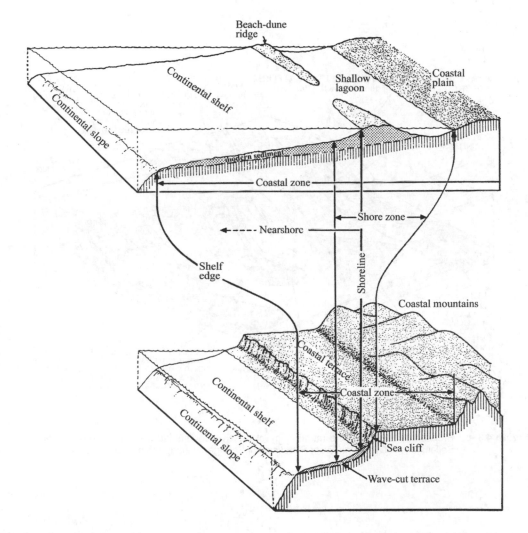

Figure 4.3–7 Coastal zone characteristics of trailing edge (passive) continental margins (top) and leading edge (active) continental margins (bottom). (From Kennett, J. P., *Marine Geology,* Prentice-Hall, Englewood Cliffs, NJ, 1982, 264. With permission.)

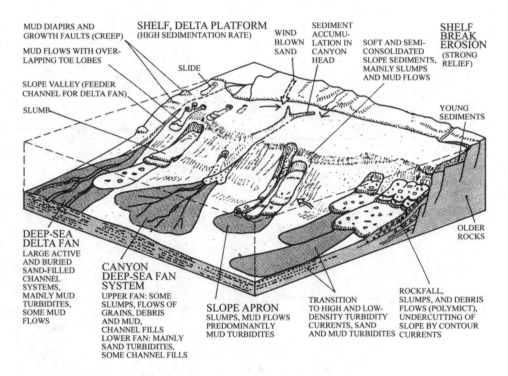

Figure 4.3–8   Processes influencing the morphology of passive continental margins. (From Einsele, G. and Seileicher, A., *Cyclic and Event Stratification,* Springer-Verlag, Berlin, 1991, 318. With permission.)

4.4 SUBMARINE CANYONS AND OCEANIC TRENCHES

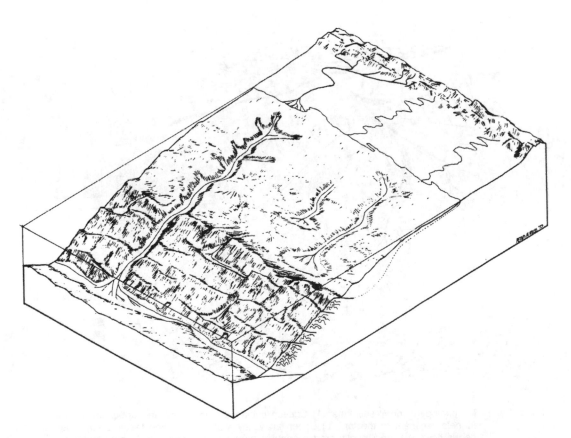

Figure 4.4–1   Diagram showing large submarine canyons as agents of sedimentary bypassing in an accretionary forearc terrain. (From Underwood, M. B. and Karig, D. E., *Geology,* 8, 432, 1980. With
permission.)

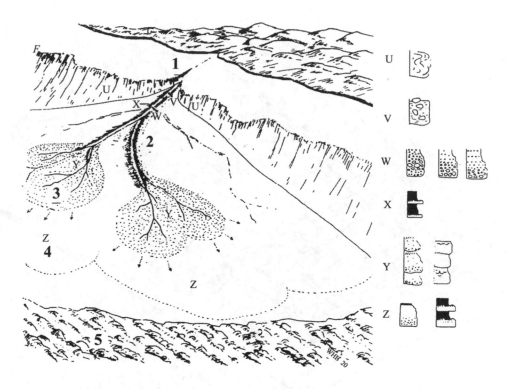

Figure 4.4–2 Morphology of deep-sea fans. (1) Canyon cutting through shelf and upper slope, traps, and funnels sediment to the fan. (2) Upper fan valley, walls with slump features (U), bottom with debris flows (V), and mostly graded coarse-grained beds. Levees with thin-bedded turbidities (X). (3) Active suprafan with distributary channel, filled with pebbly or massive sands (Y). Outer fan with classical turbidities (Z). (5) Abyssal hill region beyond the fan. Valleys between hills may have distal fan material. (From Seibold, E. and Berger, W. H., *The Sea Floor: An Introduction to Marine Geology,* 2nd ed., Springer-Verlag, Berlin, 1993, 65. With permission.)

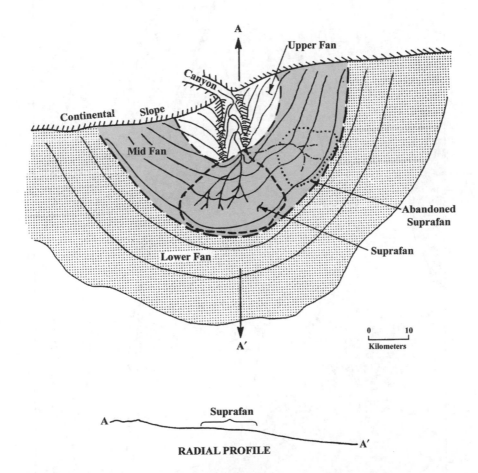

Figure 4.4–3   Schematic representation of a model for submarine fan growth emphasizing active and aban-
doned depositional lobes (suprafans). (From Normark, W.R., *Am. Assoc. Petrol. Geol. Bull.*, 54,
2170, 1970. With permission.)

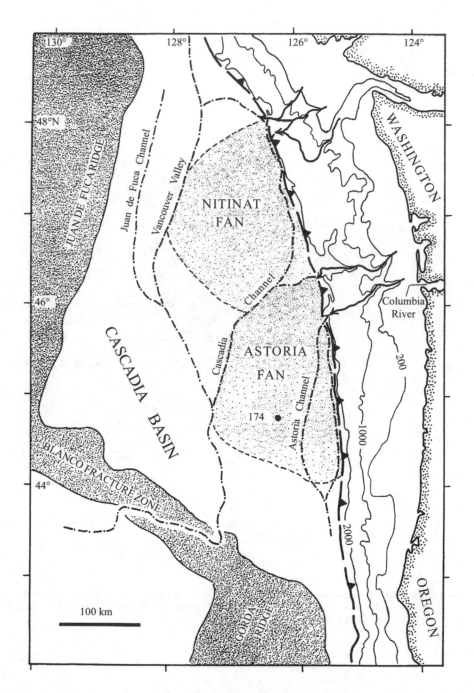

Figure 4.4–4   Astoria and Nitinat submarine fans off the Oregon and Washington coasts in the northeast Pacific
Ocean. (From Underwood, M. B., *Rev. Aquat. Sci.,* 4, 161, 1991. With permission.)

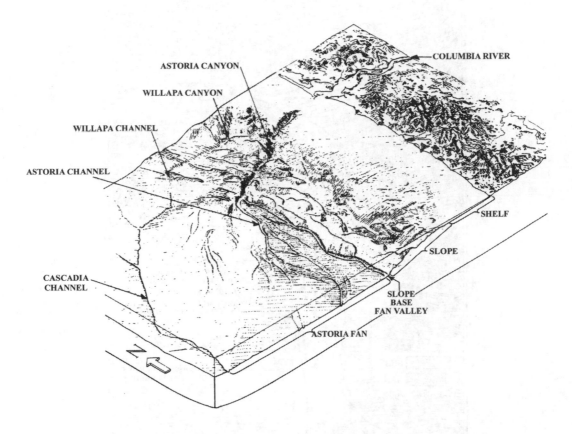

Figure 4.4–5 Detailed diagram of the Astoria submarine canyon and fan complex, offshore Oregon. (From Underwood, M. B., *Rev. Aquat. Sci.,* 4, 163, 1991. With permission.)

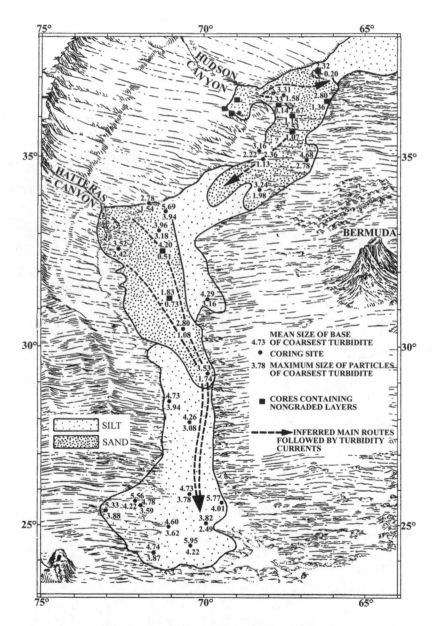

Figure 4.4–6 Sediments transported by turbidity currents through submarine canyons (Hudson and Hatteras Canyons) and accumulating as turbidities in abyssal plains off the eastern United States. (From Horn, D. R. et al., *Mar. Geol.,* 11, 287, 1971. With permission.)

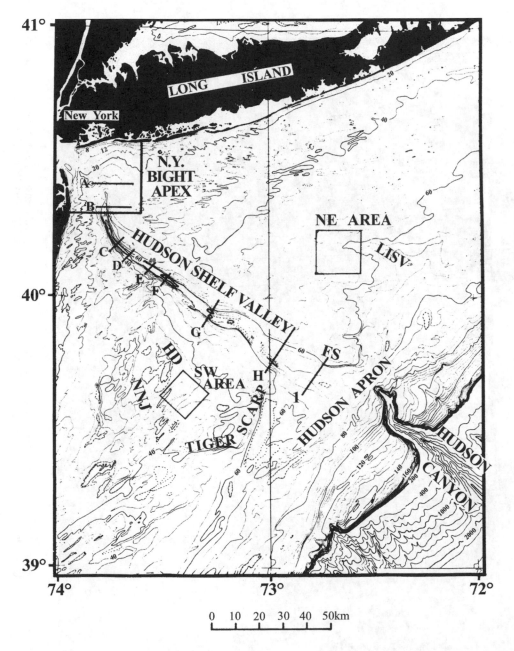

Figure 4.4–7 Bathymetric map of the Hudson Canyon and Hudson Shelf Valley. (From Uchupi, E., U.S. Geological Survey Professional Paper 5291, U.S. Geological Survey, Washington, D.C., 1970.)

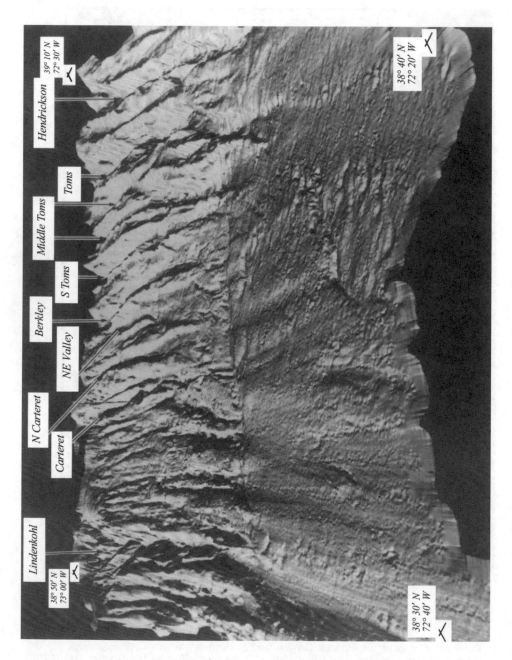

Figure 4.4–8   Image of submarine canyons along an 80-km stretch of the New Jersey continental slope from Lindenkohl Canyon to Hendrikson Canyon. (From Pratson, L. F., *Geol. Soc. Am. Bull.*, 106, 399, 1994. With permission.)

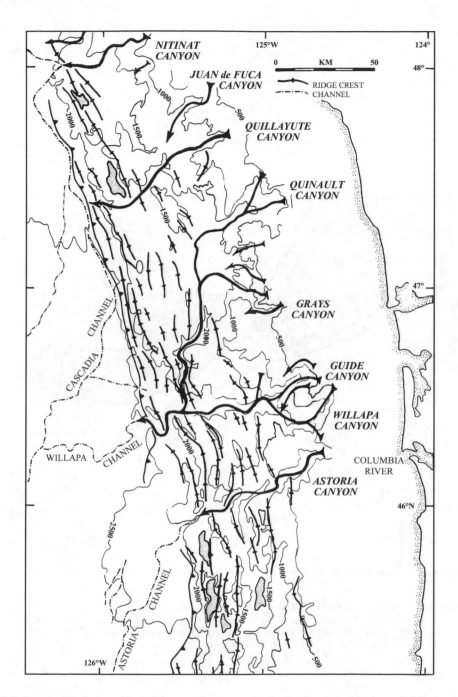

Figure 4.4–9 Simplified map of bathymetry and structure for the landward trench slope and Cascadia Basin offshore Washington and northern Oregon. Series of submarine canyons are highlighted by arrows. Stippled pattern corresponds to intraslope basins. (From Underwood, M. B., *Rev. Aquat. Sci.*, 4, 162, 1991. With permission.)

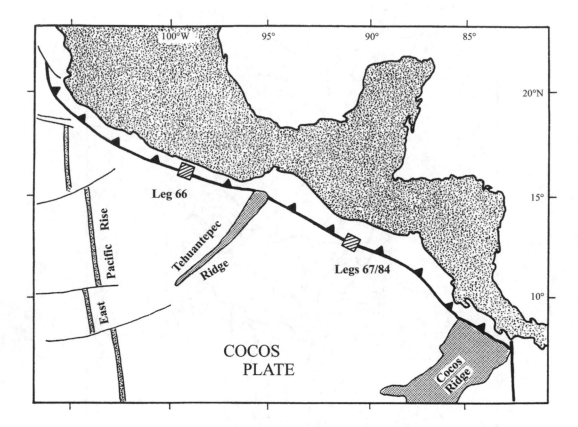

Figure 4.4–10   Geographic and tectonic setting of the Middle America Trench, east-central Pacific Ocean. The aseismic Tehuantepac Ridge marks the boundary between the continental subduction margin offshore Mexico and the subduction zone offshore Central America. (From Underwood, M. B., *Rev. Aquat. Sci.,* 4, 155, 1991. With permission.)

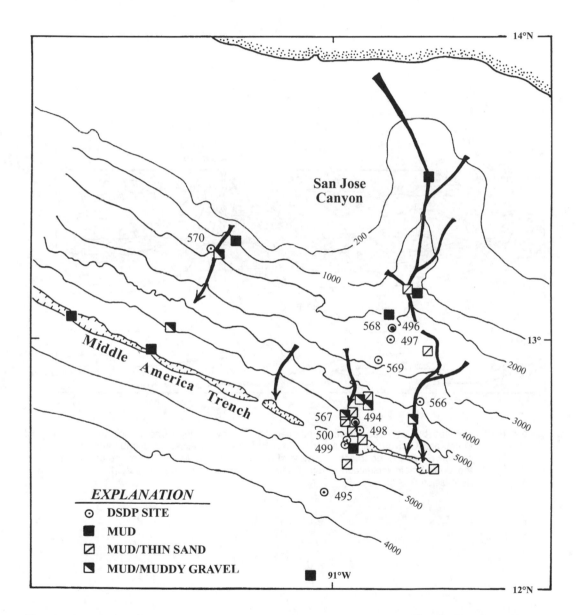

Figure 4.4–11   Map showing dominant lithologies of surface samples collected along the San Jose submarine canyon. (From Underwood, M. B., *Rev. Aquat. Sci.,* 4, 156, 1991. With permission.)

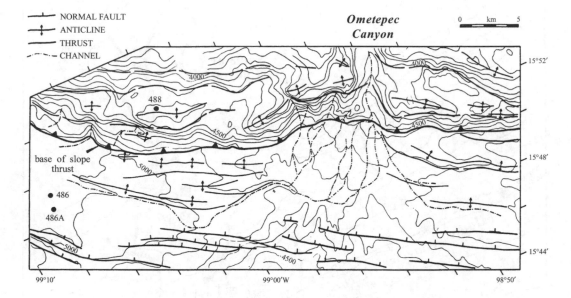

Figure 4.4–12 Detailed bathymetric and structural map for the lower slope and trench in the vicinity of Ometepec Canyon, Middle America Trench. Note the coalescing distributary channels that collectively define the trench fan. Most of the channels curve away from the canyon mouth in the down-gradient direction. (From Underwood, M. B., *Rev. Aquat. Sci.*, 4, 159, 1991. With permission.)

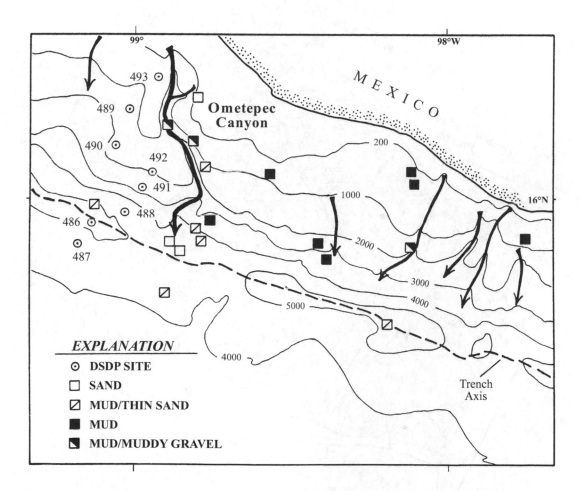

Figure 4.4–13 Map showing dominant lithologies of surface samples collected along the Ometepec Submarine Canyon and neighboring areas. Arrows highlight submarine canyons and slope gullies. Dashed line indicates position of the trench axis. (From Underwood, M. B., *Rev. Aquat. Sci.,* 4, 158, 1991. With permission.)

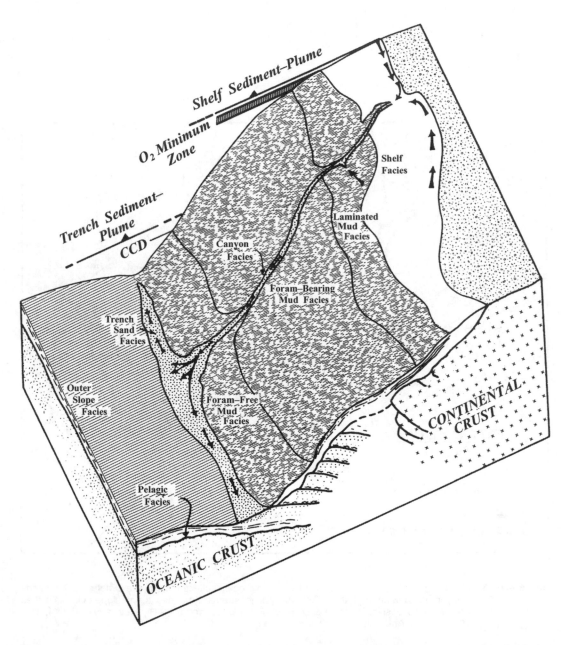

Figure 4.4–14 Sedimentary facies model for the Middle America Trench showing the influence of the Ometepec Canyon on forearc bypassing, the formation of a trench fan, and the development of a plume of suspended sediment above the trench floor. (From Underwood, M. B., *Rev. Aquat. Sci.,* 4, 160, 1991. With permission.)

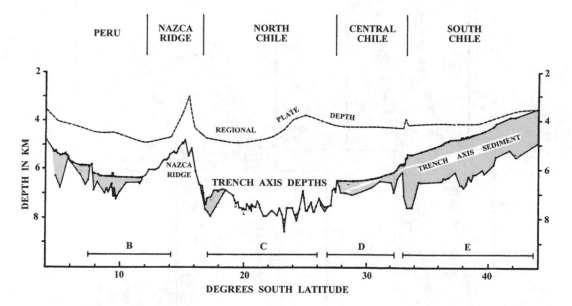

Figure 4.4–15   Axial profile of the Peru-Chile Trench, eastern Pacific Ocean. (From Schweller, W. J. et al., *Geol. Soc. Am. Mem.,* 154, 323, 1981. With permission.)

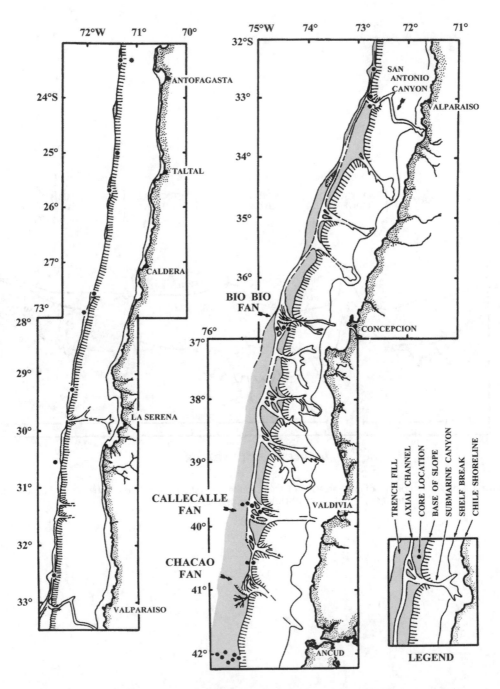

Figure 4.4–16   Map illustrating major submarine canyons, trench fans, and the axial channel of the Chile Trench.
(From Thornburg, T. M. and Kulm, L. D., *Geol. Soc. Am. Bull.,* 98, 33, 1987. With permission.)

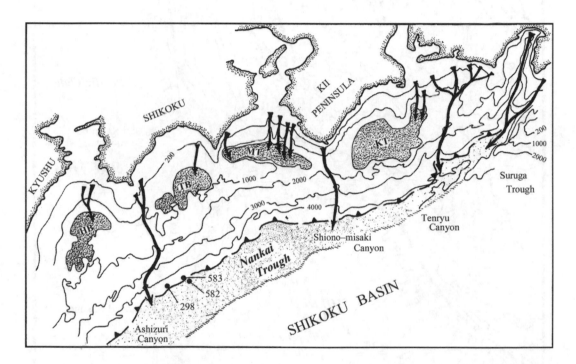

Figure 4.4–17   Simplified bathymetric map of the Nankai Trough and adjacent landward trench slope. Submarine canyons and channels are highlighted by arrows. Major forarc basins include Hyuga Basin (HB), Tosa Basin (TB), Muroto Trough (MT), and Kumano Trough (KT). Suruga Trough serves as the principal conduit for turbidity currents delivering sediment to the floor of Nankai Trough; Tenryu Canyon, Shiono-misaki, and Ashizuri Canyons also provide throughgoing pathways from the shelf edge to the trench. (From Underwood, M. B., *Rev. Aquat. Sci.,* 4, 168, 1991. With permission.)

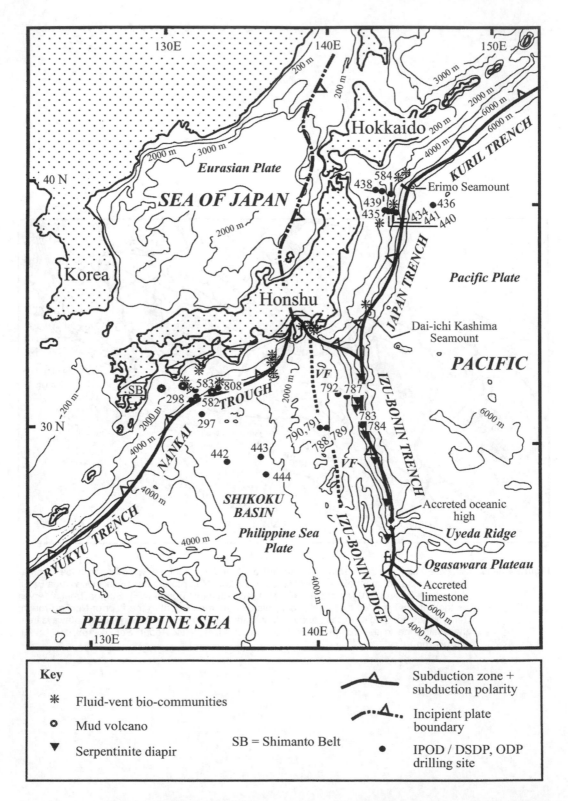

Figure 4.4–18 Map of southwest Japan and vicinity depicting major tectonic features in addition to locations of DSDP and ODP drill sites. (From Pickering K. T., in *Proceedings of the Ocean Drilling Program, Science Results,* Vol. 131, Hill, I. A., Taira, A., and Firth, J. V., Eds., Ocean Drilling Program, College Station, TX, 1993, 313. With permission.)

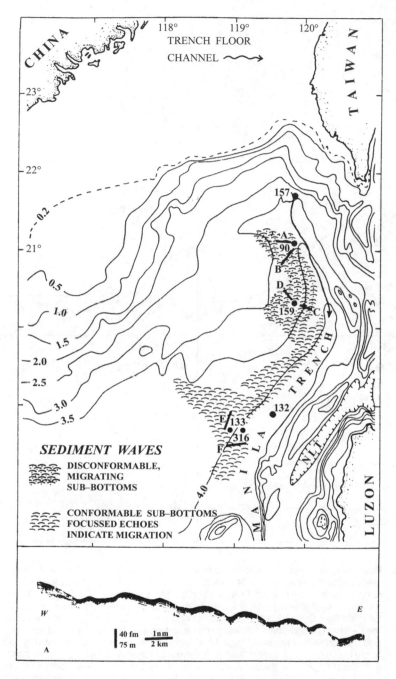

Figure 4.4–19 Map showing a field of sediment waves on the outer slope of the Manila Trench. Also shown is a seismic-reflection profile (Profile A) illustrating the geometry of the bedforms. These sediment waves were probably generated by thick turbidity currents; an impressive amount of upslope flow is also indicated by the asymmetry direction of the sediment waves and the orientation of crestlines. Bedforms decrease in size down the axial gradient of the trench. (From Damuth, J. E., *Geology,* 7, 520, 1979. With permission.)

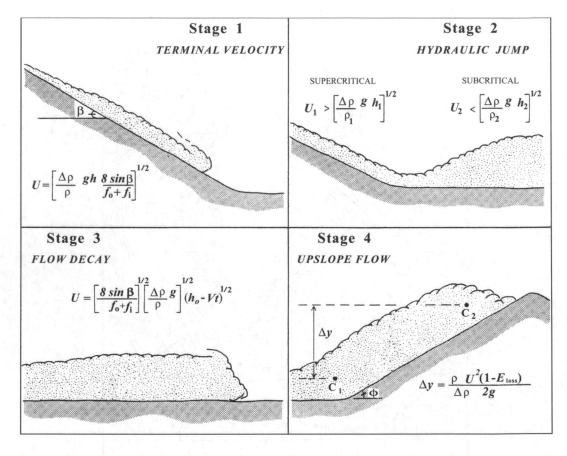

Figure 4.4–20   Four-stage evolution of a turbidity current. Stage 1 involves downslope acceleration to terminal velocity. During Stage 2, the flow encounters a break in slope and passes through a hydraulic jump (i.e., the densimetric Froude number changes from $F_r > 1$ to $F_r < 1$). Stage 3 involves gradual decay of energy on a basin floor due to frictional resistance and settling of sediment from the flow. During Stage 4, the turbidity current collides with a bathymetric obstruction, and the center of gravity shifts a finite distance upslope as kinetic energy is exchanged for potential energy. Variables are as follows: U = Mean body velocity, Δp = density contrast between turbidity current and ambient fluid, p = density of turbidity current, g = gravitational acceleration, β = slope angle, h = thickness of turbidity current, f_i = frictional coefficient at the base of the flow, f_o = frictional coefficient at the top of the flow, t = time, and v = settling rate of sedimentary particles, E_{loss} = energy loss due to frictional heating, C = center of gravity of turbidity current, Δy = vertical shift in center of gravity, ϕ = barrier angle. (From Underwood, M.B., *Rev. Aquat. Sci.*, 4, 190, 1991. With permission.)

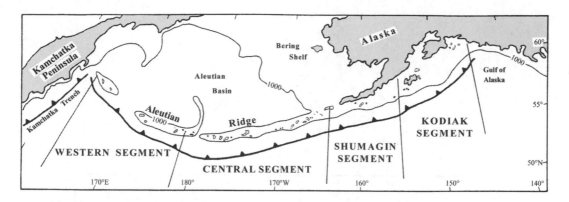

Figure 4.4–21   Map illustrating the four principal geographic segments of the Aleutian arc-trench system, North Pacific Ocean. (From Underwood, M. B., *Rev. Aquat. Sci.,* 4, 170, 1991. With permission.)

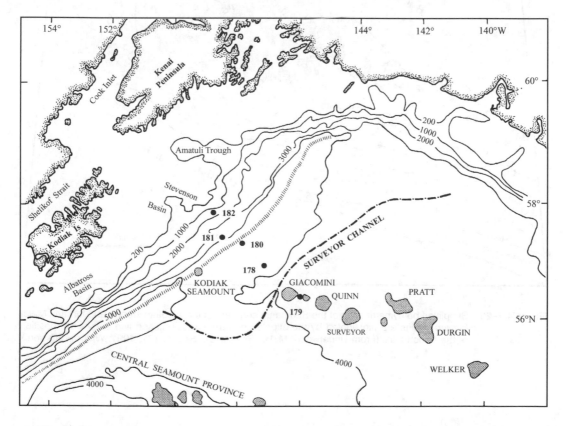

Figure 4.4–22   Simplified bathymetric map of the eastern Gulf of Alaska showing the Kodiak segment of the Aleutian forearc and trench. Also shown are major shelf basins, plus seamounts on the sub-ducting Pacific plate (stippled pattern). Cross-hatched line marks the position of the trench axis. (From Underwood, M. B., *Rev. Aquat. Sci.,* 4, 170, 1991. With permission.)

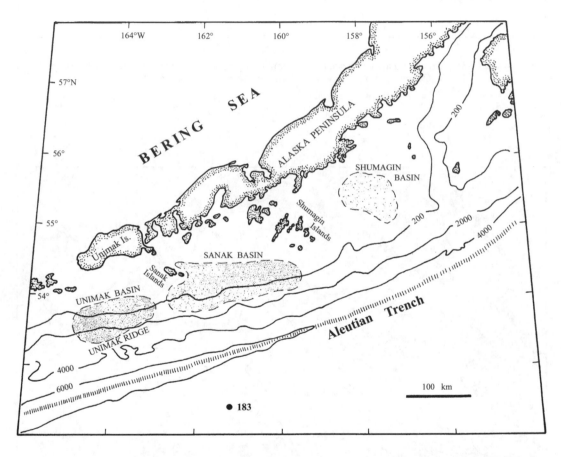

Figure 4.4–23   Simplified bathymetric map of the Shumagin segment of the Aleutian Trench and forearc. Major forearc basins are highlighted by the stippled pattern. Cross-hatched line shows the position of the trench axis. (From Underwood, M. B., *Rev. Aquat. Sci.,* 4, 171, 1991. With permission.)

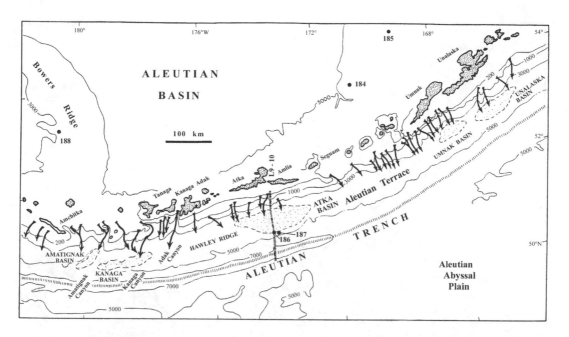

Figure 4.4–24 Simplified bathymetric map of the central segment of the Aleutian Trench and forearc. Note that the upper trench slope is highlighted by numerous submarine canyons (arrows), but none of these canyons can be traced all the way downslope to the trench floor. Also shown are major forearc basins of the mid-slope Aleutian Terrace and the trackline for USGS seismic-reflection profile L9–10. (From Underwood, M. B., *Rev. Aquat. Sci.,* 4, 173, 1991. With permission.)

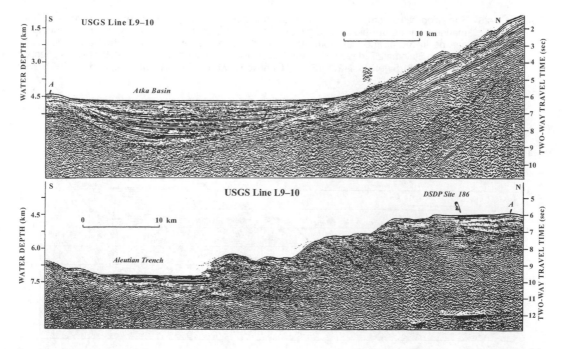

Figure 4.4–25 Unmigrated seismic-reflection profile across Atka Basin, central segment of the Aleutian forearc (see Figure 4.4–24 for location of trackline). The two segments of this profile join at point A. Note that the Atka Basin is filled almost to the brink here; other profiles show complete infilling of the Atka Basin along parallel tracklines. Accretionary ridges on the lower slope provide less than 320 m of relief to block unconfined turbidity currents. (From Underwood, M. B., *Rev. Aquat. Sci.,* 4, 174, 1991. With permission.)

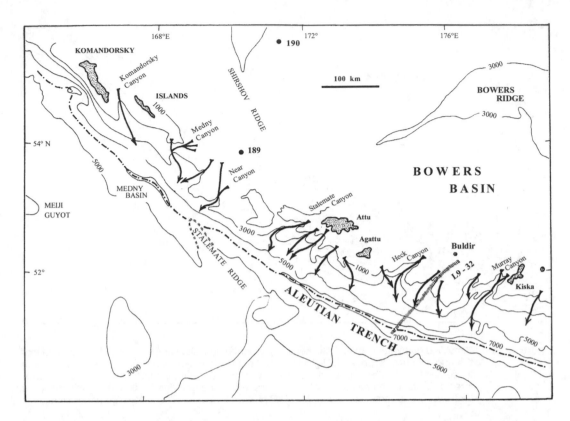

Figure 4.4–26 Simplified bathymetric map of the western segment of the Aleutian Trench and forearc. Sub-marine canyons are highlighted by arrows; as in the central segment, none of these features can be traced all the way from the arc platform to the trench floor. Note the collision zone between Stalemate Ridge and the subduction front and the trackline for USGS seismic-reflection profile L9-32 (Figure 4.4–27). (From Underwood, M. B., *Rev. Aquat. Sci.*, 4, 175, 1991. With permission.)

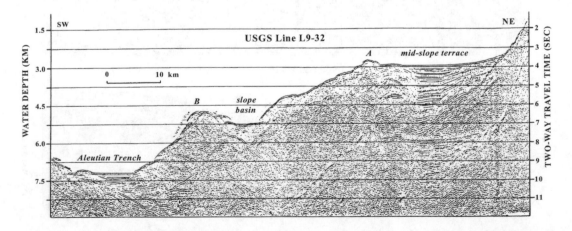

Figure 4.4–27 Unmigrated seismic-reflection profile across the western Aleutian forarc and trench (see Figure 4.4–26 for location of trackline). Note that the mid-slope basin is filled to within 225 m of the trench-slope break (Ridge A). Ridges on the lower slope display maximum relief of 540 m (e.g., Ridge B). (From Underwood, M. B., *Rev. Aquat. Sci.*, 4, 176, 1991. With permission.)

4.5 PLATE TECTONICS, MID-OCEAN RIDGES, AND OCEAN CRUST FORMATION

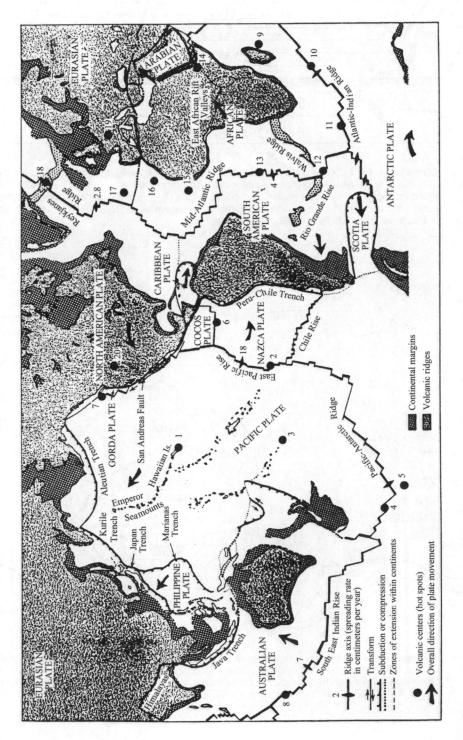

Figure 4.5–1 Major lithospheric plates and their boundaries, showing principal hot spots. Legend: 1. Hawaii, 2. Easter Island, 3. Macdonald Seamount, 4. Bellany Island, 5. Mt. Erebus, 6. Galapagos Islands, 7. Cobb Seamount, 8. Amsterdam Island, 9. Reunion Island, 10. Prince Edward Island, 11. Bouvet Island, 12. Tristan da Cunha, 13. St. Helena, 14. Afar, 15. Cape Verde Islands, 16. Canary Islands, 17. Azores, 18. Iceland, 19. Eifel, 20. Yellowstone. (From Brown, J. et al., in *Ocean Circulation*, 1989. With permission of Pergamon Press, Headington Hill Hall, Oxford, U.K.)

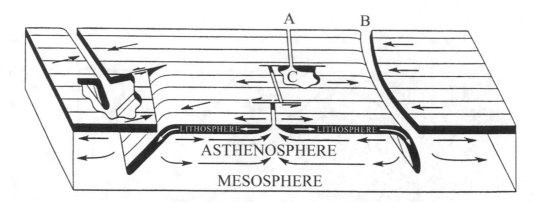

Figure 4.5–2   According to the theory of plate tectonics, the lithosphere of the earth is broken into a mosaic of a dozen or so dynamic rigid plates that move with respect to each other on a partially molten asthenosphere. The plates originate at mid-ocean ridges (A), subduct into the underlying asthenosphere at trenches (B), and slide by each other at transform faults (C). (From Isacks, B. L. et al., *J. Geophys. Res.*, 73, 5855, 1968. With permission.)

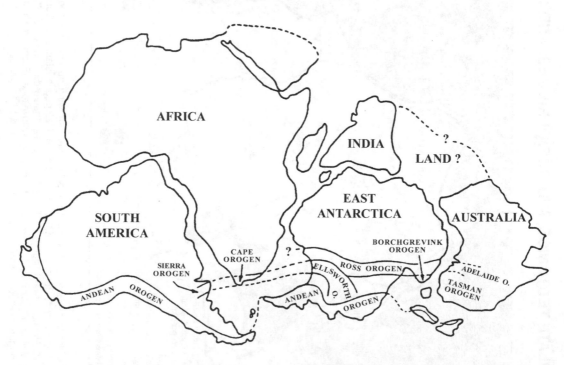

Figure 4.5–3   Illustration of the tectonic arrangement between the continents, landmasses, and orogens (including the late Mesozoic-early Cenozoic Andean Orogen) that composed the supercontinent of Gondwana before it fragmented during the Mesozoic and Cenozoic. (From Berkman, P. A., *Rev. Aquat. Sci.*, 6, 295, 1992. With permission.)

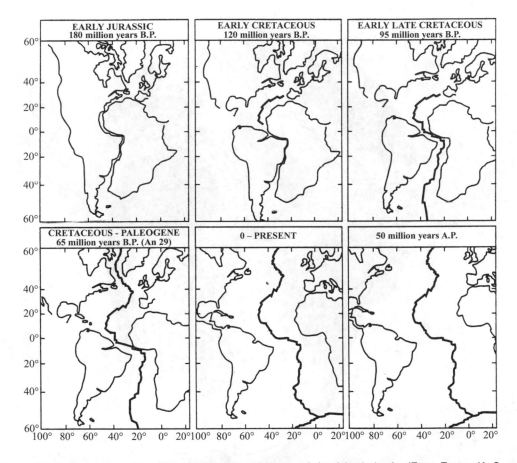

Figure 4.5–4 Formation of the Mid-Atlantic Ridge system and the Atlantic basin. (From Emery, K. O. and Uchupi, E., *The Geology of the Atlantic Ocean,* Springer-Verlag, New York, 1984. With permission.)

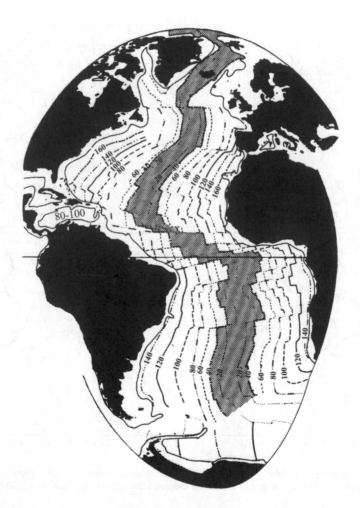

Figure 4.5–5 Age of the seafloor in the Atlantic Ocean. (From Seibold, E. and Berger, W. H., *The Sea Floor: An Introduction to Marine Geology*, 2nd ed., Springer-Verlag, Berlin, 1993, 36. With permission.)

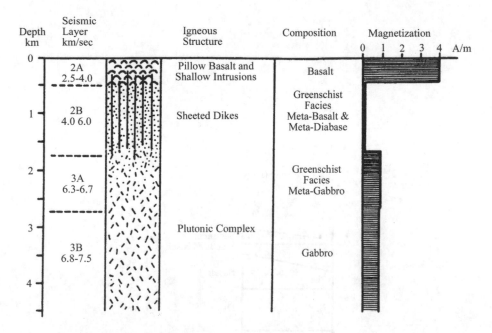

Figure 4.5–6 Model of the magnetization structure of the oceanic crust. (From Lowrie, W., in *Deep Drilling Results in the Atlantic Ocean: Ocean Crust,* Talwani, M. et al., Eds., Technical Volume, American Geophysical Union, Washington, D.C., 1979, 135. With permission.)

Table 4.5–1 Models of Oceanic Seismic Structure

Three-Layer Model			Multiple-Layer Model[a]		
Layer	Velocity V_p, km/s	Thickness, km	Layer	Velocity V_p, km/s	Thickness, km
1	~2.0	~0.5	1	1.7–2.0	0.5
2	5.07 ± 0.63	1.71 ± 0.75	2A	2.5–3.8	0.5–1.5
			2B	4.0–6.0	0.5–1.5
3	6.69 ± 0.26	4.86 ± 1.42	3A	6.5–6.8	2.0–3.0
			3B	7.0–7.7	2.0–5.0
Mantle	8.13 ± 0.24	—	Mantle	8.1	—

[a] Dividing layer 2 into layers 2A, B, and C having average velocities of 3.64, 5.19, and 6.09 km/s, respectively, has been proposed.

Source: Salisbury, M. H. et al., in *Deep Drilling Results in the Atlantic Ocean: Ocean Crust,* Talwani, M. et al., Eds., Technical Volume, American Geophysical Union, Washington, D.C., 1979, 113. With permission.

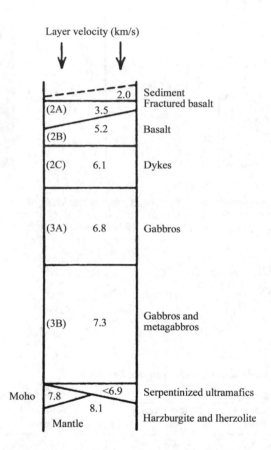

Figure 4.5–7   Seismic velocities recorded in the oceanic crust and upper mantle. (From Anderson, R. N., *Marine Geology: A Planet Earth Perspective,* John Wiley & Sons, New York, 1986, 68. Reproduced by permission of copyright owner, John Wiley & Sons, Inc.)

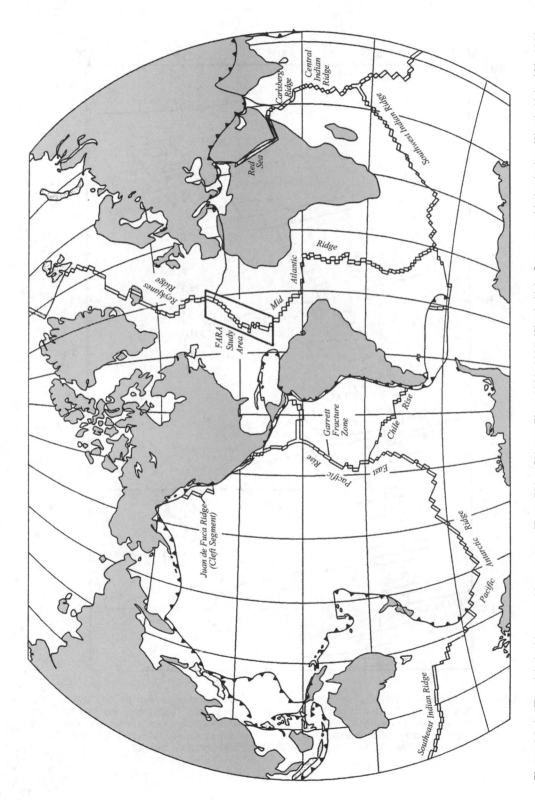

Figure 4.5–8   The global mid-ocean ridge system. (From Ridge Science Plan: 1993–1997, Woods Hole Oceanographic Institution, Woods Hole, MA, 1992. With permission.)

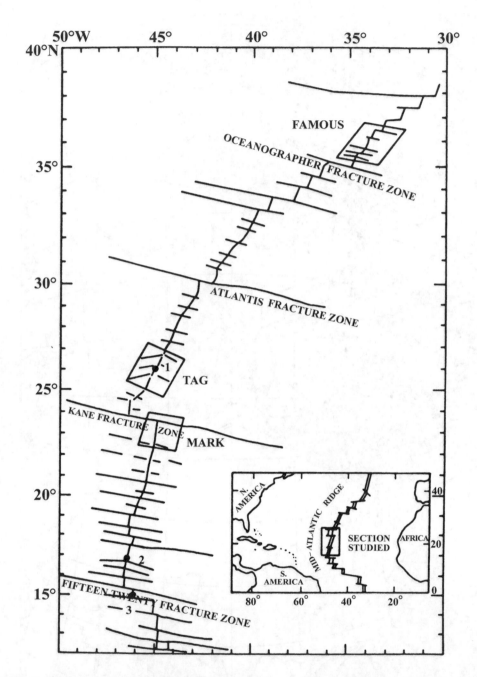

Figure 4.5–9 Map of the Mid-Atlantic Ridge between 12°N and 40°N showing major fracture zones, small-offset transform faults, and hydrothermal vent locations (e.g., TAG) along the slow-spreading ridge system. (From Rona, P. A., The Central North Atlantic Ocean Basin and Continental Margins: Geology, Geophysics, Geochemistry, and Resources, Including the Trans-Atlantic Geotraverse [TAG], NOAA Atlas 3, NOAA/ERL, Washington, D.C., 1980.)

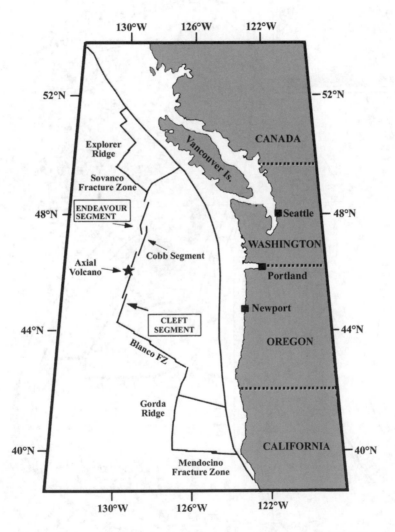

Figure 4.5–10   Seafloor spreading centers along the Explorer, Juan de Fuca, and Gorda Ridges in the northeast Pacific. (From Wright, D. and McDuff, R. E., *Ridge Events,* 9, 11, 1998. With permission.)

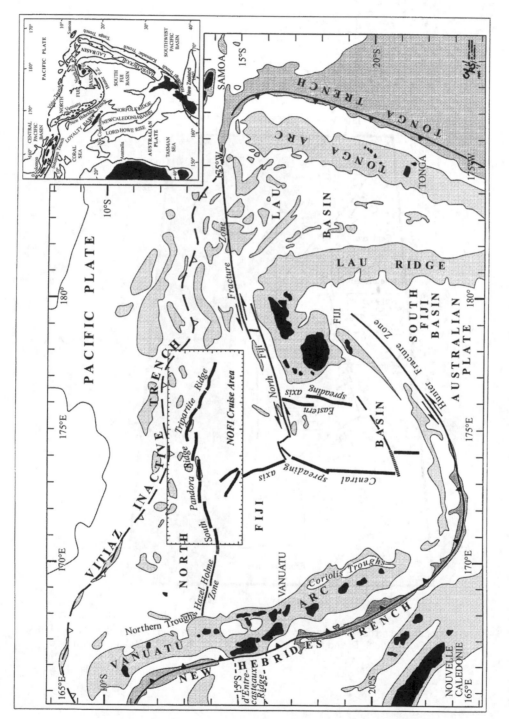

Figure 4.5–11   The North Fiji Basin in the southwest Pacific, illustrating the actively spreading axis (black lines) (SPR: South Pandora Ridge; TR: Tripartite Ridge, and HHR: Hazel Holmes Ridge). (From Lagabrielle, Y. et al., *InterRidge News*, 4, 30, 1995. With permission.)

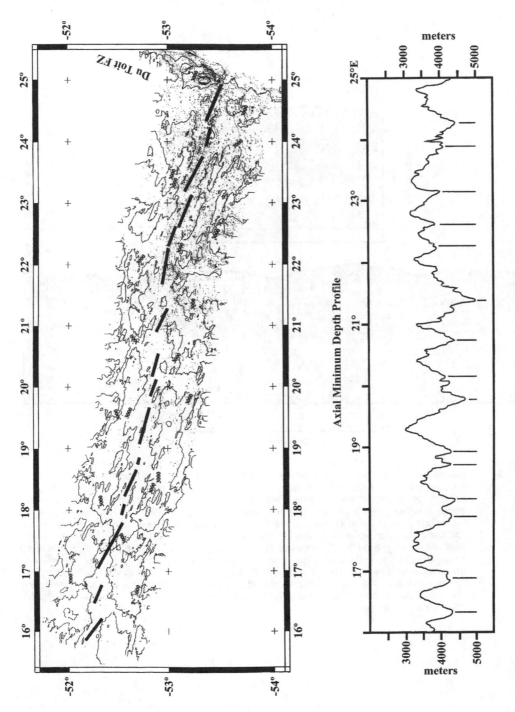

Figure 4.5–12   Bathymetric and axial minimum depth profile of the southwest Indian Ridge between 15°E and 25°E. Bathymetric contour interval is 300 m. Dark contour lines are at 1500-m intervals. Bold lines on bathymetric map show second-order ridge segments. Thin lines on axial minimum depth profile exhibit second-order ridge segment boundaries. (From Grindlay, N. R. et al., *InterRidge News*, 5, 7, 1996. With permission.)

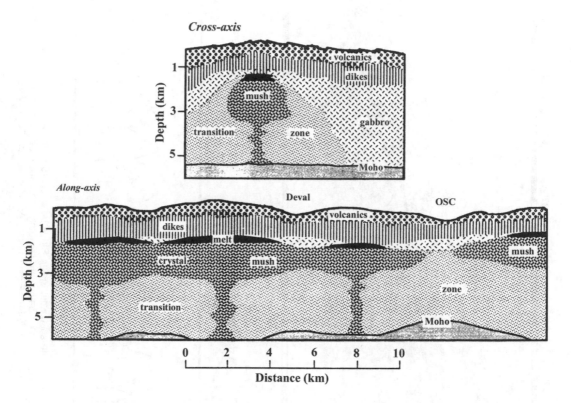

Figure 4.5–13   Interpretive model of a magma chamber along a fast-spreading (high magma supply) ridge like the East Pacific Rise based on recent geophysical and petrological constraints. The essential elements of this model are a narrow, sill-like body of melt 1 to 2 km below the ridge axis that grades downward into a partially solidified crystal mush zone, which is in turn surrounded by a transition zone to the solidified, but still hot, surrounding rock. The solidus, which defines the limit of magma, can occur anywhere from the boundary of the mush zone to the edges of the axial low-velocity zone (LVZ). Because the solidus may not be isothermal and significant lithologic variations can occur in the lower layered gabbros, isolated pockets of magma with low melt percentages can occur throughout the LVZ. Eruptions will mainly tap the molten, low-viscosity melt lens. The relative volumes of melt and mush vary along the ridge axis, particularly near ridge axis discontinuities. (From Sinton, J. M. and Detrick, R. S., *J. Geophys. Res., 97*, 197, 1992. With permission.)

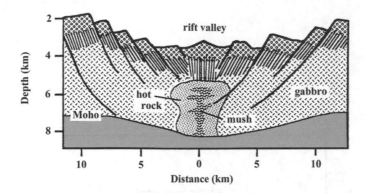

Figure 4.5–14 Interpretive model of a magma chamber beneath a slow-spreading (low magma supply) ridge like the Mid-Atlantic Ridge based on recent geophysical and petrological constraints. Such ridges are unlikely to be underlain by an eruptable magma lens in any steady-state sense. A dike-like mush zone is envisioned below the rift valley, forming small sill-like intrusive bodies that progressively crystallize to form oceanic crust. Eruptions will be closely coupled in time to injection events of new magma from the mantle. Faults bordering the rift valley may root in the brittle-ductile transition within the partially molten magma chamber. (From Sinton, J. M. and Detrick, R. S., *J. Geophys. Res.*, 97, 197, 1992. With permission.)

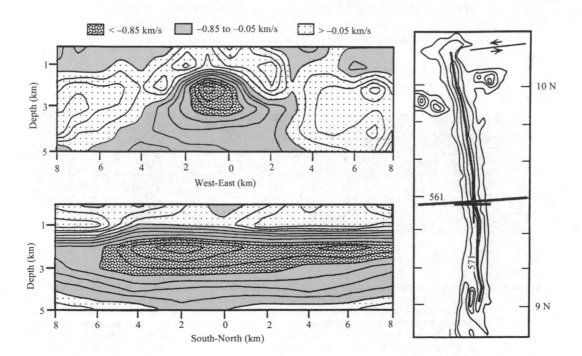

Figure 4.5–15 Vertical sections displaying the anomalous P-wave velocity structure across (top) and along (bottom) the East Pacific Rise crest near 9°30′N from the seismic tomography study of Toomey et al. (1990). The location of these sections is depicted by the solid lines in the map on the right. The largest velocity anomaly (–0.85 km/s) is confined to a narrow (<2 km-wide), thin (<1 to 1.5 km-thick) zone in the mid-crust beneath the rise axis. This body is surrounded by a broader seismic low-velocity zone (LVZ), shaded gray, which has a width of 10 to 12 km and extends to the base of the crust. The relatively small velocity anomaly associated with the bulk of this LVZ precludes the existence of a large molten magma chamber. (From Sinton, J. M. and Detrick, R. S., *J. Geophys. Res.*, 97, 197, 1992. With permission.)

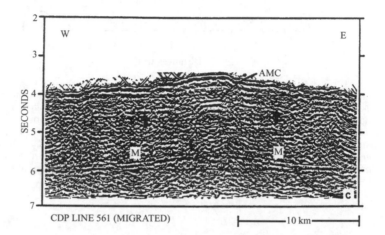

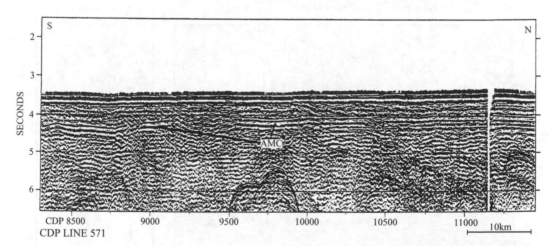

Figure 4.5–16 Multichannel seismic reflection profiles from the survey of Detrick et al. (1987) across (top) and along (bottom) of the East Pacific Rise near 9°30′N (see Figure 4.5–15 for the location). A high-amplitude, subhorizontal reflector is present beneath the rise axis. This event can be unequivocally tied to the top of the seismic LVZ shown in Figure 4.5–15 and has been interpreted as a reflection from the top of a narrow, sill-like molten body or axial magma chamber (AMC) in the crust. Note the remarkable continuity of this event along the rise axis. M is the Moho. (From Sinton, J. M. and Detrick, R. S., *J. Geophys. Res.,* 97, 197, 1992. With permission.)

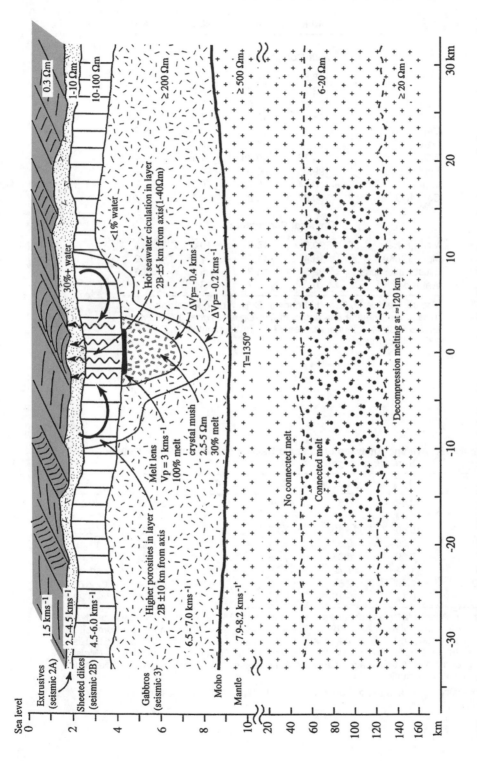

Figure 4.5–17 Magmatic anatomy of a slow-spreading ridge. Combined interpretation based on data from seismic, controlled-source electromagnetic, and magnetotelluric experiments. Seafloor bathymetry and seismic layer boundaries are quantitatively accurate based on swath bathymetry measurements and seismic modeling. Similarly, estimates of electrical resistivity, porosity, melt content, and temperature are quantitative estimates based on modeling and interpretation. Note the 10:1 break in scale at 10 km depth. (From Constable, S. et al., *InterRidge News*, 6, 21, 1997. With permission.)

Table 4.5–2   Calculated Melt Compositions and Other Physical Parameters for Melting beneath Ocean Ridges

	$P_o =$			
	40 kb	30 kb	20 kb	14 kb
Melt Compositions				
SiO_2	48.50	49.03	49.85	50.40
Al_2O_3	12.70	14.20	15.20	16.40
FeO	9.53	8.80	8.22	7.70
MgO	15.10	13.60	11.80	10.20
CaO	11.10	11.90	11.30	9.90
Na_2O	1.31	1.61	2.13	2.70
TiO_2	0.72	0.88	1.14	1.40
Sum	99.0	100.0	99.6	98.7
Mg#	0.74	0.73	0.72	0.70
CaO/Al_2O_3	0.87	0.83	0.74	0.60
Physical Parameters				
Mean F (%)	20.0	15.6	10.8	7.8
Mean P (kb)	16.4	11.7	7.3	4.9
Crustal thickness	22.5	13.8	6.8	3.6
Water depth	0	1.7	3.3	4.1

Source: Detrick, R. S. and Langmuir, C. H., The geometry and dynamics of magma chambers, in *Mid-Ocean Ridge: A Dynamic Global System,* National Academy Press, Washington, D. C., 1988, 123. With permission.

Table 4.5–3   Average Analyses of Oceanic Basalts and Basalt Glass Related to Spreading Ridges

	1	2	3	4	5	6
SiO_2	50.68	50.67	49.94	50.19	50.93	49.61
TiO_2	1.49	1.28	1.51	1.77	1.19	1.43
Al_2O_3	15.60	15.45	17.25	14.86	15.15	16.01
FeO*	9.85	9.67	8.71	11.33	10.32	11.49
MgO	7.69	8.05	7.28	7.10	7.69	7.84
CaO	11.44	11.72	11.68	11.44	11.84	11.32
Na_2O	2.66	2.51	2.76	2.66	2.32	2.76
K_2O	0.17	0.15	0.16	0.16	0.14	0.22

Note: 1, Atlantic, 51 glass analyses (Melson et al., 1975); 2, Atlantic, 155 glass analyses (data from Melson et al., 1975, Frey et al., 1974, and unpublished data from leg 37 and FAMOUS); 3, average oceanic tholeiite (Engel et al., 1965); 4, East Pacific Rise, 38 glass analyses (Melson et al., 1975); 5, Indian Ocean, 12 glass analyses (Melson et al., 1975); 6, average oceanic tholeiite (Cann, 1971). See original source for full reference citations.

Source: Bryan, W. B. et al., *J. Geophys. Res.,* 81, 4285, 1976. With permission.

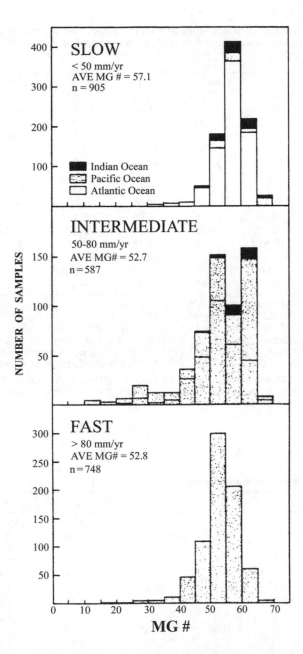

Figure 4.5–18 Histograms of over 2200 glass compositions from mid-ocean ridges at three different spreading rate ranges, keyed according to ocean. Slow-spreading centers include the Mid-Atlantic Ridge, Cayman Trough, Galapagos Rift spreading center west of about 96°W, Southwest Indian Ridge, Central Indian Ridge, and Carlsbad Ridge. Intermediate-spreading centers include the Galapagos spreading center east of 96°W, East Pacific Rise north of the Rivera Transform and west of 145°W (Pacific Antarctic plate boundary), Juan de Fuca spreading center, and Southeast Indian Ridge. All fast-spreading ridge lavas derive from the East Pacific Rise. (From Sinton, J. M. and Detrick, R. S., *J. Geophys. Res.*, 97, 197, 1992. With permission.)

Table 4.5–4   Chemical Characteristics of Some Potential Primary Ocean-Floor Basaltic
Magmas (recalculated to 100% on an anhydrous basis)

	1	2	3	4	5	6
SiO_2	49.7	49.5	49.1	50.8	50.0	50.5
TiO_2	0.72	0.81	0.62	0.62	0.65	0.76
Al_2O_3	16.4	15.7	16.5	14.4	14.5	15.6
FeO	7.89	7.45	8.78	6.89	8.26	7.3
MnO	0.12	0.15	0.15	0.10	n.d.	0.2
MgO	10.1	10.0	10.3	12.0	11.7	9.4
CaO	13.2	13.0	12.4	13.6	13.0	13.7
Na_2O	2.00	1.95	1.92	1.43	1.71	2.1
K_2O	0.01	0.17	0.07	0.07	0.03	0.4
P_2O_5	—	0.08	0.06	—	—	—
Cr_2O_3	0.07	0.14	0.06	0.10	—	—
Mg'-value	0.72	0.73	0.70	0.77	(0.72)	0.72
CaO/Al_2O_3	0.81	0.82	0.75	0.94	0.90	0.88
Ni ppm	320	249	232	—	—	—
Sm ppm	1.60	1.95	1.37	—	—	—
La/Sm	0.63	1.81	1.57	—	—	—

1. Frey et al., *J. Geophys. Res.*, 79, 5507, 1974; 3–14.
2. Rhodes (unpublished data); Chain 43 #23, 45°N.
3. Langmuir et al., *Earth Planet. Sci. Lett.,* 36, 133, 1977; FAMOUS, 527-1-1.
4. Donaldson and Brown, *Earth Planet. Sci. Lett.,* 37, 81, 1977; avg. melt inclusion in spinel.
5. Dungan and Rhodes, *Contr. Mineral. Petrol.,* 79, 1979; interpolated from melt inclusion data.
6. Watson, *J. Volcanol. Geotherm. Res.,* 1, 73, 1976; interpolated from melt inclusion data.

Source: Rhodes, J. M. and Dungan, M. A., in *Deep Drilling Results in the Atlantic Ocean: Ocean Crust,* Talwani, M. et al., Eds., Technical Volume, American Geophysical Union, Washington, D.C., 1979, 262. With permission.

Table 4.5–5   Oceanic Ultramafic Rocks (wt. %)

	1	2	3	4
SiO_2	45.4	45.9	46.07	45.95
TiO_2	0.1	0.2	0.24	0.17
Al_2O_3	1.7	3.7	3.60	3.69
(FeO)	8.3	8.3	8.35	8.29
MnO	0.1	0.1	0.11	0.11
MgO	42.9	38.6	38.42	38.53
CaO	0.7	2.3	2.17	2.34
Na_2O	0.2	0.3	0.49	0.37
K_2O	0.05	0.05	0.05	0.05
Cr_2O_3	0.3	0.3	0.3	0.3
NiO	0.2	0.2	0.2	0.2

1. Average oceanic harzburgite (75 analyses).
2. Average oceanic lherzolite (64 analyses).
3. Calculated lherzolite as the sum of 87% average harzburgite + 13% of primary tholeiitic melt under 8 to 10 kbar.
4. Calculated lherzolite as the sum of 86.6% average harzburgite + 13.4% of primary tholeiitic melt under 3 to 5 kbar.

Source: Dmitriev, L. V. et al., in *Deep Drilling Results in the Atlantic Ocean: Ocean Crust,* Talwani, M. et al., Eds., Technical Volume, American Geophysical Union, Washington, D.C., 1979, 302. With permission.

4.6 HEAT FLOW

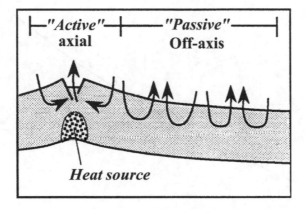

Figure 4.6–1 Illustration of convective regimes in ocean crust. High-temperature circulation in the active axial cell is driven by a magmatic or hot rock heat source, whereas lower temperature (<200°C) off-axis circulation is driven by passive cooling of the crust and lithosphere. (From Alt, J. C., in *Seafloor Hydrothermal Systems: Physical, Chemical, Biological, and Geological Interactions,* Humphris, S. E. et al., Eds., Geophysical Monograph 91, American Geophysical Union, Washington, D.C., 1995, 88. With permission.)

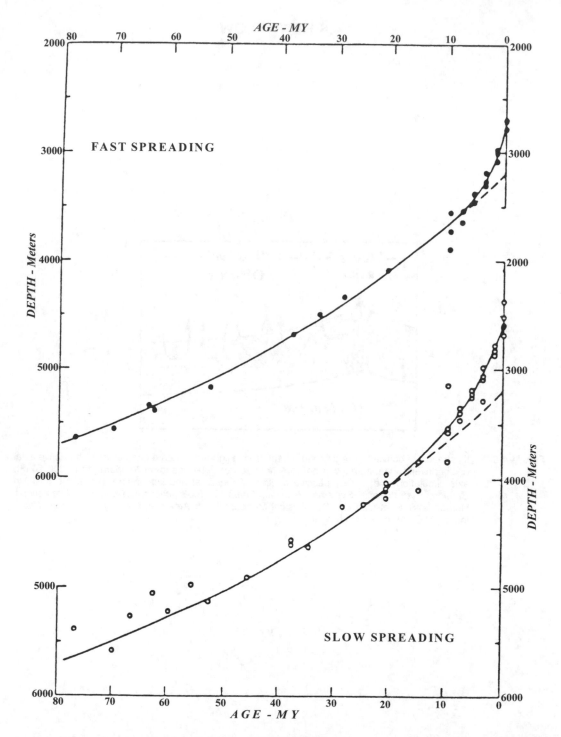

Figure 4.6–2 Both fast- and slow-spreading rate ridges subside exponentially away from the volcanic centers as the rock slowly cools. Eighty million years after formation, the lithosphere is still cooling and sinking because of thermal contraction. (From Anderson, R. N., *Marine Geology: A Planet Earth Perspective,* John Wiley & Sons, New York, 1986, 73. Reproduced by permission of copyright owner, John Wiley & Sons, Inc.)

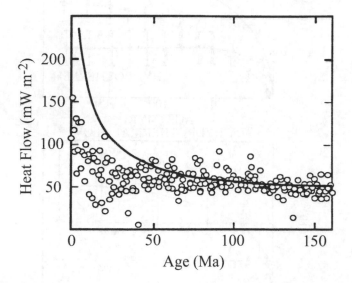

Figure 4.6–3 Oceanic heat flow vs. age of crust to ~160 Ma. Measured values (open circles) fall below theoretical conductive cooling curve (heavy line), indicating convective cooling of young ocean crust by circulating seawater. Measured values are averaged over 2 Ma intervals for the Atlantic, Pacific, and Indian Oceans. (From Stein, C. A. and Stein, S., *J. Geophys. Res.,* 99, 3081, 1994. With permission.)

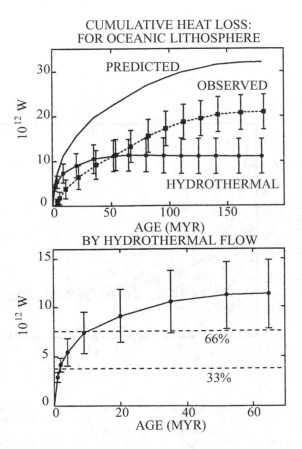

Figure 4.6–4   Top: The cumulative hydrothermal heat flux as a function of age is inferred from the difference between that predicted by the GDH1 plate cooling model and that integrated from seafloor observations. The lines connect the points whose values were computed. For clarity, the 1 Myr point is not plotted, and the observed values are offset. Error bars are one standard deviation of the data. Bottom: Cumulative inferred hydrothermal heat flux for 0 to 65 Myr values are the same as in the top panel, except for the 1 Myr estimate, which here reflects the greater uncertainty in the near-axial value resulting from incorporating uncertainties in the thermal model and crustal area. Approximately one third of the inferred hydrothermal heat flux occurs in crust younger than 1 Myr, and another one third occurs in crust older than 9 Myr. (From Stein, C. in *Seafloor Hydrothermal Systems: Physical, Chemical, Biological, and Geological Interactions,* Humphris, S. E. et al., Eds., Geophysical Monograph 91, American Geophysical Union, Washington, D.C., 1995, 428. With permission.)

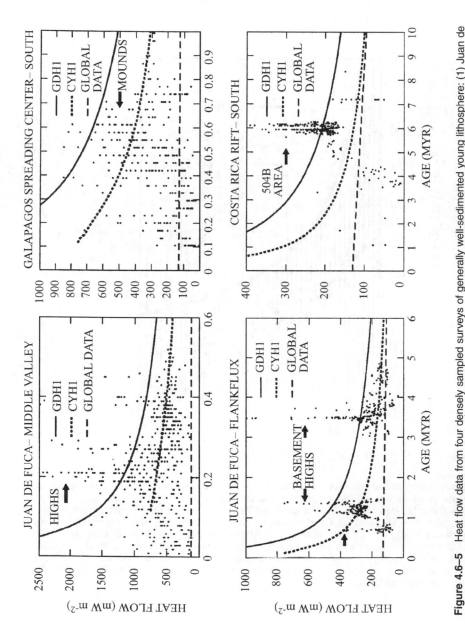

Figure 4.6—5 Heat flow data from four densely sampled surveys of generally well-sedimented young lithosphere: (1) Juan de Fuca—Middle Valley; (2) Juan de Fuca—Flankflux; (3) Galapagos Spreading Center—South; and (4) Costa Rica Rift—South. Most heat flow values are higher than the global average, presumably because of a sampling bias. In general, the measured values are significantly lower than predicted by the GDH1 model, which includes no hydrothermal cooling, suggesting that hydrothermal heat transfer is pervasive. Most values exceeding those predicted occur over high features. Also shown are the predictions of the CYH1 model, which predicts the average heat flow vs. age for young lithosphere including the effects of hydrothermal cooling. (From Stein, C., in *Seafloor Hydrothermal Systems: Physical, Chemical, Biological, and Geological Interactions*, Humphris, S. E. et al., Eds., Geophysical Monograph 91, American Geophysical Union, Washington, D.C., 1995, 436. With permission.)

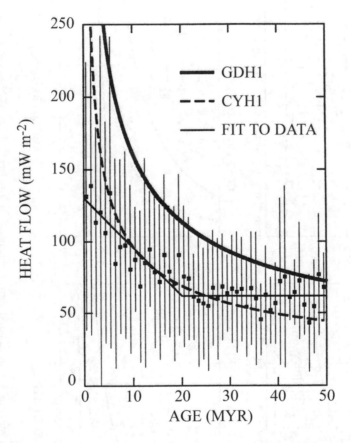

Figure 4.6–6 Globally averaged heat flow data and models for young lithosphere. Data means and standard deviations are shown for 1-Myr age bins. The means are well approximated by a simple two-stage linear fit. Also shown are the predicted heat flow for the GDH1 model, which has no hydrothermal cooling, and the CYH1 model, which includes hydrothermal cooling. CYH1, which was fit to the heat flow for ages greater than 10 Myr, predicts values higher than observed for younger ages, in accord with the expected sampling bias. (From Stein, C. et al., in *Seafloor Hydrothermal Systems: Physical, Chemical, Biological, and Geological Interactions,* Humphris, S. E. et al., Eds., Geophysical Monograph 91, American Geophysical Union, Washington, D.C., 1995, 438. With permission.)

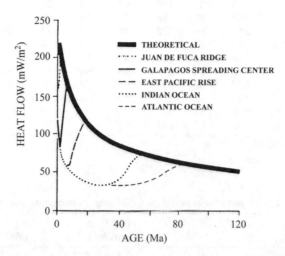

Figure 4.6–7 Schematic representation of the average heat flow as a function of age observed on the flanks of a variety of seafloor spreading ridges. Average heat flow through young seafloor is lower than expected due to the effects of hydrothermal circulation. Values approach the level predicted by simple cooling theory when sedimentation hydrologically seals the crust. The time when this occurs depends primarily on the local sedimentation rate, and in part on the roughness of the seafloor and the premeability of the sediments. (From Wright, J. A. and Louden, K. E., Eds., *CRC Handbook of Seafloor Heat Flow,* CRC Press, Boca Raton, FL, 1989. With permission.)

Table 4.6–1   Comparison of Calculated Heat-Flux Values from the Juan de Fuca Ridge

Vent Field	Calculated Heat Flux (MW)				
	1	2	3	4	5
SSR	2,096	30,240	1.6	814	580 ± 351
Axial	7,337	81,890			
NSR	2,266	28,220			
Surveyor smt	172	3,020			
Endeavour	1,133	12,600			1700 ± 1100
Total	13,004	155,970			

Note: SSR = southern symmetrical ridge, NSR = northern symmetrical ridge.

1. Model 1 line source.
2. Model 2 flow model at a velocity of 1 cm/s.
3. Detailed survey, point-source model.
4. Detailed survey, line-source model.
5. Three-dimensional survey flow model type.

Source: Tivey, M. A. and Johnson, H. P., *Rev. Aquat. Sci.,* 1, 473, 1989. With permission.

Table 4.6–2   Mean Heat Flow as a Function of Water Depth by Ocean Basin

Water Depth (m)	Pacific			Atlantic			Indian			Marginal Basins		
	q^a	σ^b	N^c	q	σ	N	q	σ	N	q	σ	N
1000–1500	45	79	3	135	117	97	116	53	4	81	30	70
1500–2000	191	116	101	88	85	63	98	79	14	170	268	81
2000–2500	233	201	212	98	73	47	104	74	28	107	85	84
2500–3000	198	156	452	104	72	106	80	66	40	80	39	94
3000–3500	109	81	306	76	73	141	51	32	77	78	39	151
3500–4000	82	59	282	67	41	223	54	44	84	75	39	120
4000–4500	58	39	357	56	35	188	57	36	134	55	51	64
4500–5000	57	35	168	53	24	219	66	41	117	65	33	65
5000–5500	58	22	135	61	35	190	61	35	113	62	37	37
5500–6000	52	16	62	57	57	215	51	27	41	64	33	33

[a] q = mean heat flow in mW/m².
[b] σ = standard deviation.
[c] N = number of values.

Table 4.6–3   Heat Flow and Age Estimates In Marginal Basins

Basin		Heat Flow (mW/m²)	Age (Ma)
Aleutian Basin		55.2 ± 0.4	117–132
Balearic Basin		92 ± 10	20–25
Caroline Basin		85 ± 31	28–36
Celebes Sea		56 ± 22	65–72
Coral Sea Basin		72 ± 11	56–64
Parece Vela Basin		88 ± 21	20–30
Shikoku Basin		82 ± 29	14–24
South China Sea	(1)	88 ± 6	27.5–33
	(2)	107 ± 4	19–23
Sulu Sea		89 ± 7	41–47
Tyrrhenian Sea	(1)	134 ± 8	7–12
	(2)	151 ± 10	5–8
West Philippine Basin		68 ± 22	39–50

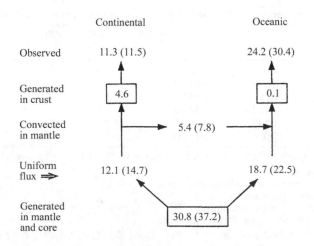

Figure 4.6–8   Heat flow bookkeeping. Note world average heat flows upward from continental and oceanic crust (in W/m²) are rather similar, although there is much more heat generation by radioactivity in continental crust, which is much thicker. (From Harper, J. F., *Rev. Aquat. Sci.,* 1, 322, 1989. With permission.)

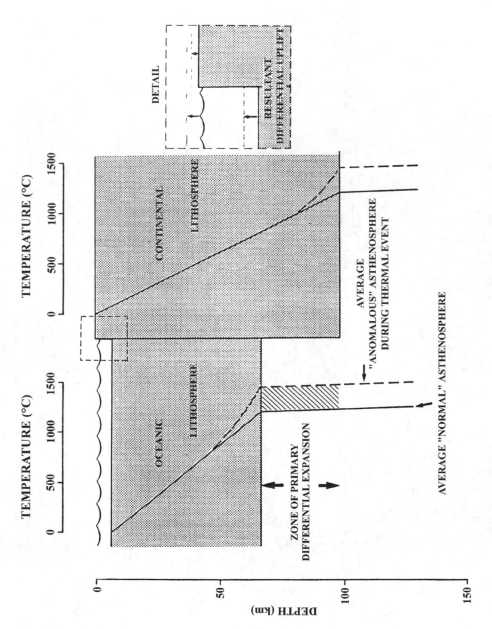

Figure 4.6–9 Schematic representation of typical continental and oceanic geotherms before and during a thermal event during which the convective heat supply in the asthenosphere, and thus the average temperature, is increased. A resultant decrease in continental freeboard will occur simultaneously with the increase in the average temperature in the asthenosphere. (From Wright, J. A. and Louden, K. E., Eds., *CRC Handbook of Seafloor Heat Flow*, CRC Press, Boca Raton, FL, 1989. With permission.)

4.7 HYDROTHERMAL VENTS

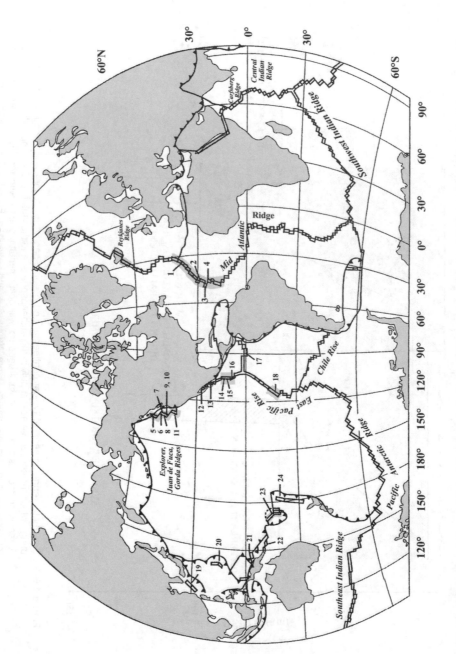

Figure 4.7–1   Comparison of the global extent of plume studies and seafloor vent fluid sampling. Shaded areas on spreading centers mark the locations of multi-segment plume mapping investigations on the Reykjanes Ridge, Mid-Atlantic Ridge, northeast Pacific ridges, Gulf of California, northern and southern East Pacific Rise, and the north Fiji Basin. Numbers mark sites of vent fluid collection and analysis along spreading centers; fluid discharge sites noted only from visual observations are not included. (From Baker, E. T. et al., in *Seafloor Hydrothermal Systems: Physical, Chemical, Biological, and Geological Interactions*, Humphris, S. E. et al., Eds., Geophysical Monograph 91, American Geophysical Union, Washington, D.C., 1995, 65. With permission.)

Table 4.7–1   Known Locations of Hydrothermal Venting in the Oceans

Explorer Ridge, northeast Pacific (McConachy and Scott, 1987)
Juan de Fuca Ridge
  Middle Valley near 48.5°N (Davis et al., 1987)
  Endeavour segment near 48°N (Tivey and Delaney, 1986; Delaney et al., 1992)
  Axial seamount near 46°N (CASM, 1985; Baker et al., 1990)
  Coaxial segment near 46°N (Baker et al., 1993)
  Cleft segment near 45°N (Normark et al., 1983, 1987)
Gorda Ridge (Baker et al., 1987; Morton et al., 1987)
East Pacific Rise
  Guaymas Basin, Gulf of California (Lonsdale and Becker, 1985)
  21°N (RISE, 1980; Ballard et al., 1981; Von Damm, et al., 1985)
  13°N (Hekinian et al., 1983a, b)
  11°N (McConachy et al., 1986)
  9–10°N (Haymon et al., 1991, 1993)
  15°S (Lupton and Craig, 1981)
Galapagos Rift near 86°W (Corliss et al., 1979; Crane and Ballard, 1980)
Mid-Atlantic Ridge
  37°N: Lucky Strike (Langmuir et al., 1993)
  29°N: Broken Spur (Murton et al., 1993)
  26°N: TAG (Rona et al., 1986; Thompson et al., 1988; Rona and Thompson, 1993)
  23°N: Snake Pit (Detrick et al., 1986; Campbell et al., 1988b)
  15°N (Rona et al., 1987, 1992)
Red Sea (Degens and Ross, 1969; Shanks and Bischoff, 1980)
Back-arc basins and marginal seas
  Okinawa Trough (Halbach et al., 1989)
  Mariana Trough (Horibe et al., 1986)
  Manus Basin (Both et al., 1986)
  Woodlark Basin
  North Fiji Basin Ridge (Auzende et al., 1991)
  Lau Basin (Fouquet et al., 1991a, b)
Hot-spot and arc volcanoes
  Loihi seamount, Hawaii (Malahoff et al., 1982; Karl et al., 1988; Sedwick et al., 1992)
  MacDonald and Teahitia Seamounts (McMurtry et al., 1989; Stuben et al., 1989; Michard et al., 1993)
  Kasuga seamounts (McMurtry et al., 1993)

Note: References in original source.

Source: Seyfried, W. E., Jr. and Mottl, M. J., in *The Microbiology of Deep-Sea Hydrothermal Vents,* Karl, D. M., Ed., CRC Press, Boca Raton, FL, 1995, 5. With permission.

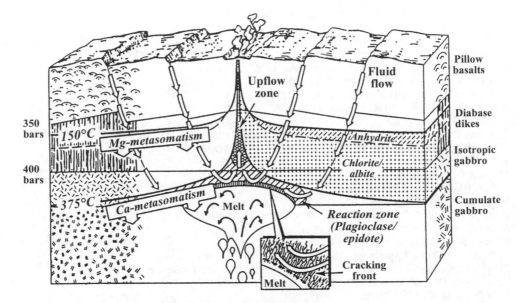

Figure 4.7–2 Schematic cross section of a mid-ocean ridge hydrothermal system showing the inferred location of the subseafloor magma chamber and the circulation path of seawater-derived hydrothermal fluid. Downflowing limbs of hydrothermal convection cells recharge the deep seated, high-temperature reaction zone immediately adjacent to the top of the magma chamber. It is in the reaction zone that seawater chemistry will change most, as it equilibrates with alteration phases formed from diabase and gabbro. A combination of high temperatures, relatively low pressures, and an abundance of fresh rock at the cracking front of the recently crystallized magma chamber enhances rock alteration and changes in fluid composition. Changes in the physical properties of the fluid, especially density and heat capacity, cause it to ascend to the seafloor along permeable fractures that characterize the upflow zone, where cooling may occur by adiabatic decompression and conduction of heat into the wall rocks. (From Seyfried, W. E., Jr. and Mottl, M. J., in *The Microbiology of Deep-Sea Hydrothermal Vents,* Karl, D. M., Ed., CRC Press, Boca Raton, FL, 1995, 6. With permission.)

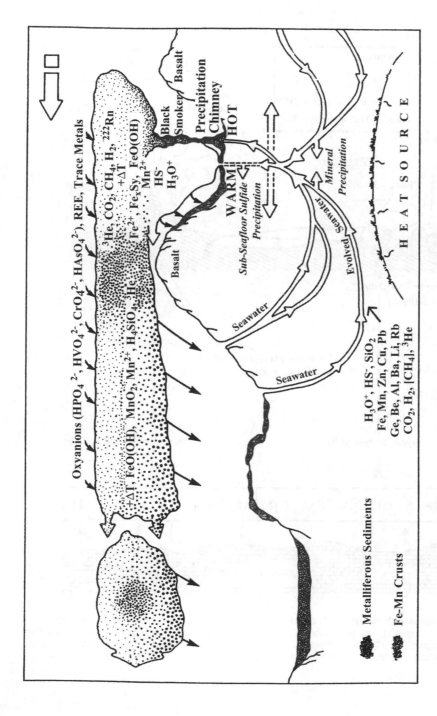

Figure 4.7-3 Diagram depicting a hydrothermal plume and underlying vent system at a seafloor spreading center. Schematic representation shows high-temperature vent sources, buoyant plume, neutrally buoyant proximal plume, and the discontinuous distal plume. (From Massoth, G. J. et al., in *Global Venting, Midwater, and Benthic Ecological Processes*, DeLuca, M. P. and Babb, I., Eds., National Undersea Research Program Report 88-4, NOAA, Rockville, MD, 1988, 29. With permission.)

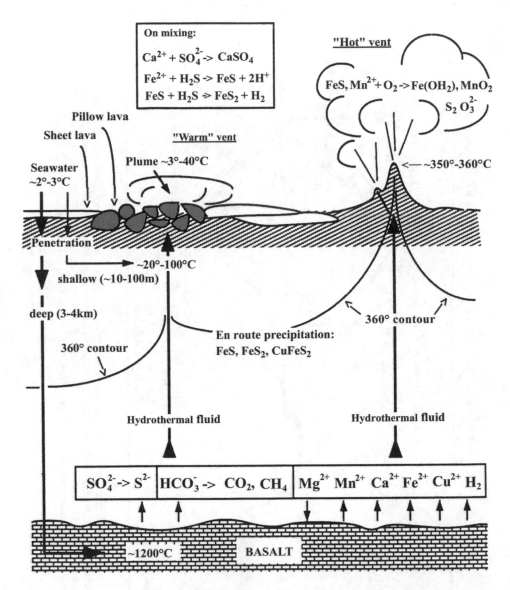

Figure 4.7–4 Major processes during the hydrothermal circulation of seawater through the oceanic crust at mid-ocean ridges. (From Jannasch, H. W., in *Seafloor Hydrothermal Systems: Physical, Chemical, Biological, and Geological Interactions,* Humphris, S. E. et al., Eds., Geophysical Monograph 91, American Geophysical Union, Washington, D.C., 1995, 275. With permission.)

Table 4.7–2 Mineral Distribution in EPR Hydrothermal Deposits[a]

		Axis Deposits							Off-Axis Deposits
		Chimneys			Particulates		Sediments		
	Mounds	Black Smokers	White Smokers	Inactive Chimneys	Black Smokers	White Smokers	Proximal	Distal	Off-Axis Deposits
Sulfides/Sulfosalts									
(Most Abundant)									
Sphalerite (Zn(Fe)S)	A	C	R-A	A	A		A		R-T
Wurtzite (Zn(Fe)S)	A	R-C	R-A	A			C		
Pyrite (Fe_2S)	A	C	A	A	R-C	C-A	A		A
Chalcopyrite ($CuFeS_2$)	C	C-A	R	R-C	T		C-A		C-A
(Less Abundant)									
Iss-isocubanite (variable $CuFe_2S_3$)	R-C	C-A	R	R-C	R				T
Marcasite (Fe_2S)	C	C	C-A	C-A			R-T		C-A
Melnicovite (FeS_{2x})	C	C	C	C-A					C-A
Pyrrhotite ($Fe_{1-x}S$)		R-C			A		T-A		R-T
Bornite-chalcocite (Cu_5FeS_4-Cu_2S)		R-C							R-T
Covellite (CuS)		R							R
Digenite (Cu_9S_5)		R					T		R-T
Idaite ($Cu_{5.5}FeS_{6.5}$)		T							
Galena (PbS)			T	T					
Jordanite ($Pb_9As_4S_{15}$)			T	T					
Tennantite $(Cu, Ag)_{10}(Fe, Zn, Cu)_2As_4S_{23}$		T		T					
Vallerite $(2(Cu, Fe)_2S\cdot3(Mg, Al)(OH)_2)$		T							
Sulfates									
Anhydrite ($CaSO_4$)		A	A						T-A
Gypsum ($CaSO_4 \cdot 2H_2O$)		R	R						R
Caminite ($MgSO_4 \cdot xMg(OH)_2 \cdot (1-2x)H_2O$)		R						T-R	R
Barite ($BaSO_4$)	R-C	T	R-C	R-C		C			
Jarosite-natrojarosite $(K, Na)Fe_3(SO_4)_2(OH)_6$	R			R					
Chalcanthite ($CuSO_4 \cdot 5H_2O$)									
Carbonate									
Magnesite ($MgCO_3$)		T							
Elements									
Sulfur (S)	R		R	C-A			R-T	T	T

(continued)

Table 4.7–2 Mineral Distribution in EPR Hydrothermal Deposits[a] (continued)

| | | Axis Deposits | | | | | Sediments | | |
| | | Chimneys | | | Particulates | | | | |
	Mounds	Black Smokers	White Smokers	Inactive Chimneys	Black Smokers	White Smokers	Proximal	Distal	Off-Axis Deposits
Oxides/Oxyhydroxides									
Goethite (FeO(OH))	C			R-C					C
Lepidocrocite (FeO(OH))	R			T-R					
Hematite (Fe_2O_3)		T-R							
Magnetite (Fe_3O_4)		T-R							
"Amorphous" Fe compounds	C-A	R-C	R-C	C-A			T-A	C-A	R-A
"Amorphous" Mn compunds								C-A	T-A
Psilomelane (Ba, $H_2O)_2Mn_5O_{10}$)									R-T
Silicates									
Amorphous silica ($SiO_2 \cdot nH_2O$)	C	R	C	C		A		A	R-A
Quartz (SiO_2)									A-T
Talc ($Mg_3Si_4O_{10}(OH)_2$)	R	R-C							
Nontronite	R			T-R				A	T-A
(Fe, Al, $Mg)_2(Si_{3.66}Al_{0.34})O_{10}(OH)_2$ Illite-smectite	R	R	R-C	R-C					R-T
Aluminosilicate gel									
Hydroxychlorides									
Atacamite ($Cu_2Cl(OH)_3$)									T-C

[a] Key: A = abundant > C = common > R = rare > T = trace.

Source: Haymon, R. M., in The Geology of North America, Vol. N, The Eastern Pacific Ocean and Hawaii, Winterer, E. et al., Eds., Geological Society of America, Boulder, CO, 1989, 173. With permission.

Table 4.7–3   Chemical Composition of Massive Sulfide Samples from the East Pacific Rise, 21°N and Galapagos Rift Spreading Centers

| | East Pacific Rise, 21°N | | | Galapagos Rift | |
	Sphalerite/Wurtzite-Rich		Silica-Rich	Sphalerite/Wurtzite-Rich	Pyrite-Rich
	(wt. %)				
Fe	14.7	26.2	16.7	15.6	44.1
Zn	34.9	20.3	41.8	46.9	0.14
Cu	0.23	1.3	0.89	0.35	4.98
Pb	0.61	0.07	0.29	0.30	<0.07
S	31.3	39.7	34.9	36.8	52.2
SO_3	<0.01	7.6	<0.01	<0.03	<0.03
SiO_2	19.0	<0.5	4.3	1.5	<0.1
Al_2O_3	0.3	0.11	0.77	0.15	<0.06
MgO	<0.03	0.07	0.02	<0.05	<0.05
CaO	<0.01	5.42	<0.01	<0.03	<0.03
Sum	101.4	100.77	99.67	101.60	101.42
	(ppm)				
Ag	241	34	202	290	<10
As	483	770	215	411	125
Au	NA	0.17	<0.2	0.13	0.05
B	<7[a]	<7[a]	<7[a]	40[a]	<7[a]
Ba	6030	65	850	19[a]	16[a]
Bi	<0.2[a]	2[a]	0.2[a]	<0.2[a]	<10[a]
Cd	120	890	790	490	<32[a]
Co	<2.0	2.5	6	24	482
Cr	8	16	<30	<8[a]	55
Cs	6.7	<5.0	6.6	<9	<3
Ga	3.3[a]	18[a]	21[a]	<20[a]	15[a]
Ge	96[a]	<1.5[a]	100[a]	270	<1[a]
Hg	2[a]	<1[a]	<1[a]	<1[a]	NA
Mn	570[a]	91[a]	500[a]	720[a]	140[a]
Mo	16	78	13	3	170[a]
Ni	2[a]	5[a]	2[a]	NA	3.1[a]
Pd	0.001	0.001	0.001	<0.002	<0.002
Pt	0.002	0.002	<0.001	<0.005	<0.005
Rh	0.002	0.0010	0.0022	0.003	<0.001
Sb	45.0	13	52.9	34	1.8
Sc	<0.4	0.2	0.25	<1	<0.3
Se	7[a]	172	10[a]	29	100
Sr	220	9	19	<10[a]	<1[a]
Te	<1[a]	2[a]	<1[a]	<1[a]	NA
Tl	40[a]	20[a]	<1[a]	10[a]	<5[a]
U	6.0	1.3	3.1	10	1.0
Y	4[a]	<1.5[a]	3[a]	<2[a]	<2[a]
W	1.0	<2.0	<3.0	<10[a]	<10[a]
Zr	9[a]	<3[a]	43[a]	28[a]	<3[a]

Note: NA = not analyzed.

[a] Semiquantitative optical emission spectroscopy. The following elements are below their respective detection limits (ppm) for all samples (by semiquantitative emission spectroscopy): Na (500), K (700), Ti (30), P (500), Be (1), Ce (40), La (10), Pr (20), Nd (10), Sm (20), Eu (8), Gd (20), Tb (100), Dy (10), Ho (8), Er (10), Yb (10), Li (20), Nb (3), Sn (1), Ta (400), Th (20).

Source: Haymon, R. M., in The Geology of North America, Vol. N, The Eastern Pacific Ocean and Hawaii, Winterer, E. et al., Eds., Geological Society of America, Boulder, CO, 1989, 173. With permission.

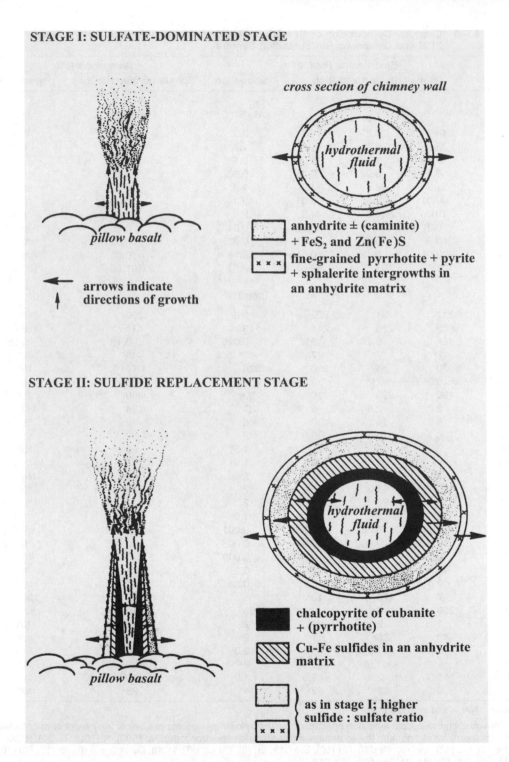

Figure 4.7–5 Formation of black smoker (hydrothermal) vents. The first stage of growth involves anhydrite (calcium sulfate) precipitation from seawater to build chimney walls. The second stage entails the sulfide replacement of anhydrite precipitated earlier in the walls. Mineral zonation in the black smoker is evident in the cross section of the chimney wall. (From Haymon, R. M., in *The Geology of North America,* Vol. N, *The Eastern Pacific Ocean and Hawaii,* Winterer, E. et al., Eds., Geological Society of America, Boulder, CO, 1989, 173. With permission.)

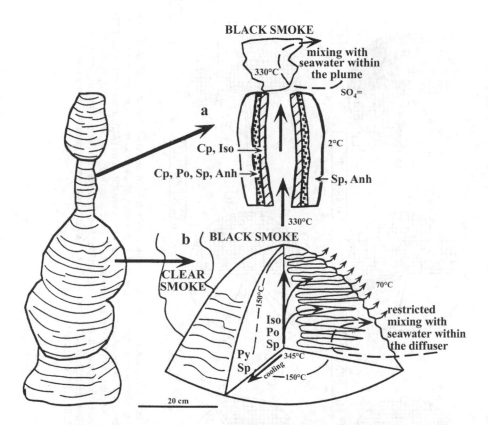

Figure 4.7–6 Schematic representations of (a) a black smoker chimney and (b) a diffuser, both from the Snakepit site on the Mid-Atlantic Ridge. Differences in mineralogical zonation, styles of mixing with seawater, and temperature zonation are evident. (From Tivey, M. K., in *Seafloor Hydrothermal Systems: Physical, Chemical, Biological, and Geological Interactions,* Humphris, S. E. et al., Eds., Geophysical Monograph 91, American Geophysical Union, Washington, D.C., 1995, 170. With permission.)

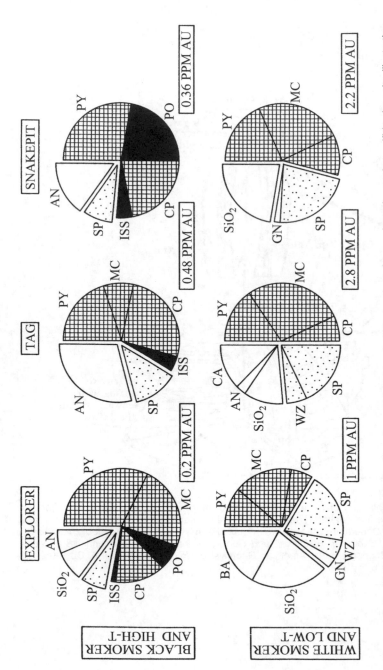

Figure 4.7-7 Comparison of the bulk mineralogy of black and white smokers from selected massive sulfide deposits illustrating relative abundances of different mineral phases. High-T = assemblages with formation temperatures >250°C; Low-T = assemblages with formation temperatures ≤205°C. Abbreviations are anhydrite (AN), amorphous silica (SiO₂), sphalerite (SP), wurtzite (WZ), isocubanite (ISS), chalcopyrite (CP), pyrrhotite (PO), marcasite (MC), pyrite (PY), galena (GN), anhydrite (CA). (From Hannington, M. D. et al., in *Seafloor Hydrothermal Systems: Physical, Chemical, Biological, and Geological Interactions,* Humphris, S. E. et al., Eds., Geophysical Monograph 91, American Geophysical Union, Washington, D.C., 1995, 127. With permission.)

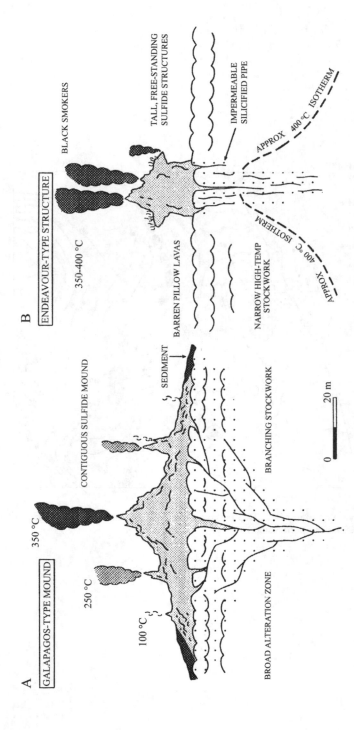

Figure 4.7–8 Features of contrasting deposit morphologies, showing the role of subseafloor permeability in controlling styles of venting: (A) Galapagos-type mound and (B) Endeavour-type structure. (From Hannington, M. D. et al., in *Seafloor Hydrothermal Systems: Physical, Chemical, Biological, and Geological Interactions*, Humphris, S. E. et al., Eds., Geophysical Monograph 91, American Geophysical Union, Washington, D.C., 1995, 135. With permission.)

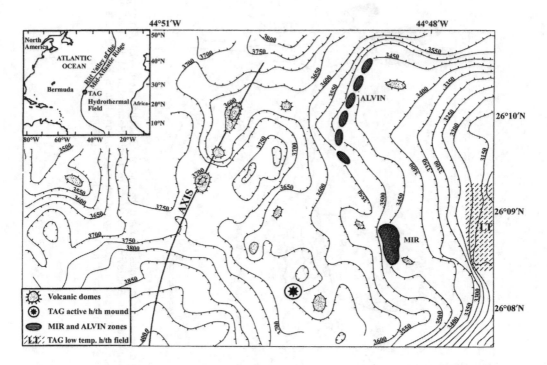

Figure 4.7–9   Seabeam bathymetric map of the TAG hydrothermal field. (From Humphris, S. E., *JOI/USSAC Newsl.*, 7, 1, 1994. With permission.)

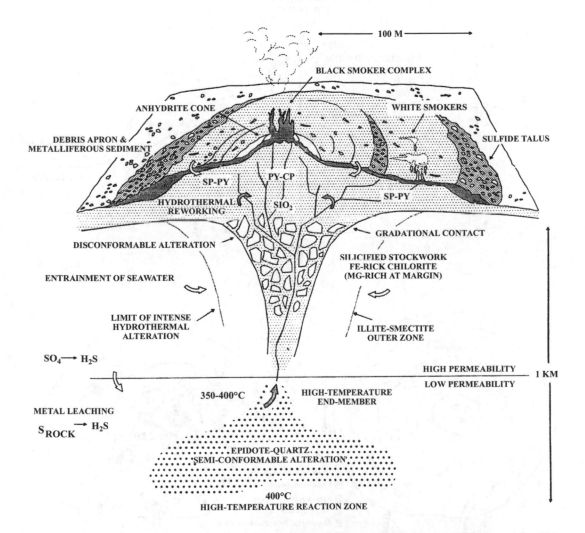

Figure 4.7–10 (a) Cross-sectional profile across the active TAG mound. (b) Structural features on the TAG mound. (c) Geologic map showing locations of different orifices on the TAG mound, dominant mineral compositions, and major morphologic features of the mound. (From Fornari, D. J. and Embley, R. W., in *Seafloor Hydrothermal Systems: Physical, Chemical, Biological, and Geological Interactions,* Humphris, S. E. et al., Eds., Geophysical Monograph 91, American Geophysical Union, Washington, D.C., 1995, 6. With permission.)

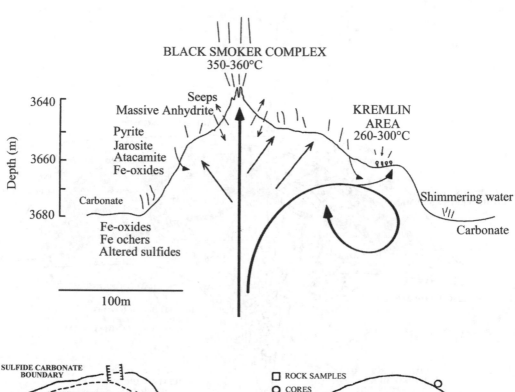

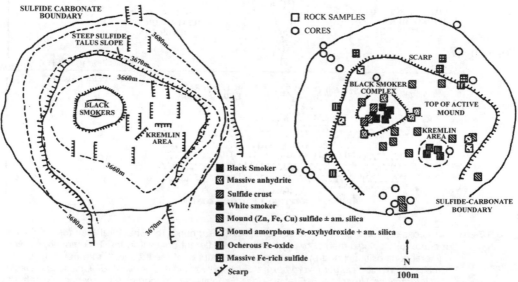

Figure 4.7–11   Schematic diagram depicting the principal components of a seafloor sulfide deposit and associated hydrothermal systems at the TAG mound. Arrows indicate fluid flow paths for seawater
(open) and hydrothermal fluids (shaded). During hydrothermal circulation of seawater, SO_4 is
precipitated as anhydrite or reduced to H_2S. Reduced sulfur in the hydrothermal fluids is derived
mainly from leaching of the rocks. (SP = sphalerite, PY = pyrite, CP = chalcopyrite). (From
Hannington, M. D., in *Seafloor Hydrothermal Systems: Physical, Chemical, Biological, and
Geological Interactions,* Humphris, S. E. et al., Eds., Geophysical Monograph 91, American
Geophysical Union, Washington, D.C., 1995, 117. With permission.)

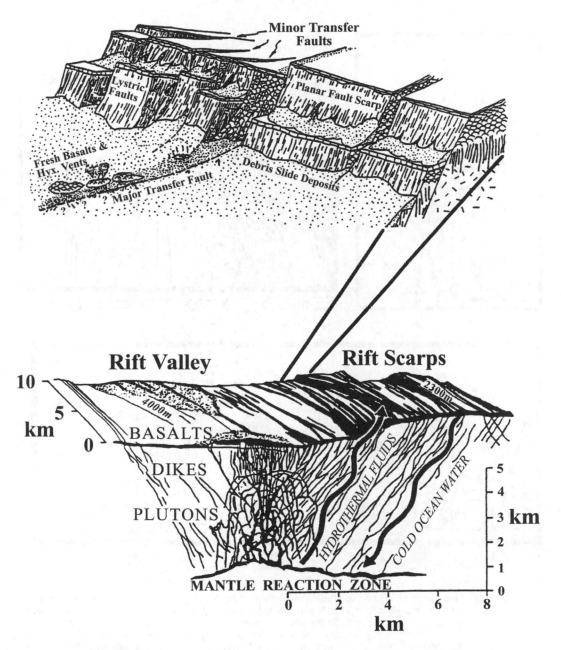

Figure 4.7–12   Perspective models of the morphology and structure of the Mid-Atlantic Ridge rift valley in the TAG region (top), and the inferred subseafloor fracture patterns and circulation system (bottom) associated with a slow-spreading mid-ocean ridge crest. (From Fornari, D. J. and Embley, R. W., in *Seafloor Hydrothermal Systems: Physical, Chemical, Biological, and Geological Interactions,* Humphris, S. E. et al., Eds., Geophysical Monograph 91, American Geophysical Union, Washington, D.C., 1995, 7. With permission.)

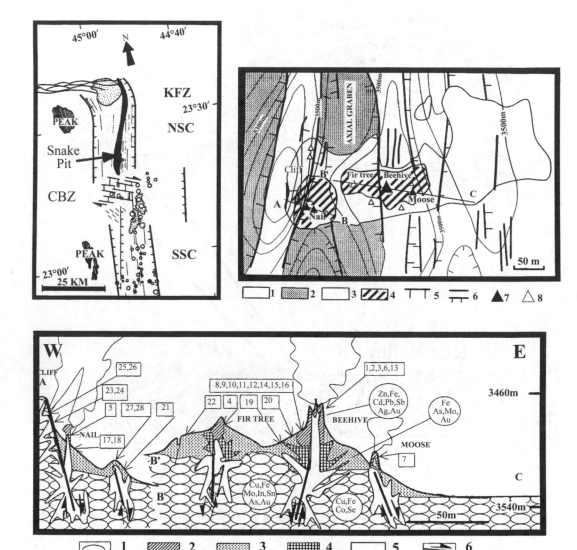

Figure 4.7–13   (Top Left) Schematic morphostructural map of the MARK area which includes the Snake Pit hydrothermal vent field. (Top Right) Geologic map of Snake Pit vent deposits: 1 = pillow lava flows; 2 = rubble; 3 = hydrothermal sediments; 4 = sulfide mounds; 5 = normal fault; 6 = fissure, minor fault; 7 = active black smoker; 8 = inactive chimneys. (Bottom) Cross section through the Snake Pit vent area showing geological and hydrothermal relationships: 1 = pillow lava; 2 = Zn-rich chimney; 3 = Fe-rich massive sulfide; 4 = Cu-rich massive sulfide; 5 = Cu-rich sulfide of stockwork and central part of chimney; 6 = normal fault. (From Fornari, D. J. and Embley, R. W., in *Seafloor Hydrothermal Systems: Physical, Chemical, Biological, and Geological Interactions,* Humphris, S. E. et al., Eds., Geophysical Monograph 91, American Geophysical Union, Washington, D.C., 1995, 8. With permission.)

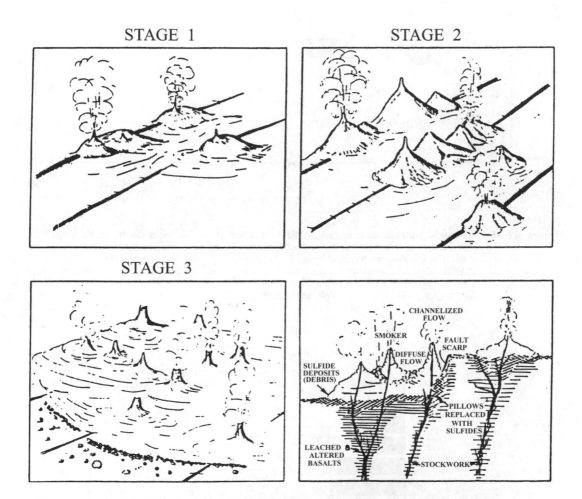

Figure 4.7–14 Sequence of development of a hydrothermal vent field at an unsedimented spreading center. (1) Initial stage showing small discrete vents over existing fissures. (2) Established stage illustrating coalescing hydrothermal vent deposits along fissures. (3) Mature vent field depicting large equidimensional deposits. (4) Profile of an active hydrothermal vent system. (From Tivey, M. A. and Johnson, H. P., *Rev. Aquat. Sci.,* 1, 473, 1989. With permission.)

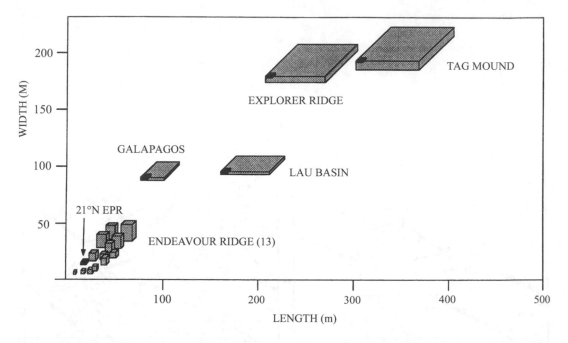

Figure 4.7–15 Schematic representation of relative sizes of major sulfide deposits on the modern seafloor (subseafloor stockwork mineralization not included). The black square in the lower left corner of each box represents a typical, small 21°N-type East Pacific Rise deposit. (From Hannington, M. D., in *Seafloor Hydrothermal Systems: Physical, Chemical, Biological, and Geological Interactions,* Humphris, S. E. et al., Eds., Geophysical Monograph 91, American Geophysical Union, Washington, D.C., 1995, 146. With permission.)

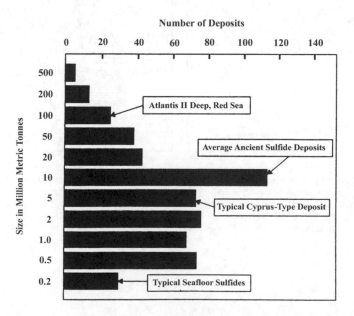

Figure 4.7–16 Histogram of tonnages of major massive sulfide deposits worldwide in comparison to modern seafloor sulfides. (From Hannington, M. D. et al., in *Seafloor Hydrothermal Systems: Physical, Chemical, Biological, and Geological Interactions,* Humphris, S. E. et al., Eds., Geophysical Monograph 91, American Geophysical Union, Washington, D.C., 1995, 147. With permission.)

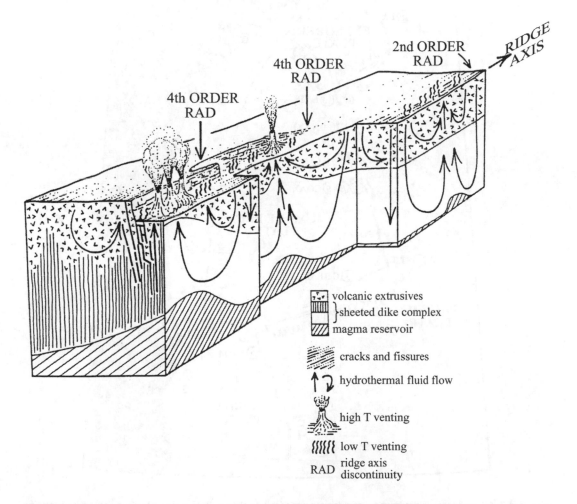

Figure 4.7–17 Schematic model of hydrothermal fluid flow, low- and high-temperature venting, and ridge axis discontinuities along a portion of the fast-spreading East Pacific Rise. (From Haymon, R. M. et al., *Earth Planet. Sci. Lett.,* 104, 513, 1991. With permission.)

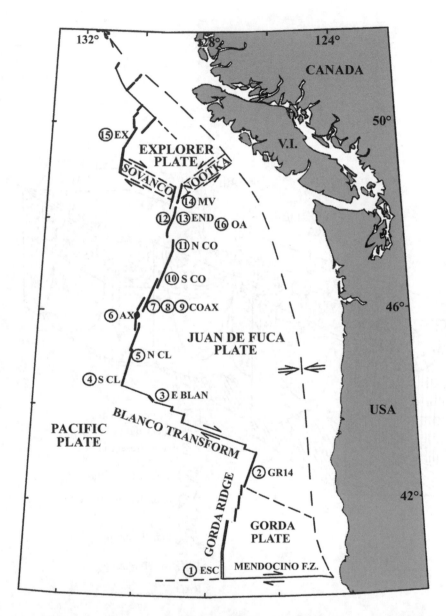

Figure 4.7–18 Schematic diagram of the ridge system in the northeast Pacific. Thick lines indicate the spreading center segments, thinner lines indicate fracture zones, the dashed line represents the Cascadia Subduction Zone, and arrows indicate the direction of plate motion. The circled numbers show the locations of hydrothermal sites: 1 = Escanaba Trough; 2 = Gorda Ridge 14; 3 = East Blanco; 4 = South Cleft; 5 = North Cleft; 6 = Axial Volcano; 7, 8, and 9 = Source, floc, and lava flow vents on the Coaxial segment; 10 = South Cobb; 11 = North Cobb; 12 and 13 = Main Endeavour vent field and High-Rise vent field on Endeavour segment; 14 = Middle Valley; 15 = Explorer Ridge; 16 = Off-axis site. (From Fornari, D. J. and Embley, R. W., in *Seafloor Hydrothermal Systems: Physical, Chemical, Biological, and Geological Interactions,* Humphris, S. E. et al., Eds., Geophysical Monograph 91, American Geophysical Union, Washington, D.C., 1995, 15. With permission.)

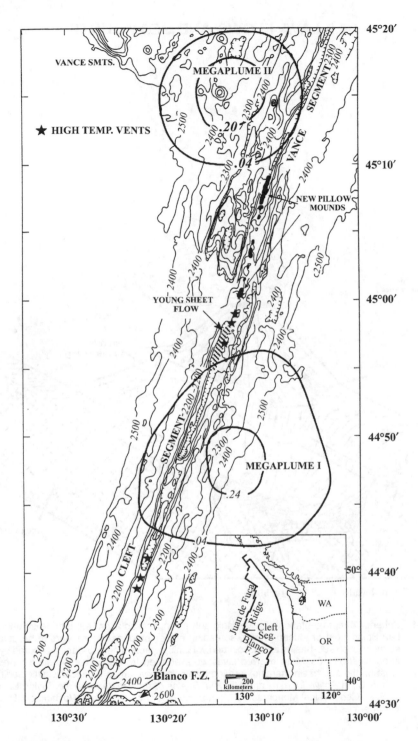

Figure 4.7–19 Bathymetric map of southern Juan de Fuca Ridge based on Sea Beam surveys. Contour interval is 100 m except for the axial valleys, where it is 20 m. Temperature anomaly contours of Megaplume I and Megaplume II are indicated. Locations of new pillow mounds erupted in the 1980s and young sheet flow are also noted. Stars show locations of high-temperature hydro-thermal vents. (From Fornari, D. J. and Embley, R. W., in *Seafloor Hydrothermal Systems: Physical, Chemical, Biological, and Geological Interactions,* Humphris, S. E. et al., Eds., Geo-physical Monograph 91, American Geophysical Union, Washington, D.C., 1995, 23. With per-mission.)

4.8 LAVA FLOWS AND SEAMOUNTS

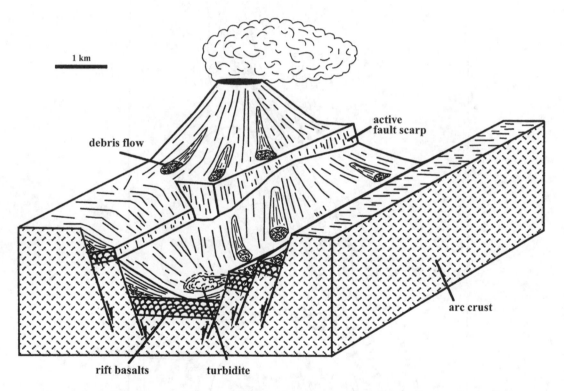

Figure 4.8–1   Schematic representation of a volcanic sub-basin and adjacent submarine seamount volcano. Lau Basin crust produced prior to spreading center propagation appears to be composed of a series of such sub-basins and associated volcanic centers. The sub-basin is floored by MORB/IAT transitional basalts. More evolved lavas were erupted from the volcano sited on extended arc crust on the sub-basin margins, forming locally large volumes of volcaniclastic sediment. (From Clift, P. D., in *Active Margins and Marginal Basins of the Western Pacific,* Taylor, B. and Natland, J., Eds., Geophysical Monograph 88, American Geophysical Union, Washington, D.C., 1995, 77. With permission.)

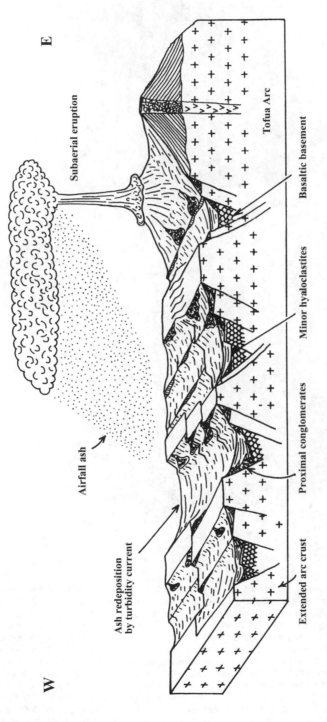

Figure 4.8–2   Schematic representation of the modern Lau Basin, depicting volcaniclastic debris from the arc volcanic front that is ponded close to the arc, except in the case of widespread airfall ash material. (From Marsaglia, K. M. et al., in *Active Margins and Marginal Basins of the Western Pacific,* Taylor, B. and Natland, J., Eds., Geophysical Monograph 88, American Geophysical Union, Washington, D.C., 1995, 303. With permission.)

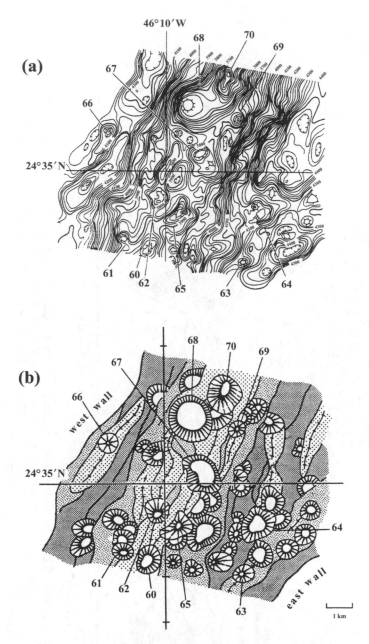

Figure 4.8–3   (a) Bathymetry of a portion of the inner valley floor of the northern Mid-Atlantic Ridge traced from Sea Beam swaths plotted at 20-m intervals. Tick Marks point downhill. Numbers show the locations of features that fully meet seamount criteria. An axial volcanic ridge is centrally located within the segment and presumed to be the primary site of crustal accretion. The scale of irregularity of the topography suggests that the axial volcanic ridge is composed of piled-up seamounts and fissure-fed pillow flows. (a) Sea Beam swaths of study area; and (b) provisional interpretation of the Sea Beam swaths. Seamounts are depicted with summit areas unshaded and slopes hatched. Some seamounts are smaller than 50 m. Contours that are rounded to form sections of circles are interpreted as seamounts partly buried by later volcanism or seamounts abutting preexisting features. Note that the seamounts range in shape from pointy to flat-topped cones. Ridges and slopes with parallel contours are interpreted as products of fissure eruptions. These are shown stippled and overlapping one another, with lighter stippling denoting the youngest eruptions. Straight lines with ticks mark the crest of the ridges. (From Smith, D. K. and Cann, J. R., *J. Geophys. Res.*, 97, 1645, 1992. With permission.)

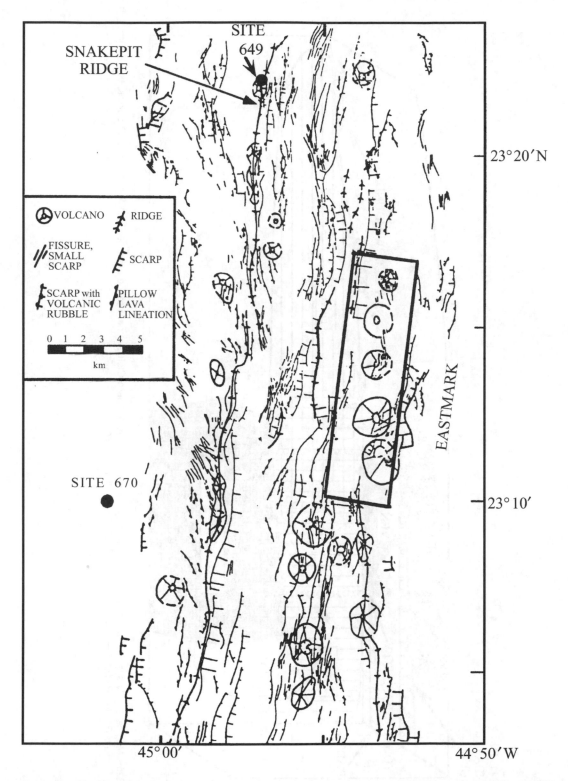

Figure 4.8–4 Map of the central valley area of the MARK area on the Mid-Atlantic Ridge based on Sea MARC imaging and Sea Beam bathymetry. Note pillow lava lineations and volcanic edifices (seamounts). (From Bryan, W. B. et al., *J. Geophys. Res.*, 99, 2973, 1994. With permission.)

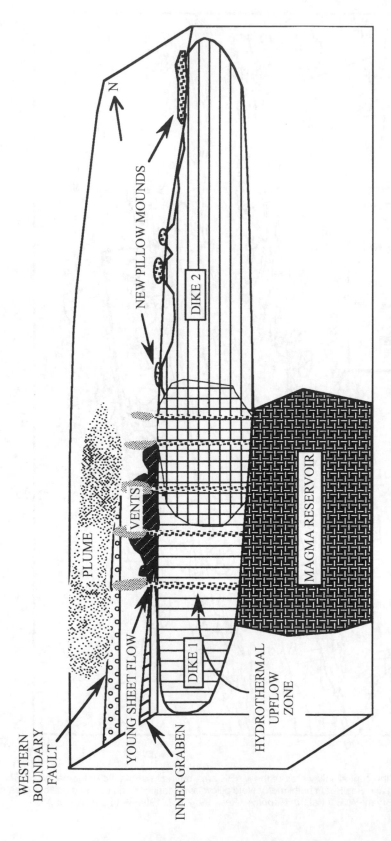

Figure 4.8–5 Magma supply system for pillow mounds and sheet flows along the northern Cleft segment of the Juan de Fuca Ridge. (From Chadwick, W. W., Jr. and Embley, R. W., *J. Geophys. Res.*, 99, 4761, 1994. With permission.)

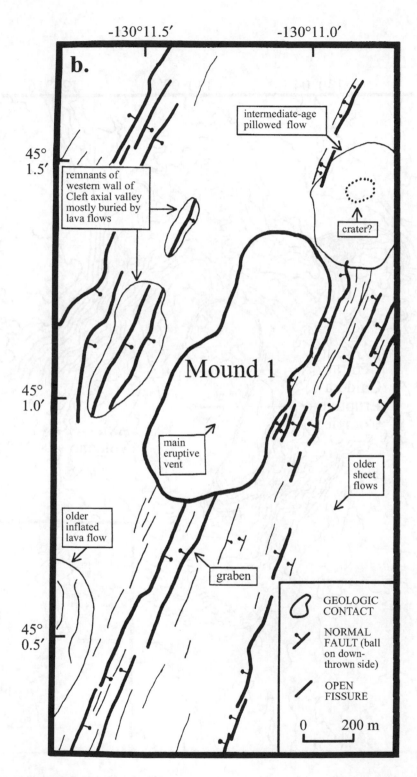

Figure 4.8–6 Map of Pillow Mound 1 on the northern Cleft segment of the Juan de Fuca Ridge based on SeaMARC I sidescan sonar imaging. (From Chadwick, W. W., Jr. and Embley, R. W., *J. Geophys. Res.*, 99, 4761, 1994. With permission.)

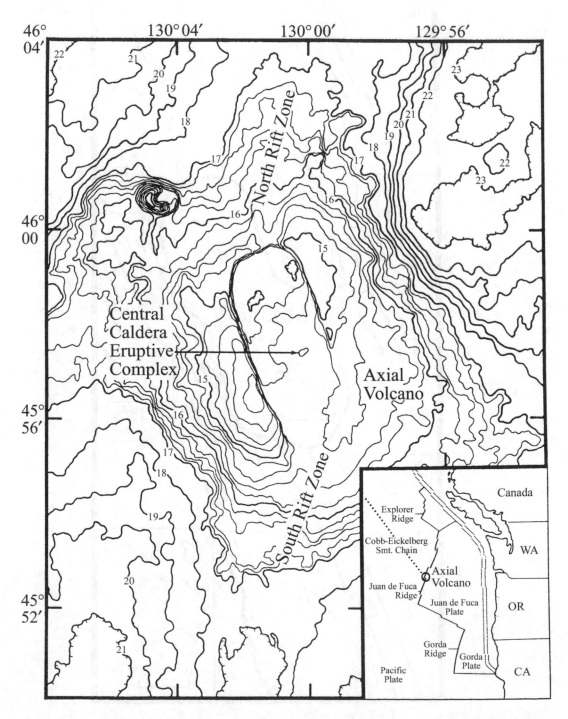

Figure 4.8–7 Map of the Central Caldera Eruptive Complex within Axial Volcano's summit caldera. Contour interval is 20 m. Inset shows location of Axial Volcano with respect to the crustal plates and plate boundary of the northeast Pacific Ocean. (From Applegate, B. and Embley, R. W., *Bull. Volcanol.*, 54, 447, 1992. With permission.)

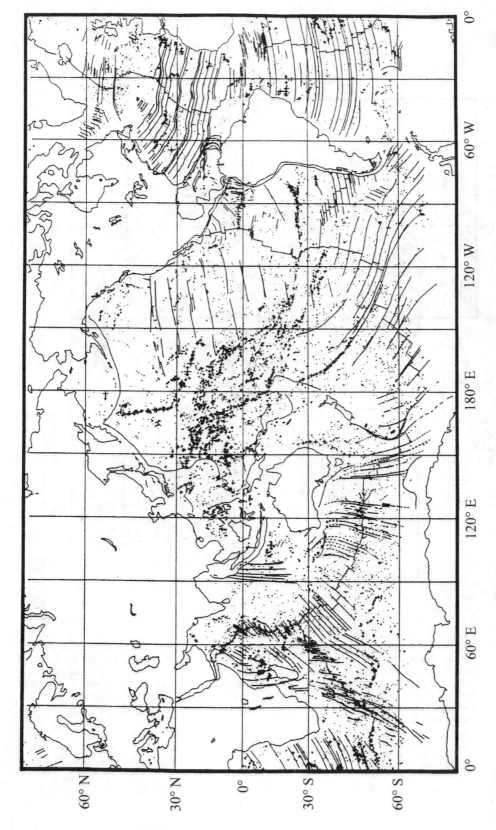

Figure 4.8-8 The global distribution of all seamounts located in SEASAT altimetry data. Seamounts are plotted as crosses with symbol size proportional to signal amplitude. Note how seamounts tend to cluster and form linear chains. (From Smith, D. K., *Rev. Aquat. Sci.*, 5, 197, 1991. With permission.)

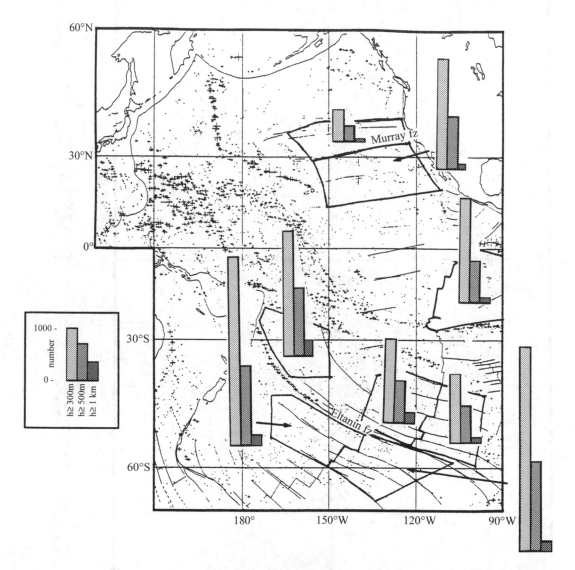

Figure 4.8–9 Map of the Pacific Ocean with histograms showing the cumulative number of seamounts per 10^6 km^2 for small-, medium-, and large-sized seamounts. Scale is shown to the left. Changes in abundances as well as size distribution are evident. In the North Pacific, a dramatic decrease in the number of seamounts is observed north across the Murray Fracture Zone. This is also true across the Eltanin Fracture Zone in the South Pacific. In addition, the oldest crust in the western Pacific has the highest number of large-sized seamounts. (From Smith, D. K., *Rev. Aquat. Sci.*, 5, 197, 1991. With permission.)

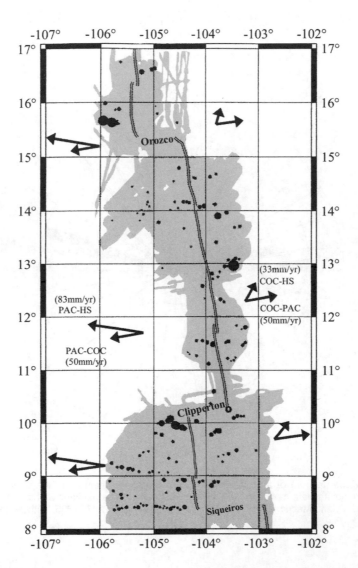

Figure 4.8–10   Near-axis seamounts along the northern East Pacific Rise and its flanks, from 8°15′N to 17°N. Complete bathymetric coverage is indicated by gray shading, and seamounts ≥200 m high are shown as filled circles scaled to the map scale. The double line is the East Pacific Rise, and the arrows show the magnitudes and directions of relative and absolute plate motion. (From Scheirer, D. S. and Macdonald, K. C., *J. Geophys. Res.*, 100, 2239, 1995. With permission.)

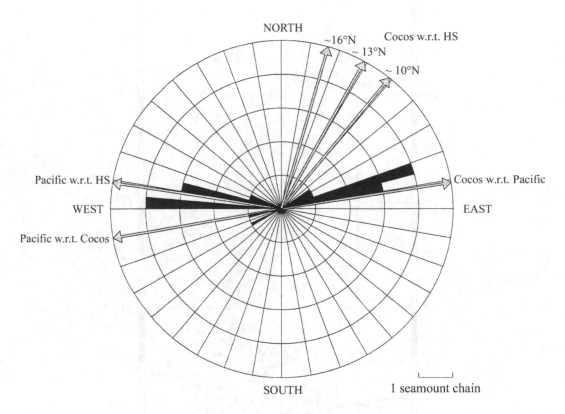

Figure 4.8–11   Rose diagram of the orientations of 21 seamount chains and the directions of relative and absolute plate motion in the northern East Pacific Rise between 8°N and 17°N. (From Scheirer, D. S. and Macdonald, K. C., *J. Geophys. Res.,* 100, 2247, 1995. With permission.)

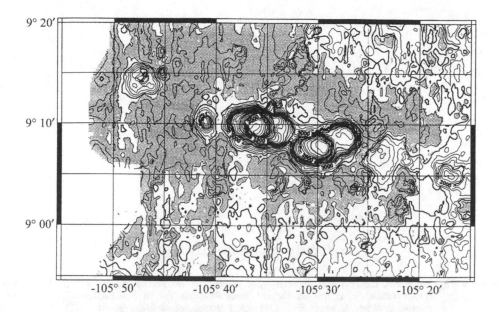

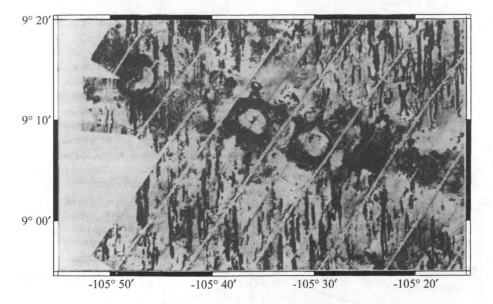

Figure 4.8–12 Bathymetry and side scan date of the P9°05 seamount chain. The bathymetry is contoured every 50 m with heavy lines every 250 m. The gray pattern indicates depths deeper than 3300 m. Dark regions in the side scan images indicate areas of high sonar backscatter, such as from seafloor slopes facing the ship or unsedimented lava; light areas denote low sonar backscatter, such as from acoustic shadows or sedimented seafloor. (From Scheirer, D. S. and Macdonald, K. C., *J. Geophys. Res.*, 100, 2242, 1995. With permission.)

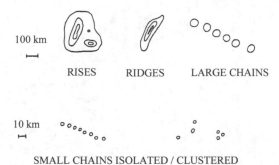

Figure 4.8–13   Schematic drawing depicting the types of volcanic landforms present in the eastern Pacific. (From Batiza, R., in *The Geology of North America,* Vol. N, *The Eastern Pacific Ocean and Hawaii,* Winterer, E. et al., Eds., Geological Society of America, Boulder, CO, 1989, 289. With permission.)

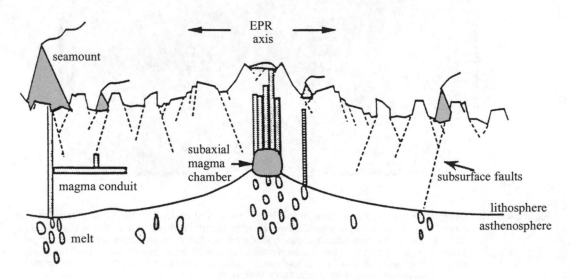

Figure 4.8–14   Formation of seamounts in proximity to the fast-spreading East Pacific Rise axis. (From Smith, D. K., *Rev. Aquat. Sci.,* 5, 197, 1991. With permission.)

4.9 MARINE MINERAL DEPOSITS

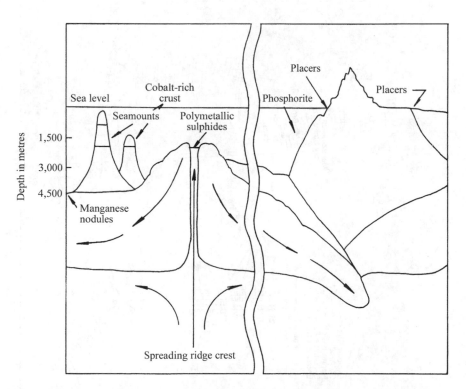

Figure 4.9–1 Physiographic setting of seafloor mineral deposits. (From Cronan, D. S., *Marine Minerals in Exclusive Economic Zones,* Chapman & Hall, London, 1992, 12. With permission.)

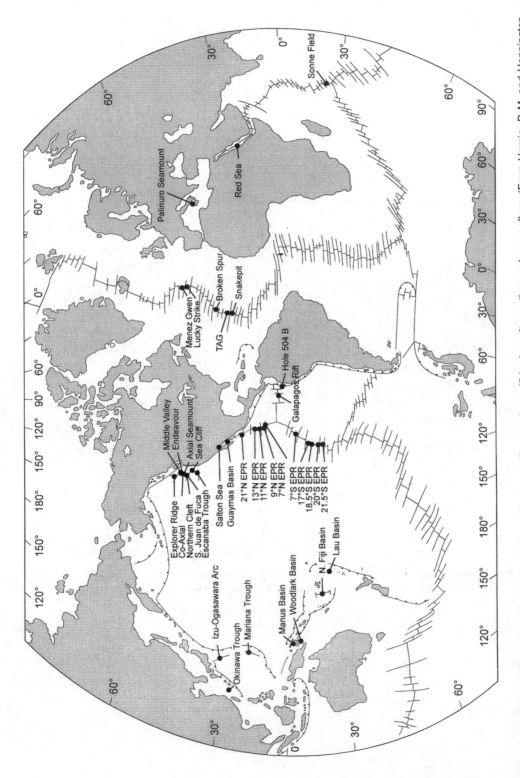

Figure 4.9-2 Location of hydrothermal systems and polymetallic massive sulfide deposits on the modern seafloor. (From Herzig, P. M. and Hannington, M. D., in *Handbook of Marine Mineral Deposits*, Cronan, D. S., Ed., CRC Press, Boca Raton, FL, 2000, 348. With permission.)

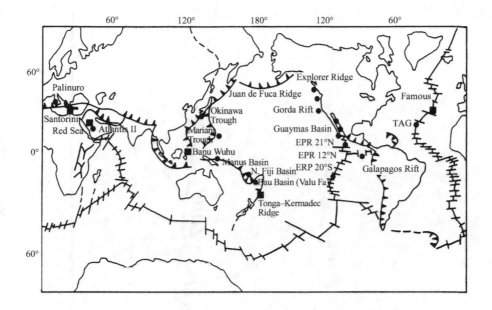

Figure 4.9–3 Examples of submarine hydrothermal mineralization at divergent and convergent plate margins. EPR = East Pacific Rise. Divergent plate boundaries and rift zones; ▲ collision and subduction zones; ■ oxidic metalliferous deposits; ● black smokers and polymetallic sulphide deposits. (From Cronan, D. S., *Marine Minerals in Exclusive Economic Zones,* Chapman & Hall, London, 1992, 134. With permission.)

Table 4.9–1 **Approximate Partial Composition of Manganese Nodules from Different Environments (in wt %)**

	Abyssal	Seamounts	Active Ridges	Continental Margins
Mn	16.78	14.62	15.51	38.69
Fe	17.27	15.81	19.15	1.34
Ni	0.540	0.351	0.306	0.121
Cu	0.370	0.058	0.081	0.082
Co	0.256	1.15	0.400	0.011

Source: Cronan, D. S., *Marine Minerals in Exclusive Economic Zones,* Chapman & Hall, London, 1992, 100. With permission.

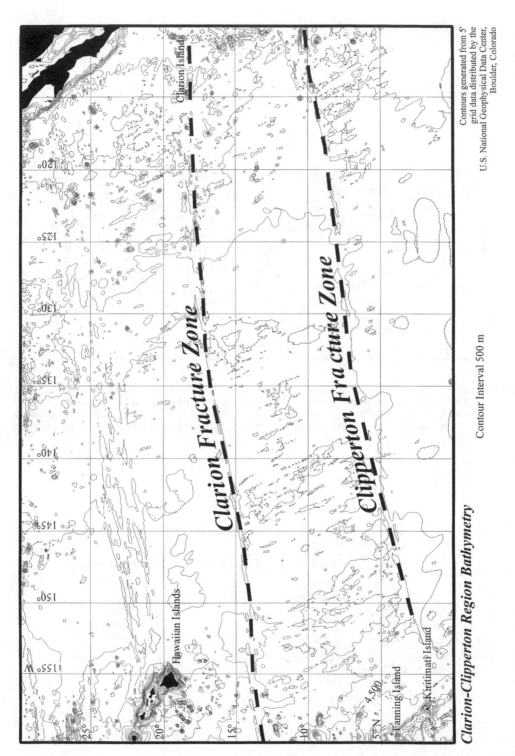

Clarion-Clipperton Region Bathymetry

Figure 4.9–4   Clarion-Clipperton Fracture Zone Region (CCZ), showing the major bathymetric features and bounding land masses. The manganese nodules with the highest concentrations of nickel and copper and with the highest densities (kg/m²) are found between these two fracture zones. (From Morgan, C. L., in *Handbook of Marine Mineral Deposits*, Cronan, D. S., Ed., CRC Press, Boca Raton, FL, 2000, 146. With permission.)

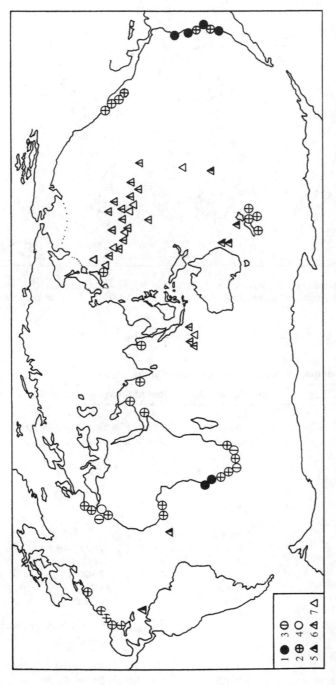

Figure 4.9–5 Phosphorite locations on the seafloor: 1–4 = phosphorites on continental margins; 5–7 = phosphorites on seamounts. Ages: 1 = Holocene; 2, 5 = Neogene; 3, 6 = Paleogene; 4, 7 = Cretaceous. (From Cronan, D. S., *Marine Minerals in Exclusive Economic Zones*, Chapman & Hall, London, 1992. With permission.)

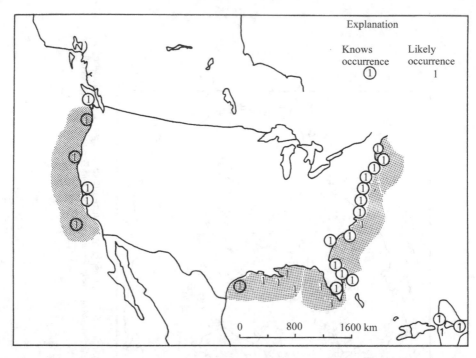

Figure 4.9–6 Aggregate deposits along the coasts of the United States. (From Cronan, D. S., *Marine Minerals in Exclusive Economic Zones,* Chapman & Hall, London, 1992, 26. With permission.)

Table 4.9–2   Principal Placer Minerals and Their Composition

Gems	Specific Gravity	Composition and Principal Element
Diamond	3.5	C
Garnet	3.5–4.27	$(CaMgFeMn)_3 (FeAlCrTi)_2 (SiO_4)_3$
Ruby	3.9–4.1	Al_2O_3
Emerald	3.9–4.1	Al_2O_3
Topaz	3.4–3.6	$Al_2F_2SiO_4$
Heavy noble metals		
Gold	20	Au
Platinum	21.5	Pt
Light heavy minerals		
Beryl	2.75–2.8	$Be_3Al_2Si_6O_{18}$ (Be)
Corundum	3.9–4.1	Al_2O_3 (Al)
Rutile	4.2	TiO_2 (Ti)
Zircon	4.7	$ZrSiO_4$ (Zr)
Chromite	4.5–4.8	$FeCr_2O_4$ (Cr)
Ilmenite	4.5–5.0	$FeOTiO_2$ (Ti)
Magnetite	5.18	Fe_3O_4 (Fe)
Monazite	5.27	$(CeLa Yt) PO_4 ThO_2 SiO_2$ (Th REE)
Scheelite	5.9–6.1	$CaWO_4$ (W)
Heavy heavy minerals		
Cassiterite	6.8–7.1	SnO_2 (Sn)
Columbite, Tantalite	5.2–7.9	$(FeMn) (NbTa)_2O_6$ (Nb, Ta)
Cinnabar	8–10	HgS (Hg)

Source: Cronan, D. S., *Marine Minerals in Exclusive Economic Zones,* Chapman & Hall, London, 1992, 32. With permission.

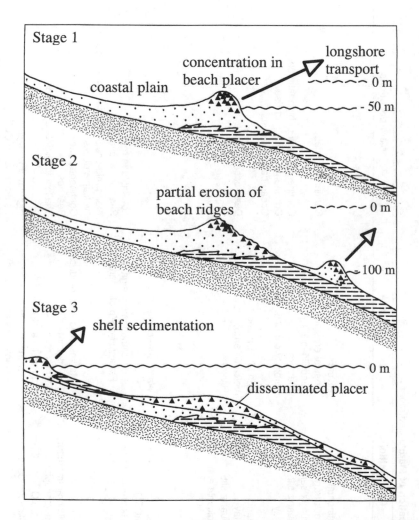

Figure 4.9–7 Conceptual model for the origin of disseminated placer deposits with light heavy minerals (indicated by triangles) on the shelf during medium, low, and high sea level. (From Kudrass, H. R., in *Handbook of Marine Mineral Deposits*, Cronan, D. S., Ed., CRC Press, Boca Raton, FL, 2000, 9. With permission.)

4.10 MARINE SEDIMENTS

Table 4.10–1 Classification of Marine Sediments

Type	Composition	Source	Main Locations
Lithogenous			
	Continental margin		
	Rock fragments	Rivers	Continental shelf
	Quartz sand	Coastal erosion	
	Quartz silt	Landslides	
	Clay	Glaciers	Continental shelf in high latitudes
		Turbidity currents	Continental slope and rise
			Ocean basin margins
	Oceanic	Wind-blown dust	
	Clay	Volcanic eruptions	Deep-ocean basins
	Quartz silt	Rivers	
	Volcanic ash		
Biogenous			
	Calcium carbonate ($CaCO_3$)	Warm surface water	Low-latitude regions
	Calcareous ooze (microscopic)	Coccolithophores (algae)	Sea floor above CCD
		Foraminifers (protozoans)	Along mid-ocean ridges and the tops of volcanic peaks
	Shell/coral fragments (macroscopic)	Macroscopic shell-producing organisms	Continental shelf
		Coral reefs	Beaches
			Shallow low-latitude regions
	Silica ($SiO_2 \cdot nH_2O$)	Cold surface water	High-latitude regions
	Siliceous ooze	Diatoms (algae)	Seafloor below CCD
		Radiolarians (protozoans)	Surface current divergence near the equator

Hydrogenous		
Manganese nodules (manganese, iron, copper, nickel, coabalt)		Abyssal plain
Phosphorite (phosphorus)	Precipitation of dissolved materials directly from seawater due to chemical reactions	Continental shelf
Oolites ($CaCO_3$)		Shallow shelf in low-latitude regions
Metal sulfides (iron, nickel, copper, zinc, silver)		Hydrothermal vents at mid-ocean ridges
Evaporites (gypsum, halite, other salts)		Shallow restricted basins where evaporation is high in low-latitude regions
Cosmogenous		
Iron-nickel spherules	Space dust	In very small proportions mixed with all types of sediment in all marine environments
Tektites (silica glass)		
Iron-nickel meteorites	Meteors	Localized near meteor impact structures
Silicate chondrites		

Source: Thurman, H. V. and Trujillo, A. P., *Essentials of Ocenography,* 6th ed., Prentice-Hall, Upper Saddle River, NJ, 1999, 103. With permission.

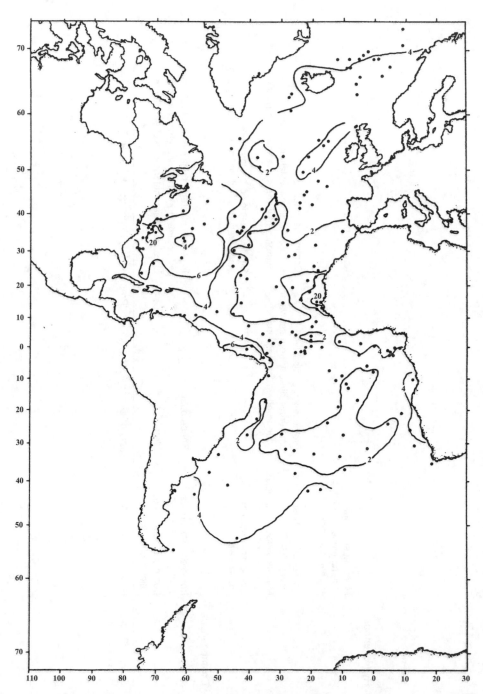

Figure 4.10–1 Sedimentation accumulation rates in the Atlantic Ocean in centimeters per 1000 years. (From Balsam, W. L. and Deaton, B. C., *Rev. Aquat. Sci.,* 4, 411, 1991. With permission.)

Table 4.10–2 Radiometric Sediment Accumulation Rates on the Northern Mid-Atlantic Ridge

Core and Location		Sedimentation Rate, cm/10³ year	
		Bulk	Carbonate Free
Near 42°N			
CH 82-24	Ridge flank, on foothill	2.7	0.95
CH 82-26	Ridge flank on plain among low hills	2.1	0.53
CH 82-30(1)	Ridge flank on small rise	1.5	0.27
CH 82-41	Foothills west of rift valley	1.3	0.31
Near 37°N, FAMOUS Area			
Core 52[a]	Interaction of FZB and RV3	0.98	0.19
Core 124[a]	Intersection of FZB and RV2	0.94	0.17
Core 125[a]	Intersection of FZB and RV2	0.50	0.10
Core 118[a]	Near FZA on scarp	0.35	0.04
Core 527–3[a]	Floor of inner valley, near center of RV2	2.90	0.78
TAG Area 26°N			
Core 4A[a]	Median valley floor	1.8	0.74
24°N			
ZEP 12	Ridge flank	—	0.45
ZEP 13[a]	Ridge crest valleys	~0.40	~0.20
ZEP 15[a]	Ridge crest valleys	~0.22	~0.05
ZEP 18[a]	Ridge crest valleys	~0.25	~0.11
ZEP 22	Ridge flank	0.25	—
20°N			
All 42–13	Ridge flank	0.50	0.17
All 42–17[a]	Ridge crest; hills adjacent to median valley	0.73	0.13
All 42–33	Ridge flank	0.47	0.31
All 42–41	Abyssal hill west of ridge	1.30	1.26

[a] Median valley cores.

Source: Scott, M. R. et al., in Deep Drilling Results in the Atlantic Ocean: Ocean Crust, Talwani, M. et al., Eds., Technical Volume, American Geophysical Union, Washington, D.C., 1979, 403. With permission.

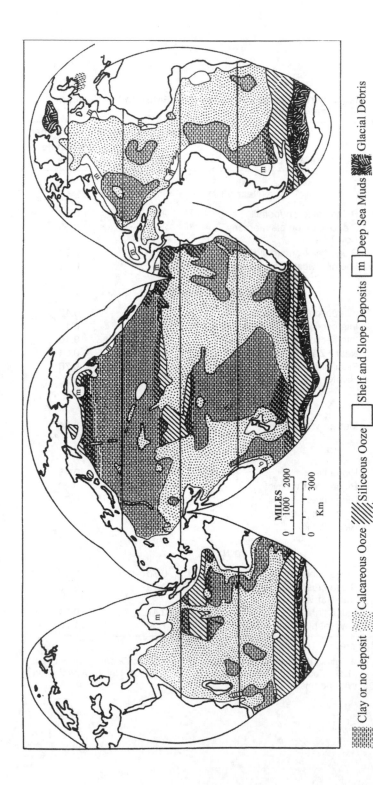

Figure 4.10–2   Sediment composition of the deep-sea floor. (From Seibold, E. and Berger, W. H., *The Sea Floor: An Introduction to Marine Geology*, Springer-Verlag, Berlin, 1993, 218. With permission.)

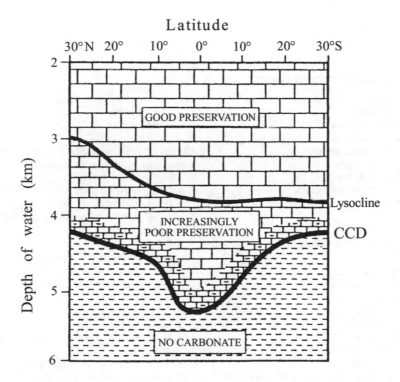

Figure 4.10–3 Depth of the lysocline and calcium carbonate compensation depth (CCD) for a north–south profile of the tropical central Pacific Ocean showing relative degrees of carbonate preservation. Note depression of the CCD due to enhanced productivity in the equatorial upwelling area. (From Grotsch, J. et al., in *Cycles and Events in Stratigraphy,* Einsele, G. et al., Eds., Springer-Verlag, Berlin, 1991, 110. With permission.)

4.11 ESTUARIES, BEACHES, AND CONTINENTAL SHELVES

Table 4.11–1   Classification of Coasts

I. Coasts Shaped by Nonmarine Processes
   (Primary Coasts)
   A. Land erosion coasts
      1. Drowned rivers
      2. Drowned glacial-erosion coasts
         a. Fjord (narrow)
         b. Trough (Wide)
   B. Subaerial deposition coasts
      1. River deposition coasts
         a. Deltas
         b. Alluvial plains
      2. Glacial-deposition coasts
         a. Moraines
         b. Drumlins
      3. Wind deposition coasts
         a. Dunes
         b. Sand flats
      4. Landslide coasts
   C. Volcanic coasts
      1. Lava flow coasts
      2. Tephra coasts
      3. Coasts formed by volcanic collapse or
         explosion
   D. Coasts shaped by earth movements
      1. Faults
      2. Folds

      3. Sedimentary extrusions
         a. Mud lumps
         b. Salt domes
   E. Ice coasts
II. Coasts Shaped by Marine Processors or Marine
    Organisms (Secondary Coasts)
    A. Wave erosion coasts
       1. Straightened coasts
       2. Irregular coasts
    B. Marine deposition coasts (prograded by
       waves, currents)
       1. Barrier coasts
          a. Sand beaches (single ridge)
          b. Sand islands (multiple ridges, dunes)
          c. Sand spits (connected to mainland)
          d. Bay barriers
       2. Cuspate forelands (large projecting points)
       3. Beach plains
       4. Mud flats, salt marshes (no breaking
          waves)
    C. Coasts formed by biological activity
       1. Coral reef, algae (in the tropics)
       2. Oyster reefs
       3. Mangrove coasts
       4. Marsh grass
       5. Serpulid reefs

Source: Thurman, H. V., *Introductory Oceanography,* 8th ed., Prentice-Hall, Upper Saddle River, NJ, 1997, 288. With permission.

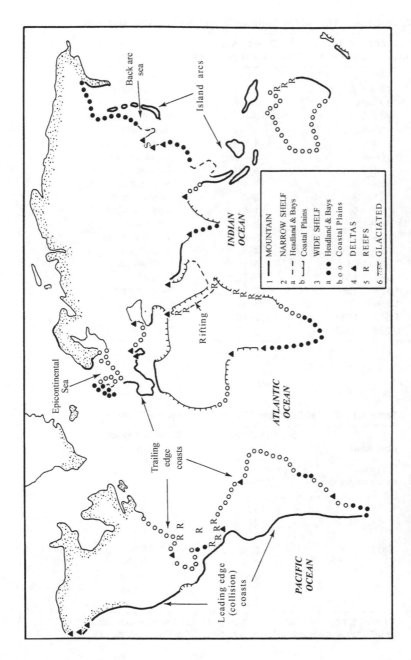

Figure 4.11–1 The tectonic classification of the world's coastlines. (From Carter, R. W. G., *Coastal Environments: An Introduction to the Physical, Ecological, and Cultural Systems of Coastlines,* Academic Press, London, 1988, 17. With permission.)

Table 4.11–2   The Udden–Wentworth Particle Size Classification

Grade Limits (diameter in mm)	Grade Limits (diameter in phi units)	Name
>256	≤ -8	Boulder
256–128	-8–-7	Large cobble
128–64	-7–-6	Small cobble
64–32	-6–-5	Very large pebble
32–16	-5–-4	Large pebble
16–8	-4–-3	Medium pebble
8–4	-3–-2	Small pebble
4–2	-2–-1	Granule
2–1	-1–0	Very coarse sand
1–1/2	0–1	Coarse sand
1/2–1/4	1–2	Medium sand
1/4–1/8	2–3	Fine sand
1/8–1/16	3–4	Very fine sand
1/16–1/32	4–5	Coarse silt
1/32–1/64	5–6	Medium silt
1/64–1/128	6–7	Fine silt
1/128–1/256	7–8	Very fine silt
1/256–1/512	8–9	Coarse clay
1/512–1/1024	9–10	Medium clay
1/1024–1/2048	10–11	Fine clay

Table 4.11–3   Settling Velocities of Sedimentary Particles

Sediment	Median Diameter (μm)	Settling Velocity (m/day)
Fine sand	250–125	1040
Very fine sand	125–62	301
Silt	31.2	75.2
	15.6	18.8
	7.8	4.7
	3.9	1.2
Clay	1.95	0.3
	0.98	0.074
	0.49	0.018
	0.25	0.004
	0.12	0.001

Source: King, C. A. M., *Introduction to Marine Geology and Geomorphology,* Edward Arnold, London, 1975. With permission.

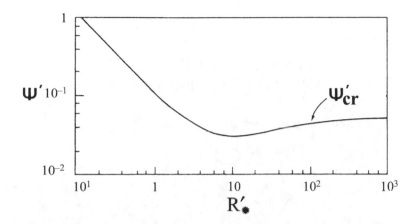

Figure 4.11–2 Shields diagram illustrating relationship between grain Reynolds number (*x* axis) and the critical Shields parameter (*y* axis) at which particle motion is initiated. (From Wright, L. D., *Morphodynamics of Inner Continental Shelves,* CRC Press, Boca Raton, FL, 1995, 141.)

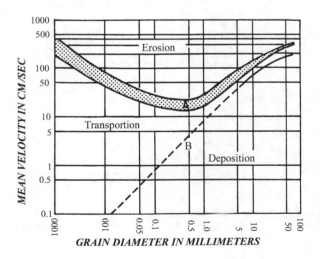

Figure 4.11–3 Hjulström's diagram showing the relationship between grain size and water velocity. (From Hjulström, F., *Upsala Univ. Geol. Inst.,* B, 25, 221, 1935. With permission.)

DYNAMIC EQUIVALENCE
AND SORTING PROCESSES

A. Settling Equivalence

B. Entrainment Equivalence (Selective Entrainment)

C. Transport Equivalence (Transport Sorting)

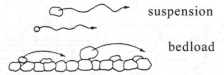

suspension

bedload

D. Dispersive – Pressure Equivalence
   (Shear Sorting)

Figure 4.11–4 Schematic illustration of sorting processes and possible dynamic equilibria acting on sediments. (From Komar, P. D., *Rev. Aquat. Sci.,* 1, 393, 1989. With permission.)

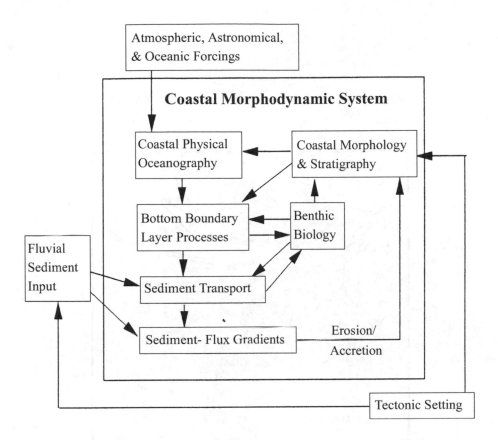

Figure 4.11–5 Flow diagram depicting the idealized elements and linkages in a coastal morphodynamic system. (From Wright, L. D., *Morphodynamics of Inner Continental Shelves,* CRC Press, Boca Raton, FL, 1995, 5. With permission.)

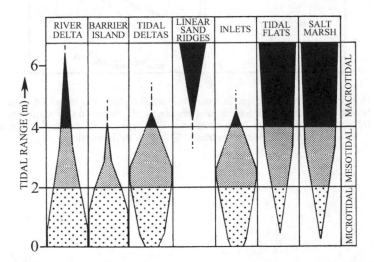

Figure 4.11–6 Variation in morphology of coastal plain shorelines with respect to differences in tidal range. (From Hayes, M. O., in *Barrier Islands from the Gulf of St. Lawrence to the Gulf of Mexico,* Leatherman, S. P., Ed., Academic Press, New York, 1979, 4. With permission.)

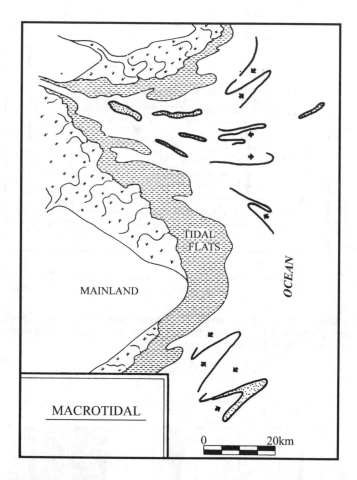

Figure 4.11–7 Morphological model of typical macrotidal coastal plain shoreline with medium wave energy. Note absence of barrier islands and presence of offshore linear sand ridges. (From Hayes, M. O., in *Barrier Islands from the Gulf of St. Lawrence to the Gulf of Mexico,* Leatherman, S. P., Ed., Academic Press, New York, 1979, 6. With permission.)

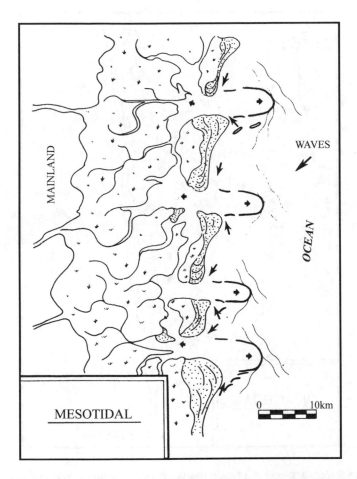

Figure 4.11–8 Morphological model of a typical mesotidal barrier island shoreline with medium wave energy. Note abundance of tidal inlets. (From Hayes, M. O., in *Barrier Islands from the Gulf of St. Lawrence to the Gulf of Mexico,* Leatherman, S. P., Ed., Academic Press, New York, 1979, 13. With permission.)

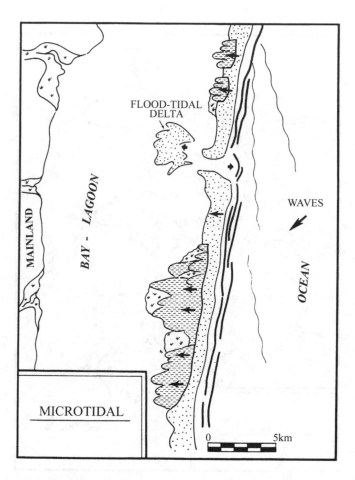

Figure 4.11–9 Morphological model of a typical microtidal barrier island shoreline with medium wave energy. Note abundance of washover areas and paucity of tidal inlets. Flood-tidal deltas tend to be considerably larger than ebb-tidal deltas. (From Hayes, M. O., in *Barrier Islands from the Gulf of St. Lawrence to the Gulf of Mexico,* Leatherman, S. P., Ed., Academic Press, New York, 1979, 19. With permission.)

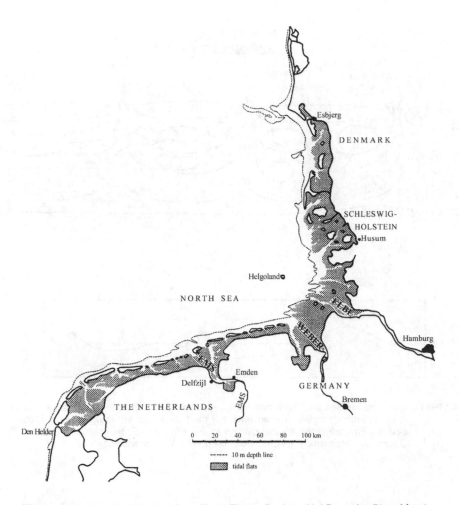

Figure 4.11–10 The Wadden Sea. (From Eisma, D., *Intertidal Deposits: River Mouths, Tidal Flats, and Coastal Lagoons,* CRC Press, Boca Raton, FL, 1998, 132. With permission.)

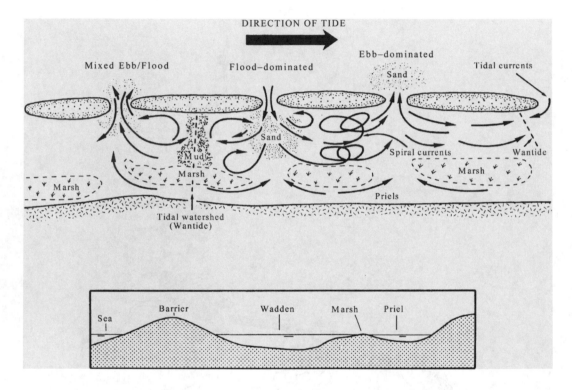

Figure 4.11–11   The Wadden Sea tidal dynamics model, displaying the types of tidal passes and the wantide (tidal watershed) zones. The movement of the tide is from left to right. (From Carter, R. W. G., *Coastal Environments: An Introduction to the Physical, Ecological, and Cultural Systems of Coastlines,* Academic Press, London, 1988, 171. With permission.)

Table 4.11–4   Selection of Features Typical for Intertidal Deposits

Unidirectional

Lower phase plane bed

Small-scale ripples—straight and sinuous: immature: linguoid: mature; Baas et al.,1993

Erosive ripple forms—Reineck and Singh, 1973, their Figure 7.8

Back-flow ripples—small-scale ripples formed in troughs of large-scale ripples through backflow; Boersma et al., 1968

Megaripples, dunes—straight-crested, undulatory/sinuous, lunate

Tidal bundle sequences—series of, ideally, about 28 megaripple foresets of increasing and decreasing thickness, formed due to the increasing and decreasing transport capacity of tidal currents during a neap-spring-neap-neap period; in intertidal environments only one (high-slack water) mud drape can be deposited

Upper phase plane bed—current lineation

Antidunes—extremely low preservation potential

Graded beds—deposited during storms

Bidirectional

Ebb caps on megaripples—Boersma and Terwindt, 1981

Flaser-linzen bedding—Reineck and Singh, 1973

Wave ripples

Combined ripples produced by the combined action of waves and currents with different directions

Interference ripples—combination of wave and/or current ripples with different direction; Reineck and Singh, 1973

Adhesion ripples—accretion of generally preexisting ripple forms due to aeolian transport during low water: van Straaten, 1953

Runoff Structures

Microdeltas in channels along shoals with runoff through small gullies

Rill marks—Reineck and Singh, 1973

Erosional Structures

Shell lags—scour marks, Figure 7.10

Tool Marks

Figure 7.11; see also Dionne, 1988

Deformation Structures

Load cast

Cavernous sand—Emery, 1945

Convolute lamination—de Boer, 1979

Mud volcanoes

Mud cracks produced by desiccation

Slumps and Faults

Formed due to slope instability, e.g., along the margins of channels

Bioturbation

U burrows (*Arenicola,* certain shrimps); Figure 7.7

Pressure structures; Figure 7.12

Biological Life

Manyfold; shell bioherms; see text and other chapters of original source

Table 4.11–4   Selection of Features Typical for Intertidal Deposits (continued)

Typical Sediment Types

Ooids; sand-sized concentric carbonate grains, formed in agitated waters in the subtidal and intertidal zone;
  Bathurst, 1971
Broken and complete skeletal carbonate of organisms typical for the intertidal (and often also subtidal) zone
Clay flakes, formed due to shrinkage of thin mud layers and/or fine-grained layers stabilized by algae and/or
  early cementation
Organic remains of inter-supratidal plants

Additional

Raindrop imprints
Foam impressions
Water-level marks
Crystal imprints in evaporite environments, etc.

Note: See original source for cited figures and reference citations.
Source: Eisma, D., *Intertidal Deposits: River Mouths, Tidal Flats, and Coastal Lagoons,* CRC Press, Boca
Raton, FL, 1998, 346. With permission.

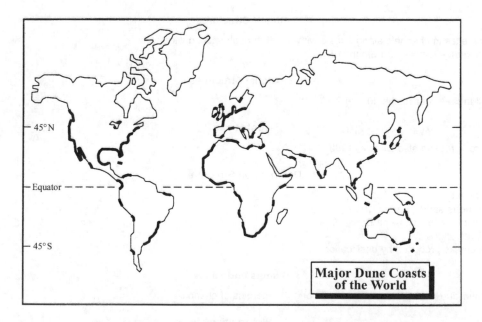

Figure 4.11–12   Distribution of the major dune systems of the world. (From Carter, R. W. G. et al., in *Coastal
                    Dunes: Form and Processes,* Nordstrom, K. F. and Psuty, N., Eds., John Wiley & Sons, New
                    York, 1990, 3. With permission.)

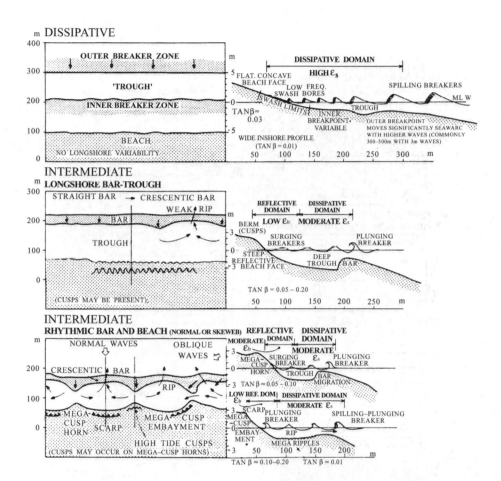

Figure 4.11–13 Six common surf zone and beach mophodynamic states. The dissipative and reflective extremes are separated by at least four intermediate states. (From Wright, L. D., *Morphodynamics of Inner Continental Shelves,* CRC Press, Boca Raton, FL, 1998, 186. With permission.)

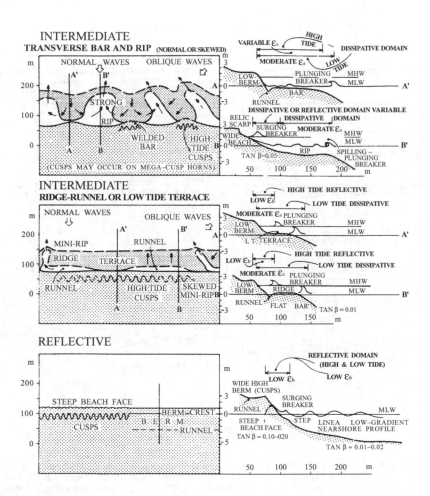

Figure 4.11–13   (continued)

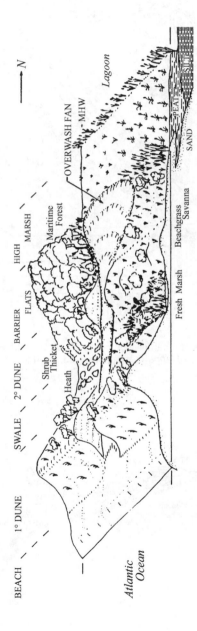

Figure 4.11–14 Barrier-beach environments, vegetation, and sediments. Sand deposits from overwash and aeolian activity produce new substrates for colonization by plants. These deposits may then be buried by peat as marshes develop, forming discrete lenses. (From Clark, J. S., *Rev. Aquat. Sci.*, 2, 509, 1990. With permission.)

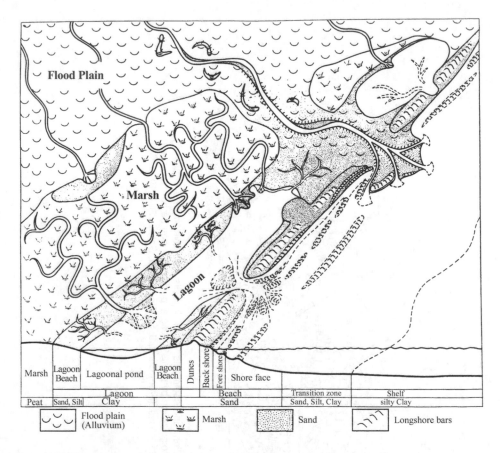

Figure 4.11–15 Typical geomorphic features of a barrier-type coast. (From Seibold, E. and Berger, W. H., *The Sea Floor: An Introduction to Marine Geology,* 2nd ed., Springer-Verlag, Berlin, 1993, 141. With permission.)

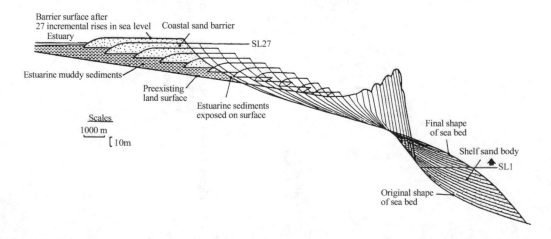

Figure 4.11–16 Example of shelf and barrier evolution in response to sea level rise and coastal transgression across complex shelf topography as predicted by a model of Cowell et al. (1991, 1992). (From Wright, L. D., *Morphodynamics of Inner Continental Shelves,* CRC Press, Boca Raton, FL, 1995, 175. With permission.)

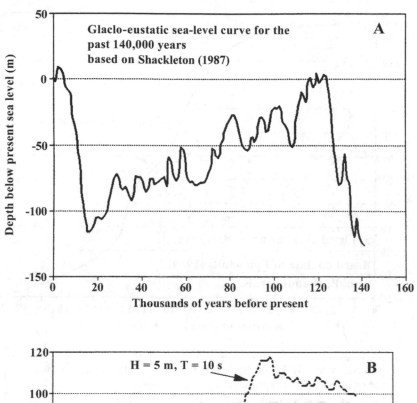

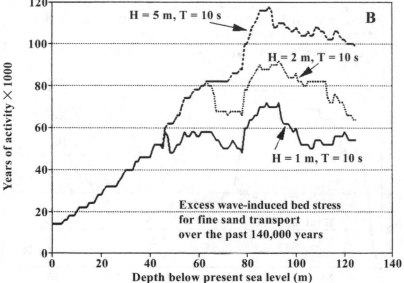

Figure 4.11–17   Sea level variation and bed level activity for the past 140,000 years. (A) Sea level curve for the past 140,000 years (based on Shackleton, 1987). (B) Predicted distribution of bed agitation by waves of different sizes over the past 140,000 years. (From Wright, L. D., *Morphodynamics of Inner Continental Shelves,* CRC Press, Boca Raton, FL, 1995, 34. With permission.)

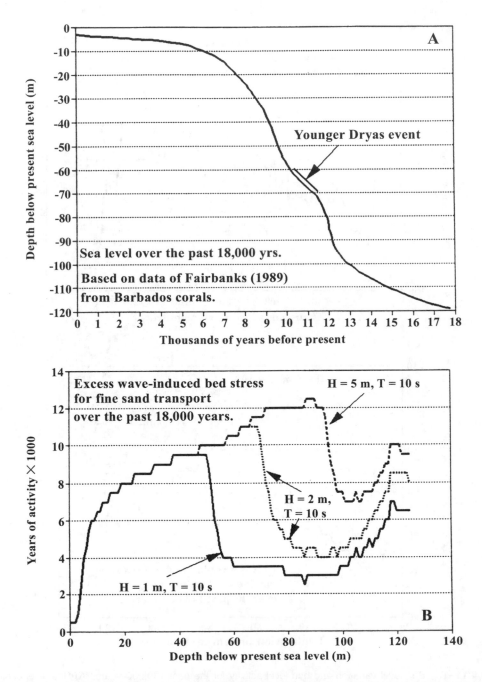

Figure 4.11–18 Sea level variation and bed level activity for the past 18,000 years. (A) Sea level curve for the past 18,000 years. (B) Predicted distribution of bed agitation by waves of different sizes over the past 18,000 years. (From Wright, L. D., *Morphodynamics of Inner Continental Shelves*, CRC Press, Boca Raton, FL, 1995, 36. With permission.)

Table 4.11–5   Average Sediment Discharge to the Ocean of the 25 Rivers with the Largest
Sediment Load

River	Average Sediment Discharge[a] (10^6/t/year)	Average Water Discharge (km^3/year)	Average Concentration (mg/l)
Amazon	1000–1300 [1000–1300]	6300	190
Yellow River (Huang He)	1100 (100) [1200]	49	22040
Ganges-Brahmaputra	900–1200	970	1720
Chang Jiang	480	900	531
Irrawaddy	260 [260]	430	619
Magdalena	220	240	928
Mississippi	210 (400) [500]	580	362
Godavari	170	92	1140
Orinoco	150 (150) [150]	1100	136
Red River (Hung Ho)	160	120	1301
Mekong	160	470	340
Purari/Fly	110	150	1040
Salween	~100	300	300
Mackenzie	199 (100) [100]	310	327
Parana/Uruguay	100	470	195
Zhu Jiang (pearl)	80	300	228
Copper	70 (70) [70]	39	1770
Choshui	66	6	11000
Yukon	60 (60) [60]	195	308
Amur	52	325	160
Indus	50 [250]	240	208
Zaire	43	1250	34
Liao He	41	6	6833
Niger	40	190	210
Danube	40 [70]	210	190

Rivers That Formerly Discharged Large Sediment Loads

Nile	0 [125]	0 (was 39)	
Colorado	>1 [125]	1 (was 20)	

Other Rivers That Discharge Large Volumes of Water

Zambesi	20	220	90
Ob	16	385	42
Yenesei	13	560	23
Lena	12	510	24
Columbia	8 [15]	250	32
St. Lawrence	3	450	7

[a] ( ) Presumed natural level: [ ] year 1890.

Adapted from Milliman and Meade (1983) and Meade (personal communication).

Source: Eisma, D., *Intertidal Deposits: River Mouths, Tidal Flats, and Coastal Lagoons,* CRC Press, Boca Raton, FL, 1998, 17. With permission.

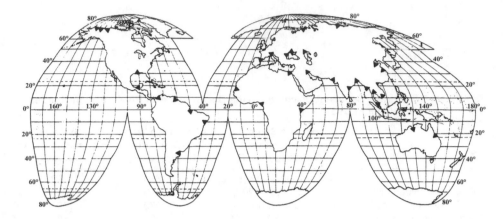

Figure 4.11–19 Worldwide distribution of major deltas. (From Wright, L. D. et al., Louisiana State University Studies Inst. Tech. Rep. 156, Louisiana State University, Baton Rouge, 1974. With permission.)

Figure 4.11–20 The distribution of continental shelves in the Northern Hemisphere. (From Walsh, J. J., *On the Nature of Continental Shelves,* Academic Press, New York, 1989. With permission.)

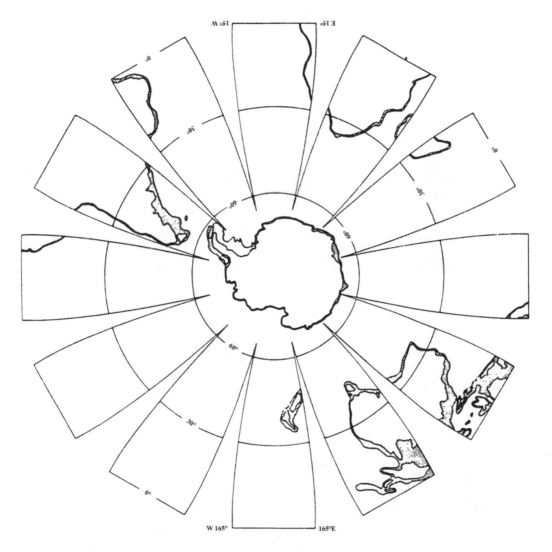

Figure 4.11–21   The distribution of continental shelves in the Southern Hemisphere. (From Walsh, J. J., *On the Nature of Continental Shelves,* Academic Press, New York, 1989. With permission.)

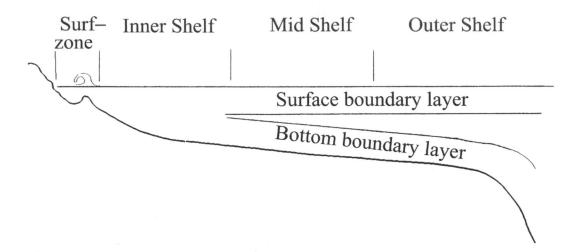

Figure 4.11–22 The inner shelf is a friction-dominated realm where surface and bottom boundary layers overlap. (From Wright, L. D., *Morphodynamics of Inner Continental Shelves,* CRC Press, Boca Raton, FL, 1995, 51. With permission.)

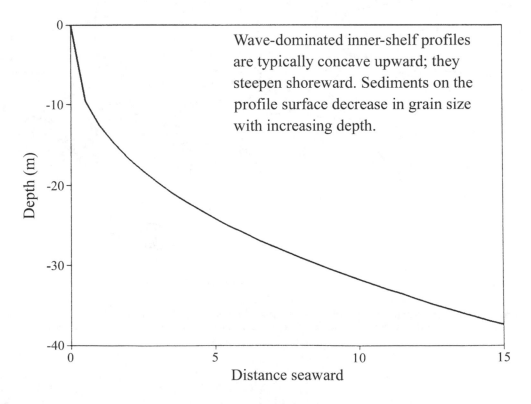

Wave-dominated inner-shelf profiles are typically concave upward; they steepen shoreward. Sediments on the profile surface decrease in grain size with increasing depth.

Figure 4.11–23 Typical concave upward inner shelf profile. (From Wright, L. D., *Morphodynamics of Inner Continental Shelves,* CRC Press, Boca Raton, FL, 1995, 14. With permission.)

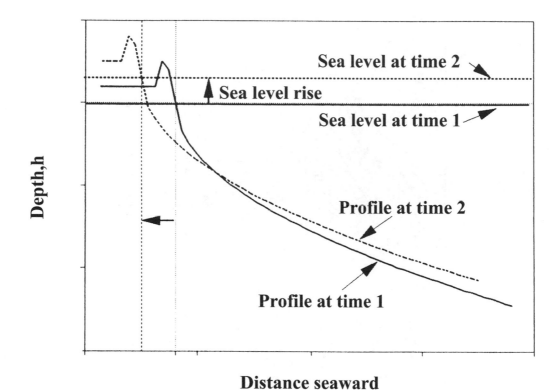

Figure 4.11–24 Concept of the "Bruun Rule" whereby a sea level rise is assumed to cause an inner shelf profile of equilibrium to experience an upward and shoreward translation. (From Wright, L. D., *Morphodynamics of Inner Continental Shelves,* CRC Press, Boca Raton, FL, 1995, 26. With permission.)

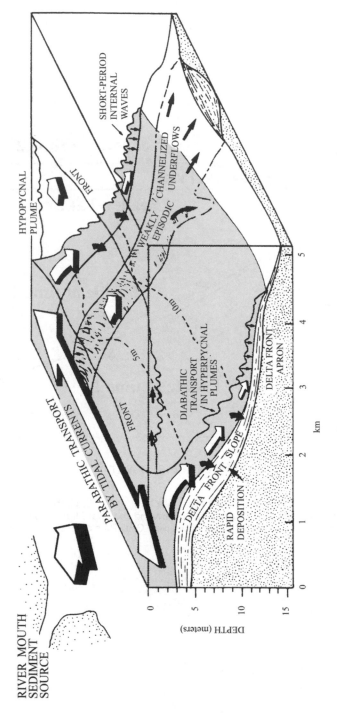

Figure 4.11–25 Schematic representation illustrating along-shelf and across-shelf transports by a combination of tidal currents and hyperpycnal flows. (From Wright, L. D., *Morphodynamics of Inner Continental Shelves*, CRC Press, Boca Raton, FL, 1995, 95. With permission.)

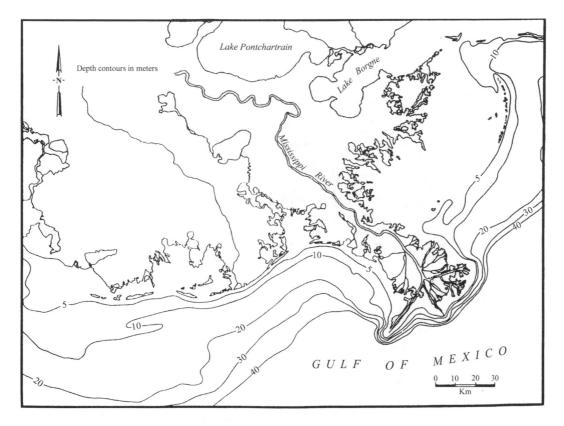

Figure 4.11–26   Bathymetric map of the Louisiana shelf. (From Wright, L. D., *Morphodynamics of Inner Continental Shelves,* CRC Press, Boca Raton, FL, 1995, 41. With permission.)

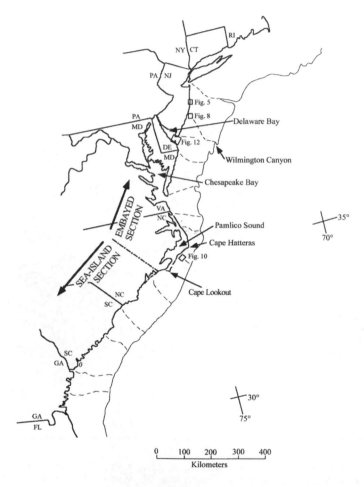

Figure 4.11–27 U.S. Atlantic continental margin. Shelf edge is indicated by the 200-m bathymetric contour. Approximate positions of major paleochannels that were known to exist during the most recent (Holocene) sea-level rise are indicated with dashed lines. (From Ashley, G. M. and Sheridan, R. E., SEPM Special Publication No. 51, 1994, 287. With permission.)

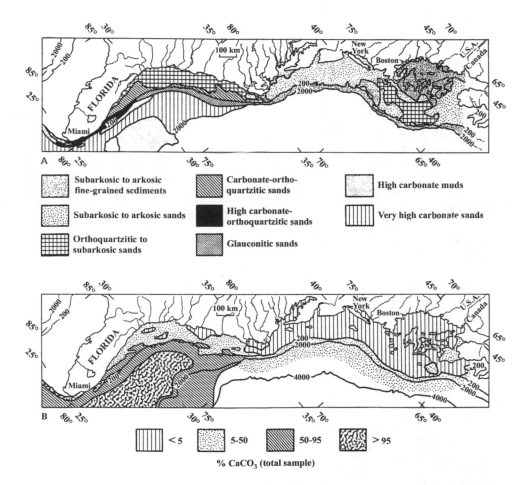

Figure 4.11–28 Continental margin sediments along the east coast of the United States. Top panel displays the sediment types; bottom panel shows calcium carbonate content. (From Milliman, J. D. et al., *Geol. Soc. Am. Bull.*, 82, 1315, 1972. With permission.)

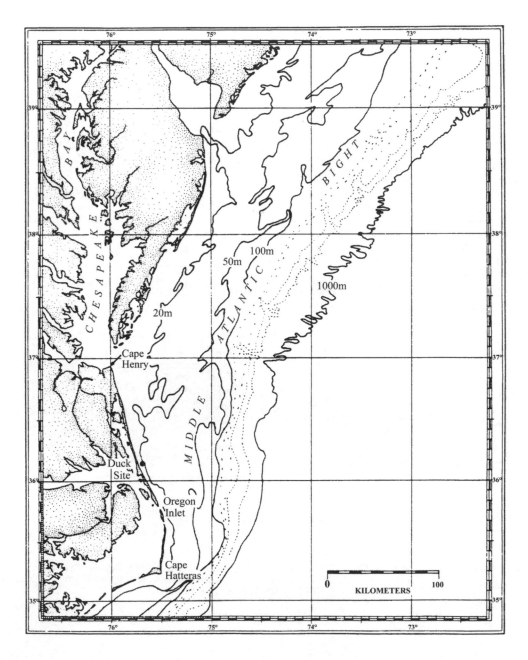

Figure 4.11–29 Bathymetric map of the Middle Atlantic Bight. (From Wright, L. D., *Morphodynamics of Inner Continental Shelves,* CRC Press, Boca Raton, FL, 1995, 37. With permission.)

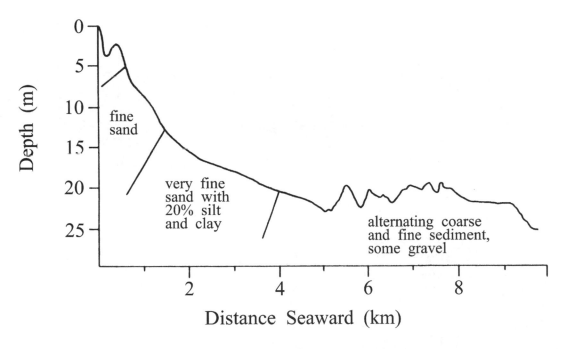

Figure 4.11–30   Profile of the southern Middle Atlantic Bight inner shelf off Duck, North Carolina. (From Wright, L. D., *Morphodynamics of Inner Continental Shelves,* CRC Press, Boca Raton, FL, 1995, 38. With permission.)

CHAPTER 5

Marine Biology

I. INTRODUCTION

Estuarine and oceanic environments support a multitude of organisms from the smallest of protozoans to the largest of mammals. The broadest division of the marine realm separates the benthic (i.e., bottom) and the pelagic (i.e., water column) environments. The benthic environment consists of supratidal (supralittoral), intertidal (littoral), subtidal (sublittoral or shelf), bathyal, abyssal, and hadal zones. The pelagic environment, in turn, comprises neritic (inshore) and oceanic (offshore) zones. The neritic zone can be further subdivided on the basis of water depth into epipelagic (0 to 200 m depth), mesopelagic (200 to 1000 m), bathypelagic (1000 to 2000 m), abyssalpelagic (2000 to 6000 m), and hadalpelagic regions (>6000 m). Organisms inhabiting the neritic and oceanic zones are logically classified as neritic (coastal ocean) and oceanic species. They can also be grouped according to their life habits into plankton (free-floating), nekton (swimming), and benthos (bottom-dwelling).

This chapter focuses on the major taxonomic groups of organisms inhabiting estuarine and marine environments. It also examines the physical, chemical, and biological factors affecting the abundance, distribution, and diversity of these organisms. In addition, it provides useful information on estuarine and marine communities and the complex and dynamic habitats that they utilize.

II. BACTERIA

Marine bacteria are microscopic, unicellular organisms less than 2 μm in diameter, which belong to the phylum Schizomycophyta. They can be broadly subdivided into autotrophic and heterotrophic forms. Autotrophic bacteria derive energy through photosynthesis (phototrophs) or through the oxidation of inorganic compounds (chemolithotrophs). Included in this category are those bacteria that use hydrogen in water as electron donors, while releasing oxygen, and those bacteria that utilize reduced substances (e.g., sulfides, molecular hydrogen, or carbon compounds) as electron donors in photoassimilation of carbon dioxide.[1] Heterotrophic bacteria, saprophytes and parasites, obtain energy from other organic compounds. Photosynthetic bacteria, phototrophs, encompass marine forms such as anoxyphotobacteria and oxyphotobacteria. In marine systems, phototrophic bacteria have rather limited significance as primary producers and remain relatively unimportant in biotic transformations.[2] Chemosynthetic bacteria (chemolithotrophs) are integral components in several geochemical cycles (e.g., nitrogen and sulfur cycles). Examples include nitrifying bacteria (family Nitrobacteriaceae), which convert ammonia to nitrate (e.g., *Nitrosomonas*) and nitrite to nitrate (e.g., *Nitrobacter*), and sulfur bacteria (i.e., sulfur-oxidizing and sulfur-granule-containing forms), which oxidize sulfide, sulfur, or thiosulfate to sulfate. Recently, chemosynthetic bacteria that oxidize sulfur and other inorganic compounds at hydrothermal vent and cold-water sulfide/methane seep environments in the deep sea have received much attention because the energy derived from their oxidation supports lavish biotic communities.[3,4] These communities have been the focus of a number of comprehensive investigations.[5-8]

Five microbial habitats are delineated in estuarine and marine environments: planktonic, neustonic, epibiotic, benthic, and endobiotic types. Free-floating bacteria (i.e., bacterioplankton) assimilate soluble organic matter in the water column, removing this material in microgram-per-liter or nanogram-per-liter concentrations.[9] Although bacterioplankton attain relatively high numbers in some coastal waters, they may be less abundant than bacterial neuston populations that are highly responsive to the greater concentrations of fixed carbon and nutrients accumulating at the air–seawater interface. Epibiotic bacteria, in turn, colonize the surfaces of marine substrates, where they serve as a food source for protozoans and other heterotrophs. An extremely large number of bacteria inhabit seafloor sediments, living in both aerobic and anaerobic zones. Other bacteria enter into commensalistic, mutualistic, and parasitic relationships with other organisms. The symbiotic activity of certain endobacteria (i.e., parasites) has been well chronicled, largely because of the diseases and other problems they inflict on marine plants and animals.

Marine bacteria attain peak numbers in estuarine waters ($\sim 10^6$ to 10^8 cells/ml), and gradually decline in abundance and production from the coastal ocean (1 to 3×10^6 cells/ml) to neritic (10^4 to 10^6 cells/ml) zones.[1,10-15] Bacterial biomass likewise decreases from more than 10 μg C/l in estuarine waters to 5 to 10 μg C/l and 1 to 5 μg C/l in coastal oceanic and neritic waters, respectively.[16] Similar trends are evident in bottom sediments, with highest bacterial cell counts (10^{11} cells/cm^3) observed in salt marsh and mudflat sediments along estuarine shores, and lower counts recorded in subtidal shelf and deep-sea sediments.[14] Bacterial cell counts, as well as bacterial production, also diminish with increasing depth within the top 20 cm of the sediment column. Rublee,[17] for example, reported that bacterial cell counts of salt marshes dropped from 1 to 20×10^9 cells/cm^3 in surface sediments to 1 to 3×10^9 cells/cm^3 at a depth of 20 cm. Bacterial numbers typically peak near the sediment–water interface within the upper 2 cm of seafloor sediments, and then they decline at progressively greater depths. The aforementioned trends in bacterial abundance in estuarine and oceanic environments closely correlate with the concentration of organic matter present in the water column and bottom sediments. Highest bacterial cell counts and production occur in sediments enriched in particulate organic matter or enhanced by high dissolved organic carbon concentrations (e.g., salt marsh, mangrove, and seagrass biotopes).

The density of bacteria on sediment surfaces ranges from ~ 1 cell/0.3 μm^2 to 1 cell/400 μm^2.[18,19] The highest numbers exist in sediments with elevated concentrations of organic matter. More bacteria attach to fine sediments (clay and silt) than to coarse sediments (sand and gravel), concentrating in cracks, crevices, and depressions in the grains. Hence, only a relatively small area of a given particle is colonized by the microbes at any time. Attached bacteria are generally larger (~ 0.6 μg in diameter) than free-living (bacterioplankton) forms (<0.4 μg in diameter).[20,21] The abundance of attached relative to free-living bacteria increases from the open ocean to nearshore oceanic and estuarine environments.

The role of bacteria is vital to the health of estuarine and marine ecosystems. For example, bacteria are critical to the decomposition of organic matter, cycling of nutrients and other substances, and energy flow in food webs. A large fraction of marine bacteria are saprobes, obtaining their nourishment from dead organic matter. Bacterial-mediated decomposition of organic matter proceeds largely by hydrolysis of extracellular enzymes released by the microoganisms, followed by uptake and subsequent incorporation of solubilized compounds into microbial biomass.[14,22] Bacterial mineralization of the organic matter releases inorganic chemical constituents. The traditional view of bacteria as decomposers in marine systems is that they colonize detrital material, assimilate nutrients, and convert particulate organic matter into dissolved organic matter to meet their energy requirements. They then serve as a food source for microfauna (e.g., ciliates and flagellates) and macrofauna (e.g., detritivores).

Bacterial decomposition of organic matter takes place both in aerobic and anaerobic environments. Aerobic heterotrophic bacteria occur in three distinct habitats, that is, in the water column either suspended or attached to detrital particles, in the top layer of seafloor sediments, and in living and dead tissues of plants and animals. Anaerobic bacteria characteristically inhabit deeper, anoxic sediment layers, anoxic seas characterized by very poor circulation (e.g., bottom waters of fjord-type estuaries), and some polluted regions. Much microbial decomposition and mineralization of organic matter take place in sediments of saltmarshes and other wetlands, as well as in estuarine and coastal oceanic bottom sediments only a few millimeters or centimeters below the sediment–water interface. The depth to the anaerobic zone in seafloor sediments is a function of physical–chemical properties and biological processes occurring in the water column and sediments. For example, light intensity at the sediment surface, the degree of turbulence on the seafloor, permeability of the sediment, bioturbation, and content of organic matter affect the position of the anaerobic zone.[23] All are influenced by other physical, chemical, and biological factors. The balance between the downward diffusion of oxygen in the sediments and its consumption determines the depth of the transition zone separating the oxidized and reduced layers.

Anaerobic bacteria have more specific roles in the transformation of organic matter. Although aerobic bacteria use a wide range of natural substrates, the anaerobes utilize a more-restricted group of compounds. On the basis of their specialized biochemistry, anaerobic forms have been differentiated into four broad types: (1) fermenting bacteria; (2) dissimilatory sulfate–reducing bacteria; (3) dissimilatory nitrogenous oxide–reducing bacteria; and (4) methanogenic bacteria. Among these forms, dissimilatory sulfate–reducing bacteria use a restricted group of low-molecular-weight compounds (e.g., lactate) and generate biomass, carbon dioxide, and soluble end products. Methanogenic bacteria likewise require a specific group of compounds (i.e., carbon dioxide, acetate, formate, and methanol), with methane as their end product. Fermenting and dissimilatory nitrogenous oxide–reducing bacteria employ a wider array of carbon sources; their metabolic products include biomass, carbon dioxide, and low-molecular-weight fermentation. Anaerobic microbial metabolism is much more complex than aerobic microbial metabolism, which principally transforms particulate organic carbon and dissolved organic carbon to biomass and carbon dioxide.[24]

Because bacteria are responsible for most organic matter decomposition in estuarine and marine environments, they also play a vital function in the cycling of nutrients. During the summer months, nutrients released during the breakdown and mineralization of organic matter may be completely assimilated by autotrophs. Mineralization processes can be considered a counterpart of photosynthesis and chemosynthesis.

Aside from their unequivocal importance in the transformation of organic matter, bacteria are key biotic components in trophic energetics. As stated by Alongi (p. 2),[22] "The energetics of the pelagic food web is dominated by the microbial loop—a complex network of autotrophic and heterotrophic bacteria, cyanobacteria, protozoa, and microzooplankton—within which a substantial share of fixed carbon and energy is incorporated into bacteria and subsequently dissipated as it is transferred from one consumer to another." The concept of the microbial loop has been revolutionizing the traditional view of pelagic food webs for more than a decade.[25] The classic perception of the pelagic food web is one of essentially a linear pathway of energy flow from phytoplankton through zooplankton to higher-order consumers (e.g., fish and mammals). The revised view of the pelagic food web, as conveyed by Azam et al.,[25] includes a microbial loop involving the cycling of organic matter through microbes before entering the classic food web.[1]

Marine microheterotrophic activity now appears to be tightly coupled to that of primary producers, with bacteria utilizing dissolved organic matter from living phytoplankton as well as dead phytoplankton remains.[26] As much as 10 to 50% of phytoplankton production is converted to dissolved organic matter and assimilated by bacterioplankton;[27] phagotrophic protists consume the bacterioplankton and, in turn, are ingested by microzooplankton or larger zooplankton.[28–31] Thus, a substantial amount of primary production may cycle through bacteria, much of the dissolved organic matter being recovered through a "microbial loop" rather than lost through remineralization as occurs in the traditional food chain. The structure and function of the microbial links in pelagic food webs are subjects of ongoing investigations, but remain uncertain. Clearly, opinions differ in terms of the relative importance of bacterioplankton as primarily remineralizers of the nutrients fixed by phytoplankton or as a biomass source for the macroplanktonic food chain.[21]

Questions have been raised regarding the applicability of the microbial loop concept to benthic food chains.[22] A benthic microbial loop, analogous to that proposed by Azam et al.,[25] may be even more profound in seafloor sediments, where bacteria consume as much as 80% of the organic inputs.[21] While larger zooplankton ineffectively graze on bacterioplankton in pelagic waters, meta-zoan bacterivores (e.g., meiofauna and benthic macrofauna) as well as protozoans efficiently ingest the microbes in bottom substrates. Trophic interactions among macrofaunal food webs in seafloor sediments, however, often are poorly understood.[32] Indeed, sediment-ingesting benthic macrofauna consume bacteria, protozoans, and meiofauna, potentially obfuscating energy flow from lower trophic levels. Protozoans, especially microflagellates,[33] remove most of the bacterial production in the sediments, macrofaunal grazing on protozoan bacterivores indirectly accounting for only slight cropping of bacteria. Less than 10% of bacterial production is consumed directly by benthic macrofauna in most marine sediments, the highest recorded values ($\sim$40% of production) restricted to intense feeding in surface sediments. Bacterial production in sediments may serve principally as a sink for energy and nutrient flow.[21]

Alongi[22] stressed that the role of sediment bacteria in benthic food chains and nutrient cycles is essentially one of aerobic trophic links and anaerobic nutrient sinks. In surface aerobic sediments where most consumer organisms live, bacteria are linked trophically. This is not the case in anaerobic sediments below, where anaerobic bacteria and some protozoa constitute most forms of life. The concept of microbial loops in pelagic and perhaps benthic food webs and the view of bacteria as competitors for food and nutrients of other estuarine and marine organisms continue to stimulate much interest in the interplay of microbial groups, material cycles, and the factors affecting them.

III. PHYTOPLANKTON

The principal primary producers of the world's oceans are microscopic, free-floating plants (i.e., phytoplankton), which inhabit surface waters, including those under ice in polar seas. These unicellular, filamentous, or chain-forming species encompass a wide diversity of photosynthetic organisms. The principal taxonomic groups include diatoms (class Bacillariophyceae), dinoflagellates (class Pyrrophyceae), coccolithophores (class Prymnesiophyceae), and silicoflagellates (class Chysophyceae). In estuaries, lagoons, and coastal embayments, other taxonomic groups may locally predominate, such as euglenoid flagellates (class Euglenophyceae), green algae (class Chlorophyceae), blue-green algae (class Cyanophyceae), and brown-colored phytoflagellates (class Haptophyceae).

Phytoplankton are subdivided into four major size classes: (1) nanoplankton (2 to 20 μm); (2) microplankton (20 to 200 μm); (3) mesoplankton (0.2 to 20 mm); and (4) macroplankton ($>$2 cm). Most phytoplankton assemblages consist of nanoplankton (mainly diatoms, coccolithophores, and silicoflagellates) and microplankton (diatoms and dinoflagellates). Phycologists often differentiate net plankton and nanoplankton based on the nominal aperture size of plankton nets deployed in the field. Phytoplankton retained by fine-mesh nets (i.e., $\sim$64 μm apertures) constitute the net plankton, and those passing through the nets comprise the nanoplankton.

Phytoplankton play a critical role in initiating the flow of energy in a useful form through oceanic ecosystems. They are responsible for at least 90% of the photosynthesis in marine waters, the remaining 10% largely ascribable to benthic macroalgae and vascular plants (i.e., salt marsh grasses, mangroves, and seagrasses) in intertidal and subtidal environments. Although most of this production is coupled to microscopic forms, macroscopic floating algae (e.g., *Sargassum*) are locally significant, as evidenced in the Sargasso Sea.

A. Major Taxonomic Groups

1. Diatoms

Among the most important constituents of phytoplankton communities worldwide are diatoms, the dominant taxonomic group in the oceans. These highly productive, diminutive plants, typically ranging in size from 10 to 200 μm, secrete an external siliceous skeleton (pectin impregnated with silica) called a frustule, which is composed of two minute valves in a "pillbox" arrangement. Diatoms occur as single cells or chains of cells floating in the water column or attached to surfaces.[1] The planktonic forms are principally centric diatoms characterized by circular or spherical tests, and the benthic forms, primarily pennate diatoms typified by oblong or elongate tests.

Although ubiquitous in the oceans, diatoms attain greatest abundance in high-latitude polar regions, in the neritic zone of boreal and temperate waters, and in nutrient-rich upwelling areas. Sedimented frustules underlying waters of high fertility often form extensive seafloor deposits termed *diatomaceous oozes*.[34] In some waters, the diatoms are so abundant that they may comprise more than 90% of the suspended silica in the water column.[35]

2. Dinoflagellates

These widespread unicellular, biflagellated planktonic algae are second to the diatoms in total marine abundance. However, in some areas, blooms of dinoflagellates exceed diatom numbers.[36] Dinoflagellates frequently dominate phytoplankton communities in subtropical and tropical waters, and they are major components of temperate and boreal autumnal assemblages.[37]

Many dinoflagellates are not strict autotrophs. About 50% of them lack chloroplasts and carry out heterotrophic production. Others are mixotrophs obtaining energy from both autotrophic and heterotrophic processes. Some species are parasitic or symbiotic.

Dinoflagellates generally range in size from ~5 to 100 μm. When conditions are favorable, they reproduce rapidly. Dinoflagellate blooms exceeding 10^6 cells/l commonly develop in estuaries and coastal lagoons during the warmer months of the year. These blooms impart a reddish-brown color to the water, generating the so-called red tides. In systems with poor water circulation, dinoflagellate blooms can contribute to severe oxygen depletion, leading to anoxia or hypoxia, which threaten entire biotic communities.

Some dinoflagellates (e.g., certain species of *Alexandrium, Gymnodinium,* and *Pyrodinium*) produce neurotoxins that commonly cause mass mortality of fish, shellfish, and other living resources. Toxic red-tide blooms can be particularly devastating to estuarine and marine organisms.[38] Saxitoxin, a potent neurotoxin, is responsible for paralytic shellfish poisoning, a neurological disorder in humans resulting from the consumption of contaminated shellfish. This disorder is often fatal.

3. Coccolithophores

A substantial fraction of the nanoplankton in open ocean waters consists of coccolithophores, unicellular flagellated algae characterized by an external covering of small calcareous plates (i.e., coccoliths). Most coccolithophores are less than 25 μm in size. Highest abundances occur in subtropical and tropical waters, although a few species reach peak numbers in colder regions. The skeletal remains of coccolithophores are a major component of calcareous oozes in the deep sea.

4. Silicoflagellates

Another source of siliceous particles on the ocean floor are the skeletal remains of silicoflagellates. These unicellular, uniflagellate organisms, which range from ~10 to 200 μm in size, secrete an internal skeleton of opaline silica. Although found in seafloor sediments of all the major ocean basins, silicoflagellates are most numerous in cold, nutrient-rich regions. However, they generally do not constitute a major fraction of siliceous oozes.

B. Primary Productivity

Annual phytoplankton production rates average ~50 g C/m²/year in open ocean waters compared with ~100 g C/m²/year in coastal marine systems and ~300 g C/m²/year in upwelling areas.[1,39] Lowest productivity values in the oceans occur in the convergent gyres.[34] Estuaries exhibit a wide range of production rates from ~5 g C/m²/year in turbid waters to ~530 g C/m²/year in clearer systems.[40–42] In a comprehensive review, Boynton et al.[43] showed that the mean annual phytoplankton productivity for 45 estuaries was 190 g C/m²/year. Coastal and upwelling regions benefit from greater nutrient supplies, and thus have higher levels of productivity.

The dominant physical–chemical factors controlling phytoplankton production in marine waters are solar radiation and nutrient availability.[1,34,36,37,39] Four aspects of solar radiation influence phytoplankton production: (1) the intensity of incident light; (2) changes in light on passing from the atmosphere into the water column; (3) changes in light with increasing water depth; and (4) the utilization of radiant energy by phytoplankton. A portion of incident light is lost by scattering and reflection at the sea surface.[44–46] The angle of the sun, the degree of cloud cover, and the roughness of the sea surface modulate the amount of light reflected at the sea surface. Absorption and scattering of light by water molecules, suspended particles, and dissolved substances further attenuate light in the water column.

The amount of solar radiation reaching the sea surface is strongly latitude dependent, being lowest at the poles and highest in the tropics. Seasonal variations in illumination with latitude contribute to different seasonal phytoplankton production patterns in tropical, subtropical, temperate, boreal, and polar regions. However, other factors, particularly essential nutrient availability, also influence these patterns. For example, low nutrient concentrations in the euphotic zone of tropical waters cause relatively low phytoplankton productivity despite high light intensity year-round. In contrast, reasonably high nutrient availability in polar waters is offset by reduced solar radiation, which also limits phytoplankton productivity. Mid-latitude regions, typified by more favorable light and nutrient conditions, generally display peak annual productivity levels.[1,34]

Other physical factors must also be considered. For instance, vertical mixing greatly affects the distribution of nutrients in the water column, and hence phytoplankton productivity. The development of a thermocline during spring in temperate regions isolates the mixed layer in the photic zone, leading to phytoplankton blooms. Increased light triggers these seasonal events.

The major nutrient elements include nitrogen, phosphorus, and silicon. Of these three elements, nitrogen (as nitrate, NO_3^-) and phosphorus (as phosphate, PO_4^{3-}) have the greatest impact on primary productivity; both are necessary for survival of autotrophs, yet exist in very small concentrations in seawater. For example, nitrate levels in seawater amount to ~1 μg-atom/l or less and rarely exceed 25 μg-atom/ℓ, whereas phosphate values usually range from 0 to 3 μg-atom/ℓ.[47] Silicon, when present in very low concentrations,[48] represses metabolic activity of the cell and can limit phytoplankton production. It represents an essential element for the skeletal growth of diatoms, as well as radiolarians and certain sponges. Elements other than nitrogen and phosphorus also are required by autotrophs, but their availability usually does not limit growth. These encompass the major elements (e.g., calcium, carbon, magnesium, oxygen, and potassium), minor and trace elements (e.g., cobalt, copper, iron, molybdenum, vanadium, and zinc), and organic nutrients (e.g., biotin, cobalamine, and thiamine).[36] Some trace metals (e.g., copper and zinc), however, can be toxic to phytoplankton even at low concentrations and, consequently, may hinder their productivity.[49]

IV. ZOOPLANKTON

As primary herbivores in the sea, zooplankton serve an essential role in estuarine and marine food chains as an intermediate link between primary producers and secondary consumers. They comprise a highly diverse group of passively drifting animals. While many are strict herbivores consuming phytoplankton, others are carnivores, detritivores, or omnivores. Some species obtain nutrition by direct uptake of dissolved organic constituents. Most zooplankton gather food via filter feeding or raptorial feeding.[36] Despite ingesting large volumes of food, zooplankton assimilate only a portion of it; the remainder is egested. Zooplankton fecal pellets are important components in detrital food webs of many estuarine and shallow coastal marine systems.[50]

Zooplankton are predominantly minute, passively drifting organisms, although some forms (e.g., jellyfish) grow to several meters in size. All zooplanktan have only limited mobility, and thus are easily entrained in currents and occasionally transported considerable distances. Water circulation, therefore, plays a significant role in the distribution of these important lower-trophic-level consumers.

A. Zooplankton Classifications

Zooplankton are classified based on their taxonomy, size, and length of planktonic life. Several phyla generally dominate zooplankton communities in marine waters, including protozoans, cnidarians, mollusks, annelids, arthropods, echinoderms, chaetognaths, and chordates.[51] Four major size categories of zooplankton are delineated: nanozooplankton, microzooplankton, mesozooplankton, and macrozooplankton. Three groups of zooplankton are recognized based on duration of planktonic life: holoplankton, meroplankton, and tychoplankton.

1. Classification by Size

Zooplankton forms that pass through a plankton net with a mesh size of 202 μm constitute the nanozooplankton and microzooplankton, and those forms retained by the net comprise the mesozooplankton. Still larger individuals collected by plankton nets with a mesh size of 505 μm include the macrozooplankton. Zooplankton less than ~60 μm in size consist mainly of protozoans, especially foraminiferans, radiolarians, and tinitinnids. In the open oceans, the tests of foraminiferans and radiolarians produce thick accumulations of *Globigerina* and radiolarian oozes in some regions (see Chapter 4). In shallow coastal marine waters and estuaries, however, copepod nauplii and the meroplankton of benthic invertebrates become increasingly important among these smaller size groups.

Copepods are by far the most important group of mesozooplankton in the sea in terms of absolute abundance and biomass, although rotifers, cladocerans, and larger meroplankton are also significant. Calanoid, cyclopoid, and harpacticoid copepods are abundant in estuaries worldwide. In some systems, these diminutive forms completely dominate the mesozooplankton. As such, they are critical components in the flow of energy through these valuable coastal ecosystems, serving as a vital link between phytoplankton and larger consumers.

Among the macrozooplankton, three groups are particularly notable. These include the jellyfish group (i.e., hydromedusae, comb jellies, and true jellyfish), crustaceans (e.g., amphipods, isopods, mysid shrimp, and true shrimp), and polychaetes. Members of the jellyfish group often reproduce rapidly in estuaries and coastal marine waters during certain seasons. Many crustacean populations follow a similar explosive pattern. Krill are particularly abundant in the highly productive pelagic waters of the Antarctic.[52] These shrimplike euphausiids, which measure ~2 to 5 cm in length, are a main staple of whales and fish.

2. Classification by Length of Planktonic Life

a. Holoplankton

Permanent residents of the plankton are termed holoplankton. Excluding protozoans, ~5000 holoplanktonic species have been described.[34] Many of these species belong to several holoplanktonic groups such as the medusae, siphonophores, ctenophores, chaetognaths, heteropods, pteropods, amphipods, ostracods, copepods, euphausiids, and salps. Among the protozoans, commonly encountered holoplankton in the sea include the dinoflagellates, ciliates, foraminiferans, and radiolarians.

In many estuaries and coastal marine waters, the principal holoplanktonic groups are copepods, cladocerans, and rotifers. Copepods often predominate, with calanoids outnumbering cyclopoids and harpacticoids. Species of *Acartia, Eurytemora, Pseudodiaptomus,* and *Tortanus* inhabit many estuaries. Distinct spatial distributions are evident. For example, *Eurytemora* spp. typically occur in the upper reaches, and *Acartia* spp. (*A. bifilosa, A. discaudata, A. hudsonica,* and *A. tonsa*) in the middle reaches. Species of *Centropages, Oithona, Paracalanus,* and *Pseudocalanus* often proliferate in the lower estuary. Although estuarine species generally dominate in the upper estuary, marine forms are most abundant in the lower estuary. Species diversity tends to increase downestuary because of the influx of marine forms.[34, 53, 54]

b. Meroplankton

Many benthic invertebrates, benthic chordates, and fish remain planktonic for only a portion of their life cycle, and hence comprise the meroplankton. About 70% of all benthic marine invertebrate species have a meroplanktonic life stage.[34] The eggs and larvae of some of these populations often dominate the meroplankton in estuaries and shallow coastal marine waters. Fish eggs and fish larvae (i.e., icthyoplankton) are also numerous in these systems. Most estuarine and marine fishes produce planktonic eggs and/or larvae.[48] Leiby[55] discusses the methods by which the eggs and larvae of estuarine and marine fishes enter the planktonic community.

The abundance of meroplankton in the water column is coupled to reproductive cycles of adults which, in turn, are often modulated by environmental conditions. Meroplankton appear year-round in tropical waters, but seasonally in higher latitudes. Warmer temperatures trigger spawning of benthic invertebrates in temperate and cold-water inshore regions. Reproductive activity may also be tied to increased food supply. A similar relationship of fish reproduction to environmental change is clearly evident.[34]

Most benthic marine invertebrates and teleostean fishes exhibit high fecundity. A female American oyster (*Crassostrea virginica*), for example, can produce more than 10^8 eggs in a single spawn,[56] and a female hard clam (*Mercenaria mercenaria*), more than 10^7 eggs per spawning season.[57] Female plaice (*Pleuronectes platessa*), haddock (*Melanogrammus aeglefinus*), and cod (*Gadus morrhua*) release ~2.5 × 10^5, ~5 × 10^5, and >1 × 10^6 eggs, respectively.[34] The meroplankton of benthic marine invertebrates and the ichthyoplankton of marine fishes are susceptible to the vagaries of environmental conditions that severely deplete their numbers. Early life mortality for many forms may be as much as 99.999%. Temperature, salinity, turbidity, circulation, and seafloor characteristics, as well as other physical and chemical factors, influence larval development, distribution, and survivorship. Biological factors, including predation, food availability, and seasonal abundance of adults and larvae, also affect meroplankton success.[58] Predation alone causes larval mortality to exceed 90%.[36]

The duration of meroplankton larvae in the plankton generally ranges from several minutes to a few months depending on the species and type of larval development. For benthic marine invertebrates, five types of larval development patterns are apparent. These include planktotrophy, lecithotrophy, viviparity, demersal development, and direct development. Shallow coastal marine and estuarine benthic invertebrates typically have planktonic larval stages. Deep-sea species, however, mainly show direct development or brood protection of young.[34,59,60]

In estuaries and shallow coastal marine waters, meroplankton abundance fluctuates markedly because of different reproductive strategies of benthic invertebrate and fish populations and the responses of their meroplankton to variable environmental conditions. Spawning does not occur synchronously and the residency of the plankton in the water column differs substantially among these populations. Pulses of meroplankton often do not overlap. As a result, the timing of peak numbers of major taxonomic groups in the plankton can deviate considerably.

c. Tychoplankton

Bioturbation, dredging, bottom shear stresses associated with wave and current action, and other processes that roil seafloor sediments promote the upward translocation of demersal zooplankton into the plankton.[61,62] These tychoplanktonic organisms, aperiodically inoculated into the water column, provide forage for carnivorous zooplankton and planktivorous fishes. Various species of amphipods, isopods, cumaceans, and mysids are representative forms. Although these organisms augment the food supply of resident populations in the water column, they rarely comprise a quantitatively significant fraction of the zooplankton community.

Zooplankton have preferred depth ranges, occupying distinct vertical zones (epi-, meso-, bathy-, and abyssopelagic waters) in the ocean. Some species, the pleuston, live at the sea surface and others, the neuston, inhabit the upper tens of millimeters of the surface water. Still other forms adapt to the vertical gradients of temperature, light, and pressure found at greater depths in the water column. The morphology and behavior of zooplankton in deeper zones differ from those of epipelagic species. Mesopelagic and bathypelagic zooplankton generally have a wider distribution than epipelagic species, which are commonly associated with specific water masses. The species diversity of oceanic zooplankton is higher in low latitudes, although the abundance of individuals is relatively low in these waters. In contrast, fewer species occur in higher latitudes, but the number of individuals in each species tends to be higher.[34,36,37,39]

More locally, zooplankton display a patchy distribution. A number of biological or physical factors may be responsible for the patchiness, including the aggregation of zooplankton for reproduction, the interactions between zooplankton and their food, and responses to circulation patterns and chemical gradients (e.g., salinity). Zooplankton patches may persist for periods of days to many months or even years.[34,37]

Many zooplankton species migrate vertically in the water column. Three types of migration patterns have been ascertained: nocturnal, twilight, and reverse migration.[63] Nocturnal migration is the most common type among estuarine and marine zooplankton. In this case, zooplankton begin to ascend the water column near sunset, reach a minimum depth between sunset and sunrise, and then descend to a maximum depth during the day. In contrast to the daily ascent and descent of a nocturnal migration, twilight migration consists of two ascents and descents per day. Zooplankton undergoing twilight migration commence their ascent of the water column about sunrise; however, upon attaining a minimum depth, they later descend at night in a migration termed the *midnight sink*. At sunrise, they ascend once again, but subsequently descend to the daytime depth. Reverse migration involves the ascent of zooplankton to a minimum depth during the day followed by a descent to a maximum depth at night. This pattern is opposite to that of nocturnal migration and remains the least common type of zooplankton migration observed in the sea.

The distance traveled by zooplankton during vertical migration periods can be substantial. Many migrate several hundred meters per day, and some of the larger, stronger forms can cover distances up to a kilometer or more. During such migrations, the zooplankton may experience significant changes in physical and chemical conditions.

Light appears to be the key environmental cue that triggers zooplankton diel vertical migrations. More specifically, the rate and direction of change of light intensity from the ambient level (adaptive intensity) are deemed to be of paramount importance for initiating the vertical movements.[64] Factors such as an absolute amount of change in light intensity, a relative rate of light intensity change, a change in depth of a particular light intensity, a change in the polarized light pattern, and a change in underwater spectra can all elicit vertical migration of zooplankton populations.[65-67]

Diel vertical migrations confer several advantages on zooplankton. Among the most notable benefits are predator avoidance, greater transport and dispersal, increased food intake and utilization, and maximization of fecundity.[37,39,68,69] These migrations accelerate the transfer of organic materials from the euphotic zone to deeper waters. They may also facilitate genetic exchange by the mixing of individuals of a given population.[34]

Factors other than light affect zooplankton dynamics. For example, temperature and salinity can alter diel vertical migration patterns. The organism's age, physiological condition, and reproductive stage also influence these movements. Temperature has likewise been shown to affect the growth, fecundity, longevity, and other life processes of these diminutive animals.[47,70]

V. BENTHOS

Hard-bottom and soft-bottom substrates in the sea provide a wide array of habitats that support numerous species of benthic marine plants and animals. About 200,000 benthic species occupy these habitats compared to ~5000 species of larger zooplankton, >20,000 species of pelagic fish, and 140 species of marine mammals.[34] These organisms live attached to or move on or in seafloor sediments.[15] Some remain attached to hard substrates (e.g., rocks, shells of organisms, and anthropogenic structures) throughout their lives. Others (i.e., mobile species) move freely on or in bottom substrates in search of food, shelter, or refuges from predators. Most benthic species inhabit soft-bottom substrates (i.e., unconsolidated clay, silt, and sand). The structure of benthic communities depends greatly on physical–chemical conditions in the environment—particularly temperature, salinity, and light—and the character of the substrate (i.e., sediment composition, grain size, sorting, etc.).

In estuarine and continental shelf environments, physical and chemical conditions can fluctuate substantially along the seabed. Most extreme conditions are observed in intertidal and supratidal zones, where bottom habitats are subject to periodic immersion and exposure, and in surf zones, where sediment flux is considerable due to wave action. Much greater stability exists in deep-sea benthic environments (excluding hydrothermal vent systems), which are characterized by much more uniform temperature and salinity, as well as other factors. In addition, biotic interactions (e.g., competition and predation) on the seafloor can vary markedly in different physiographic regions of the marine realm.

Greater environmental stability in deep-sea and nearshore tropical waters has favored the evolution of more highly diverse benthic communities.[36] In contrast, shallow-water systems in temperate latitudes are typified by more variable environmental conditions and benthic communities with lower species diversity. Here, however, the total abundance of individuals within each species tends to be high, which differs from the low abundances found in deep-sea populations.

A. Benthic Flora

A widely diverse group of bottom-dwelling plants inhabits intertidal and shallow subtidal zones of estuaries and the coastal ocean. Light is a major factor controlling the distribution of these plants, which generally occur in waters less than ~30 m depth.[37] Many species attach directly to the seabed or, more conspicuously, to rocky shores and anthropogenic structures. However, the most productive plant communities are those found in fringing salt marshes and mangroves, as well as shallow subtidal seagrass communities.

The benthic flora are subdivided into microphyte and macrophyte components. The microphytes (or microscopic plants) consist of diatoms, dinoflagellates, and blue-green algae. They commonly inhabit mudflats and sand flats in intertidal habitats, growing on sediment grains or forming mats on sediment surfaces. Significant numbers also exist in saltmarsh and other wetland habitats, as well as in shallow subtidal areas and in coral reef sediments. Others live on stones and rocks as lithophytes and on anthropogenic structures. Motile forms, particularly naviculoid diatoms, migrate vertically in sediments in response to changing light conditions.

Microalgae growing on the surfaces of leaves and stems of macrophytes (i.e., epiphytes) form a furlike covering known as periphyton or *Aufwuchs* that provides a rich food supply for many grazing herbivores, such as mud snails, fiddler crabs, and herbivorous fish. The group also represents an important food source for suspension and deposit feeders in shallow-water systems.[71] In addition to their importance in the food web, microalgae can influence water chemistry by regulating oxygen and nutrient fluxes across the sediment–water interface.[72,73]

Primary production of benthic microalgae amounts to ~25 to 2000 g C/m²/year in marine environments.[1] Lower production values are frequently caused by insufficient light impinging on the sediment surface. This is commonly observed in more turbid systems.[74]

Benthic macroalgae (seaweeds) are common inhabitants of rocky intertidal zones in temperate and subtropical regions, where they attach to hard substrates via rootlike holdfasts or basal disks. These plants are much less common on mud and sand substrates, being limited by turbidity, sedimentation, and shading effects. Some species are free-floating forms in coastal and open ocean (e.g., *Sargassum*) waters. Although species of Chlorophyta (green algae), Rhodophyta (red algae), and Phaeophyta (brown algae) are all represented, members of Phaeophyta dominate many shore regions.[75] Rockweeds (Fucales), which primarily inhabit rocky intertidal zones in cool temperate latitudes, and kelps (Laminariales), which live subtidally, provide examples.[22] Other macroalgae (e.g., *Cladophora, Enteromorpha, Ulva*) are usually less extensive. Some taxa (e.g., *Blidingia minima* var. *subsalsa, Enteromorpha clathrata,* and *Vaucheria* spp.) attain peak abundances in estuaries.

Macroalgal species often exhibit a conspicuous zonation pattern on many rocky shores. Several factors are largely responsible for this, notably species competition, grazing pressure, and physiological stresses associated with emersion. For example, the lower limits of fucoid algae in the intertidal zone appear to be controlled by species competition, and the upper limits by physiological stresses associated with desiccation. On the rocky shores of New England, the red alga *Chondrus crispus* dominates the lower intertidal zone outcompeting the brown alga *Fucus vesiculosus,* which predominates farther up the shore. The brown alga is more resistant to desiccation and therefore dominates the middle intertidal zone.[75]

Primary production of seaweeds can be considerable, at times exceeding those of all other macrophytes. Subtidal kelp beds (i.e., *Ecklonia, Laminaria,* and *Macrocystis*), for example, have annual primary production values ranging from 400 to 1900 g C/m²/year. Their biomass, in turn, can be greater than several metric tons per square meter.[75] Under certain conditions, blooms of macroalgae develop in subtidal waters, which can have devastating impacts on a system by reducing dissolved oxygen levels and eliminating valuable benthic habitat areas (e.g., seagrass beds).[76] In contrast, harsh environmental conditions in intertidal zones commonly limit the standing crop biomass and productivity of benthic macroalgae. Fluctuations of temperature, salinity, light intensity, nutrient availability, sediment stability, as well as competition and grazing pressure, greatly affect the abundance and distribution of macroalgal populations. Periodic submergence and emergence associated with tidal action create stressful conditions for floral assemblages inhabiting intertidal zones. Consequently, production estimates for intertidal habitats are quite variable.

Some of the most extensively developed benthic macroflora occur in salt marsh, mangrove, and seagrass systems. The principal vegetation in these biotopes are flowering plants, which flourish in soft, fine sediments. Saltmarsh plants dominate the plant communities of intertidal zones in mid- and high-latitude regions, although they are most luxuriant in temperate latitudes. Mangroves replace salt marshes as the dominant coastal vegetation at ~28° latitude, and the two communities co-occur between ~27° and 38° latitude.[22] Seagrasses have a wider distribution, occupying shallow subtidal waters of all latitudes except the most polar.[77] These vascular plants are highly productive throughout the range.

1. Salt Marshes

Along protected marine shores and embayments as well as the margins of temperate and subpolar estuaries, saltmarsh communities (i.e., halophytic grasses, sedges, and succulents) develop on muddy substrates at and above the mid-tide level.[22,36,37,39,75,77–79] Most of the macroflora belong to a few cosmopolitan genera (i.e., *Spartina, Juncus,* and *Salicornia*) that are broadly distributed, and species diversity is relatively low compared to other plant communities. The flora is more variable in Europe than North America.[75] Along the Atlantic Coast of North America, for example, the diversity of flora is lower than that along the North Atlantic Coast of Europe, with the cordgrass *Spartina alterniflora* dominating between mean sea level and mean high water and grading landward into species of *Juncus* and *Salicornia,* as well as *Distichlis spicata* and *S. patens.* Plant zonation also appears to be more conspicuous in North American salt marshes. In Europe, marshes bordering the Atlantic Ocean, the English Channel, and the North Sea show distinct compositional variations. The Atlantic marshes are dominated by the genera *Festuca* and *Puccinella;* the south coast of England, by *S. anglica* and *S. townsendii;* and the North Sea, by species of *Armeria, Limonium, Spergularia,* and *Triglochin,* in addition to the sea plantain *Plantago maritima.*[39]

In North America, salt marshes can be divided into three geographic units: (1) Bay of Fundy and New England marshes; (2) Atlantic and Gulf coastal plain marshes; and (3) Pacific marshes. Compared with the lush marsh vegetation of the Bay of Fundy, New England, Atlantic, and Gulf regions, the Pacific Coast is rather depauperate in saltmarsh grasses. However, the diversity of salt marsh flora on the Pacific Coast tends to be greater than that observed on the Atlantic Coast. The zonation and succession patterns of the flora are also more complex.[80]

Six types of salt marshes have been documented, namely, estuarine, Wadden, lagoonal, beach plain, bog, and polderland varieties.[81] Differences among the saltmarsh types have been attributed to several major factors:

1. The character and diversity of the indigenous flora;
2. The effects of climatic, hydrographic, and edaphic factors upon this flora;
3. The availability, composition, mode of deposition, and compaction of sediments, both organic and inorganic;
4. The organism–substrate interrelationships, including burrowing animals and the prowess of plants in affecting marsh growth;
5. The topography and areal extent of the depositional surface;
6. The range of tides;
7. The wave and current energy; and
8. The tectonic and eustatic stability of the coastal area.[82]

Ideal sites of saltmarsh development are sheltered coastal areas where erosion is minimal and sediments are regularly deposited.[37,75,77] Initial colonization occurs on mudflat surfaces between the levels of mean high water neap (MHWN) and mean high water.[80] Saltmarsh growth proceeds as sedimentation produces a surface above the MHWN level. Gradual accretion of sediments (~3 to 10 mm/year) promotes maturation of the marsh. Successional development of the marsh may be arrested by higher rates of sedimentation, which can also restrict species richness.

Tidal salt marshes grade into tidal freshwater marshes farther inland as salinity decreases to low levels. Tidal freshwater marshes supplant tidal salt marshes as the principal benthic floral habitat where the average annual salinity amounts to ~0.5‰ or less in tidally affected areas. This change often takes place considerable distances from bays and open coastal waters. Nontidal freshwater wetlands replace tidal freshwater marshes farther upstream.

Primary production of saltmarshes varies greatly. As reviewed by Nybakken,[83] the annual net primary production of saltmarshes in different regions of the United States is as follows: New Jersey (325 g C/m²/year), Georgia (1600 g C/m²/year), Gulf Coast (~300 to 3000 g C/m²/year), California (50 to 1500 g C/m²/year), and the Pacific northwest (100 to 1000 g C/m²/year). In European salt marshes, production values generally range from ~250 to 500 g C/m²/year.[75] Belowground production approaches or even exceeds aboveground production.[84,85] The belowground biomass of roots and rhizomes in *Spartina* marshes may be two to four times greater than the aboveground biomass of shoots and standing dead matter.[22,84]

Salt marsh vegetation not only serves as a food source for organisms, but also anchors sediment and reduces erosion. Much of the production enters a complex web of decomposer food chains. Saltmarsh habitats filter many contaminants released from nearby watersheds.[86] In addition, they represent a source or sink of nutrients at various times, and therefore can strongly influence the production of adjacent waters.

2. Seagrasses

Among the most highly productive, widespread communities of vascular plants in shallow temperate, subtropical, and tropical seas are the seagrasses, a group of monocotyledonous angiosperms occurring from the lower intertidal zone down to depths of ~50 m. The most extensive beds or meadows of seagrasses develop in shallow subtidal estuarine and inshore marine waters less than ~5 m in depth. They typically appear as isolated patches to thick carpets of grasses, which can completely blanket the bottom. Seagrass colonization is greatest on soft sediments, but some plants also inhabit hard bottoms.

About 50 species of seagrasses exist worldwide; they belong to 12 genera (i.e., *Amphibolis, Cymodocea, Enhalus, Halodule, Halophila, Heterozostera, Posidonia, Phyllospadix, Syringodium, Thalassia, Thalassodendron,* and *Zostera*). More taxa inhabit tropical waters than temperate regions. In North America, *Zostera* (eelgrass) is the dominant genus in temperate waters and also has a broad distribution in boreal systems along both the Atlantic and Pacific Coasts. *Thalassia* (turtlegrass) predominates in subtropical and tropical regions. *Halodule* and *Phyllospadix* are two other abundant genera in North American waters.

Seagrass distribution depends on several physical–chemical factors, particularly light, temperature, salinity, turbidity, wave action, and currents. Extensive seagrass meadows tend to form in areas of low water movement, whereas mounds or patches of grasses commonly appear in areas exposed to high water motion.[83] As seagrasses grow, they naturally dampen wave action and current flow, thereby promoting the deposition of sediments and the expansion of the beds.

Seagrasses are morphologically similar. A network of roots and rhizomes anchors the plants to the substrate, and a dense arrangement of stems and straplike leaves grow above the substrate, in some cases reaching the water surface. The stems and blades are ecologically important because they provide subhabitats for epibenthic flora and fauna. Many commercially and recreationally important finfish and shellfish species utilize seagrass habitats as feeding and reproductive grounds, as well as nursery areas.

Aside from their value as habitat formers, seagrass meadows contribute significantly to the primary production of estuarine waters. Seagrass annual primary production estimates vary from ~60 to 1500 g C/m²/year.[87] As in the case of saltmarsh habitats, belowground production of roots and rhizomes in seagrass meadows can rival that of leaf and shoot production.[22,88–90] Most production estimates are for the two most intensely studied species, specifically *Zostera marina* and *Thalassia testudinum.*

3. Mangroves

Mangroves or mangals consist of inshore communities of halophytic trees, palms, shrubs, and creepers that are physiologically adapted to grow in saline conditions along subtropical and tropical coastlines of the world between ~28°N and 25°S latitudes. These communities grow as forests or dense thickets on unconsolidated sediments, ranging locally from the highest tide mark down nearly to mean sea level.[39,91] They are most extensively developed on intertidal and shallow subtidal zones of protected embayments, tidal lagoons, and estuaries, where wave energy is reduced and sheltered conditions foster sediment accretion. Mangroves heavily colonize the tropics, fringing up to 75% of the coastline in this region.[83]

Mangrove trees are stabilized in bottom sediments by an array of shallow roots. Prop or drop roots extend from the trunk and branches of the trees and terminate only a few centimeters in the sediments. Cable roots extend horizontally from the stem base and support air roots (i.e., pneumatophores) that project vertically upward to the surface. The pneumatophores enable the roots to receive oxygen despite being surrounded by anoxic muds. Anchoring and feeding types of roots form on prop, drop, and cable roots.[91] The root systems mitigate erosion, enhance bank stabilization, and protect the shoreline.

There are about 80 species of monocots and dicots belonging to 16 genera that have been described in mangrove communities, with at least 34 species in 9 genera believed to be true mangroves.[77,92] *Avicennia, Bruguiera, Rhizophora,* and *Sonneratia* appear to be the dominant genera.[83] In the United States, Florida has the most well-developed mangroves. Here the black mangrove (*A. germinans*), red mangrove (*R. mangle*), and white mangrove (*Languncularis racemosa*) dominate the communities. Mangroves line more than 1.7×10^5 ha of the Florida coastline.

Mangrove vegetation generally grows in a zoned pattern due to different species tolerances to salinity, tidal inundation, and other factors. In South Florida, for example, *Conocarpus erecta* (or buttonwood) occasionally is encountered in the upper intertidal zone, but usually comprises part of the sand/strand vegetation. *Languncularis racemosa* predominates in the middle and upper intertidal zone, whereas *A. germinans* occupies sites in the lower intertidal. *Rhizophora mangle* extends seaward of *A. germinans*, building a fringe of vegetation in shallow subtidal waters. The zonation of mangroves is usually more pronounced in other geographic regions, such as the Indo-Pacific, where more species (~30 to 40) are present.

Annual primary production of mangroves ranges from ~350 to 500 g C/m²/year.[39] Wood production accounts for ~60% of mangrove net primary production, and leaves, twigs, and flowering parts are responsible for the remainder.[22,93] Most of this production goes ungrazed and enters detritus food chains. The standing crops of mangrove forests, in turn, are typically great, with aboveground biomass values commonly exceeding 2000 g dry weight.[22] Most of this biomass is due to the growth of stems and prop roots and less to the growth of branches and leaves.

B. Benthic Fauna

Based on where they live relative to the substrate, benthic animals are broadly subdivided into epifauna, which live on the surface of soft sediments and hard bottoms, and infauna, which live within soft sediments.[78,83,94] Most of the benthic fauna (~80%) consist of epifaunal populations. Commonly occurring infauna are burrowing clams, polychaete worms, and various gastropods. Important epifauna include barnacles, corals, mussels, bryozoans, and sponges. Another group of organisms live in close association with the substrate, but also swim above it (e.g., crabs, prawns, and flatfish).[34]

Benthic fauna are also subdivided by size into micro-, meio-, macro-, and megafauna.[37,75] The microfauna constitute those individuals smaller than 0.1 mm in size.[23] Protozoans—mainly ciliates and foraminifera—largely comprise this group. The largest protozoans, the Xenophyophoria, are abundant in the hadal zone. The meiofauna, are larger animals retained on sieves of 0.1- to 1.0-mm mesh. Two categories are differentiated: (1) temporary meiofauna, which are juvenile stages of the meiofauna; and (2) permanent meiofauna, i.e., gastrotrichs, kinorhynchs, nematodes, rotifers, archiannelids, halacarines, harpacticoid copepods, ostracods, mystacocarids, and tardigrades as well as representatives of the bryozoans, gastropods, holothurians, hydrozoans, oligochaetes, polychaetes, turbellarians, nemertines, and tunicates. Still larger animals (1 mm to ~20 cm) captured on 1-mm mesh sieves comprise the macrofauna. Many species of bivalves, gastropods, polychaete worms, and other taxa provide examples. The largest benthic fauna, exceeding 20 cm in size, are the megafauna.

Abundance of benthic organisms decreases from estuaries to the deep sea. In estuaries and on the continental shelf, microfaunal densities in bottom sediments exceed 10^7 individuals/m^2. Meiofaunal densities range from ~10^4 to 10^7 individuals/m^2 in these environments, and decline to ~10^4 to 10^5 individuals/m^2 in deep-sea sediments. Abundance of benthic macrofauna, especially opportunistic species (e.g., *Mulinia lateralis, Pectinaria gouldi,* and *Capitella capitata*), can be greater than 10^5 individuals/m^2 in some estuarine and shallow coastal marine systems. However, the density of the macrofauna drops to ~30 to 200 individuals/m^2 in the deep sea. Here, the biomass averages only ~0.002 to 0.2 g/m^2.[36]

1. *Spatial Distribution*

Bottom habitats in marine environments differ greatly in physical characteristics, such as sediment properties, presence of hard substrates, depth, temperature, light, wave action, currents, and degree of exposure and desiccation (in intertidal and supratidal zones). Biotic interactions (e.g., competition, predation, and grazing) also vary from one type of habitat to another. Hence, both physical and biological factors must be considered when assessing the spatial distribution of benthic fauna in these environments.

A conspicuous feature of many rocky, sandy, and muddy shores is animal zonation.[75] On the rocky intertidal shores of England, for example, two faunal groups are documented. Barnacles (*Chthamalus stellatus*), gastropods (*Littorina neritoides*), and isopods (*Ligia* sp.) predominate in the upper intertidal zone, and barnacles (*Balanus balanoides*), limpets (*Patella vulgata*), and mussels (*Mytilus edulis*) extend from the middle intertidal zone to the subtidal fringe.

On sandy shores of England, amphipods (*Talitrus saltator* and *Talorchestia* sp.) are most abundant in the upper intertidal zone. Lugworms (*Arenicola marina*), cockles (*Cardium edule*), and a number of crustaceans (i.e., species of *Bathyporeia, Eurydice,* and *Haustorius*) occur in the middle intertidal zone. Bivalves (*Cardium edule, Ensis ensis,* and *Tellina* sp.) dominate toward low water.

On muddy shores of England, lugworms (*A. marina*), amphipods (*Corophium volutator*), and polychaetes (*Nereis diversicolor*) occupy the middle intertidal zone, although *A. marina* and *C. volutator* extend into upper intertidal areas. Bivalves (i.e., species of *Cardium, Macoma,* and *Tellina*) are found toward low water. They often attain high abundances in this area.

Benthic communities generally exhibit patchy distribution patterns resulting from the responses of organisms to physical, biological, or chemical factors. The patchiness is both temporally and spatially variable. Clustering of benthic fauna is often apparent even over very small areas of a few centimeters.[95] The nature of bottom sediments alone may account for much patchiness. For instance, the polychaetes. *Polydora ligni* and *Scrobicularia plana* live in muddy sediments, and high densities of the worms commonly occur in local areas containing high concentrations of organic matter. In contrast, the polychaetes *Ophelia* spp. and the amphipods *Bathyporeia* spp. and *Haustorius* spp. prefer sandy bottoms.[96] Clumping of these populations may reflect larval settlement patterns.

Some populations (e.g., periwinkles) have adapted gregarious behavior to increase their probability of successful reproduction.[34] The clustering of other fauna is ascribable to predation and competition. This may be most evident along rocky, intertidal shores.[83] The proximity of many epifaunal predators elicits responses from infaunal prey that commonly leads to a repositioning of individuals within the sediment column. Predators directly modulate the abundance of their prey by consuming larvae, juveniles, and adults; alternatively, they indirectly influence the survivorship of their prey by burrowing through sediments, disturbing the sediment surface, and reducing larval settlement.[97]

Competitive interactions between species in soft-bottom benthic communities can be mitigated by vertical partitioning of sediments.[98] Feeding and burrowing activities govern the vertical distribution and abundance of certain infaunal populations.[99] Constraints on burrowing depth due to body size, rather than resource partitioning of competitors, have been shown to be critical in regulating the position of the infauna in sediments of some estuaries.[97]

The geochemistry of bottom sediments exerts some control on the vertical distribution of benthic fauna that can lead to the clumping of individuals at various points below the sediment–water interface. The infauna, for example, are responsive to vertical gradients of dissolved oxygen. The macrofauna concentrate in the oxygenated surface sediments and decline in abundance with increasing depth and decreasing oxygen levels. Some representatives of the microfauna and meiofauna, however, peak in abundance in the deeper sediment layers.[100] Other factors influencing the vertical zonation of the fauna include food availability, amount of organic matter, and sediment grain size.[101]

Random perturbations or stochastic events associated with physical disturbances of the seafloor are capable of restructuring benthic communities.[102] Wave-induced disturbances, dredging and dredged material disposal, and sediment erosion during high-magnitude storms cause major aperiodic density changes. The successional pattern of benthic communities hinges on the frequency and nature of such disturbances.[103,104] In areas subjected to frequent physical perturbations, pioneering species of infauna—inhabiting near-surface sediments—tend to dominate the community. The pioneering forms feed near the sediment surface or from the water column. Habitats devoid of physical disturbances harbor higher-order successional stages or equilibrium stages of benthos dominated by bioturbating infauna, which feed at greater depths within the bottom sediments. Disturbances of the seafloor appear to be principally responsible for the spatial mosaic patterns observed in many soft-bottom benthic communities.[103,105]

2. Reproduction and Larval Dispersal

Most benthic macrofaunal populations, except those in the high Arctic and deep sea, have high fecundities and a planktonic larval phase to maximize dispersal.[23] These populations experience an extremely high wastage of numbers in the plankton.[106,107] In contrast, meiofaunal populations typically produce only a few gametes, with many individuals undergoing direct development. Although the reproductive output in terms of the total fecundity is small, several behavioral processes minimize reproductive losses. First, meiofaunal taxa may reproduce continuously through the year.[107,108] Second, relatively short generation times characterize numerous species. Third, life-history development can be delayed by some populations through resting eggs or larval stage delay.[108–110] These adaptations promote larger population densities.

Levin and Bridges[111] proposed a classification scheme for larval development of marine invertebrates composed of four categories:

1. Mode of larval nutrition (i.e., planktotrophy; facultative planktotrophy; maternally derived nutrition—lecithotrophy, adelphophagy, and translocation; osmotrophy; and autotrophy—photoautotrophy, chemoautotrophy, somatoautotrophy);
2. Site of development (i.e., planktonic; demersal; benthic—aparental and parental);
3. Dispersal potential (i.e., teleplanic, actaeplanic, anchiplanic, and aplanic); and
4. Morphogenesis (i.e., indirect and direct).

The dispersal potential of the larvae is coupled to the length of time spent in the plankton. This depends on the mode of development, environmental factors, and chemical or physical cues that induce the larvae to settle and metamorphose. Planktotrophic larvae, because of their higher abundance and longer pelagic existence, have greater dispersal capability than their lecithotrophic counterparts.[112,113] For most intertidal and subtidal benthic invertebrates, the larvae reside in the plankton for periods of minutes to months.[114] For the long-life planktotrophic forms, currents can transport the larvae long distances to populate remote habitats.

While the longevity of the planktonic larval stage and horizontal advective processes largely determine the potential for dispersal, the behavior of the larvae affects the degree of dispersal.[59,115,116] Invertebrate larvae exhibit different abilities to control the direction, frequency, and speed of swimming. For example, decapod crustacean larvae are strong swimmers that generally exert greater control over their horizontal movements in the water column. Others, such as ciliated bivalve larvae, move in helical paths and spin while swimming. These larvae are not capable of swimming strongly enough in a horizontal plane to greatly influence their distribution.[117] Therefore, the swimming behavior of the larvae in the water column can maximize or minimize horizontal advective processes and the magnitude of their dispersal.

The orientation and position of larvae in the water column are influenced by multiple cues. Pressure, salinity, temperature, light, gravity, and currents elicit specific larval responses including barokinesis, halokinesis, thermokinesis, geotaxis, phototaxis, and rheotaxis, respectively.[117] Certain chemical or physical cues trigger metamorphosis and settlement of larvae. Chemicals released by adults of the same species induce larvae to metamorphose, which also affects population distributions along the seafloor.

Clumped distribution patterns of benthic marine invertebrates commonly arise from the dynamics of larval settlement. Larval recruitment (i.e., settlement, attachment, and metamorphosis) of many benthic populations reflects gregarious behavioral patterns mediated by adult-derived chemical cues. Pheromones, for example, elicit behavioral responses in the larvae that foster gregarious settlement. Biochemical control of larval recruitment to the benthos has been demonstrated among arthropods, bryozoans, ascidian chordates, coelenterates, echinoderms, and mollusks.[118]

Clumped distributions of benthic marine invertebrates also result from factors other than gregarious settlement of larvae to chemical cues. For example, suitable substrates for larval settlement may themselves have a patchy distribution.[118] In addition, substrates covered with algae, bacterial coatings, organic matter, and other substances are in some cases the principal attractant to larval settlement.[115]

3. Feeding Strategies, Burrowing, and Bioturbation

Based on the mode of obtaining food, five types of benthic fauna are recognized: suspension feeders, deposit feeders, herbivores, carnivores, and scavengers.[37] Suspension and deposit feeders consist mainly of benthic macrofauna. They tend to occur in sediments of different composition. For example, suspension feeders predominate in sandy substrates where there are lower amounts of particles available in the water column to clog their filtering apparatus. They obtain most of their nutrition from phytoplankton, although some species also consume bacteria, small zooplankton, and detritus. Examples include bivalves (e.g., *Mercenaria mercenaria* and *Mytilus edulis*), polychaetes (e.g., *Sabella pavonina*), and ascidians (e.g., *Ciona intestinalis*). Deposit feeders, in contrast, are most numerous in soft, organic-rich muddy sediments. Some deposit feeders (i.e., nonselective feeders) ingest sediment and organic particles together, with little if any selectivity. They appear to obtain most of their nutrition from bacteria attached to the particles, voiding the sediment and nondigestible matter. Other deposit feeders (i.e., selective feeders) actively separate their food from the sediment particles. Examples of deposit feeders can be found among the amphipods (e.g., *Corophium volutator*), bivalves (e.g., *Tellina tenuis*), gastropods (e.g., *Ilyanassa obsoleta*), and polychaetes (e.g., *Arenicola marina*).

Browsing herbivores graze on plants or animals present on substrate surfaces. For example, microalgae growing on rock and wooden surfaces are often consumed by grazing sea urchins (e.g.,

Arbacia punctulata) and mud snails (e.g., *Hydrobia ulvae*). Carnivores take a much more aggressive role in obtaining food, seizing and capturing their prey. When live prey is not available, many also act as scavengers, consuming dead or decaying matter.[37] Among well-known carnivores in benthic communities are starfish (e.g., *Asterias forbesi*), polychaetes (e.g., *Glycera americana*), gastropods (*Busycon carica, Polinices duplicatus,* and *Urosalpinx cinerea*), and crustaceans (e.g., *Callinectes sapidus* and *Carcinus maenus*).

Feeding and burrowing activities of benthic fauna alter the texture of bottom sediments. Intense, deep-vertical burrowing (i.e., 20 to 30 cm) by benthic infauna facilitates homogeneity of the sediment column. The formation of animal tubes, pits and depressions, excavation and fecal mounds, crawling trails, and burrows along the sediment–water interface exacerbate bed roughness, thereby affecting fluid motion and sediment erosion and transport in the benthic boundary layer.[119,120] Bioturbation (i.e., biogenic particle manipulation and pore water exchange) also influences inter-particle adhesion, water content of sediments, and the geochemistry of interstitial waters.

Both the stabilization and destabilization of bottom sediments have been attributed to biogenic activity of the benthos. These organisms enhance the exchange of gases across the sediment–water interface and nutrient mixing in the sediments. Among bioturbating organisms, pioneering species consisting primarily of tubicolous or sedimentary forms rework sediments most intensely in the upper 2 cm. Sedimentary effects ascribable to these organisms are (1) subsurface deposit feeding, which blankets the substratum with fecal pellets; (2) fluid bioturbation, which pumps water into and out of the bottom through vertically oriented tubes; and (3) construction of dense tube aggre-gations, which may influence microtopography and bottom roughness. In contrast, benthic com-munities dominated by high-order successional stages rework sediments to greater depths. Infaunal deposit feeders and deeply burrowing errant or tube-dwelling forms that feed head down (i.e., conveyer-belt species) rework sediments at depths below 2 cm.[103,121]

4. Biomass and Species Diversity

a. Biomass

Benthic fauna decrease in biomass from shallow waters to the deep sea. For example, benthic macrofaunal biomass declines from ~200 g dry weight/m^2 on the continental shelf to only 0.2 g dry weight/m^2 below 3 km.[75] This gradient reflects in large part the lower density of populations and the generally smaller size of organisms comprising the deep-sea benthos.[36] The abundance of bottom-dwelling organisms is not uniform in the deep abyss. The number of benthic organisms in abyssal sediments beneath high productivity waters of the Antarctic and Arctic exceeds that of benthic organisms in abyssal sediments beneath the less productive temperate waters.[75]

Exceptions to the aforementioned deep-sea benthic biomass patterns are evident at deep-sea hydrothermal vent sites along mid-ocean ridges, and in bottom sediments of deep-sea trenches. Deep-sea hydrothermal vent communities appear as an oasis of life supported by high chemosyn-thetic primary production with biomass values that rival the most productive shallow-water benthic communities.[6] They stand in stark contrast to the depauperate biomass typically observed in the deep sea. Estimates of wet weight biomass of common species in active vent fields have ranged from ~8 to 30 kg/m^2.[122–124] These values are 500 to 1000 times greater than those registered on nonvent deep-sea assemblages.[5,125] Much lower wet weight biomass figures (~100 to 500 g/m^2) are also characteristic of estuarine ecosystems.[126] Macrobenthic biomass decreases exponentially along a depth gradient from shallow coastal ocean waters to abyssal regions.[127–129]

The elevated benthic biomass values in deep-sea trenches relative to abyssal regions may be the result of the greater accumulation of organic matter, which can support more organisms. Deep-sea trenches are commonly located in close proximity to continents. The influx of organic matter from these landmasses appears to be significant.[37]

b. Diversity

There are two components of species diversity: the number of species in an area and their patterns of relative abundance.[83] During the past 30 years, considerable work has been conducted on the relationship of large-scale diversity patterns to water depth and latitude. Two conspicuous gradients in species richness of marine benthic species have been documented in the sea. When plotting species richness of benthic fauna vs. the depth gradient of the ocean, a parabolic pattern is discerned, with the number of species being relatively low on the continental shelf, increasing at upper-continental rise depths, and then decreasing again at abyssal depths.[130-134] Depth gradients in biological and physical properties have been invoked to explain the parabolic patterns.[135] In regard to latitude gradients, species richness of benthic fauna is highest in the tropics, intermediate in temperate waters, and lowest in the Antarctic and Arctic.[75]

The deep sea exhibits remarkably high species richness. An estimated 5×10^5 to 1×10^6 benthic macrofaunal species exist there.[136-140] A number of hypotheses have been advanced to explain high species diversity in the deep sea and the aforementioned latitude and shelf, deep-sea gradients in species richness.[141,142] As summarized by Valiela,[1] these include:

- The time hypothesis
- The spatial heterogeneity hypothesis
- The competition hypothesis
- The environmental stability hypothesis
- The productivity hypothesis
- The predation hypothesis

Not all deep-sea environments have high species diversity, low biomass, and low population densities. Deep-sea hydrothermal vents, for example, are characterized by relatively low species diversity but high biomasses and population densities. Worldwide, more than 20 new families, 90 new genera, and 400 new species have been identified at deep-sea hydrothermal vents since 1977.[143] At any particular hydrothermal vent field, however, only a few species typically dominate the benthic communities, such as vestimentiferan tube worms (*Riftia pachyptila*), giant white clams (*Calyptogena magnifica*), mussels (*Bathymodiolus thermophilus*), and "eyeless" caridean shrimp (*Chorocaris chacei* and *Rimicaris exoculata*). Since 1977, when deep-sea hydrothermal vents were initially discovered at a depth of 2500 m along the Galapagos Rift spreading center, other hydrothermal vent communities have been found along the East Pacific Rise (e.g., 9°N, 11°N, 13°N, and 21°N), in Guaymas Basin, along the Gorda, Juan de Fuca, and Explorer Ridges, in the Mariana back-arc spreading center, along the Mid-Atlantic Ridge (e.g., TAG and Snake Pit sites), and elsewhere.[5-8,125,144] Moreover, comparable fauna have been collected at cold seep localities at the base of the Florida Escarpment, the Gulf of Mexico slope off Louisiana, Alaminos Canyon, and along the Oregon subduction zone. Biological processes (e.g., growth rates) of some vent organisms (e.g., *Calyptogena magnifica*) have been shown to proceed at rates that are extremely rapid for a deep-sea environment and comparable to those from some shallow-water environments.[145] Hydrothermal vents are highly ephemeral systems in the deep sea, and, as a result, the spectacular benthic communities inhabiting them develop rapidly but are subject to mass extinction when the heated fluids cease to flow on the seafloor.

C. Coral Reefs

Among the most spectacular benthic habitats in the marine environment are coral reefs, which occupy $\sim 1.9 \times 10^8$ km² or less than 1% of the world's oceans.[34] These wave-resistant structures, found in shallow warm (23 to 25°C) subtropical and tropical seas between ~25°N and 25°S latitude, originate from the skeletal construction of hermatypic corals, calcareous algae, and other calcium carbonate-secreting organisms.[36,37,78] Coral reefs are best developed in clear open marine waters less than ~20 m in depth, with the rate of calcification declining with increasing depth.[75] Zooxanthellae, symbiotic photosynthetic algae (dinoflagellates), live in endoderm cells of the coral, providing the animals with various photosynthetic by-products. Major controls on coral reef growth and production include light, temperature, salinity, depth, turbidity, nutrients, local hydrodynamics, and predation.[22] Because of a combination of bottom topography and depth, and different degrees of wave action and exposure, all reefs display distinctive zonation patterns.[34]

Coral reefs are highly productive and characterized by great species richness. As many as 3000 animal species may inhabit a single reef.[34] Estimates of reef production range from 300 to 5000 g C/m^2/year.[75] In addition to the zooxanthellae, other major primary producers are calcareous algae, filamentous algae, and marine grasses. Aside from providing primary production, the zooxanthellae confer other advantages on corals, notably the enhancement of calcification, lipogenesis, and nutrition.[37]

Three categories of coral reefs are delineated: atolls, barrier reefs, and fringing reefs. Atolls predominate in the tropical Pacific, and barrier reefs and fringing reefs occur in coral-reef zones of all oceans. The Indo-Pacific reefs contain the highest diversity of coral species, with the Atlantic reefs being impoverished in comparison.[34] Although these reefs exhibit highly complex and variable morphologies, all are similar in that they have been formed entirely by biological activity.[83]

VI. NEKTON

The highest-trophic-level organisms of estuaries and oceans are those capable of sustained locomotion, actively swimming through the water in search of prey. Included here are finfish, marine mammals, marine reptiles, and some birds. The nekton are ecologically important, primarily acting as major predators in biotic communities. They are also commercially important, supplying food for livestock, poultry, pets, and humans. Various species serve as sources of fur and other commodities. Because of their great economic value, many fish, cetacean, and pinniped species have been subject to considerable harvesting pressure, in some cases leading to excessive depletion of population numbers. Occasionally, species have been placed on threatened or endangered species lists to ensure their survival and future viability.

A. Fish

Fish comprise the largest group of nekton in the sea. They are subdivided taxonomically into three classes: Agnatha, Chondrichthyes, and Osteichthyes. The Agnatha or primitive forms contain ~50 species of fish, including the primitive hagfish and lampreys. They are principally parasites and scavengers. The Chondrichthyes or elasmobranch fish lack scales and have a cartilaginous skeleton. This class, which has ~300 species, encompasses the sharks, skates, and rays. The Osteichthyes or bony fish (teleosts) represent the majority of marine species (>20,000). Members of the class feed at all trophic levels; the smallest and most abundant species (e.g., anchovies, herring, and sardines) occupy lower trophic levels mainly consuming plankton, whereas the largest, piscivorous forms (e.g., bluefish, jackfish, and tunas) occupy the upper trophic levels. Some species (e.g., summer flounder, weakfish, and cod) consume both benthic invertebrates and other fish.

Other classifications of fish are based on ecological criteria. For example, species may be separated into stenohaline and euryhaline or stenothermal and eurythermal categories based on salinity and temperature tolerances, respectively. McHugh[146] used breeding, migratory, and ecologial criteria to categorize estuarine fish into five distinct groups:

1. Freshwater fishes that occasionally enter brackish waters;
2. Truly estuarine species that spend their entire lives in the estuary;

3. Anadromous and catadromous species;
4. Marine species that pay regular visits to the estuary, usually as adults;
5. Marine species that use the estuary largely as a nursery ground, spawning and spending much of their adult life at sea, but often returning seasonally to the estuary; and
6. Adventitious visitors that appear irregularly and have no apparent estuarine requirements.

Using similar criteria, Moyle and Cech[147] subdivided fish populations into five broad classes: (1) freshwater; (2) diadromous; (3) true estuarine; (4) nondependent marine; and (5) dependent marine fishes.

One of the broadest ecological classification schemes for marine fish differentiates species according to the environments they inhabit. The oceanic realm has two major divisions: the pelagic and benthic environments. Fishes inhabiting these environments are also differentiated from the resident forms living in estuaries.

1. Representative Fish Faunas[147–154]

a. Estuaries

Common fishes of estuaries include the anchovies (Engraulidae), killifish (Cyprinodontidae), silversides (Atherinidae), herrings (Clupeidae), mullet (Mugilidae), pipefish (Syngnathidae), drums (Sciaenidae), flounders (Bothidae, Pleuronectidae), eels (Anguillidae), and gobies (Gobiidae).

b. Pelagic Environment

i. Neritic zone—Characteristic fishes in the neritic zone are herrings (Clupeidae), eels (*Anguilla*), mackerels (Scombridae), bluefish (*Pomatomus*), butterfishes (Stromateidae), tunas (*Thunnus*), marlin (*Makaira*), snappers (Lutjanidae), grunts (Pomadasyidae), porgies (Sparidae), sea trout (*Cynoscion*), barracudas (Sphyraenidae), and sharks.

ii. Epipelagic zone—Occupants of the epipelagic zone are some albacores, bonitos, and tunas (Scombridae), dolphins (Coryphaena), mantas (Mobulidae), marlin (*Makaira*), sailfish (*Istiophorus*), molas (*Mola*), and lanternfish (Myctophidae).

iii. Mesopelagic zone—In the mesopelagic zone, fish examples include deep-sea eels (*Synaphobranchus*), the deep-sea swallower (*Chiasmodus*), lanternfishes (Myctophidae), stalkeyed fish (*Idiacanthus*), and stomiatoids.

iv. Bathypelagic zone—Examples of fishes found in the bathypelagic zone are the deep-sea swallower (*Chiasmodus*), deep-water eels (e.g., *Cyema*), stomiatoids (e.g., *Chauliodus* and *Malacosteus*), scorpionfishes (Scorpaenidae), dories (Zeidae), gulpers (*Eurypharynx*), and swallowers (*Saccopharynx*).

v. Abyssopelagic zone—Some of the fishes encountered in the abyssopelagic zone are deep-water eels (*Cynema*), deep-sea anglers (*Borophryne* and *Melanocetus*), gulpers (*Eurypharynx*), and stomiatoids (*Chauliodus*).

c. Benthic Environment

i. Supratidal zone—Only a few species have established niches in this environment. Some of these include gobies (Gobiidae), eels (*Anguilla*), and clingfishes (Gobiesocidae).

ii. Intertidal zone—Representative fishes of the intertidal zone include stingrays (Dasyatidae), flounders (Bothidae, Pleuronectidae), soles (Soleidae), eels (*Anguilla*), morays (Muraenidae), clingfishes (Gobiesocidae), sculpins (Cottidae), searobins (Triglidae), blennies (Blenniidae), gobies (Gobiidae), pipefishes and seahorses (Syngnathidae), and cusk-eels (Ophidiidae).

iii.   Subtidal zone—Fishes commonly occurring in the inner subtidal zone of the continental shelf (to a depth of ~50 m) are skates (Rajidae), stingrays (Dasyatidae), flounders and soles (Pleuronectiformes), searobins (Triglidae), dogfish sharks (Squalidae), bonefish (Albulidae), eels (*Anguilla*), morays (Muraenidae), seahorses and pipefishes (Syngnathidae), croakers, kingfish, and drums (Sciaenidae), hakes (Gadidae), wrasses (Labridae), butterflyfishes and angelfishes (Chaetodontidae), Parrotfishes (Scaridae), trunkfishes (Ostraciidae), puffers (Tetraodontidae), and blennies (Blenniidae). In the outer subtidal zone of the continental shelf (from 50 to ~200 m depth) common fishes include cod (*Gadus*), haddock (*Melanogrammus*), hakes (*Merluccius* and *Urophycis*), halibuts (*Hippoglossus*), chimaeras (*Chimaera*), hagfishes (Myxinidae), eels (*Anguilla*), and pollock (*Pollachius*).

iv.   Bathyal zone—Some typical bathyal fishes are halibuts (*Hippoglossus*), chimaeras (Chimaera), cod (*Gadus*), and hagfishes (Myxinidae).

v.   Abyssal zone—Abyssal fishes include rattails or grenadiers (Macrouridae), eels (*Synaphobranchus*), brotulas (Brotulidae), and relatives of the lanternfish (e.g., *Bathypterois* and *Ipnops*).

vi.   Hadal zone—Examples of fishes inhabiting the hadal region are rattails (Macrouridae), deep-water eels (*Synaphobranchus*), and brotulids (*Bassogigas*).

Environmental conditions greatly influence fish assemblages found at specific areas in the sea. Of particular importance are temperature, salinity, light, and currents. Biological factors of significance include food supply, competition, and predation.

Estuaries rank among the most physically unstable areas for fish. Populations must deal with temperature as well as salinity gradients.[151–153] In temperate and boreal regions, seasonal temperature changes usually have a marked effect on the structure of fish assemblages. The migratory patterns of many species, for example, are strongly coupled to seasonal temperature levels. Changing thermal gradients can act as barriers to certain species (e.g., bluefish), thereby affecting their migratory behavior. To maximize survivorship in estuarine habitats, fishes thermoregulate behaviorally, avoiding or selecting environmental temperatures. However, the observed distribution of a species in an estuary reflects its response to other factors as well, such as food availability, nutritional state, competition, predation, and habitat requirements.

A common problem encountered in some estuaries is reduced dissolved oxygen. In severe cases, when dissolved oxygen concentrations decrease below 4 ml/l and approach 0 ml/l, fish populations are impacted. Migration routes may be effectively blocked by oxygen-depleted water masses that spread over broad areas, persisting for months. Schooling fish entering waters devoid of oxygen can be trapped, occasionally culminating in mass mortality.

B. Crustaceans and Cephalopods

Among the invertebrates, crustaceans and cephalopods constitute important members of the nekton. Pelagic swimming crabs, shrimp, euphausiids, cuttlefish, octopods, and squid are examples. Shrimp and squid are of greatest commercial interest because they serve as major sources of food in many countries. Although some squid (e.g., *Architeuthis*) exceed 20 m in length,[34] most are no longer than 50 to 60 cm.[15] They are carnivores, feeding on crustaceans (e.g., crabs and shrimp), cephalopods (e.g., other squid), and fish.

C. Marine Reptiles

There are two major reptilian representatives in the oceans—sea snakes (~50 species) and sea turtles (5 species). Both groups primarily inhabit warm tropical waters. While marine turtles return to sandy beaches on land to lay their eggs above the high-tide mark, sea snakes remain in the ocean where they bear their live young.[34] Sea turtles are harvested for their meat, and their shells are sought for decorative purposes. Because of human predation pressure, many conservation groups have expressed concern for the long-term health of these reptiles. Their total numbers have been dramatically reduced throughout the world in recent years.[83]

D. Marine Mammals

Some of the most spectacular members of the oceanic nekton are marine mammals. Four orders of mammals inhabit the sea: (1) the Cetacea (whales, porpoises, and dolphins); (2) the Pinnipedia (seals, sea lions, and walruses); (3) the Sirenia (manatees and dugongs); and (4) the Carnivora (sea otters). Approximately 140 mammalian species are represented. The Cetacea is the largest order, containing more than 75 species. Members of the Cetacea give birth to their young at sea. They regularly traverse great distances in search of food, which consists of a variety of prey. The larger whales (i.e., baleen whales) feed on zooplankton, benthic invertebrates, or fish. These animals commonly exceed 10 m in length and include the largest mammal on earth (i.e., the blue whale *Balaenoptera musculus* is >30 m long). The toothed whales (including porpoises and dolphins) are major predators; they consume a wide variety of fish.

In contrast to the cetaceans, the pinnipeds give birth to their young on land or on floating ice.[83] They are similar to the toothed whales in that they largely prey on fish. There are 32 species of pinnipeds inhabiting marine waters worldwide. Many have been heavily exploited for their fur, oil, or ivory.[34] As a consequence, they have been the target of many conservation efforts leading to the relaxing of hunting pressure in many regions. These changes have had a positive effect on the revitalization of various pinniped species.

The sirenians feed lower on the food chain, consuming larger plants. These herbivorous mammals inhabit rivers, estuaries, and shallow coastal marine waters in low latitudes. Only three species of manatees and one dugong species belong to this order. As in the case of the pinnipeds, the sirenians have been hunted in the past for their meat and oil, causing drastic reductions in their abundance. During the past few decades, efforts have been expended to protect these mammals from further exploitation and other anthropogenic impacts.

Many species of marine mammals are long-lived. For example, sperm whales and fin whales may live for 80 years. Bottlenose porpoises, gray seals, and harbor seals often exceed 30 years in age.[83] The low fecundity, long development times, and long life spans characterizing the marine mammals make them vulnerable to human exploitation.[34]

E. Seabirds

Many species of seabirds (>250 species) utilize estuarine and marine environments, frequenting numerous habitats in search of food. Three broad ecological groups are recognized based on avian behavior and feeding. These include (1) the coastline birds that occasionally move inland (e.g., cormorants, gulls, and coastal terns); (2) the divers that catch fish below the sea surface (e.g., gannets and penguins); and (3) the ocean-going forms that spend most of their lives over and on the sea surface (e.g., albatrosses, petrels, and shearwaters).[75] Most of these birds consume fish as part of their diet.[15] Auks, albatrosses, penguins, petrels, and gannets are the most highly adapted birds to the marine environment.[34]

Seabirds exhibit several modes of feeding. Most actively pursue food in the uppermost areas of the water column. While cormorants, gannets, murres, pelicans, penguins, puffins, and terns dive or swim underwater to obtain food, other species such as gulls, petrels, and skimmers skim the neuston at the sea surface.[34] The highest abundances of seabirds occur where the food supply is greatest along coastal areas, in upwelling zones, and at oceanic fronts.

Seabirds nest and breed on land, often in dense colonies. In some cases, the birds fly many kilometers to their breeding grounds. It is during breeding periods that seabirds are most susceptible to predation. Nesting colonies are commonly attacked by rats, racoons, foxes, domestic animals, and other predators. Eggs and young of the species are particularly vulnerable prey to these intruders. Human disturbance of nesting sites also can have devastating impacts on the bird populations.

Many bird species other than seabirds frequent estuarine and neighboring coastal environments in search of suitable habitats for foraging, breeding, and nesting. Among these groups of birds are waterfowl (e.g., ducks, geese, mergansers, and swans), waders (e.g., egrets, ibises, and herons), and shorebirds (i.e., plovers and true shorebirds). Waterfowl and waders generally comprise the most abundant bird populations along estuaries. Many migrating waterfowl and shorebirds in North America nest in the tundra regions of Canada and Alaska, but overwinter or stop briefly along estuaries far to the south where they feed. The estuaries are critical to the long-term health and viability of these avifauna.

In the Delaware Bay area, shorebirds gain as much as 50% of their body weight in fat over a 10- to 14-day foraging period, consuming large amounts of horseshoe crab (*Limulus polyphemus*) eggs along beaches.[86] Delaware Bay is a major staging area for shorebirds migrating from South America. More than a million of these birds use the beaches and coastal marshes in the Delaware Bay area each spring. A number of species inhabit tidal marshes and mudflats of the estuary year-round. Many other temperate estuaries provide valuable habitat for migrating shorebirds as well.

Shorebirds, seabirds, and waterfowl play significant roles in coastal food webs.[155–160] Many of these birds exert considerable predation pressure on some benthic and pelagic fauna. The predation effect is often most evident in benthic macroinvertebrate communities inhabiting tidal flats. Avifauna are particularly susceptible to a wide range of human impacts (e.g., pollution, habitat destruction, and hunting) in the coastal zone.[49]

REFERENCES

1. Valiela, I., *Marine Ecological Processes,* 2nd ed., Springer-Verlag, New York, 1995.
2. Wolff, W. J., Biotic aspects of the chemistry of estuaries, in *Chemistry and Biogeochemistry of Estuaries,* Olausson, E. and Cato, I., Eds., John Wiley & Sons, Chichester, 1980, 263.
3. Wirsen, C. O., Jannasch, H. W., and Molyneaux S. J., Chemosynthetic microbial activity at Mid-Atlantic Ridge hydrothermal vent sites, *J. Geophys. Res.,* 98, 9693, 1993.
4. Jannasch, H. W., Microbial interactions with hydrothermal fluids, in *Seafloor Hydrothermal Systems: Physical, Chemical, Biological, and Geological Interactions,* Humphris, S. E., Zierenberg, R. A., Mullineaux, L. S., and Thomson, R. E., Eds., Geophysical Monograph 91, American Geophysical Union, Washington, D.C., 1995, 273.
5. Tunnicliffe, V., The biology of hydrothermal vents: ecology and evolution, *Oceanogr. Mar. Biol., Annu. Rev.,* 29, 319, 1991.
6. Lutz, R. A. and Kennish, M. J., Ecology of deep-sea hydrothermal vent communities: a review, *Rev. Geophys.,* 31, 211, 1993.
7. Van Dover, C. L., Ecology of Mid-Atlantic Ridge hydrothermal vents, in *Hydrothermal Vents and Processes,* Parson, L. M., Walker, C. L., and Dixon, D. R., Eds., Geological Society Special Publication No. 87, The Geological Society of London, London, 1995, 257.
8. Shank, T. M., Fornari, D. J., Von Damm, K. L., Lilley, M. D., Haymon, R. M., and Lutz, R. A., Temporal and spatial patterns of biological community development at nascent deep-sea hydrothermal vents (9°50′N, East Pacific Rise), *Deep-Sea Res. II,* 45, 465, 1998.
9. Pomeroy, L. R., Microbial processes in the sea: diversity in nature and science, in *Heterotrophic Activity in the Sea,* Hobbie, J. E. and Williams, P. J. le B., Eds., Plenum Press, New York, 1984, 1.
10. Sieburth, J. M., Bacterial substrates and productivity in marine ecosystems, *Annu. Rev. Ecol. Syst.,* 7, 259, 1976.
11. Fenchel, T., Suspended marine bacteria as a food source, in *Flows of Energy and Materials in Marine Ecosystems: Theory and Practice,* Fasham, M. J. R., Ed., Plenum Press, New York, 1984, 301.
12. Fenchel, T., Ecology of heterotrophic microflagellates. IV. Quantitative occurrence and importance as consumers of bacteria, *Mar. Ecol. Prog. Ser.,* 9, 35, 1985.
13. Ducklow, H. W. and Carlson, C. A., Oceanic bacterial production, *Adv. Microb. Ecol.,* 12, 113, 1992.
14. Ducklow, H. W. and Shiah, F.-K., Bacterial production in estuaries, in *Aquatic Microbiology: An Ecological Approach,* Ford, T. E., Ed., Blackwell Scientific Publications, Boston, 1993, 261.
15. Pinet, P. R., *Invitation to Oceanography,* Jones and Bartlett Publishers, Boston, 1998.

16. Williams, P. J. le B., Bacterial production in the marine food chain: the emperor's new suit of clothes? in *Flows of Energy and Materials in Marine Ecosystems: Theory and Practice,* Fasham, M. J. R., Ed., Plenum Press, New York, 1984, 271.

17. Rublee, P. A., Bacteria and microbial distribution in estuarine sediments, in *Estuarine Comparisons,* Kennedy, V. S., Ed., Academic Press, New York, 1982, 159.

18. Rublee, P. and Dornseif, B. E., Direct counts of bacteria in the sediments of a North Carolina salt marsh, *Estuaries,* 1, 188, 1978.

19. Dale, N., Bacteria in intertidal sediments: factors related to their distribution, *Limnol. Oceanogr.,* 19, 509, 1974.

20. Azam, F. and Hodson, R. E., Size distribution and activity of marine microheterotrophs, *Limnol. Oceanogr.,* 22, 492, 1977.

21. Kemp, P. F., The fate of benthic bacterial production, *Rev. Aquat. Sci.,* 2, 109, 1990.

22. Alongi, D. M., *Coastal Ecosystem Processes*, CRC Press, Boca Raton, FL, 1998.

23. Fenchel, T. M., The ecology of micro- and meiobenthos, *Annu. Rev. Ecol. Syst.,* 9, 99, 1978.

24. Pomeroy, L. R., Bancroft, K., Breed, J., Christian, R. R., Frankberg, D., Hall, J. R., Maurer, L. G., Wiebe, W. J., Wiegert, R. G., and Wetzel, R. L., Flux of organic matter through a salt marsh, in *Estuarine Processes,* Vol. 2, Wiley, M., Ed., Academic Press, New York, 1977, 270.

25. Azam, F., Fenchel, T., Field, J. G., Gray, J. S., Meyer-Reil, L.-A., and Thingstad, T. F., The ecological role of water column microbes in the sea, *Mar. Ecol. Prog. Ser.,* 19, 257, 1983.

26. Biddanda, B. A. and Pomeroy, L. R., Microbial aggregation and degradation of phytoplankton-derived detritus in seawater. I. Microbial succession, *Mar. Ecol. Prog. Ser.,* 42, 79, 1988.

27. Turner, J. T., Tester, P. A., and Ferguson, R. L., The marine cladoceran *Penilia avirostris* and the "microbial loop" of pelagic food webs, *Limnol. Oceanogr.,* 33, 245, 1988.

28. Fenchel, T., The ecology of heterotrophic microflagellates, *Adv. Microb. Ecol.,* 9, 57, 1986.

29. Sherr, E. B., Sherr, B. F., Fallon, R. D., and Newell, S. Y., Small, aloricate ciliates as a major component of the marine heterotrophic nanoplankton, *Limnol. Oceanogr.,* 31, 177, 1986.

30. Sherr, E. B., Sherr, B. F., and Paffenhöfer, G.-A., Phagotrophic protozoa as food for metazoans: a "missing" trophic link in marine pelagic food webs? *Mar. Microb. Ecol.,* 1, 61, 1986.

31. Michaels, A. F. and Silver, M. W., Primary production, sinking fluxes, and the microbial food web, *Deep-Sea Res.* Part A, 35, 473, 1988.

32. Alongi, D. M., Microbial-meiofaunal interrelationships in some tropical intertidal sediments, *J. Mar. Res.,* 46, 349, 1988.

33. Nygaard, K., Borsheim, K. Y., and Thingstad, T. F., Grazing rates on bacteria by marine heterotrophic microflagellates compared to uptake rates of bacterial-sized monodisperse fluorescent latex beads, *Mar. Ecol. Prog. Ser.,* 44, 159, 1988.

34. Lalli, C. M. and Parsons, T. R., *Biological Oceanography: An Introduction,* Pergamon Press, Oxford, England, 1993.

35. Kennett, J. P., *Marine Geology,* Prentice-Hall, Englewood Cliffs, NJ, 1982.

36. Gross, M. G., *Oceanography: A View of the Earth,* 3rd ed., Prentice-Hall, Englewood Cliffs, NJ, 1990.

37. Levinton, J. S., *Marine Ecology,* Prentice-Hall, Englewood Cliffs, NJ, 1982.

38. Smayda, T. J. and Shimizu, Y., Eds., *Toxic Phytoplankton Blooms in the Sea,* Elsevier Scientific Publishing, New York, 1993.

39. Mann, K. H., *Ecology of Coastal Waters: A Systems Approach,* University of California Press, Berkeley, 1982.

40. Joint, I. R. and Pomeroy, A. J., Primary production in a turbid estuary, *Estuarine Coastal Shelf Sci.,* 13, 303, 1981.

41. Cole, B. E. and Cloern, J. E., Significance of biomass and light availability to phytoplankton productivity in San Francisco Bay, *Mar. Ecol. Prog. Ser.,* 17, 15, 1984.

42. Day, J. W., Jr., Hall, C. A. S., Kemp, W. M., and Yáñez-Arancibia, A., *Estuarine Ecology,* John Wiley & Sons, New York, 1989.

43. Boynton, W. R., Kemp, W. M., and Keefe, C. W., A comparative analysis of nutrients and other factors influencing estuarine phytoplankton production, in *Estuarine Comparisons,* Kennedy, V. S., Ed., Academic Press, New York, 1982, 69.

44. Campbell, J. W. and Aarup, T., Photosynthetically available radiation at high latitudes, *Limnol. Oceanogr.,* 34, 1490, 1989.

45. Helbling, E. W., Villafañe, V., Ferrario, M., and Holm-Hansen, O., Impact of natural ultraviolet radiation on rates of photosynthesis and on specific marine phytoplankton species, *Mar. Ecol. Prog. Ser.,* 80, 89, 1992.

46. Kirk, J. T. O., The nature and measurement of the light environment in the ocean, in *Primary Productivity and Biogeochemical Cycles in the Sea,* Falkowski, P. G. and Woodhead, A. D., Eds., Plenum Press, New York, 1992, 9.

47. Raymont, J. E. G., *Plankton and Productivity in the Oceans,* 2nd ed., Vol. 1, Pergamon Press, Oxford, 1980.

48. Omori, M. and Hamner, W. M., Patchy distribution of zooplankton: behavior, population assessment and sampling problems, *Mar. Biol.,* 72, 193, 1982.

49. Kennish, M. J., *Ecology of Estuaries: Anthropogenic Effects,* CRC Press, Boca Raton, FL, 1992.

50. Turner, J. T., Zooplankton feeding ecology: do co-occurring copepods compete for the same food, *Rev. Aquat. Sci.,* 5, 101, 1991.

51. Omori, M. and Ikeda, T., *Methods in Marine Zooplankton Ecology,* John Wiley & Sons, New York, 1984.

52. Knox, C. A., *The Biology of the Southern Ocean,* Cambridge University Press, New York, 1994.

53. Perkins, E. J., *Biology of Estuaries and Coastal Waters,* Academic Press, London, 1974.

54. Grindley, J. R., Estuarine plankton, in *Estuarine Ecology: With Particular Reference to Southern Africa,* Day, J. H., Ed., A. A. Balkema, Rotterdam, 1981, 117.

55. Leiby, M. M., Life history and ecology of pelagic fish eggs and larvae, in *Marine Plankton Life Cycle Strategies,* Steidinger, K. A. and Walker, L. M., Eds., CRC Press, Boca Raton, FL, 1984, 121.

56. Galtsoff, P. S., The American oyster *Crassostrea virginica* Gmelin, *Fish. Bull.* (U.S.), Vol. 64, 1964.

57. Bricelj, V. M. and Malouf, R. E., Aspects of reproduction of hard clams (*Mercenaria mercenaria*) in Great South Bay, New York, *Proc. Natl. Shellfish. Assoc.,* 70, 216, 1980.

58. Norcross, B. L. and Shaw, R. F., Oceanic and estuarine transport of fish eggs and larvae: a review, *Trans. Am. Fish. Soc.,* 113, 153, 1984.

59. Day, R. and McEdward, L., Aspects of the physiology and ecology of pelagic larvae of marine benthic invertebrates, in *Marine Plankton Life Cycle Strategies,* Steidinger, K. A. and Walker, L. M., Eds., CRC Press, Boca Raton, FL, 1984, 93.

60. McEdward, L., Ed., *Ecology of Marine Invertebrate Larvae,* CRC Press, Boca Raton, FL, 1995.

61. Marcus, N. H., Recruitment of copepod nauplii into the plankton: importance of diapause eggs and benthic processes, *Mar. Ecol. Prog. Ser.,* 15, 47, 1984.

62. Marcus, N. H. and Schmidt-Gengenbach, J., Recruitment of individuals into the plankton: the importance of bioturbation, *Limnol. Oceanogr.,* 31, 206, 1986.

63. Forward, R. B., Jr., Diel vertical migration: zooplankton photobiology and behavior, *Oceanogr. Mar. Biol. Annu. Rev.,* 26, 361, 1988.

64. Forward, R. B., Jr., Cronin, T. W., and Stearns, D. E., Control of diel vertical migration: photoresponses of a larval crustacean, *Limnol. Oceanogr.,* 29, 146, 1984.

65. Stearns, D. E. and Forward, R. B., Jr., Photosensitivity of the calanoid copepod *Acartia tonsa, Mar. Biol.,* 82, 85, 1984.

66. Stearns, D. E., Copepod grazing behavior in simulated natural light and its relation to nocturnal feeding, *Mar. Ecol. Prog. Ser.,* 30, 65, 1986.

67. Forward, R. B., Jr., A reconsideration of the shadow response of a larval crustacean, *Mar. Behav. Physiol.,* 12, 99, 1986.

68. Heinle, D. R., Zooplankton, in *Functional Adaptations of Marine Organisms,* Vernberg, F. J. and Vernberg, W. B., Eds., Academic Press, New York, 1981, 85.

69. Davis, C. C., Planktonic Copepoda (including Monstrilloida), in *Marine Plankton Life Cycle Strategies,* Steidinger, K. A. and Walker, L. M., Eds., CRC Press, Boca Raton, FL, 1984, 67.

70. Heinle, D. R., Temperature and zooplankton, *Chesapeake Sci.,* 10, 186, 1969.

71. Miller, D. C., Geider, R. J., and MacIntyre, H. L., Microphytobenthos: the ecological role of the "secret garden" of unvegetated, shallow water marine habitats. II. Role in sediment stability and shallow water foods webs, *Estuaries,* 19, 202, 1996.

72. Rizzo, W. M., Lackey, G. J., and Christian, R. R., Significance of euphotic subtidal sediments to oxygen and nutrient cycling in a temperate estuary, *Mar. Ecol. Prog. Ser.,* 86, 51, 1992.

73. Reay, W. G., Gallagher, D. L., and Simmons, J., Sediment-water column oxygen and nutrient fluxes in nearshore environments of the lower Delmarva Peninsula, USA, *Mar. Ecol. Prog. Ser.,* 118, 215, 1995.

74. Cahoon, L. B. and Cooke, J. E., Benthic microalgal production in Onslow Bay, North Carolina, USA, *Mar. Ecol. Prog. Ser.,* 84, 185, 1992.

75. Meadows, P. S. and Campbell, J. I., *An Introduction to Marine Science,* 2nd ed., Blackie and Son Ltd., Glasgow, 1985.

76. Valiela, I., McClelland, J., Hauxwell, J., Behr, P. J., Hersh, D., and Foreman, K., Macroalgal blooms in shallow estuaries: controls and ecophysiological and ecosystem consequences, *Limnol. Oceanogr.,* 42, 1105, 1997.

77. Boaden, P. J. S. and Seed, R., *An Introduction to Coastal Ecology,* Blackie and Son Ltd., Glasgow, 1988.

78. Barnes, R. S. K. and Hughes, R. N., *An Introduction to Marine Ecology,* 2nd ed., Blackwell Scientific Publications, Cambridge, 1988.

79. Ford, T. E., Ed., *Aquatic Microbiology: An Ecological Approach,* Blackwell Scientific Publications, Cambridge, 1993.

80. Macdonald, K. B., Plant and animal communities of Pacific North American salt marshes, in *Wet Coastal Ecosystems,* Chapman, V. J., Ed., Elsevier, Amsterdam, 1977, 167.

81. Beeftink, W. G., Salt-marshes, in *The Coastline,* Barnes, R. S. K., Ed., John Wiley & Sons, Chichester, 1977, 93.

82. Frey, R. W. and Basan, P. B., Coastal saltmarshes, in *Coastal Sedimentary Environments,* 2nd ed., Davis, R. A., Ed., Springer-Verlag, New York, 1985, 225.

83. Nybakken, J. W., *Marine Biology: An Ecological Approach,* 2nd ed., Harper & Row, New York, 1988.

84. Shubauer, J. P. and Hopkinson, C. S., Above- and below-ground emergent macrophyte production and turnover in a coastal marsh ecosystem, Georgia, *Limnol. Oceanogr.,* 29, 1052, 1984.

85. Hackney, C. T. and de la Cruz, A. A., Below-ground productivity of roots and rhizomes in a giant cordgrass marsh, *Estuaries,* 9, 112, 1986.

86. Kennish, M. J., Ed., *Estuary Restoration and Maintenance: The National Estuary Program,* CRC Press, Boca Raton, FL, 1999.

87. Phillips, R. C. and McRoy, C. P., Eds., *Handbook of Seagrass Biology: An Ecosystem Perspective,* Garland STPM Press, New York, 1980.

88. Charpy-Robaud, C. and Sournia, A., The comparative estimation of phytoplankton, microphytobenthic, and macrophytobenthic primary production in the oceans, *Mar. Micro. Food Webs,* 4, 31, 1990.

89. Hillman, K., Walker, D. I., Larkum, A. W. D., and McComb, A. J., Productivity and nutrient limitation, in *Biology of Seagrasses,* Larkum, A. W. D., McComb, A. J., and Shepard, S. A., Eds., Elsevier, Amsterdam, 1989, Chap. 19.

90. Vermaat, J. E., Agawin, N. S. R., Duarte, C. M., Fortes, M. D., Marba, N., and Uri, J. S., Meadow maintenance, growth, and productivity of a mixed Philippine seagrass bed, *Mar. Ecol. Prog. Ser.,* 124, 215, 1995.

91. Dawes, C. J., *Marine Botany,* John Wiley & Sons, New York, 1981.

92. Lin, P., *Mangrove Vegetation,* Springer-Verlag, New York, 1988.

93. Twilley, R. R., Chen, R. H., and Hargis, T., Carbon sinks in mangroves and their implications to carbon budget of tropical coastal ecosystems, *Water Air Soil Pollut.,* 64, 265, 1992.

94. Jumars, P. A., *Concepts in Biological Oceanography: An Interdisciplinary Primer,* Oxford University Press, Oxford, 1993.

95. Eckman, J. E., Small-scale patterns and processes in a soft substratum intertidal community, *J. Mar. Res.,* 37, 437, 1979.

96. Wolff, W. J., Estuarine benthos, in *Estuaries and Enclosed Seas,* Ketchum, B. H., Ed., Elsevier, Amsterdam, 1983, 151.

97. Ambrose, W. G., Jr., Role of predatory infauna in structuring marine soft-bottom communities, *Mar. Ecol. Prog. Ser.,* 17, 109, 1984.

98. Hines, A. H. and Comtois, K. L., Vertical distribution of infauna in sediments of a subestuary of central Chesapeake Bay, *Estuaries,* 8, 296, 1985.

99. Commito, J. A., The importance of predation by infaunal polychaetes in controlling the structure of a soft-bottom community in Maine, USA, *Mar. Biol.,* 68, 77, 1982.

100. Fenchel, T. and Jansson, B. O., On the vertical distribution of the microfauna in the sediments of a brackish-water beach, *Ophelia,* 3, 161, 1966.

101. Coull, B. C. and Bell, S. S., Perspectives of marine meiofaunal ecology, in *Ecological Processes in Coastal Marine Systems,* Livingston, R. J., Ed., Plenum Press, New York, 1979, 189.

102. Nichols, F. H. and Thompson, J. K., Time scales of change in the San Francisco Bay benthos, *Hydrobiologia,* 129, 121, 1985.

103. Rhoads, D. C. and Boyer, L. F., The effects of marine benthos on physical properties of sediments: a successional perspective, in *Animal-Sediment Relations: The Biogenic Alteration of Sediments,* McCall, P. L. and Tevesz, M J. S., Eds., Plenum Press, New York, 1982, 3.

104. Rhoads, D. C. and Germano, J. D., Interpreting long-term changes in benthic community structure: a new protocol, *Hydrobiologia,* 142, 291, 1986.

105. Brenchley, B. A., Disturbance and community structure: an experimental study of bioturbation in marine soft-bottom environments, *J. Mar. Res.,* 39, 767, 1981.

106. Rumrill, S. S., Natural mortality of marine invertebrate larvae, *Ophelia,* 32, 163, 1990.

107. Morgan, S. G., Life and death in the plankton: larval mortality and adaptation, in *Ecology of Marine Invertebrate Larvae,* McEdward, L., Ed., CRC Press, Boca Raton, FL, 1995, 279.

108. Warwick, R. M., Population dynamics and secondary production of benthos, in *Marine Benthic Dynamics,* Tenore, K. R. and Coull, B. C., Eds., University of South Carolina Press, Columbia, 1980, 1.

109. Woombs, M. and Laybourn-Parry, J., Growth, reproduction, and longevity in nematodes from sewage treatment plants, *Oecologia,* 64, 168, 1984.

110. Coull, B. C. and Dudley, B. W., Delayed nauplier development of meiobenthic copepods, *Biol. Bull.,* 150, 38, 1976.

111. Levin, L. A. and Bridges, T. S., Pattern and diversity in reproduction and development, in *Ecology of Marine Invertebrate Larvae,* McEdward, L., Ed., CRC Press, Boca Raton, FL, 1995, 1.

112. Jablonski, D. and Lutz, R. A., Larval ecology of marine benthic invertebrates: paleobiological implications, *Biol. Rev.,* 58, 21, 1983.

113. Jablonski, D., Larval ecology and macroevolution in marine invertebrates, *Bull. Mar. Sci.,* 39, 565, 1986.

114. Shanks, A. L., Mechanisms of cross-shelf dispersal of larval invertebrates and fish, in *Ecology of Marine Invertebrate Larvae,* McEdward, L., Ed., CRC Press, Boca Raton, FL, 1995, 323.

115. Crisp, D. J., Overview of research on marine invertebrate larvae, 1940–1980, in *Marine Biodeterioration: An Interdisciplinary Study,* Costlow, J. D. and Tipper, R. C., Eds., Naval Institute Press, Annapolis, MD, 1984, 103.

116. Scheltema, R. S., On dispersal and planktonic larvae of benthic invertebrates: an eclectic overview and summary of problems, *Bull. Mar. Sci.,* 39, 290, 1986.

117. Young, C. M., Behavior and locomotion during the dispersal phase of larval life, in *Ecology of Marine Invertebrate Larvae,* McEdward, L., Ed., CRC Press, Boca Raton, FL, 1995, 249.

118. Chia, F. S. and Rice, M. E., Eds., *Settlement and Metamorphosis of Marine Invertebrates,* Elsevier/North Holland, New York, 1978.

119. Wright, L. D., Benthic boundary layers of estuarine and coastal environments, *Rev. Aquat. Sci.,* 1, 75, 1989.

120. Wright, L. D., *Morphodynamics of Inner Continental Shelves,* CRC Press, Boca Raton, FL, 1995.

121. Rhoads, D. C., Organism-sediment relations on the muddy seafloor, *Oceanogr. Mar. Biol. Annu. Rev.,* 12, 263, 1974.

122. Hessler, R. R. and Smithey, W. M., Jr., The distribution and community structure of megafauna at the Galapagos Rift hydrothermal vents, in *Hydrothermal Processes at Seafloor Spreading Centers,* Rona, P. A., Boström, K., Laubier, L., and Smith, K. L., Jr., NATO Conf. Ser. 4, Vol. 12, Plenum Press, New York, 1983, 735.

123. Somero, G. N., Siebenaller, J. F., and Hochachka, P. W., Biochemical and physiological adaptations of deep-sea animals, in *The Sea,* Vol. 8, *Deep-Sea Biology,* Rowe, G. T., Ed., John Wiley & Sons, New York, 1983, 261.

124. Grassle, J. F., The ecology of deep-sea hydrothermal vent communities, *Adv. Mar. Biol.,* 23, 301, 1986.

125. Tunnicliffe, V., Hydrothermal-vent communities of the deep sea, *Am. Sci.,* 80, 336, 1992.

126. Kennish, M. J., *Ecology of Estuaries,* VoL 2, Biological Aspects, CRC Press, Boca Raton, FL, 1990.

127. Sokolova, M. N., Trophic structure of deep-sea macrobenthos, *Mar. Biol.,* 16, 1, 1972.

128. Rowe, G. T., Benthic biomass and surface productivity, in *Fertility of the Sea,* Vol, 2, Costlow, J. D., Ed., Gordon and Breach, New York, 1971, 147.

129. Rowe, G. T., Biomass and production of the deep-sea macrobenthos, in *The Sea,* Vol. 8, *Deep-Sea Biology,* Rowe, G. T., John Wiley & Sons, New York, 1983, 97.

130. Rex, M. A., Community structures in the deep-sea benthos, *Annu. Rev. Ecol. Syst.,* 12, 331, 1981.

131. Rex, M. A., Geographical patterns of species diversity in the deep-sea benthos, in *The Sea,* Vol. 8, Rowe, G. T., Ed., John Wiley & Sons, New York, 1983, 453.

132. Rex, M. A., Stuart, C. T., Hessler, R. R., Allen, J. A., Sanders, H. L., and Wilson, G. D. F., Global-scale latitudinal patterns of species diversity in the deep-sea benthos, *Nature,* 365, 636.

133. Rex, M. A., Stuart, C., and Etter, R. J., Large-scale patterns of species diversity in the deep-sea benthos, in *Marine Biodiversity: Causes and Consequences,* Ormond, R. J. A., Gage, J. D., and Angel, M. V., Eds., Cambridge University Press, Cambridge, 1997, 94.

134. Paterson, G. L. J. and Lambshead, P. J. D., Bathymetric patterns of polychaete diversity in the Rockall Trough, northeast Atlantic, *Deep-Sea Res.,* 42, 1199, 1995.

135. Pineda, J., Boundary effects on the vertical ranges of deep-sea species, *Deep-Sea Res.,* 40, 2179, 1993.

136. Grassle, J. F. and Maciolek, N. J., Deep-sea species richness: regional and local diversity estimates from quantitative bottom samples, *Am. Nat.,* 139, 313, 1992.

137. May, R. M., Bottoms up for the oceans, *Nature,* 357, 278, 1992.

138. Poore, G. C. B. and Wilson, G. D. F., Marine species richness, *Nature,* 361, 597, 1993.

139. Poore, G. C. B., Just, J., and Coven, B. F., Composition and diversity of Crustacea Isopoda of the southeastern Australian continental slope, *Deep-Sea Res.,* 41, 677, 1994.

140. Gage, J. D., Why are there so many species in deep-sea sediments? *J. Exp. Mar. Biol. Ecol.,* 200, 257, 1996.

141. Sanders, H. L., Marine benthic diversity: a comparative study, *Am. Nat.,* 102, 243, 1968.

142. Rhode, K., Latitudinal gradients in species diversity: the search for the primary cause, *Oikos,* 65, 514, 1992.

143. Lutz, R. A., Desbruyères, D., Shank, T. M., and Vrijenhoek, R. C., A deep-sea hydrothermal vent community dominated by Stauromedusae, *Deep-Sea Res.,* 45, 329, 1998.

144. Lutz, R. A., The biology of deep-sea vents and seeps, *Oceanus,* 34, 75, 1991.

145. Lutz, R. A., Shank, T. M., Fornari, D. J., Haymon, R. M., Lilley, M. D., Von Damm, K. L., and Desbruyères, D., Rapid growth at deep-sea vents, *Nature,* 371, 663, 1994.

146. McHugh, J. L., Estuarine nekton, in *Estuaries,* Lauff, G. H., Ed., Publ. 83, American Association for the Advancement of Science, Washington, D.C., 1967, 581.

147. Moyle, P. B. and Cech, J. J. Jr., *Fishes: An Introduction to Ichthyology,* Prentice-Hall, Englewood Cliffs, NJ, 1982.

148. Stickney, R. R., *Estuarine Ecology of the Southeastern United States and Gulf of Mexico,* Texas A&M University Press, College Station, Texas, 1984.

149. Evans, D. H., Ed., *The Physiology of Fishes,* 2nd ed., CRC Press, Boca Raton, FL, 1996.

150. Pitcher, T. J., Ed., *The Behavior of Teleost Fishes,* Johns Hopkins University Press, Baltimore, MD, 1986.

151. Day, J. H., Blaber, S. J. M., and Wallace, J. H., Estuarine fishes, in *Estuarine Ecology: With Particular Reference to Southern Africa,* Day, J. H., Ed., A. A. Balkema, Rotterdam, 1981, 197.

152. Haedrich, R. L., Estuarine fishes, in *Estuaries and Enclosed Seas,* Ketchum, B. H., Ed., Elsevier Scientific, Amsterdam, 1983, 183.

153. Yáñez-Arancibia, A., Linares, F. A., and Day, J. W., Jr., Fish community structure and function in Terminos Lagoon, a tropical estuary in the southern Gulf of Mexico, in *Estuarine Perspectives,* Kennedy, V., Ed., Academic Press, New York, 1980, 465.

154. Lager, K. F., Bardach, J. E., and Miller, R. R., *Ichthyology,* John Wiley & Sons, New York, 1962.

155. McLusky, D. S., *The Estuarine Ecosystem,* Halsted Press, New York, 1981.

156. Burger, J. and Olla, B. L., Eds., *Shorebirds: Breeding Behavior and Populations,* Vol. 5, Plenum Press, New York, 1984.

157. Burger, J. and Olla, B. L., Eds., *Shorebirds: Breeding Behavior and Populations,* Vol. 6, Plenum Press, New York, 1984.

158. Feare, C. J. and Summers, R., Birds as predators on rocky shores, in *The Ecology of Rocky Coasts,* Moore, P. G. and Seed, R., Eds., Hodder and Stoughton, Sevenoaks, England, 1985.

159. Perry, M. C., Waterfowl of Chesapeake Bay, in *Contaminant Problems and Management of Living Chesapeake Bay Resources,* Majumdar, S. K., Hall, L. W., Jr., and Austin, H. M., Eds., Pennsylvania Academy of Science, Easton, 1987, 94.

160. Perry, M. C. and Uhler, F. M., Food habits and distribution of wintering canvasbacks, *Aythya valisineria,* on Chesapeake Bay, *Estuaries,* 11, 57,

5.1 MARINE ORGANISMS:
MAJOR GROUPS AND COMPOSITION

Table 5.1–1   Typical Cosmopolitan Oceanic Species

Siphonophora	Mollusca	Copepoda
Physophora hydrostatica	*Euclio pyramidata*	*Rhincalanus nasutus*
Agalma elegans	*Euclio cuspidata*	*Eucalanus elongatus*
Dimophyes arctica	*Diacria trispinosa*	*Pleuromamma robusta*
Lensia conoidea	*Pneumodermopsis ciliata*	*Euchirella rostrata*
Chelopheys appendiculata	*Taonidium pfefferi*	*Euchirella curticaudata*
Sulculeolaria biloba	*Tracheloteuthis risei*	*Oithona spinirostris*

Medusae	Polychaeta	Other Crustacea
Cosmetira pilosella	*Travisiopsis lanceolata*	*Lepas* sp.
Laodicea undulata	*Vanadis formosa*	*Munnopsis murrayi*
Halicreas sp.	*Rhynchonerella angelini*	*Brachyscelus crusulum*
Periphylla periphylla	*Tomopteris septentrionalis*	

Thaliacea	Chaetognatha	
Salpa fusiformis	*Sagitta serratodentata f. tasmanica*	*Meganyctiphane norvegica*
Dolioletta gegenbauri	*Sagitta hexaptera*	*Euphausia krohni Anchialus agilis*

Table 5.1–2   Some Planktonic Species Typical of Deep Water

Gaetaenus pileatus	*Amalopenaeus elegans*	*Sagitta macrocephala*
Arietellus plumifer	*Hymenodora elegans*	*S. zetesios*
Pontoptilus muticus	*Boreomysis microps*	*Eukrohnia fowleri*
Centraugaptilus rattrayi	*Eucopia unguiculata*	*Nectonemertes miriabilis*
Augaptilus megalaurus and many	*Cyphocaris anonyx*	*Spiratella helicoides*
other copepods	*Scina* sp.	*Histioteuthis boneltiana*

Table 5.1–3   Species Composition of the Five World Distributional Zones of Planktonic Foraminifera

Northern and Southern Cold-Water Regions

1.   Arctic and antarctic zones
      Globigerina pachyderma (Ehrenberg): Left-coiling variety; right-coiling in subarctic and subantarctic
      zones
2.   Subarctic and subantarctic zones
      Globigerina quinqueloba (Natland)
      Globigerina bulloides (d'Orbigny)
      Globigerinita bradyi (Wiesner)
      Globorotalia scitula (Brady)

Transition Zones

3.   Northern and south transition zones between cold-water and warm-water regions
      Globorotalia inflata (d'Orbigny): With mixed occurrences of subpolar and tropical–subtropical species

Warm-Water Region

4.   Northern and southern subtropical zones
      Globigerinoides ruber (d'Orbigny): Pink variety in Atlantic Ocean only
      Globigerinoides conglobatus (Brady): Autumn species
      Hastigerina pelagica (d'Orbigny)
      Globigerinita glutinata (Egger)
      Globorotalia truncatulinoides (d'Orbigny)
      Globorotalia hirsuta (d'Orbigny)   Winter species
      Globigerina rubescens (Hofker)   Winter species
      Globigerinella aequilateralis (Brady)   Prefer outer margins of subtropical central
      Orbulina universa (d'Orbigny)   water masses and into transitional zone
      Globoquadrina dutertrei (d'Orbigny)
      Globigerina falconensis (Blow)
      Globorotalia crassaformis (Galloway and Wissler)

5.   Tropical Zone
      Globigerinoides sacculifer (Brady): Including *Sphaeroidinella dehiscens* (Parker and Jones)
      Globorotalia menardii (d'Orbigny)
      Globorotalia tumida (Brady)
      Pulleniatina obliquiloculata (Parker and Jones)
      Candeina nitida (d'Orbigny)
      Hastigerinella digitata (Rhumbler)
      Globoquadrina conglomerata (Schwager)   Restricted to Indo-Pacific
      Globigerinella adamsi (Banner and Blow)   Restricted to Indo-Pacific
      Globoquadrina hexagona (Natland)   Restricted to Indo-Pacific

Note: The species are listed under the zone where their highest concentrations are observed, but they are
          not necessarily limited to these areas. Most species listed under the subtropical zones are also
          common in the tropical waters.

[a] Usually located in central water masses between 20°N and 40°N, or between 20°S and 40°S latitude.

Table 5.1–4 Areas of North Pacific in Which Listed Species Have Been Shown to Occur

Organism	Subarctic	Transitional	Central	Equatorial	Eastern Tropic Pacific	Warm-Water Cosmopolites	Comments
PROTOZOA							
Foraminifera							
Globigerina quinqueloba	+						
Globigerinoides minuta	+						
Globigerina pachyderma	+						
Globorotalia truncatulinoides			+				
Pulleniatina obliquiloculata				+			
Sphaeroidinella dehiscens				+			
Globigerina conglomerata				+			
Globorotalia tumida				+			
Globorotalia hirsuta				+			
Globigerinella aequilateralis						+	Pure
Globigerinoides conglobata						+	Pure
Globigerinoides rubra						+	Pure
Orbulina universa						+	Pure
Globigerinoides sacculifera						+	Peak at equator
Globorotalia menardii						+	Peak at equator
Globigerina eggeri						+	Edge effect
Hastigerina pelagica						+	Edge effect
Radiolaria							
Castanidium apsteini	+						Doubtful; may also be deep central
Castanidium variabile	+						Doubtful; may also be deep central
Haeckeliana porcellana	+						
Castanea amphora			+				
Castanissa brevidentata			+				
Castanella thomsoni			+				T. zone w/upwelled water?
Castanea henseni			+				T. zone w/upwelled water?
Castanea globosa			+				T. zone w/upwelled water?
Castanidium longispinum				+			
Castanella aculeata				+			

(continued)

CHAETOGNATHA
Sagitta elegans
Eukrohnia hamata
Sagitta scrippsae
Sagitta pseudoserratodentata
Sagitta californica — Crossing W.T.P.
Sagitta ferox
Sagitta robusta — Patchy
Sagitta regularis
Sagitta hexaptera
Sagitta enflata — Peak at equator
Pterosagitta draco — Peak at equator
Sagitta pacifica — Peak at equator
Sagitta minima — Edge effect

ANNELIDA
Tomopteris septentrionalis
Tomopteris pacifica
Poeobius meseres

ARTHROPODA
Copepoda
Calanus pacificus
Calanus plumchrus
Calanus tonsus
Calanus cristatus
Eucalanus bungii bungii — May be T. zone
Eucalanus elongatus hyalinus
Eucalanus bungii californicus — South Pacific also
Clausocalanus pergens
Clausocalanus lividus

Table 5.1–4 Areas of North Pacific in Which Listed Species Have Been Shown to Occur (continued)

Organism	Subarctic	Transitional	Central	Equatorial	Eastern Tropic Pacific	Warm-Water Cosmopolites	Comments
Eucalanus subcrassus				+			
Rhincalanus cornutus				+			
Eucalanus inermis					+		
Eucalanus crassus						+	Patchy, "pure"
Rhincalanus nasutus						+	Very patchy, almost pure equatorial
Eucalanus attenuatus						+	Peak at equator; some edge effect
Eucalanus subtenuis				+			Patchy, peak at equator; some edge effect
Clausocalanus arcuicornis							
Eucalanus longiceps							
Rhincalanus gigas				+			
Clausocalanus laticeps							
Euphausiacea							
Thysanoessa longipes	+						
Euphausia pacifica	+						
Thysanopoda acutifrons		+					
Thysanoessa gregaria		+					
Euphausia gibboides		+					
Nematoscelis difficilismegalops		+					
Nematoscelis atlantica			+				
Euphausia brevis			+				
Euphausia hemigibba			+				
Euphausia gibba			+				
Euphausia mutica			+				
Stylocheiron suhmii			+				
Euphausia diomediae				+			
Euphausia distinguenda				+			
Nematoscelis gracilis				+			Crossing in W.T.P.
Euphausia distinguenda					+		
Euphausia eximia					+		
Euphausia lamelligera					+		
Euphausia tenera						+	Peak at equator
Stylocheiron abbreviatum						+	Avoids E.T.P.
Euphausia superba							
Amphipoda							
Parathimisto pacifica	+						

Taxon							Remarks
MOLLUSCA							
Pteropoda							
Limacina helicina	+						
Clio polita	+						
Corolla pacifica		+					
Clio balantium		+					
Cavolinia inflexa			+				Crossing in W.T.P.
Clio pyramidata			+				Crossing in W.T.P.
Styliola subula			+				Crossing in W.T.P.
Limacina lesueuri							
Clio n.sp.							
Cavolinia uncinata				+			Very patchy
Limacina trochiformis				+	+		
Limacina inflata						+	
Cavolinia longirostris						+	Very patchy; almost pure equatorial
Cavolinia gibbosa						+	Very patchy, avoids E.T.P.
Hyalocylix striata						+	
Creseis virgula						+	Edge effect
Creseis acicula						+	
Cavolinia tridentata						+	Peak at equator
Diacria trispinosa						+	Peak at equator
Limacina bulimoides						+	Avoids E.T.P.
Clio antarctica						+	
Heteropoda							
Caranaria japonica			+				
Gymnosomata							
Clione limacina		+					

Note: E.T.P., Eastern Tropical Pacific; W.T.P., Western Tropical Pacific.

Source: McGowan, J. A., in *The Micropaleontology of Oceans*, Cambridge University Press, Cambridge, 1971, 14. With permission.

Table 5.1–5　Elemental Ratios (by atoms) in Aquatic Organisms or Biomass[a]

Organism	C	N	P
Marine phytoplankton	106	16	1
Marine bacteria	47	7	1
	45	9	1
Tropical zooplankton	144	29	1
Aloricate ciliates	242–56	16	1
Oceanic seston (<200 μm)	122	15	1
(<3 μm)	181	20	1
(Total, 0–250 m)	120	16	
(Total, surface)	132	15	1
Atoll lagoon seston	490–55	64–9	1
Sedimenting detritus	410	29	1
Dissolved organics	300–400	19	1
Turf algae (prokaryote)	432	36	1
Macroalgae (eukaryotic)	696	36	1
Seagrasses (leaves)	458	21	1
(Roots and rhizomes)	596	12	1
Mangrove leaves (live)	1133	29	1
Mangrove litterfall	4567	24	1
Mangrove creek DOM	700–250	35–18	1
Nonmarine emergent macrophytes			
Cyperus papyrus (live)		45	1
(papyrus detritus)	4775	577	1
Typha domingensis (live)	715	24	1
T. domingensis (dead)	1107	4	1
Lepironia articulata	1920	16	
Freshwater nonemergent macrophytes			
Utricularia flexuosa	480	16	
Rainforest leaves (live)	516	20	1
	516–401	38–26	1
Rainforest litterfall	678	25	1
Floating macrophytes	480	27	1
Salvinia molesta (in sewage lagoon)		14–10	1
Azolla pinnata	329	40	1
Paspalum repens	880–416	16	
Eichhornia crassipes	267–166	16	
Amazon floodplain lake seston	253	27	1
Macrophyte epiphytes	230–85	25–12	1

[a] C:N and N:P ratios were calculated relative to N = 16 and P = 1 where data for full C:N:P ratios could not be obtained.

Source: Connell, D. W. and Hawker, D. W., Eds., *Pollution in Tropical Aquatic Systems,* CRC Press, Boca Raton, FL, 1992, 32. With permission.

Table 5.1–6   Total Carbon, Carbonate Carbon, and Organic Carbon Concentration in Planktonic Marine Organisms

Organism	Total Carbon (% DW)	Carbonate Carbon (% DW)	Organic Carbon (% DW)	Organic Carbon (% of DW)
Cnidaria				
Cyanea capillata	13.8	—	13.8	36.0
Physalia physalis	31.4	—	31.4	62.8
Pelagia noctiluca	12.9	—	12.9	26.0
Pelagia noctiluca	15.9	—	15.9	31.2
Aequorea vitrina	26.8	—	26.8	52.5
Average	17.5	—	17.5	41.7
Ctenophora				
Mnemiopsis sp.	6.4	—	6.4	20.6
Arthropoda				
Euphausia krohnii	35.8	—	35.8	43.9
Centropages hamatus	36.3	—	36.3	46.2
C. typicus 1:1				
Calanus finmarchicus	41.7	—	41.7	50.5
Meganyctiphanes norvegica	42.0	—	42.0	51.6
Lophogaster sp.	46.8	—	46.8	57.4
Centropages sp.	38.5	—	38.5	49.7
Centropages sp.	38.7	—	38.7	50.0
Sagitta elegans 1:1				
Idotea metallica	33.2	2.36	30.8	48.0
Calanus finmarchicus	39.8	—	39.8	48.3
Mixed copepods	35.6	—	35.6	46.0
Calanus finmarchicus (a small admixture of euphausiids and shell-less pteropods)	37.8	—	37.8	46.0
Average	38.3	—	38.0	48.9
Mollusca				
Limacina sp.	28.3	2.74	25.6	56.0
Ommastrephes sp.	45.1	—	45.1	48.8
Sthenoteuthis sp.	37.2	—	37.2	40.4
Clione limacina	26.3	—	26.3	39.4
Illex illecebrosus	39.2	—	39.2	42.6
Squid eggs (Loligo)	21.7	—	21.7	45.0
Average	33.1	—	32.7	45.4
Chordata				
Salpa sp.	10.6	—	10.6	46.1
Salpa sp.	9.6	—	3.6	39.0
Salpa fusiformis	7.8	—	7.8	33.9
Pyrosoma sp.	9.4	—	9.4	41.0
Average	9.4	—	9.4	40.0
Mixed Samples				
Mixed copepods and phytoplankton	29.8	—	29.8	38.5
Copepods and phytoplankton	25.2	—	25.2	48.0
Phytoplankton and fish	4.8	1.54	3.3	32.7
Phytoplankton and copepods	6.6	—	6.6	56.0
Copepods and phytoplankton	14.3	—	14.3	48.0
Mixed zooplankton	28.4	—	28.4	48.6
Average	18.2	—	17.9	38.8

Note: All entries are % of dry weight (DW) except last column, which is ash-free (AF) dry weight. Where no results are listed, the inorganic carbonate was not detectable.

Source: Curl, H. S., Jr., *J. Mar. Res.*, 20(3), 185, 1962. With permission.

Table 5.1–7   Organic Content of Copepods and Sagittae Based on Dry Weights

	Protein %	Fat %	Carbohydrate %	Ash %	P$_2$O$_5$ %	Nitrogen %
Copepods	70.9–77.0	4.6–19.2	0–4.4	4.2–6.4	0.9–2.6	11.1–12.0
Sagittae	69.6	1.9	13.9	16.3	3.6	10.9

Table 5.1–8   Classification of Organic Material in Aquatic Systems

Size Category	Size Range (diameter)
CPOM—Coarse particulate organic material (or matter)	>1 mm
FPOM—Fine particulate organic material (or matter)	<1 mm but >0.45 μm
DOM—Dissolved organic material (or matter)	<0.45 μm

Source: Wotton, R. S., Ed., *The Biology of Particles in Aquatic Systems,* CRC Press, Boca Raton, FL, 1990, 4. With permission.

5.2 BIOLOGICAL PRODUCTION IN THE OCEAN

Table 5.2–1 Global Primary Production in Marine Ecosystems

Area	Primary Production (g C/m²/year)	World Ocean Area (km²)	World Ocean Area (%)	Total Primary Production (metric tons of carbon/year)
Upwellings	640	0.36×10^6	0.1	0.23×10^9
Coasts	160	54×10^6	15.0	8.6×10^9
Open oceans	130	307×10^6	85.0	39.9×10^9

Source: Smith, S. and Hollibaugh, J., *Rev. Geophys.,* 31, 75, 1993. With permission.

Table 5.2–2 Gross Primary Production: Marine and Terrestrial Ecosystems

Ocean Area	Range (g C/m²/year)	Average (g C/m²/year)	Land Area	Amount (g C/m²/year)
Open ocean	50–160	130 ± 35	Deserts, grasslands	50
Coastal ocean	100–250	160 ± 40	Forests, common crops, pastures	25–150
Upwelling zones	200–500	300 ± 100	Rain forests, moist crops, intensive agriculture	150–500
Upwelling zones	300–800	640 ± 150	Sugarcane and sorghum	500–1500
Saltmarshes	1000–4000	2471		

Source: Smith, S. and Hollibaugh, J., *Rev. Geophys.,* 31, 75, 1993. With permission.

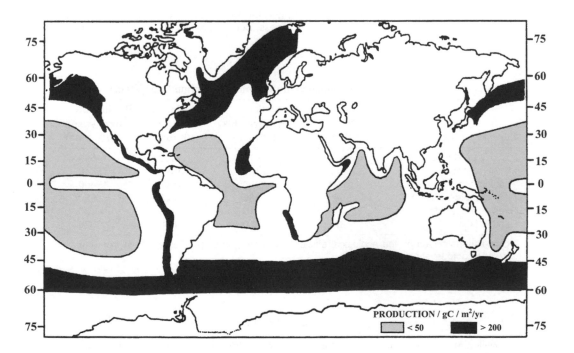

Figure 5.2–1 Geographic distribution of primary productivity in the oceans. (From Millero, F. J., *Chemical Oceanography,* 2nd ed., CRC Press, Boca Raton, FL, 1998, 317. With permission.)

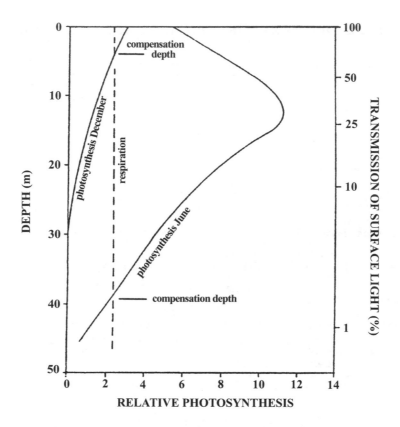

Figure 5.2–2 Integrated daily rate of primary productivity in the sea as a function of depth. (From Millero, F. J., *Chemical Oceanography,* 2nd ed., CRC Press, Boca Raton, FL, 1998, 317. With permission.)

Table 5.2–3   Annual Production Estimates in the Antarctic Marine Ecosystem[a]

Species Group	Minimum Production Carbon (10^8 tons)	Minimum Production Total Weight[b] (10^6 tons)	Maximum Production Carbon (10^6 tons)	Maximum Production Total Weight[b] (10^6 tons)
Primary production	661	6610	4450	44500
Microbial loop	215	2150	1584	15840
Zooplankton	209	2090	1540	15400
Fish + squid	58	580	428	4280
Birds + mammals	1	10	8	80
Benthic flux	82	820	602	6020

[a] Calculations based on the ecological efficiency of energy transfer (Table 5.2–4) from the next lower trophic level as described.[1]
[b] Assumed relation between carbon and total weight is 1:10.[2]

Source: Berkman, P. A., *Rev. Aquat. Sci.,* 6, 306, 1992. With permission.

REFERENCES

1. Huntley, M. E., Lopez, M. D. G., and Karl, D. M., Top predators in the Southern Ocean: a major leak in the biological carbon pump, *Science,* 253, 64, 1991.
2. Ryther, J. H., Photosynthesis and fish production in the sea, *Science,* 166, 72, 1969.

Table 5.2–4   Ecological Efficiencies in the Antarctic Marine Ecosystem

Species Group	Relative Ingestion (*I*)	Relative Growth (*G*)	Ecological Efficiency[a] (G_n/I_n)
Microbial loop (M)			
Phytoplankton (P)	0.125		
Total M	0.125	0.044	0.352
Zooplankton (Z)			
P	0.875		
M[b]	0.075		
Total Z	0.950	0.325	0.342
Fish + squid (F)			
Z	0.234		
Total F	0.234	0.065	0.278
Bird + mammals (B)			
Z	0.091		
F	0.065		
Total B	0.156	0.003	0.019
Organic flux to the benthos (f)[c]			
Z			0.285
F			0.023
B			0.034
Total f			0.342

Note: Mean ingestion (I) and growth (G) data calculated relative to one unit of fixed carbon.[1]

[a] Values correspond to gross growth efficiencies in Reference 1.
[b] Microbial growth plus egestion were ingested by the zooplankton.
[c] Based on estimates of egestion from Z, F, and B.

Source: Berkman, P. A., *Rev. Aquat. Sci.,* 6, 306, 1992. With permission.

REFERENCES

1. Huntley, M. E., Lopez, M. D. G., and Karl, D. M., Top predators in the Southern Ocean: a major leak in the biological carbon pump, *Science,* 253, 64, 1991.
2. Ryther, J. H., Photosynthesis and fish production in the sea, *Science,* 166, 72, 1969.

Table 5.2–5   Categorization of the World's Continental Shelves Based on Location, Major Rivers, and Primary Productivity

Latitude (°)	Region	Major River	Primary Production (g C m²/year)
Eastern Boundary Currents			
0–30	Ecuador–Chile	—	1000–2000
	Southwest Africa	—	1000–2000
	Northwest Africa	—	200–500
	Baja California	—	600
	Somali coast	Juba	175
	Arabian Sea	Indus	200
30–60	California–Washington	Columbia	150–200
	Portugal–Morocco	Tagus	60–290
Western Boundary Currents			
0–30	Brazil	Amazon	90
	Gulf of Guinea	Congo	130
	Oman/Persian Gulfs	Tigris	80

Table 5.2–5   Categorization of the World's Continental Shelves Based on Location,
               Major Rivers, and Primary Productivity (continued)

Latitude (°)	Region	Major River	Primary Production (g C/m²/year)
	Bay of Bengal	Ganges	110
	Andaman Sea	Irrawaddy	50
	Java/Banda Seas	Brantas	110
	Timor Sea	Fitzroy	100
	Coral Sea	Fly	20–175
	Arafura Sea	Mitchell	150
	Red Sea	Awash	35
	Mozambique Channel	Zambesi	100–150
	South China Sea	Mekong	215–317
	Caribbean Sea	Orinoco	66–139
	Central America	Magdalena	180
	West Florida shelf	Appalachicola	30
	South Atlantic Bight	Altamaha	130–350

Mesotrophic Systems

Latitude (°)	Region	Major River	Primary Production (g C/m²/year)
30–60	Australian Bight	Murray	50–70
	New Zealand	Waikato	115
	Argentina–Uruguay	Parana	70
	Southern Chile	Valdivia	90
	Southern Mediterranean	Nile	30–45
	Gulf of Alaska	Fraser	50
	Nova Scotia–Maine	St. Lawrence	130
	Labrador Sea	Churchill	24–100
	Okhotsk Sea	Amur	75
	Bering Sea	Kuskokwim	170

Phototrophic Systems

Latitude (°)	Region	Major River	Primary Production (g C/m²/year)
60–90	Beaufort Sea	Mackenzie	10–20
	Chukchi Sea	Yukon	40–180
	East Siberian Sea	Kolyma	70
	Laptev Sea	Lena	70
	Kara Sea	Ob	70
	Barents Sea	Pechora	25–96
	Greenland–Norwegian Seas	Tjorsa	40–60
	Weddell–Ross Seas	—	12–86

Eutrophic Systems

Latitude (°)	Region	Major River	Primary Production (g C/m²/year)
30–60	Mid-Atlantic Bight	Hudson	300–380
	Baltic Sea	Vistula	75–150
	East China Sea	Yangtze	170
	Sea of Japan	Ishikari	100–200
	North–Irish Sea	Rhine	100–250
	Adriatic Sea	Po	68–85
	Caspian Sea	Volga	100
	Black Sea	Danube	50–150
	Bay of Biscay	Loire	120
	Louisiana/Texas shelf	Mississippi	100

Source: Alongi, D. M., *Coastal Ecosystem Processes,* CRC Press, Boca Raton, FL, 1998, 256. With permission.

Table 5.2–6 Comparison of Production (g C/m²/Year) in Various Regions of the North American Shelf and the North Sea

	NS	GM	GB	MA	North Sea
Primary production	102–128	162–364	450	150–200	103
Zooplankton	19.3	20.5	20	13.6	12.8
Macrobenthos	8.1	9.7	9.2	17.9	2–5
Fish	3.4	3.2	6.8	3.2	1.4

Note: NS = Nova Scotia, GM = Gulf of Maine, GB = Georges Bank, MA = Mid-Atlantic Bight.

Source: Alongi, D. M., *Coastal Ecosystem Processes,* CRC Press, Boca Raton, FL, 1998, 270. With permission.

Table 5.2–7 Nitrogen Soures and Resulting Rates of Primary Production in the Gulf of Maine

Nitrogen Source	Primary Production (g C/m²/year)
New nitrogen	
Winter convective overturn	25.2
Eastern Gulf plume	36.6
Vertical eddy diffusion	32.2–108
Upwelling	
Coastal Maine	
Estuarine	8.0
Eckman	8.0
Southwest Nova Scotia	36.6
Recycled nitrogen	16–110
Total primary production	162–364

Source: From Townsend, D. W., *Rev. Aquat. Sci.,* 5, 211, 1991. With permission.

Table 5.2–8 Comparison of Primary Production for Coastal Ecosystems

	g C/m²/year
Coastal waters	
Ocean waters	5–50
Upwelling zones	50–220
Shallow shelf	30–150
Coastal bays	50–120
Surf zone	20–30
Subtidal	
Seaweeds	800–1500
Coral reefs	1700–2500
Seagrasses	120–350
Intertidal	
Rockweeds	100–250
Mollusks	10
Sandy beaches	10–30
Estuarine flats	500–750
Commercial oyster beds	400

Table 5.2–8   Comparison of Primary Production for
Coastal Ecosystems (continued)

Supratidal	
Salt marshes (temperate)	700–1300
Salt marshes (Arctic)	100–150
Mangals	350–1200
Sand dunes (fore)	400–500
Sand dunes (rear)	150–175

Note: All figures relate to *net* production (gross production
minus respiratory losses).

Source: Carter, R. W. G., *Coastal Environments: An Introduction
to the Physical, Ecological, and Cultural Systems of Coastlines,*
Academic Press, London, 1988, 21. With permission.

Table 5.2–9   Primary Production of Estuarine Habitats (values expressed in g C/m²/year
or g dry wt/m²/year)

Plant Type	Location	g C/m²/year	g dry wt/m²/year
Phytoplankton	Baltic Sea	48–94	
	St. Margaret's Bay, Canada	190	
	Cochin Backwater, India	124	
	Ems Estuary, Netherlands	13–55	
	Grevelingen, Netherlands	130	
	Wadden Sea, Netherlands	100–200	
	Loch Etive, Scotland	70	
	Lynher, U.K.	81.7	
	Alewife Cove, U.S.	162	
	Barataria Bay, U.S.	210	
	Beaufort, U.S.	52.5	
	Bissel Cove, U.S.	56	
	Charlestown River, U.S.	42	
	Core Sound, U.S.	67	
	Duplin River, U.S.	248	
	Flax Pond, U.S.	60	
	Hempstead Bay, U.S.	177	
	Jordan Cove, U.S.	66	
	Long Island Sound, U.S.	205	
		308	
	Narragansett Bay, U.S.	242	
	Niantic River, U.S.	72	
	North Inlet, U.S.	346	
Microbenthic algae	Danish fjords	116	
	Wadden Sea, Netherlands	115–178	
		101 ± 39	
	Grevelingen, Netherlands	25–37	
	Ythan Estuary, U.K.	31	
	Lynher, U.K.	143	
	Alewife Cove, U.S.	45	
	Barataria Bay, U.S.	240	
	Bissel Cove, U.S.	52	
	Charlestown River, U.S.	41	
	Delaware, U.S.	160	
	False Bay, U.S.	143–226	
	Flax Pond, U.S.	52	
	Hempstead Bay, U.S.	62	
	Jordan Cove, U.S.	41	
	Niantic River, U.S.	32	
	Sapelo Island, U.S.	180	

Table 5.2–9   Primary Production of Estuarine Habitats (values expressed in g C/m²/year or g dry wt/m²/year) (continued)

Plant Type	Location	g C/m²/year	g dry wt/m²/year
Seagrasses			
Halodule wrightii	North Carolina, U.S. (intertidal)	70–240	
	South Florida—Gulf Coast, U.S. (intertidal)	70–240	
Thalassia testudinum	South Florida—Gulf Coast—Texas, U.S. (subtidal)	580–900	
Zostera marina	Denmark		
	Leaves[a]		856
	Roots, rhizomes[b]		241
	Alaska, U.S.	19–552	
	Beaufort, U.S.	350	
	North Carolina, U.S. (intertidal)	330	
	Little Egg Harbor, U.S.		466
	Pacific Coast, U.S. (subtidal)	90–540	
	Puget Sound, U.S.	58–330	
		58–1500	116–680
			10–1200
Saltmarsh grasses			
Carex spp.	Arctic Waters	10–120[a]	
Distichlis spicata	Atlantic Coast, U.S.		1070–3400[b]
	Gulf Coast, U.S. (Louisiana)	1600[a]	
	Pacific Coast, U.S.	300–600[a]	
Juncus gerardi	Maine, U.S.		485[a]
	Atlantic Coast, U.S.		1620–4290[b]
Juncus roemerianus	North Carolina, U.S.		754[a]
	Georgia, U.S.		2200[a]
	Gulf Coast, U.S. (Louisiana)	1700[a]	
	Gulf Coast, U.S.		1360–7600[b]
Puccinellia phrygondes	Arctic waters	25–70[a]	
Salicornia virginica	Atlantic Coast, U.S.		430–1430[b]
Salicornia spp.	Norfolk, U.K.		867[a]
	Pacific Coast, U.S.	325–1000[a]	
Spartina alterniflora	Massachusetts, U.S.		3500[b]
	Atlantic Coast, U.S.	200–800[a]	
	Atlantic Coast, U.S.	220–1680[a]	
	Atlantic Coast, U.S.		220–3500[b]
	Atlantic and Gulf Coasts, U.S.	200–2000[a]	
	Gulf Coast, U.S. (Louisiana)	1300[a]	279–6000[a]
Spartina foliosa	Pacific Coast, U.S.	400–850[a]	
Spartina patens	Maine, U.S.		5163[a]
	Atlantic Coast, U.S.		310–3270[b]
	Atlantic and Gulf Coasts, U.S.	500–700[a]	
	Gulf Coast, U.S. (Louisiana)	3000[a]	
Mixed species	Netherlands	100–500[a]	200–1000[a]
	Europe		11–1100[a]
	Barataria Bay, U.S.	590[a]	1175[a]
	Gulf Coast, U.S.	500[a]	
	North America	100–1700[b]	
	North America		300–4000[a]
Rhizophora mangle	South Florida, U.S.	400	
		300	
			100
			470–730
			(leaf fall)

[a] Above ground.
[b] Below ground.

Source: Kennish, M., *Ecology of Estuaries,* Vol. 1, CRC Press, Boca Raton, FL, 1986, 163. With permission.

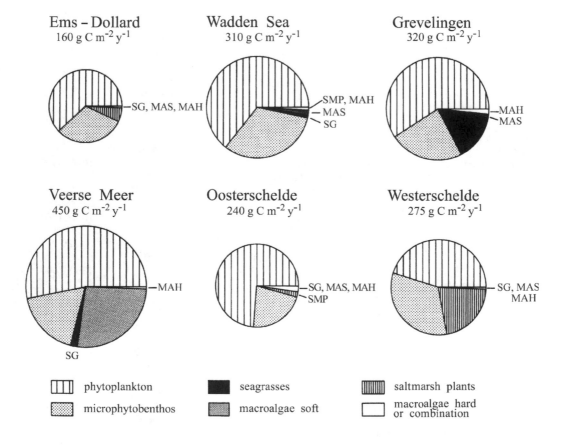

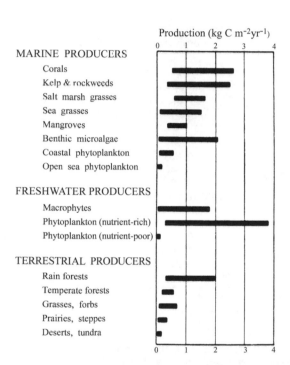

Figure 5.2–4 Annual net production rates of marine, freshwater, and terrestrial producers. (From Valiela, I., *Marine Ecological Processes,* 2nd ed., Springer-Verlag, New York, 1995. With permission.)

Table 5.2–10   Percentage of Primary Production Consumed by Herbivores in Marine and Terrestrial Environments[a]

Coastal Environments	Percentage of Production Eaten by Herbivores	Number of Trophic Steps	Ref.
Vascular plants			
Eelgrass, North Sea	4	3	Nienhuis and van Ierland (1978)
Salt marsh, Georgia	4.6	3–4	Teal (1962)
Salt marsh, North Carolina	58		Smith and Odum (1981)
Mangrove swamp, Florida	9–27[b]		Onuf et al. (1977)
Phytoplankton			
Long Island Sound	73[c]	4	Riley (1956)
Narragansett Bay	0–30[d]	4	Martin (1970)
Cochin Backwater, India	10–40		Qasim and Odum (1981)
Beaufort Sound	1.9–8.9		Williams et al. (1968)
Offshore California	7–52 (average 23)		Beers and Stewart (1971)
Peruvian upwelling	92, 54–61	3	Walsh (1975), Whiteledge (1978)
Open seas (all phytoplankton)			
Georges Bank	50–54	4	Riley (1963), Cohen et al. (1981)
North Sea	75–80	4–6	Crisp (1975)
Sargasso Sea	100	5	Menzel and Ryther (1971)
Eastern Tropical Pacific	39–140 (average 70)[e]	5	Beers and Stewart (1971)

Note: See original source for reference citations.

[a] Annual consumption except where indicated otherwise. These values are rough, but best possible estimates based on many assumptions and extrapolations.

[b] Leaves and buds only.

[c] This is an estimate of consumption of organic matter in the water column. Larger zooplankton consume about 20%, microzooplankton and bacteria an additional 43%. In the bottom, benthic animals use an estimated 31% of net primary production.

[d] Of standing stock of algae.

[e] Includes only microzooplankton that passed through a 202-μm mesh. The biomass of these small species was about 24% of that of the larger zooplankton. Total consumption could easily be larger than reported if any of the larger species are herbivorous.

Source: Valiela, I., *Marine Ecological Processes,* Springer-Verlag, New York, 1984. With permission.

Table 5.2–11 Estimates of Zooplankton Production

Organism or Group	Area and Period	Production mg C/m³/day	Daily Production Ratios x/y Standing Crop	Production Net Primary Production Ratios x/y
Calanus finmarchicus	E. Barents Sea; year	7.8	0.002	0.03
C. cristatus	N.W. Pacific; summer	5.6	0.012	—
C. plumchrus	N.W. Pacific; summer	4.6	0.010	—
Eucalanus bungii	N.W. Pacific; summer	3.5	0.014	—
C. glacialis	N. Bering Sea; year	0.7	—	0.005
C. plumchrus	W. Bering Sea; year	3.1	—	0.012
C. cristatus	W. Bering Sea; year	3.8	—	0.015
E. bungii	W. Bering Sea; year	7.3	—	0.03
Diaptomus salinus	Aral Sea; year	0.66	0.007	—
Acartia clausi	Black Sea; year	0.38	0.035	0.001
Centropages kröyeri	Black Sea; summer	0.19	0.077	0.0002
Euphasia pacifica	N.E. Pacific; year	0.9	0.008	0.0048
Acartia tonsa	Chesapeake Bay estuary; summer	77	0.50	0.05
A. clausi	Black Sea bay; June	15	0.17	—
A. clausi	Black Sea, open sea; June	6.6	0.23	0.08
Calanus helgolandicus	Black Sea, open sea; June	28	0.15	0.07
Zooplankton	Georges Bank; year	200	0.03	0.25
Zooplankton	English Channel; year	75	0.10	0.30
Zooplankton	Long Island Sound; year	166	0.17	0.30
Zooplankton	N. North Sea; April–Sept.	180	0.048	0.58
Herbivorous copepods	North Sea; Jan.–June	4.9	0.08	0.14
Copepods (mainly Calanus)	North Sea; March–June	46[a]	0.10[a]	0.20[a]
Zooplankton	Gulf of Panama; Jan.–April	70 or 234	0.29 or 0.98	0.09 or 0.31

[a] Author's calculation.

Source: Mullin, M., *Oceanogr. Mar. Biol. Annu. Rev., 7,* 308, 1970. With permission by George Allen and Unwin Ltd., London.

Table 5.2–12   Bacterial Production in Sediment (mg C/m/day) to 1 cm Depth,
                as Measured by Tritiated Thymidine Incorporation

Production	Total Measured Depth	Habitat Description
2.7	1	Freshwater river sand
3.7	15	Beach sand
2.5	0.4	Beach sand
40.5	25	Nearshore sand
16.5	25	Offshore sand
800	2	Mangrove intertidal
200–700	1	Mangrove creek bank
230	1	Mangrove creek bank
300–925	0.4	Coral reef sand
20.1	20	Salt marsh
450	0.5	Salt marsh
85	2	Seagrass bed
40	0.3	Seagrass bed
60	0.5	Seagrass bed
1200	1	Saline pond
250–500	1	Aquaculture ponds

Source: Kemp, P. F., *Rev. Aquat. Sci.,* 2, 109, 1990. With permission.

5.3 BACTERIA AND PROTOZOA

Table 5.3–1 Bacterial Abundance in Some Intertidal and Subtidal Sediments[a]

Location/Site Description	No. Samples	Depth (cm)	Cell Numbers × 10^9		Reference
			per g dry weight	per cm^3	
Marsh Sediments					
Newport River Estuary, NC; *Spartina alterniflora* marsh, yearly mean	13	0–1	13.8 (8.5–22.0)	8.5 (5.3–13.7)	Source
	13	5–6	5.7 (3.2–7.7)	5.4 (3.3–8.1)	
	13	10–11	2.9 (1.8–4.4)	3.0(1.7–4.2)	
	13	20–21	1.8 (1.1–30)	1.9 (1.0–2.9)	
Newport River Estuary, NC; transect across *Spartina* marsh	4	0–1	19.7 (12.0–34.1)	9.4 (8.4–10.9)	Rublee and Dornseif (1978)
	4	5–6	7.2 (4.6–9.8)	6.1 (5.1–7.1)	
	4	10–11	3.2 (2.3–5.0)	3.5 (2.4–4.4)	
	4	20–21	2.1 (1.6–2.3)	2.3 (2.2–2.6)	
Rhode River Estuary, MD; *Typha angustifolia/Scirpus* spp. low marsh	3	0–1	66.3 (57.2–84.6)	16.0 (13.8–20.4)	Source
	3	5–6	35.6 (24.2–48.0)	9.7 (6.6–13.1)	
	3	10–11	29.9 (27.9–31.9)	3.0 (7.6–8.1)	
	2	20–21	8.8 (7.4–10.3)	3.0 (2.5–3.5)	
Rhode River Estuary, MD; three high marsh sites, *Spartina cynusoroides, S. patens, Distichlis spicata, Scirpus* spp., *Hibiscus* spp.	3	0–1	60.1 (37.5–76.6)	12.0 (11.3–12.4)	Source
	3	5–6	51.2 (36.9–69.1)	9.5 (7.2–11.0)	
	3	10–11	37.0 (30.8–43.2)	6.1 (4.3–9.3)	
	3	20–21	23.8 (19.9–25.9)	3.9 (2.8–5.1)	
	1	30–31	31.7	3.6	
Great Sippewissett marsh, MA; tall *Spartina alterniflora*, annual mean	12	0–1	38.0 (10.0–64.0)		J. E. Hobbie and J. Helfrich (personal communication)
	12	2–3	28.0		
	12	5–6	35.0		
	12	9–10	42.0		
	12	19–20	24.0		
	12	29–30	16.0		
Great Sippewissett marsh, MA; short *S. alterniflora*, annual mean	12	0–1	54.0 (24.0–80.0)		J. E. Hobbie and J. Helfrich (personal communication)
	12	2–3	29.0		
	12	5–6	29.0		
	12	9–10	28.0		
	12	19–20	18.0		
	12	29–30	12.0		

(continued)

Table 5.3–1 Bacterial Abundance in Some Intertidal and Subtidal Sediments[a] (continued)

Location/Site Description	No. Samples	Depth (cm)	Cell Numbers × 10^9 per g dry weight	per cm^3	Reference
Great Sippewissett marsh, MA; High marsh, *S. patens*, annual mean	12	0–1	49.0		J. E. Hobbie and J. Helfrich (personal communication)
	12	2–3		(16.0–120.0)	
	12	5–6	35.0		
	12	9–10	30.0		
	12	19–20	29.0		
	12	29–30	18.0		
			19.0		
Intertidal/Subtidal Mudflats and Sand Flats					
Sapelo Island, GA; intertidal sand flat adjacent to *Spartina* marsh, winter samples	3	0–1		1.0 (0.6–1.2)	S. Y. Newell (personal communication)
Newport River Estuary, NC; subtidal sand, June–November	18	0–1	3.3 (1.6–5.9)	2.7 (1.2–4.8)	Shelton (1979)
	18	5–6	2.4 (1.0–5.9)	1.9 (0.8–4.8)	
Newport River Estuary, NC; subtidal mud, June–November—"sulfuretum"	18	0–1	8.8 (3.1–22.9)	4.2 (2.0–8.3)	Shelton (1979)
	18	5–6	5.8 (1.3–20.5)	2.7 (1.1–7.4)	
Rhode River, MD; subtidal mudflat	3	0–1	17.4 (9.6–26.8)	9.1 (5.0–14.0)	Source
	2	2–3	16.0 (11.7–19.7)	8.4 (6.1–10.3)	
	3	5–6	13.0 (11.8–14.2)	8.7 (7.9–9.5)	
	3	10–11	8.0 (6.7–9.1)	6.2 (5.2–7.1)	
	2	20–21	4.7 (3.3–6.2)	3.6 (2.5–4.7)	
Lowe's Cove Damariscotta Estuary, ME; March–August	6	sfc	2.8 (1.2–4.7)		M. DeFlaun (personal communication)
Halifax, Nova Scotia; high intertidal mudflats, April–November	13	sfc	6.0 (2.2–8.6)	3.5 (2.2–4.8)	L. M. Cammen (personal communication)
	13	1	5.7 (3.7–7.9)	4.5 (3.5–5.5)	
	13	5	3.5 (2.3–5.2)	3.6 (2.9–4.3)	
Halifax, Nova Scotia; low intertidal mudflat, April–November	13	sfc	4.2 (1.5–7.0)	2.8 (1.4–4.3)	L. M. Cammen (personal communication)
	13	1	4.0 (2.8–5.2)	3.5 (2.1–4.1)	
	13	5	4.0 (2.7–5.2)	3.8 (2.2–5.1)	
Petpeswick Inlet, Nova Scotia, seven intertidal mud- and sand flat stations, May–September	17	sfc	2.8 (0.3–10.0)		Dale (1974)
	2	1	2.9 (2.1–3.6)		
	1	2	6.3		
	16	5	2.5 (0.1–7.0)		
	14	10	1.8 (0.1–4.7)		

Note: See original source for reference citations.

[a] Values given as mean (and range) of observations; sfc = surface.

Source: Rublee, P. A., in *Estuarine Comparisons*, Kennedy, V. S., Ed., Academic Press, New York, 1982, 159. With permission.

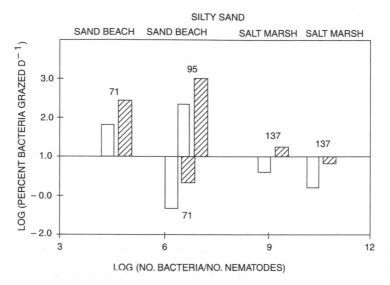

Figure 5.3–1 Predicted percent of bacterial abundance grazed by nematodes per day, assuming all nematodes were bacterivorous. Open bars are based on minimum grazing rates (0.07 μg C per individual per day), filled bars on maximum grazing rates (0.3 μg C per individual per day). Adjacent numbers are references from which nematode and bacterial abundances were taken. (From Kemp, P. F., *Rev. Aquat. Sci.*, 2, 109, 1990. With permission.)

Table 5.3–2 Direct Counts of Bacteria in Some Tropical Mangrove and Coral Reef Sediments

Location	Habitat and Season	Depth (cm)	Cell Numbers × 10⁹
Kaneohe Bay, Hawaii	Fine coral sand	0–1	3.7(2.4–5.0)
	Coral rubble	0–1	0.5(0.3–0.7)
	Coarse coral sand	0–1	0.5(0.3–0.7)
Majuro atoll, Marshall Islands	Lagoon sands	0–1	2.2(0.8–4.3)
	Dead coral	Surface	3.9(1.9–7.2)
Heron Island, GBR, Australia	Reef flat, summer	0–1	2.8(2.3–3.2)
Lizard Island, GBR, Australia	Reef flat, summer	0–0.5	0.5(0.36–0.60)[a]
	Reef flat, winter	0–0.5	0.72
Hamilton Harbor, Bermuda	Carbonate silt	0–1	2.6[b]
		3	2.0
			1.5
		7	1.3
		9	1.4
		11	1.1
Davies Reef, Australia	Reef front, winter	0–1	0.9(0.3–1.5)
	Reef flat, winter	0–1	1.5(1.4–1.6)
	Shallow lagoon, winter	0–1	1.2(0.7–1.9)
	Shallow lagoon, summer	0–1	0.6(0.2–1.0)[c]
	Deep lagoon, winter	0–1	1.7(1.2–2.4)
	Deep lagoon, summer	0–1	1.0(0.7–2.1)[c]
Umtata, Southern Africa	Mangrove mud, winter	0–1	0.9(0.8–1.0)
Cienaga Grande, Colombia	Tropical lagoon, mangrove muds, winter	Surface	0.5(0.4–1.3)[d]
Hinchinbrook Island, Australia	Low intertidal mangroves	0–0.1	1.30 (±70)[e]
Bowling Green Bay, Australia	Mangrove creek bank	0–1	2.4(2.33–2.46)
Cape York peninsula, Australia	Mangroves, winter	0–2	150(11–359)
	Mangroves, summer	0–2	77(20–344)
Hinchinbrook Island, Australia	Mangroves, autumn	0–2	23(20–25)
	Mangroves, winter	0–2	153(51–225)
	Mangroves, spring	0–2	170(105–297)
	Mangroves, summer	0–2	163(125–225)

Table 5.3–2   Direct Counts of Bacteria in Some Tropical Mangrove and Coral Reef Sediments (continued)

Location	Habitat and Season	Depth (cm)	Cell Numbers × 10⁹
	Low intertidal mangroves, autumn	0–2	16.4(8.7–24.0)
		2–4	29.8(12.9–46.7)
		4–6	7.7(5.7–9.6)
		6–8	7.0(5.9–8.1)
		8–10	3.1(2.3–3.8)
		0–2	16.3(14.6–18.0)
		2–4	10.5 (10.2–10.7)
		4–6	2.6(2.4–2.7)
		6–8	3.0(2.3–3.6)
		8–10	4.6(4.1–5.0)
Chunda Bay, Australia	Low intertidal mangroves	0–5	11.3(0.7–34.3)
	High intertidal mangroves	0–5	8.1(0.1–30.1)

Note: Values are given as mean (and range) of cell numbers per gram weight of sediment.

[a] Estimated from biomass measurements using 20 fg cell⁻¹.

[b] Mean values only, as estimated from graphs in Figure 2 in Hines, M. E., *Proc. 5th Intl. Coral Reef Congr.,* Vol. 3, Moorea, French Polynesia, 1985, 427.

[c] Unpublished data (Alongi).

[d] Converted from cells ml⁻¹ using dry wt conversion; standard deviation in parentheses.

[e] Converted from fatty acids assuming 10 fg carbon cell⁻¹.

Source: Alongi, D. M., *Rev. Aquat. Sci.,* 1, 243, 1989. With permission.

Table 5.3–3   Some Ingestion Rates (bacteria cells/hour) of Ciliates and Flagellates

Organism	Ingestion Rate	Method Used
Benthic ciliate sp.	180–600	Emptying of vacuoles
Benthic ciliates	59–410	FL beads, bacteria
	128	FL bacteria
	414–1881	Decreasing numbers
Pelagic ciliate sp.	100–800	Latex beads
	4.3–237 × 10³	FL beads
Pelagic ciliates	33–76	FL beads
	60–120	FL bacteria
	731–2150	FL bacteria
Benthic flagellates	9–12	Decreasing numbers
Pelagic flagellates	9–35	FL bacteria
	23,117	Decreasing numbers
	0–144	Decreasing numbers
	50–210	Decreasing numbers
	27–254	Decreasing numbers
	10–75	Decreasing numbers
	20–80	Decreasing numbers

Note: Flagellate ingestion rates based on microbead ingestion are not included as many are poten-tially biased by preservation artifacts. FL = fluorescently labeled.

Source: Kemp, P. F., *Rev. Aquat. Sci.,* 2, 109, 1990. With permission.

Table 5.3–4   Protozoan Abundances in Sediment Where
Abundance Was Reported per Unit Volume

Number/ml		
Ciliates	Flagellates	Habitat
6–50	75	Mangrove
120	90	Beach sand
4.6–49	19–43	Subtidal
9	700	Lagoon
7–260	6–180	Mangrove
3340	18500	River mud
~50	10^6	Tundra pond
187–637[a]		Sheltered beach
448–1056	6880[c]	Salt marsh
7616	1.2×10^6	Saline pond
300–6200[b]		Sand beach

[a] Reported as cm^{-2}; most ciliates were at 0–1 cm depth.
[b] Total protozoa reported were "mostly ciliates."
[c] Kemp, unpublished data.
Source: Kemp, P. F., Rev. Aquat. Sci., 2, 109, 1990. With permission.

Table 5.3–5   Estimates of Free-Living Protozoans (Excluding Foraminifera) in Some Tropical
Mangrove and Coral Reef Sediments

Location	Habitat and Season	Depth (cm)	Cell Numbers $\times 10^5$
Kaneohe Bay, Hawaii	Reef rubble	0–1	24.0 (14–37)[a]
Isla Providencia, Colombia	Reef sand	0–1	9.0–23.0[a]
Moorea Island, French Polynesia	Trahura reef	0–3	0.2(0–2)[a]
Umtata, South Africa	Mngazana River mangrove muds		
	High intertidal sites	0–30	3.0 (0.6–7.8)[a]
	Mid intertidal sites	0–15	2.0 (0.3–6.2)[a]
	Low intertidal sites	0–15	0.30(0.1–0.5)[a]
Davies Reef, Great Barrier Reef	Shallow lagoon, autumn	0–2	Ciliates: 1.8(±1.6) Flagellates: 14.0(±4.0)[b]
	Reef front, winter	0–5	Ciliates: 8.0(3.0–13.1) Flagellates: 97.1(44.1–147.2)
	Reef flat, winter	0–5	Ciliates: 2.0(2.0–4.0) Flagellates: 13.1(3.2–21.0)
	Shallow lagoon, winter	0–5	Ciliates: 3.1(1.1–4.0) Flagellates: 94.0(5.0–131.3)
	Deep lagoon, winter	0–5	Ciliates: 3.3(3.0–4.4) Flagellates: 151.3(100.0–202.4)
Cape York peninsula, Qld., Australia	Mangroves, winter	0–5	Ciliates: 3.8(0.6–11.0) Flagellates: 3.4(0.6–10.0)
	Mangroves, summer	0–5	Ciliates: 15.0(2.8–26.0) Flagellates:11.4(6.4–16.0)
Hinchinbrook Island, Australia	Mangroves, autumn	0–5	Ciliates: 2.3(2.1–24) Flagellates:15.5(12.3–19.0)
	Mangroves, winter	0–5	Ciliates: 2.3(1.8–3.0) Flagellates: 3.2(3.1–3.3)
	Mangroves, spring	0–5	Ciliates: 7.5(6.2–10.4) Flagellates: 4.9(4.7–5.2)
	Mangroves, summer	0–5	Ciliates: 4.0(2.1–5.7) Flagellates: 2.3(2.0–3.1)

Table 5.3–5   Estimates of Free-Living Protozoans (Excluding Foraminifera) in Some Tropical
Mangrove and Coral Reef Sediments (continued)

Location	Habitat and Season	Depth (cm)	Cell Numbers $\times$ 10⁵
Chunda Bay, Qld., Australia	Low intertidal mangroves	0–5	Ciliates: 11.9(0.1–32.3) Flagellates: 10.3(0.2–28.7)
	High intertidal mangroves	0–5	Ciliates: 5.2(0.0–15.5) Flagellates: 7.0(0.0–30.2)

Note: Values are given as mean and range of cell numbers per square meter.

[a] Ciliates enumerated only.

[b] Standard deviations in parentheses.

Source: Alongi, D. M., *Rev. Aquat. Sci.,* 1, 250, 1989. With permission.

Table 5.3–6   Relative Biomasses of Bacteria, Macrofauna, Meiofauna, and Protozoa in g C/m² (using
the conversion g C = 50% g dry weight where necessary)

Bacteria						
To 1 cm	Total	Macrofauna	Meiofauna	Protozoa	Habitat Description	Ref.
0.06					Intertidal sand	136
	0.04	0.2	0.5	0.12	Exposed beach sand	21
	0.04	0.9[a]	0.8	0.17	Exposed beach sand	21
	2.5	5.0	0.5		Subtidal sand	95
0.036	0.54	1.25			Exposed beach sand	52
0.6	1.2			0.002[b]	Salt marsh	64
0.8	1.6			0.02[b]	Saline pond	64
	5.25	1.9	1.6		Sand beach	71
	18.0	3.2	2.2	0.2	Exposed sand beach	72
	3.6	0	0.3	0.08	Very exposed beach	72
0.14	0.28		0.073		Intertidal sand	89
1.2–2.3			0.7–0.8		Salt marsh	57
31.84			1.24		Salt marsh	137
0.025	0.067				Subtidal, oil seep	90
0.013	0.035				Subtidal, normal	90
	2.8		1.0		Coral sand	36
	1.03	6.6	0.05		Cont. shelf sand	74
	0.93	0.6	0.036		Cont. slope	74
	0.12	0.014	0.0033		Abyssal plain	74
1.5	14.0				Salt marsh	24
	0.52	1.5	1.2		Intertidal mud	138

Note:

[a] Excluding filter-feeding bivalves.

[b] Protozoan biomass includes ciliates only.

Source: Kemp, P. F., *Rev. Aquat. Sci.,* 2, 109, 1990. With permission.

5.4 MARINE PLANKTON

Table 5.4–1   Characteristics of Dominant Planktonic Organisms in the Sea

Type	Typical Size (μm)	Skeletal Material	Where Dominant
Bacteria Producers (plants)	<5	None	Sediments and surfaces, sunlit surface waters only
Blue-green algae	5	None	Nowhere, but tend to grow on surfaces
Coccolithophores	3–10	Calcium carbonate	Warm open ocean (tropical and subtropical)
Silicoflagellates	5–40	Silica	Cool open ocean (polar and subpolar)
Diatoms	20–80	Silica	Cool, nutrient-rich (upwelling, polar, and coastal)
Dinoflagellates	10–50	Cellulose or none	Warm quiet waters, wherever the others are scarce
Consumers (animals)			
Protozoans			
Radiolarians	50–500	Silica	Surface waters and sediments
Foraminifera	100–1000	Calcium carbonate	Surface waters and sediments

Source: Stowe, K., *Exploring Ocean Science,* 2nd ed., John Wiley & Sons, New York, 1996, 293. With permission.

Table 5.4–2   Classification of Plankton

Size Category	Size Range
Megaplankton	20–200 cm
Macroplankton	2–20 cm
Mesoplankton	0.2–20 mm
Microplankton	20–200 μm
Nanoplankton	2–20 μm
Picoplankton	0.2–2 μm
Femtoplankton	0.02–0.2 μm

Source: Wotton, R. S., Ed., *The Biology of Particles in Aquatic Systems,* CRC Press, Boca Raton, FL, 1990, 4. With permission.

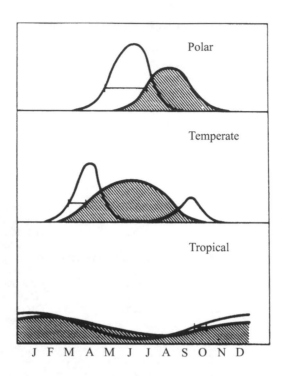

Figure 5.4–1 Seasonal cycles of phytoplankton in Arctic, temperate-boreal, and tropical waters. Stippled area delineates zooplankton seasonal cycles in these regions. The horizontal bar indicates the lag period between the increases of phytoplankon and zooplankton. (From Dring, M. J., *The Biology of Marine Plants,* Edward Arnold, London, 1982. With permission.)

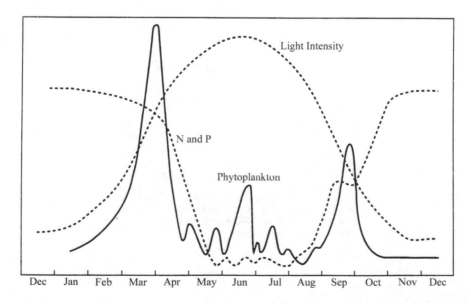

Figure 5.4–2 Seasonal variation of phytoplankton, nutrients, and light in a typical northern temperate sea. (From Millero, F. J., *Chemical Oceanography,* 2nd ed., CRC Press, Boca Raton, FL, 1996, 316. With permission.)

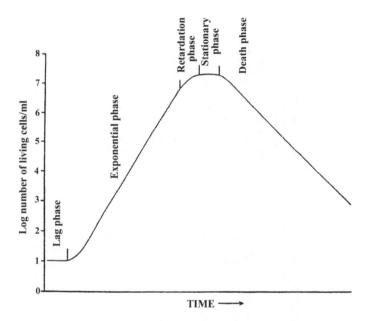

Figure 5.4–3 Idealized phytoplankton growth curve. (From Millero, F. J., *Chemical Oceanography,* 2nd ed., CRC Press, Boca Raton, FL, 1996, 315. With permission.)

Table 5.4–3   Amount of Plankton (cm³/1000 m³) at Various Depths and in Different Parts of the North Atlantic

Depth (m)	Gulf Stream 40–43°N	Continental Slope of North America 38–41°N	Gulf Stream 35–37°N	Sargasso Sea 20–37°N	North Equatorial Current 16–19°N	African Littoral 15–22°N	Canaries Current 27–36°N	32° 29′N 20° 0.9′W	32° 34′N 16° 19′W
0–50	74.5	199.0	54.9	64.7	89.6		78.6		
50–100	42.3	94.0	35.9	59.9	80.7	214.0	55.0		
100–200	26.2	35.1	23.3	32.1	30.6	94.7	27.0	26.0	28.0
200–500	25.4	30.8	11.1	10.2	15.9	59.9	15.6		
500–1,000	15.7	19.0	6.0	4.1	5.4	27.2	6.5		
1,000–2,000	5.0	7.8	2.8	1.2	1.6	—	2.1	5.0	4.0
2,000–3,000	—	—	2.1	0.6	1.0	—	0.8	1.0	0.7
3,000–4,000	—	—	—	0.3	0.3	—	0.2	0.6	—
4,000–5,000	—	—	—	0.1	—	—	—	—	—

Table 5.4–4 Changes in the Relative Numbers of Calanoida with Depth in Various Regions of the Ocean

| | Open Ocean | | | | Isolated Basin | | | |
Depth (m)	Northwestern Pacific Ocean (spring)	Tropical Zone Pacific Ocean	Tropical Zone Indian Ocean	Norwegian Sea (annual average)	Central North Polar Basin	Mediterranean Sea (Ionian Sea) Winter	Summer	Sea of Japan (winter)
0–50								
50–100	63.5	93.1	—	29.2	92.5	91.5	53.0	58.4
100–200	30.6	67.0	64.5	33.6	105	71.5	49.3	35.0
200–500	80.5	52.3	31.0	33.6	27.1	37.1	26.1	137.0
500–1,000	16.2	40.0	61.0	160	27.1	8.8	27.0	56.4
1,000–2,000	59.0	28.5	18.7	5.1	5.8	3.5	4.7	3.1

Note: Numbers in each layer are expressed in % of numbers in the overlying layer.

Table 5.4–5 Salinity Ranges of Various Copepod Species Observed in Nature

Copepod species	0	10	20	30	40	50	60	70	80
Acartia (Paracartia) africana				——					
A. (Paracartia) longipatella		————————							
A. (Arcartiella) natalensis		—————————————————							
Calanoides carinatus				—					
Centropages brachiatus				—					
C. chierchiae				—					
C. furcatus				–					
Clausidium sp.				–					
Clausocalanus furcatus				—					
Corycaeus spp.				–					
Ctenocalanus vanus				–					
Euterpina acutifrons				–					
Halicyclops spp.			—————————————————					
Harpacticus? gracilis				————————				
Hemicyclops sp.				————————————				
Nannocalanus minor				——					
Oithona brevicornis/nana		———————————————————————						
O. plumifera				—					
O. similis				——					
Paracalanus aculeatus				——					
P. crassirostris		————————							
P. parvus			———						
Porcellidium sp.				–					
Pseudodiaptomus stuhlmanni	————————————————————————————							
P. hessei	————————————————————————————							
P. nudus				–					
Rhincalanus nasutus				–					
Saphirella stages				———					
Temora turbinata				–					
Tortanus capensis				———					
Tegastes sp.				—					

Source: Grindley, J. R., *Estuarine Ecology: With Particular Reference to Southern Africa,* Day, J. H., Ed., A. A. Balkema, Rotterdam, 1981, 117. With permission.

Table 5.4–6 Assimilation of Phytoplankton by Zooplankton

Species	Food	Method	Assimilation (%)	Reference
Calanus finmarchicus	Various diatoms and flagellates	Tracer-isotope [^{32}P]	15–99	Marshall and Orr (1961)
	Skeletonema costatum	Tracer-isotope [^{14}C]	60–78	Marshall and Orr (1955b)
Temora longicornis	*S. costatum*	Tracer-isotope [^{32}P]	50–98	Berner (1962)
Euphausia pacifica	*Dunaniella primolecta*	Tracer-isotope [^{14}C]	85–99	Lasker (1960)
Ostrea edulis (larvae)	*Isochrysis galbana*	Tracer-isotope [^{32}P]	13–50	Walne (1965)
Calanus helgolandicus	Natural particulate matter	Chemical analyses	74–91	Corner (1961)
Metridia lucens	*T. nordenskioldii* *Ditylum* spp. *Artemia nauplii*	"Ratio method"	50–84	
			35	Haq (1967)
			59	
C. hyperboreus	*Exuviella* spp.	"Ratio method"	39.0–85.6	
		Chemical analyses	54.6–84.6	
Natural zooplankton	*Natural particulate material*	"Ratio method"	32.5–92.1	Conover (1966a)
C. finmarchicus	*Skeletonema costatum*	"Ratio method"	53.8–64.4	Corner et al. (1967)
	S. costatum	Chemical analyses	57.5–67.5	

Note: See original source for reference citations.

Source: Heinle, D. R., in *Functional Adaptations of Marine Organisms,* Vernberg, F. J. and Vernberg, W. B., Eds., Academic Press, New York, 1981, 85. With permission.

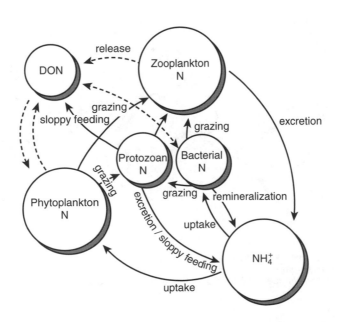

Figure 5.4–4 Idealized depiction of dissolved nitrogen cycling between phytoplankton, zooplankton, and members of the microbial loop. (From Alongi, D. M., *Coastal Ecosystem Processes,* CRC Press, Boca Raton, FL, 1998, 193. With permission.)

5.5 BENTHIC FLORA

Figure 5.5–1   Geographic distribution of genera of marine seagrasses. *Zostera* (\\\\\), *Posidonia* (∴∵∴), *Thalassia,* and *Halophila* (≡), *Cymodocea, Syringodium, Thalassia, Enhalus, Halodule,* and *Cymodocea* (/////) . (From Thayer, G. W. et al., in *Wetland Functions and Values.* Greeson. P., et al., Eds., Proc. Natl. Symp. Wetlands, American Water Resources Association, Minneapolis, 1979, 235. With permission.)

Figure 5.5–2   Geographic distribution of salt marshes (bold, stippled area) and mangroves (crosshatched area). (From Ferguson, R. L. et al., *Functional Adaptation of Marine Organisms,* Vernberg, F. J. and Vernberg, W. B., Eds., Academic Press, New York, 1981, 9. With permission.)

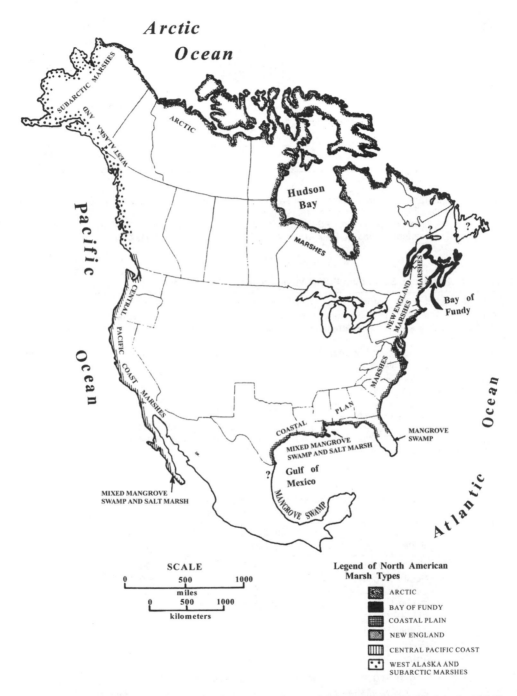

Figure 5.5–3   Schematic distribution of the three major groups of salt marshes in North America: (1) Bay of Fundy and New England marshes, (2) Atlantic and Gulf coastal plain marshes, and (3) Pacific marshes. (From Frey, R. W. and Basan, P. B., *Coastal Sedimentary Environments,* 2nd ed., Davis, R. A., Ed., Springer-Verlag, New York, 1985, 225. With permission.)

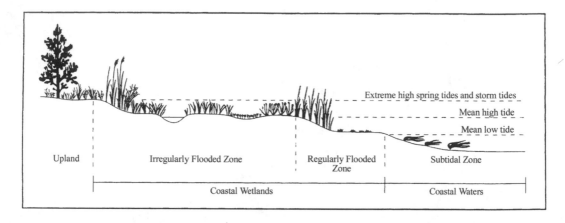

Figure 5.5–4   Diagram showing tidal marsh zones based on frequency of tidal flooding. Low marsh is flooded at least once daily, and high marsh is flooded less often. (From Tiner, R. W. and Burke, D. G., *Wetlands of Maryland,* U.S. Fish and Wildlife Service, Region 5, Hadley, MA, and Maryland Department of Natural Resources, Annapolis, 1995. With permission.)

Table 5.5–1   Known or Suspected Public Values of Intertidal Marshes

1. Shoreline protection from storm surge, winds, and waves.
2. Habitat: simultaneous food and cover for numerous fishes, crustaceans, mollusks, birds, and other vertebrates and invertebrates, many of commercial or recreational importance, some threatened or endangered.
3. Part of a set of estuarine and nearshore habitats of varying scale and location that ensure a diversity of productive fish and wildlife at all life history stages.
4. Buffer of estuarine food supply.
5. Buffer of estuarine water quality.
6. Contributor to attractive coastal image that encourages development.

Source: Montague, C. L. and Odum, H. T., Introduction: the intertidal marshes of Florida's Gulf Coast, in *Ecology and Management of Tidal Marshes: A Model from the Gulf of Mexico,* Coultas, C. L. and Hsieh, Y.-P., Eds., St. Lucie Press, Delray Beach, FL, 1997, 2. With permission.

Table 5.5–2   Intertidal Wetland Area (Marshes and Mangroves) in the Southeastern United States

State	Marsh Area (ha)	Coastline Length (km)[a]
North Carolina	64,300	500
South Carolina	204,200	350
Georgia	159,000	160
Total	**427,500**	**1,010**
Florida		
Atlantic	77,800	660
Gulf	359,700	1,340
Total	**437,500**	**2,000**

[a] Measured as a boat would most likely travel if 5 km offshore.

Source: Montague, C. L. and Odum, H. T., in *Ecology and Management of Tidal Marshes: A Model from the Gulf of Mexico,* Coultas, C. L. and Hsieh, Y.-P., Eds., St. Lucie Press, Delray Beach, FL, 1997, 10. With permission.

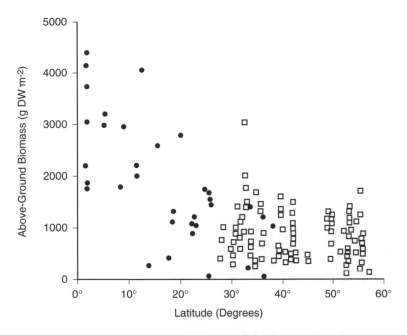

Figure 5.5–5 Global trends in aboveground biomass estimates of salt marshes (open squares) and mangrove forests (closed circles) with latitude. (From Alongi, D. M., *Coastal Ecosystem Processes*, CRC Press, Boca Raton, FL, 1998, 44. With permission.)

Table 5.5–3 Estimates of Above- and Belowground Net Primary Production (NPP) of Some Selected Species/Community Types of Saltmarsh Grasses

Community Type	Location	Aboveground NPP (g dry weight/ m²/year)	Belowground NPP (g dry weight/ m²/year)
Marsh Grasses			
Distichlis spicata	Pacific Coast	750–1500	—
	Atlantic Coast	—	1070–3400
Juncus roemerianus	Georgia	2200	—
	Gulf Coast	4250	1360–7600
Spartina alterniflora	Atlantic Coast	500–2000	550–4200
	Gulf Coast	3250	279–6000
Spartina patens	Gulf Coast	7500	—
	Atlantic Coast	—	310–3270
Mangroves			
Rhizophora apiculata	Southeast Asia	1900–3900	—
Mixed *Rhizophora* spp.	New Guinea	1750–3790	—
	Indonesia	990–2990	—
Bruguiera sexangula	China	3500	—

Source: Alongi, D. M., *Coastal Ecosystem Processes*, CRC Press, Boca Raton, FL, 1998, 48. With permission.

Table 5.5–4   *Spartina alterniflora* Above- and Belowground Primary Production along the Atlantic Coast of North America (kg/m²/year)

Area	Form	Above	Below	Total
Nova Scotia	NR	0.8	1.1	1.9
Maine	Short	0.8	0.2	1.0
Massachusetts	Tall	0.4	3.5	3.9
New York	NR	0.6	0.9	1.5
New Jersey	Short	0.5	2.3–3.2	2.8–3.7
North Carolina	Short	0.6	0.5–0.6	1.1–1.2
	Tall	1.3	0.5	1.8
South Carolina	Short	1.3	5.4	6.7
	Tall	2.5	2.4	4.9
Georgia	Short	1.4	2.0	3.4
	Medium	2.8	4.8	7.6
	Tall	3.7	2.1	5.8

Note: NR = not reported.

Source: Dame, R. F., *Rev. Aquat. Sci.,* 1, 639, 1989. With permission.

Table 5.5–5   Belowground Carbon Budget for Short-Form *Spartina alterniflora* in the Great Sippewissett Marsh, Massachusetts

Carbon Form Measured	Carbon Flux	
CO_2 evolution	720–804	(86–89%)
Carbon burial	88.8	(9–11%)
CH_4 emission	1.2–3.6	(<0.5%)
Volatile C–S compound emission	0–3.6	(<0.5%)
Calculated DOC export	0–36	(0–4%)
Total carbon input	810–936	

Note: Units are g C/m²/year. Ranges in parentheses are percentages of carbon input. DOC = dissolved organic carbon.

Source: Alongi, D. M., *Coastal Ecosystem Processes,* CRC Press, Boca Raton, FL, 1998, 75. With permission.

Table 5.5–6   Organic Matter Accumulation in *Spartina* Marshes

Area	TNPP Accumulating in Sediments (%)	Sedimentation Rate (mm/year)
Massachusetts	5.3	1
New York	37	2–6.3
New York	0	−9.5–37
North Carolina	75	>1.2
South Carolina	0	1.3

Source: Dame, R. F., *Rev. Aquat. Sci.,* 1, 639, 1989. With permission.

Table 5.5–7 **Carbon Budget from the Major Organic Decomposition Processes in the Sediments of a New England Salt Marsh and Short Belowground *Spartina* Production at the Same Site**

Process	Carbon Fluxes (g C/m^2/year)	BNPP (%)
Decomposition		
Aerobic respiration	361	31.7
Fermentation + sulfate reduction	432	37.9
Nitrate reduction	5	0.4
Methanogenesis	6	0.5
Burial	89	7.8
Export (as DOC)	36	3.2
Total losses	929	
Belowground production	1140	

Note: Decomposition was measured in terms of carbon dioxide.

Source: Dame, R. F., *Rev. Aquat. Sci.,* 1, 639, 1989. With permission.

Table 5.5–8   Flux Estimates from a Number of Marsh-Estuarine Systems Dominated by *Spartina alterniflora*

System	Type	POC	DOC	TOC	NH_4	NN	DON	PN	TN	PO_4	TP
Crommet Creek	I	68			2.1	0.3				0.6	3.2
Flax Pond	I		−8	53	−2.0	1.0				−1.4	−0.3
Canary Creek	I	−62	−38	−100	0.7	1.9	−0.9	−2.9	−1.2	−0.1	
Bly Creek	I	38	−260	−222	−0.8	0.3	−14.6	2.8	−12.5	−0.2	−0.7
Sippewissett	II	−76			−4.2	−3.9	−9.8	−6.7	−24.6	−0.6	
Gott's Marsh	II	−7			−0.4	−0.9	−2.1	−0.3	−3.7		−0.3
Rhode River	II			31		−3800			−8400		−2200
Ware Creek	II	−35	−80	−115	−2.9	2.2	−2.3		−2.8	−0.1	0.7
North Inlet	II	−128	−328	−456	−6.3	−0.9			−42.7		
Dill Creek	II	−303									
Sapelo Island	II	−208	−108	−316							−6.4
Carter Creek	III	−116	−25	−141	−0.3	0.3	−9.2	4.6	−4.0	−0.6	

Note: All values are in grams per square meter per year of marsh.

Source: Dame, R. F., *Rev. Aquat. Sci.*, 1, 639, 1989. With permission.

Table 5.5–9   Contribution of Different Autotrophs to Annual Net Primary Production in Some Salt Marshes and an Australian Mangrove Forest

Component	South Carolina	New York	Massachusetts	Australia
Marsh/mangrove	1707	692	1056	2969
Macroalgae	200	75	—	—
Microalgae	200	50	120	104
Phytoplankton	100	50	24	150
Epiphytes and neuston	100	12	25	260
Total	2307	879	1225	3483

Note: Units are g C/m²/yr.

Source: Alongi, D. M., *Coastal Ecosystem Processes,* CRC Press, Boca Raton, FL, 1998, 49. With permission.

Table 5.5–10   Nitrogen Budgets of an American Salt Marsh and an Australian Mangrove Ecosystem by the Various Processes and Their Contribution to Inputs, Outputs, and Net Exchange

	Great Sippewissett Salt Marsh	Missionary Bay Mangrove Forest
Inputs		
Precipitation	380	30
Groundwater flow	6120	30
N_2 fixation	3280	36,830
Tidal exchange	26,200	168,600
Other	10	0
Total	**35,990**	**205,490**
Outputs		
Tidal exchange	31,600	192,430
Denitrification	6940	2900
Sedimentation	1300	?
Other	30	?
Total	**39,870**	**195,330**
Net exchange	**−3880**	**10,160**

Note: Units are kg N/year for each entire system. Losses are shown as negative numbers on the net exchange line.

Source: Alongi, D. M., *Coastal Ecosystem Processes,* CRC Press, Boca Raton, FL, 1998, 84. With permission.

Table 5.5–11   Annual Exchanges of the Major Forms of Nitrogen for the Great Sippewissett Marsh and the Mangrove Forests of Missionary Bay

Form of N	Great Sippewissett Salt Marsh			Missionary Bay Mangrove Forest		
	Input	Output	Net Exchange	Input	Output	Net Exchange
NO_3–N	3420	1220	2200	5300	6300	−1000
NH_4–N	3150	3550	−400	11,140	5360	5780
DON	19,200	18,500	700	152,220	104,450	47,770
Particulate N	6750	8200	−1460	<<<1	76,320	−76,320
N_2	3280	6940	−3660	36,830	2900	33,930
Total	35,800	38,410	−2620	205,490	195,330	10,160

Note: Units are kg N/year. DON = dissolved organic nitrogen.

Source: Alongi, D. M., *Coastal Ecosystem Processes,* CRC Press, Boca Raton, FL, 1998, 85. With permission.

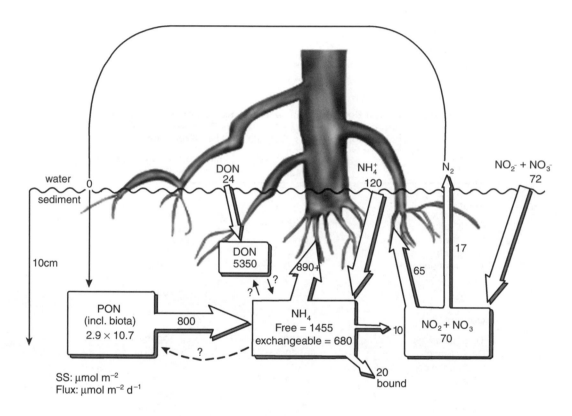

Figure 5.5–6   Nitrogen cycling in muds (10-cm depth) of mid-intertidal *Rhizophora* forests of Missionary Bay, Hinchinbrook Island, Australia. Units are mmol N m⁻² for standing stocks and mmol N m⁻² d⁻¹ for fluxes. (From Alongi, D. M., *Coastal Ecosystem Processes,* CRC Press, Boca Raton, FL, 1998, 79. With permission.)

Figure 5.5–7   A mature mangrove forest in Galley Reach, Papua New Guinea, where *Rhizophora apiculata* trees reach 30 m in height. (Photograph courtesy of B. Clough. From Alongi, D. M., *Coastal Ecosystem Processes,* CRC Press, Boca Raton, FL, 1998, 46. With permission.)

Table 5.5–12   A List of Families and Genera of Mangroves

Families and Genera	Number of Species	Indian Ocean West Pacific	Pacific America	Atlantic America	West Africa
Dicots					
Avicenniaceae					
Avicennia	11	6	3	2	1
Bombacaceae					
Camptostemon	2	2	0	0	0
Chenopodiaceae					
Suaeda[a]	2	0	0	1	1
Combretaceae					
Conocarpus[b]	1	0	1	1	1
Languncularia	1	0	1	1	1
Lumnitzera	2	2	0	0	0
Euphorbiaceae					
Exoecaria[b]	1	1	0	0	0
Leguminosae					
Machaerium[b]	1	0	1	1	1
Meliaceae					
Xylocarpus	10	8	?	2	1
Myrsinaceae					
Aegiceras	2	2	0	0	0
Myrtaceae					
Osbornia	1	1	0	0	0
Plumbaginaceae					
Aegiatilis	2	2	0	0	0
Rhizophoraceae					
Bruguiera	6	6	0	0	0
Ceriops	2	2	0	0	0
Kandelia	1	1	0	0	0
Rhizophora	7	5	2	3	3
Rubiaceae					
Scyphiphora	1	1	0	0	0
Sonneratiaceae					
Sonneratia	5	5	0	0	0
Sterculiaceae					
Heritiera[b]	2	2	0	0	0
Theaceae					
Pellicera	1	0	1	0	0
Tiliaceae					
Brownlowia[b]	17	17	0	0	0
Monocots					
Arecaceae					
Nypha[a]	1	1	0	0	0
Pandanaceae					
Pandanus[b]	1	1	0	0	0
Total	**80**	**65**	**9**	**11**	**9**

[a] *Suaeda* typically is a small- to medium-sized bush, but can become a small tree.
[b] At least some of the species in these genera are more typical of freshwater swamps behind the mangrove coastal swamp and are not considered to be true mangroves by some botanists.

Source: Dawes, C. J., *Marine Botany,* John Wiley & Sons, New York, 1981. With permission.

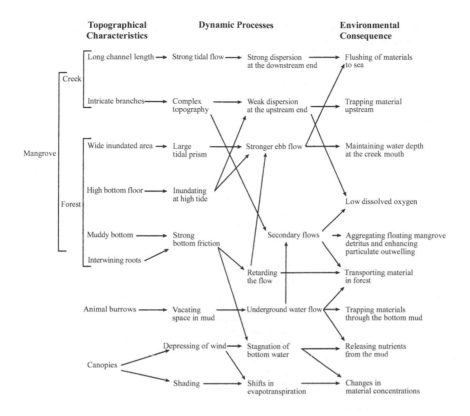

Figure 5.5–8 Summary of connections among physical, chemical, and biological processes in mangroves. (From Wolanski, E., et al., in *Tropical Mangrove Ecosystems,* Robertson, A. I. and Alongi, D. M., Eds., American Geophysical Union, Washington, D.C., 1992, chap. 3. With permission.)

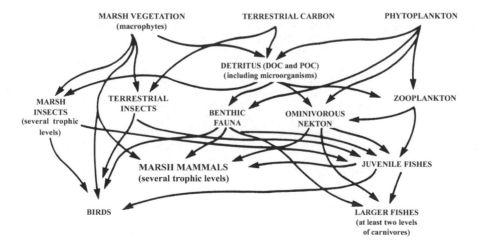

Figure 5.5–9 Hypothetical energy flow diagram for a tidal freshwater marsh. (From Odum, W. E., et al., The Ecology of Tidal Freshwater Marshes of the United States East Coast: A Community Profile, FWS/OBS-83/17, U.S. Fish and Wildlife Service, 1984.)

Table 5.5–13   Rates of Net Primary Production for Some Selected Seagrasses and Seaweeds from Various Locations

Genus/Species	Location	Net Primary Production (g C/m²/day)
Kelps		
Laminaria	North Atlantic	0.3–65.2
Macrocystis	South America, New Zealand, South Africa	1.0–4.1
Ecklonia	Australia, South Africa	1.6–3.2
Rocky intertidal/subtidal macroalgae		
Various seaweeds	Europe	0.5–9.0
Enteromorpha	Hong Kong, U.S.	0.1–2.9
Ascophyllum	U.S., Europe	1.1
Distyopteris	Caribbean	0.5–2.5
Fucus	North America	0.3–12.0
Sargassum	Caribbean	1.4
Ulva	Europe	0.6
Gracilaria	Europe	0.3
Cladophora	Europe	1.6
Seagrasses		
Zostera marina	U.S., Europe, Australia	0.2–8.0
Thalassia	U.S., Caribbean, Australia, Southeast Asia	0.1–6.0
Halodule	U.S., Caribbean	0.5–2.0
Cymodecea	Mediterranean, Australia	3.0–18.5
Posidonia	Mediterranean, Australia	2.0–6.0
Enhalus	Southeast Asia	0.3–1.6
Amphibolis	Australia	0.9–1.9

Source: Alongi, D. M., *Coastal Ecosystem Processes,* CRC Press, Boca Raton, FL, 1998, 95. With permission.

Table 5.5–14   Contribution of Various Primary Producers to Total Net Primary Production in Selected Seagrass Meadows and Kelp Beds[a]

Habitat	Location	Macrophyte	Epiphyte	Phytoplankton	Other
Halodule	Indonesia	0.64	0.23	—	—
Syringodium	Fiji	12.0	11.5	—	11.4
Halodule	Texas	7.2–96.0	0–98.4[b]	4.8[b]	—
Zostera	North Carolina	0.9	0.2	—	—
	Massachusetts	0.4–1.0	0.2–0.8	—	—
Seagrass spp.	Denmark	0.24	0.04	0.14	—
	Philippines	1.4[b]	1.8[b]	—	—
Laminaria	Nova Scotia	1.7	—	0.5	—
Ecklonia, Laminaria	South Africa	6.5	0.7	6.9	3.3

[a] Units are g C/m²/day, except where noted.
[b] Units are g AFDW/m²/day.

Source: Alongi, D. M., *Coastal Ecosystem Processes,* CRC Press, Boca Raton, FL, 1998, 98. With permission.

Table 5.5–15   Net Exchange of Particulate Materials between Representative Seagrass and
Seaweed Systems and Adjacent Coastal Areas

Habitat/Location	Net Exchange	Rate/Volume (g DW/m²/day)	% of Net Primary Production (NPP)
Seagrass/	Export	0.02	1% (*Thalassia*)
Virgin Islands		0.18–0.36	60–100% (*Syringodium*)
Seagrass/Fiji	Export	6.96	88% (*Syringodium*)
Kelp/South Africa	Export	0.32[a]	10% (mixed kelps)
Seagrass/	Export	0.01–0.26[b]	1–8% (sheltered *Zostera*)
North Carolina		0.23–0.57[b]	10–30% (exposed *Zostera*)
Seagrass/Florida	Export	0.14–0.38	8–22% (*Syringodium*)
Seagrass/	Export	0.05–0.36[a]	35–55% (exposed *Posidonia*)
Mediterranean			
Embayment/	Import	0.45	8% (mangroves)
Philippines	Export	0.76–1.3	6.5–6.7% (mixed seagrasses)
Coral Island	Export	0.08	35.5% (*Cymodocea*)
Indonesia			1.6% (*Enhalus*)
			22.9% (*Thalassia*)

[a] Units are g C/m²/day.
[b] Units are g AFDW/m²/day.

Source: Alongi, D. M., *Coastal Ecosystem Processes,* CRC Press, Boca Raton, FL, 1998, 136.
With permission.

Table 5.5–16   Productivity of Seaweeds

Species	Locality	Productivity
		(mg C/g/h)
Jointed calcareous algae (e.g., *Corallina, jania, Petrocelis*)	SW U.S.A. and Mexico	0.045 (mean)
Galaxaura squalida	Canary Islands	0.20
Botryocladia pseudodichotoma	California	0.30
Codim fragile	Mexico	0.57
Thick leathery algae	SW U.S.A. and Mexico	0.76 (mean)
(e.g., *Fucus, Sargassum, Macrocystis, Laminaria, Padina*)		
Coarsely branched algae	SW U.S.A. and Mexico	1.30 (mean)
(e.g., *Laurentia, Gelidium, Codium, Gigartina, Pterocladia*)		
Gigartina canaliculata	California	1.6
Filamentous algae	SW U.S.A. and Mexico	2.47 (mean)
(e.g., *Cladophora, Ceramium, Chaetomorpha*)		
Gelidium robustum	California	2.5
Sheetlike algae	SW U.S.A. and Mexico	5.16 (mean)
(e.g., *Ulva, Eneromorpha, Porphyra*)		
Gracilaria verrucosa	Florida	4.10–9.60
Enteromorpha compressa	Finland	0.02–10.8
Cladophora glomerata	Sweden	1.10-12.8
		(mg C/g/day)
E. compressa	Finland	47
C. glomerata	Sweden	20–180
		(g C/m²/day)
Caulerpa prolifera	Canary Islands	1.0
Pterocladia capillacea	California	1.3
Egregia laevigata	California	1.7
Gigartina canaliculata	California	2.0
Sargassum agardianum	California	2.2
Ulva californica	California	2.8
Gelidium pusillum	California	3.1

Table 5.5–16   Productivity of Seaweeds (continued)

Species	Locality	Productivity
Nereocystis luetkeana	British Columbia	5.4
Macrocystis integrifolia	British Columbia	6.7
M. pyrifera	Southern California	0.1–7.0
	California	9.5
Fucus vesiculosus	Massachusetts	20
Codium fragile	Connecticut	22
		(g C/m²/year)
Gigartina exaspertata	Washington	439–758
Laminaria hyperborea	Scotland	377–775
Iridaea cordata	Washington	144–1,012
L. longicruris	Nova Scotia, Canada	1,750
Coralline algae	Hawaii	2,080
Sargassum bed	Curaçao	2,550
Codium fragile	Connecticut	4,700

Source: Vymazal, J., *Algae and Element Cycling in Wetlands,* Lewis Publishers, Boca Raton, FL, 1995, 90. With permission.

Table 5.5–17   Examples of Potential Indicator Species for Halinity Regimes for Tidal Wetlands in the United States

Halinity Regime	Indicator Species
Hyperhaline to Euhaline	*Salicornia* spp., *Batis maritima, Lycium carolinanum, Spergularia marina, Monanthochloe littoralis,* plus stunted forms of several euhaline/polyhaline species
Euhaline	*Jaumea carnosa*, Salicornia virginica*, Lilaeopsis occidentalis**
Euhaline/polyhaline	*Spartina alterniflora, S. patens, Distichlis spicata, Iva frutescens, Juncus gerardii, Suaeda maritima, Fucus* spp., (algae), *Sporobolus virginicus, Spergularia marina, Limonium* spp., *Borrichia frutescens*
Mesohaline	*Carex lyngbyei*, Potentilla anserina*, Triglochin maritimum*, Scirpus americanus*, Cotula coronopifolia**
Mesohaline to fresh	*Typha latifolia, Phragmites australis*, Hibiscus moscheutos, Scirpus americanus, Amaranthus cannabinus, Eupatorium serotinum*
Oligohaline	*Hippuris tetraphylla*, Deschampsia cespitosa*, Grindelia integrifolia*, Hordeum brachyantherum*, Potentilla pacifica**
Oligohaline to fresh	*Nuphar luteum, Pontederia cordata, Peltandra virginica, Scirpus validus, Zizania aquatica, Zizaniopsis miliacea, Panicum hemitomom, Erianthus giganteus, Juncus canadensis, Cladium jamaicense, Rhynchospora macrostachya, Sagittaria lancifolia Hymenocallis crassifolia, Dichromena colorata, Aster novi-belgii, Polygonum punctatum, Boltonia asteroides, Sium suave, Myrica cerifera, Taxodium distichum*
Fresh	*Acorus calamus, Sparganium* spp., *Typha latifolia, Juncus effusus, Impatiens capensis, Polygonum pennsylvanicum, Rumex verticillatus, Saururus cernuus, Ludwigia palustris, Aster vimineus, Bidens laevis, Rosa palustris, Cyrilla racemiflora, Rhododendron viscosum, Cephalanthus occidentalis,* and most woody species (except mangroves and limited halophytic shrubs)

Note: Western species (including Alaska) are marked with an asterisk (*); others are eastern species. Many species exhibit a range of tolerance, yet when several of the indicator species predominate, the salinity of the area may be reasonably predicted.

Source: Tiner, R. W., *Wetland Indicators: A Guide to Wetland Identification, Delineation, Classification, and Mapping,* Lewis Publishers, Boca Raton, FL, 1999, 84. With permission.

Table 5.5–18   Tidal and Nontidal Water Regimes Defined by Cowardin et al. (1979)

Group	Type of Water	Water Regime	Definition
Tidal	Saltwater and brackish areas	Subtidal	Permanently flooded tidal waters
		Irregularly exposed	Exposed less often than daily by tides
		Regularly flooded	Daily tidal flooding and exposure to air
		Irregularly flooded	Flooded less often than daily and typically exposed to air
	Freshwater areas	Permanently flooded-tidal	Permanently flooded by tides and river or exposed irregularly by tides
		Semipermanently flooded-tidal	Flooded for most of the growing season by river overflow but with tidal fluctuation in water levels
		Regularly flooded	Daily tidal flooding and exposure to air
		Seasonally flooded-tidal	Flooded irregularly by tides and seasonally by river overflow
		Temporarily flooded-tidal	Flooded irregularly by tides and for brief periods during growing season by river overflow
Nontidal	Inland freshwater and saline areas	Permanently flooded	Flooded throughout the year in all years
		Intermittently exposed	Flooded year-round except during extreme droughts
		Semipermanently flooded	Flooded through the growing season in most years
		Seasonally flooded	Flooded for extended periods in growing season, but surface water is usually absent by end of growing season
		Saturated	Surface water is seldom present, but substrate is saturated to the surface for most of the season
		Temporarily flooded	Flooded for only brief periods during growing season, with water table usually well below the soil surface for most of the season
		Intermittently flooded	Substrate is usually exposed and only flooded for variable periods without detectable seasonal periodicity (not always wetland; may be upland in some situations)
		Artificially flooded	Duration and amount of flooding is controlled by means of pumps or siphons in combination with dikes or dams.

Source: Cowardin, L. M., et al., *Classification of Wetlands and Deepwater Habitats of the U.S., Fish and Wildlife Service,* U.S. Fish and Wildlife Service, Washington, D.C., 1979.

Table 5.5–19   Examples of Wetland Definitions Used by State Regulatory Programs in the United States

State	Definition
Connecticut	"Wetlands are those areas which border on or lie beneath tidal waters, such as, but not limited to banks, bogs, salt marshes, swamps, meadows, flats, or other low lands subject to tidal action, including those areas now or formerly connected to tidal waters, and whose surface is at or below an elevation of 1 ft above local extreme high water." (State's tidal wetland definition; the definition also includes a list of indicator plants.)
	"Wetlands mean land, including submerged land, which consists of any of the soil types designated as poorly drained, very poorly drained, alluvial, and floodplain by the National Cooperative Soils Survey, as may be amended from time to time, of the Soil Conservation Service of the United States Department of Agriculture." (State's inland wetland regulatory definition; it focuses on soil types.)
Florida	"Wetlands mean those areas that are inundated or saturated by surfacewater or groundwater at a frequency and a duration sufficient to support, and under normal circumstances do support, a prevalence of vegetation typically adapted for life in saturated soils. Soils present in wetlands generally are classified as hydric or alluvial, or possess characteristics that are associated with reducing soils conditions. The prevalent vegetation in wetlands generally consists of facultative or obligate hydrophytic macrophytes that are typically adapted to areas having soil conditions described above. These species, due to morphological, physiological, or reproductive adaptations, have the ability to grow, reproduce, or persist in aquatic environments or anaerobic soil conditions. Florida wetlands generally include swamps, marshes, bayheads, bogs, cypress domes and strands, sloughs, wet prairies, riverine swamps and marshes, mangrove swamps, and other similar areas. Florida wetlands generally do not include longleaf or slash pine flatwoods with an understory dominated by saw palmetto." (Statewide definition includes tidal and nontidal wetlands.)
Maryland	Tidal wetlands are "all state and private tidal wetlands, marshes, submerged aquatic vegetation, lands, and open water affected by the daily and periodic rise and fall of the tide within the Chesapeake Bay and its tributaries, the coastal bays adjacent to Maryland's coastal barrier islands, and the Atlantic Ocean to a distance of 3 m offshore of the low water mark." (State's tidal wetland definition; it includes deepwater areas.)
Massachusetts	"Salt Marsh means a coastal wetland that extends landward up to the highest high tide line, that is, the highest spring tide of the year, and is characterized by plants that are well adapted or prefer living in saline soils. Dominant plants within salt marshes are salt meadow cord grass (*Spartina patens*) and/or salt marsh cord grass (*Spartina alterniflora*). A salt marsh may contain tidal creeks, ditches, and pools." (State's coastal wetland definition; it was published in the first state wetland law in the nation in 1962.)
	"Bordering vegetated wetlands are freshwater wetlands which border on creeks, rivers, streams, ponds, and lakes. The types of freshwater wetlands are wet meadows, marshes, swamps, and bogs. They are areas where the topography is low and flat, and where the soils are annually saturated. The ground and surface water regime and the vegetational community which occur in each type of freshwater wetland are specified in the Act." (State's inland wetland definition; it only pertains to wetlands that border waterbodies.)
New Jersey	Coastal wetlands are "any bank, marsh, swamp, meadow, flat, or other lowland subject to tidal action in the Delaware Bay and Delaware River, Raritan Bay, Sandy Hook Bay, Shrewsbury River, including Navesink River, Shark River, and the coastal inland waterways extending southerly from Manasquan Inlet to Cape May Harbor, or any inlet, estuary or those areas now or formerly connected to tidal waters whose surface is at or below an elevation of 1 ft above local extreme high water, and upon which may grow or is capable of growing some, but not necessarily all of the following: [19 plants listed]." Coastal wetlands exclude "any land or real property subject to the jurisdiction of the Hackensack Meadowlands Development Commission...." (State's tidal wetland definition; it contains a geographic exclusion.)
New Jersey	"Freshwater wetland means an area that is inundated or saturated by surface water or groundwater at a frequency and duration sufficient to support, and that under normal circumstances does support, a prevalence of vegetation typically adapted for life in saturated soil conditions, commonly known as hydrophytic vegetation; provided, however, that the department, in designating a wetland, shall use the three-parameter approach ...developed by the U.S. Environmental Protection Agency, and any subsequent amendments thereto...." (State's inland or freshwater wetland definition.)

Table 5.5–19   Examples of Wetland Definitions Used by State Regulatory Programs in the United States (continued)

State	Definition
New York	"Freshwater wetlands mean lands and waters of the state as shown on the freshwater wetlands map which contain any or all of the following: (a) lands and submerged lands commonly called marshes, swamps, sloughs, bogs, and flats supporting aquatic or semi-aquatic vegetation of the following types: [lists indicator trees, shrubs, herbs, and aquatic species]; (b) lands and submerged lands containing remnants of any vegetation that is not aquatic or semi-aquatic that has died because of wet conditions over a sufficiently long period…provided further that such conditions can be expected to persist indefinitely, barring human intervention; (c) lands and waters substantially enclosed by aquatic or semi-aquatic vegetation…the regulation of which is necessary to protect and preserve the aquatic and semi-aquatic vegetation; and (d) the waters overlying the areas set forth in (a) and (b) and the lands underlying (c)." (State's inland wetland definition; the definition includes lists of indicator species for each wetland type; the state generally regulates wetlands 12.5 acres or larger.)
Rhode Island	"Coastal wetlands include salt marshes and freshwater or brackish wetlands contiguous to salt marshes. Areas of open water within coastal wetlands are considered a part of the wetland. Salt marshes are areas regularly inundated by salt water through either natural or artificial water courses and where one or more of the following species predominate: [eight plants listed]. Contiguous and associated freshwater or brackish marshes are those where one or more of the following species predominate: [nine plants listed]." (State's coastal wetland definition; the definition includes lists of indicator species.)
	Freshwater wetlands are defined to include, "but not be limited to marshes, swamps, bogs, ponds, river and stream flood plains and banks; areas subject to flooding or storm flowage; emergent and submergent plant communities in any body of fresh water including rivers and streams and that area of land within fifty feet (50´) of the edge of any bog, marsh, swamp, or pond." (State's inland wetland definition; various wetland types are further defined based on hydrology and indicator plants.)
Vermont	"Wetlands means those areas of the state that are inundated by surface or groundwater with a frequency sufficient to support significant vegetation or aquatic life that depend on saturated or seasonally saturated soil conditions for growth and reproduction. Such areas include but are not limited to marshes, swamps, sloughs, potholes, fens, river and lake overflows, mud flats, bogs and ponds, but excluding such areas as grow food or crops in connection with farming activities."
Wisconsin	"An area where water is at, near, or above the land surface long enough to be capable of supporting aquatic or hydrophytic vegetation and which has soils indicative of wet conditions."

Note: Some states use the federal regulatory definition for inland wetland programs (e.g., Maryland, Oregon, and Pennsylvania).

Source: Tiner, R. W., *Wetland Indicators: A Guide to Wetland Identification, Delineation, Classification, and Mapping,* Lewis Publishers, Boca Raton, FL, 1999, 12. With permission.

Table 5.5–20   Common Species of Vascular Plants Occurring in the Tidal Freshwater Habitat

Species	General Characteristics	Habitat Preference	Salinity Tolerance	Associated Species
Acorus calamus (sweet-flag)	Grows in dense colonies propagating mainly by rhizome; stemless plants up to 1.5 m with stiff, narrow basal leaves; cylindrical inflorescence emerges from side of stem (open spadix); aromatic	Shallow water or wet soil; channel margins	Fresh	*Peltandra virginica Polygonum* spp. *Impatiens capensis*
Alternanthera philoxeroides (alligatorweed)	Hollow stems with simple branches bearing opposite, lance-shaped leaves; forms dense mats; flowers on long panicles; perennial	Extremely adaptable; often emersed	Fresh to oligohaline	—
Amaranthus cannabinus (water hemp)	Erect, fleshy and stout; up to 2 m; leaves lanceolate with blades as long as 20 cm; not conspicuous until midsummer when it towers above other marsh forbs	Common to levee sections of the tidal marsh habitat; tolerates periodic inundation	Fresh to mesohaline	*Peltandra virginica Polygonum* spp. *Bidens* spp.
Asclepias incarnata (swamp milkweed)	Tall, leafy, pink-flowered herb growing solitary or in small, loose groups; lance-shaped, opposite leaves; reproduces via seeds or rhizomes	Cosmopolitan; grows in many wetland situations; high marsh species	Fresh to oligohaline	High marsh herbs
Bidens coronata B. laevis (burmarigold)	Annual plants up to 1.5 m tall, solitary or in small scattered groups; loosely branched above with opposite leaves; leaf shape variable but generally toothed or lanceolate; impressive yellow bloom late in the growing season	Cosmopolitan, growing in the upper two thirds of the intertidal zone on wet mud or in shallow water	Fresh	*Polygonum* spp. *Amaranthus cannabinus* Other *Bidens* spp.
Calamagrostis canadensis (reed-bentgrass)	Slender grass up to 1.5 m, generally forming dense colonies; long, flat leaves; loose, ovoid panicle with purplish color; perennial	Wet meadows and thickets	Fresh?	*Typha* spp. *Acorus calamus*
Carex spp. (sedges)	Grasslike sedges, culms mostly three-angled, bearing several leaves with rough margins; up to 2 m tall and usually in groups, perennial from long, stout rhizomes	Low areas with frequent flooding or damp soil	Fresh	—
Cephalanthus occidentalis (buttonbush)	Branched shrub up to 1.5 m tall with leathery smooth opposite leaves and white flowers crowded into dense, spherical, stalked heads; flowers June through August; leaf petioles reddish	Upland margins and raised hummocks of tidal freshwater marshes; wet soil	Fresh to oligohaline	*Hibiscus* spp. *Cornus amommum*
Echinochloa walteri (water's millet)	Grass up to 2 m, solitary or in small groups; long, moderately wide leaf blades; flowers in a terminal panicle which is ovoid; greenish purple, and appears in July/August	Shallow water; moist areas, disturbed sites	Fresh to oligohaline	—
Hibiscus spp. *Kosteletzkya virginica* (mallows)	Shrubform herbs up to 2 m, scattered or in large colonies; leaves wedge-shaped or rounded and alternate; large, showy pink or white flowers appearing in midsummer; perennial	Freshwater marshes or the upland margin of saline marshes with freshwater seepage	Fresh to mesohaline	*Typha* spp., *Spartina cynosuroides Polygonum* spp. *Impatiens capensis*
Eleocharis palustris E. quadrangulata (spike-rushes)	Perennials with horizontal roots-tocks; culms stout, slender, and cylindrical or squarish with a basal sheath; flowers crowded onto terminus of spikelet; between 0.5 and 1.5 m	Channel margins or stream banks in shallow water; muddy, organic substrates	Fresh to oligohaline	*Pontederia cordata Scirpus* spp. *Juncus* spp., *Leersia oryzoides*

Table 5.5-20   Common Species of Vascular Plants Occurring in the Tidal Freshwater Habitat (continued)

Species	General Characteristics	Habitat Preference	Salinity Tolerance	Associated Species
Impatiens capensis (jewelweed)	Annual plants up to 2 m with succulent, branched stems with swelling at the joints; colonial; leaves alternate and ovate or elliptic with toothed margins; flowers orange and funnel-like, appearing in July/August	Same as *Bidens* spp.; also grows in shaded portions of marshes	Fresh	*Bidens* spp. *Typha* spp. *Polygonum* spp.
Iris versicolor (blue flag)	Flat, swordlike leaves arising from a stout creeping rhizome; large, purplish-blue flowers emerge in spring from a stiff upright stem; perennial	High, shaded portions of the intertidal zone in damp soil; will not tolerate long inundations	Fresh	None in particular
Leersia oryzoides (rice cut-grass)	Weak slender grass growing in dense, matted colonies; leaf sheaths and blades very rough; emerges from creeping rhizomes and often sprawls on other vegetation	Mid-intertidal zones of marshes; high diversity vegetation patches	Fresh to oligohaline	Many; none in particular
Lythrum salicaria *Decodon verticillatus* (loosestrife)	Shrubform herbs forming large, dense colonies; aggressive; up to 1.5 m in height with lanceolate leaves opposite or whorled; upper axils branched with small purplish-pink flowers; terminal spikes pubescent; annual	Moist portions of marshes; high intertidal or upland areas	Fresh to oligohaline	*Hibiscus* spp. *Convolvulus* spp.
Mikania scandens (climbing hempweed)	Long, herbaceous vine forming matted tangles over other emergent plants; heart-shaped leaves; dense, pinkish flower clusters; slender stem; propagates by both seed and rhizome; perennial	Open, wooded swamps and marshes; shrub thickets	Fresh to oligohaline	—
Myrica cerifera (wax-myrtle)	Compact, tall, evergreen shrub with leathery alternate leaves; spicy aroma; waxy, berrylike fruits; forms extensive thickets	Most all coastal habitats; border between intertidal zone and uplands	Fresh	*Acer rubrum* *Nyssa* spp. *Taxodium distichum*
Nuphar lutecum (*N. advena*) (spatterdock)	Plant with floating or emergent leaves and flowers attached to flexible underwater stalks; rises from thick rhizomes imbedded in benthic muds; flowers deep yellow, appearing throughout the summer	Constantly submerged areas up to 1.5 m depth, or, if tidal, near or below mean low water in deep organic muds	Fresh	Usually in pure stand
Nyssa sylvatica N. aquatica (gum)	Medium-sized tree (10 m) with numerous horizontal, crooked branches; leaves crowded at twig ends turning scarlet in fall; flowers appear in April/May	Marsh/upland borders	Fresh	*Acer rubrum* *Myrica cerifera Alnus* spp.
Panicum virgatum (switch-grass)	Perennial grass 1–2 m in height in large bunches with partially woody stems; nest of hairs where leaf blade attaches to sheath; large, open, delicately branched seed head produced in late summer; rhizomatous	Dry to moist sandy soils or the mid-intertidal portions of tidal freshwater marshes; disturbed areas	Fresh to mesohaline	*Hibiscus* spp. *Scirpus* spp. *Eleocharis palustris*
Peltandra virginica (arrow-arum)	Stemless plants, 1–1.5 m tall, growing in loose colonies; several arrowhead-shaped leaves on long stalks; emerge in rather dense clumps from a thick subsurface tuber; flowers from May to June	Grows predominantly as an emergent on stream margins or intertidal marsh zones on rich, loose silt	Fresh to oligohaline	*Pontederia cordata* *Zizania aquatica* Many other species

(continued)

Table 5.5–20   Common Species of Vascular Plants Occurring in the Tidal Freshwater Habitat (continued)

Species	General Characteristics	Habitat Preference	Salinity Tolerance	Associated Species
Phragmites australis (common reed)	Tall, coarse grass with a feathery seed head; 1–4 m in height; grows aggressively from long, creeping rhizomes; perennial; flowers from July to September	Extremely cosmopolitan, growing in tidal and nontidal marshes and often associated with disturbed areas	Fresh to mesohaline	*Spartina cynosuroides* *Zizania aquatica*
Polygonum arifolium *P. sagittatum* (tearthumbs)	Plants with long, weak stems up to 2 m tall, usually leaning on other vegetation; leaves sagitate in shape and alternate; leaf midribs and stems armed with recurved barbs; flowers small and appearing in late summer; annual	Shallow water or damp soil; middle to upper intertidal zone	Fresh to oligohaline	*Bidens* spp. *Hibiscus* spp. *Impatiens capensis*
Polygonum punctatum *P. densiflorum* *P. hydropiperoides* (smartweeds)	Upright plants growing from a fibrous tuft of roots; narrowly to widely lanceolate leaves with stalks basally enclosed within a membranous sheath; up to 1 m; flowers at spike at end of stalk	Upper three quarters of intertidal zone in freshwater marshes on wet or damp soil	Fresh to oligohaline	Many species
Pontederia cordata (pickerelweed)	Rhizomatous perennial growing in dense or loose colonies; plants up to 2 m tall; fleshy, heart-shaped leaves with parallel veins and emerging from spongy stalks; flowers dark violet-blue, appearing June to August	Lower intertidal zone of tidal fresh-water marshes	Fresh to oligohaline	*Nuphar luteum* *Peltandra virginica* *Sagittaria latifolia*
Rosa palustris (swamp rose)	Shrub up to 2 m growing in loose colonies; stems lack prickles except for those occurring at bases of leaf stalks; pinnately compound leaves with fine serrate margins; showy, pink flowers appearing July/August	High intertidal zones or wet meadows	Fresh to oligohaline	*Cephalanthus occidentalis*
Rumex verticillatus (water dock)	Erect, robust annual with dark-green, lance-shaped leaves; stem swollen at nodes; attains heights over 1.5 m and grows solitary or in loose colonies; flower head is evident in late spring and can be 50 cm in length	Wet meadows or pond margins on mud or in shallow water	Fresh to oligohaline	—
Sagittaria latifolia (duck potato) *S. falcata* (bultongue)	Perennial herbs; stemless, up to 2 m in height and emerging from fibrous tubers; leaves arrowhead shaped or lanceolate with white flowers in whorls appearing on a naked stalk in July/August	Borders of rivers or marshes in low intertidal zones on organic, silty mud	Fresh to oligohaline	*Peltandra virginica* *Pontederia cordata*
Scirpus validus (soft-stem bulrush) *S. cyperinus* (woolgrass) *S. americanus* (common three square)	Medium to large rushes with cylindrical or triangular stems; inconspicuous leaf sheaths; usually grow in small groups; bear seed clusters on end or side of stem; perennial	Brackish to fresh shallow water or low to middle intertidal zones on organic clay substrates	Fresh to mesohaline	Other rushes *Typha* spp.
Sparganium eurycarpum (great burread)	Stout upright forbs up to 1 m with limp, underwater, emergent leaves attached basally and alternating up the stem; toward the terminus, stems zigzag bearing spherelike clusters of pistillate and staminate flowers	Partially submerged, shallow-water marsh areas; lower to middle intertidal zones	Fresh	*Zizania aquatica* *Leersia oryzoides* *Polygonum* spp.

Table 5.5–20 Common Species of Vascular Plants Occurring in the Tidal Freshwater Habitat (continued)

Species	General Characteristics	Habitat Preference	Salinity Tolerance	Associated Species
Spartina cynosuroides (big cordgrass)	Perennial grass attaining heights in excess of 3 m, having long, tapering leaves and growing from vigorous underground rhizomes; found in dense monospecific or mixed stands	Channel and creek margins in tidal oligohaline marshes	Fresh to mesohaline	*Phragmites australis* *Typha* spp.
Taxodium distichum (bald cypress)	Tall tree with straight trunk (40 m), conifer-like but deciduous; light porous wood covered by stringy bark; unbranched shoots originating from roots as knees	Marsh/upland borders	Fresh?	*Nyssa* spp. *Acer rubrum*
Typha latifolia *T. domingensis* *T. angustifolia* (cattails)	Stout, upright reeds up to 3 m forming dense colonies; basal leaves, long and swordlike, appearing before stems; yellowish male flower disintegrates leaving a thick, velvety-brown swelling on the spike; rhizomatous; perennial	Very cosmopolitan, occurring in shallow water or upper intertidal zones; some disturbed areas	Fresh to mesohaline	Many associates
Zizania aquatica (wildrice)	Annual or perennial aquatic grass, 1–4 m tall, usually found in colonies; short underground roots, stiff hollow stalk, and long, flat, wide leaves with rough edges; male and female flowers separate along a large terminal panicle in late summer	Fresh to slightly brackish marshes and slow streams, usually in shallow water; requires soft mud and slowly circulating water	Fresh to oligohaline	*Peltandra virginica* Many other species
Zizaniopsis mileacea (giant cutgrass)	Perennial by creeping rhizome; culms 1–4 m high; long, rough-edged leaves geniculate at lower nodes; large, loose terminal panicles appearing in midsummer; aggressive	Swamps and margins of tidal streams	Fresh?	—

Source: Odum, W. E. et al., The Ecology of Tidal Freshwater Marshes of the United States East Coast: A Community Profile, FWS/OBS-83/17, U.S. Fish and Wildlife Service, Washington, D.C., 1984.

Table 5.5–21   Representative Benthic Macroinvertebrates of Tidal Freshwater Marshes of Mid-Atlantic Region (U.S.)

Sponges
   Spongilla lacustris and other species
Hydra
   Hydra americana
   Protohydra spp.
Bryozoans
   Barentsia gracilus
   Lophopodella sp.
   Pectinatella magnitica
Leeches
   Families Glossiphoniidae, Piscicolidae
Oligochaetes
   Families Tubificidae, Naididae
Insects
   Dipteran larvae (especially family Chironomidae)
   Larvae of *Ephemeroptera, Odonata, Trichoptera,* and *Coleoptera*
Amphipods
   Hyallela azteca
   Gammarus fasciatus
   Lepidactylus dytiscus (southeastern states)
Crustaceans
   Crayfish
   Blue crab, *Callinectes sapidus*
   Caridean shrimp, *Palaemonetes paladosus*
Mollusks
   Fingernail clam, *Pisidium* spp.
   Asiatic clam, *Corbicula fluminea* (formerly *C. manilensis*)
   Brackishwater clam, *Rangia cuneata*
   Pulmonate snails (at least six families)

Source: Odum, W. E. et al.,The Ecology of Tidal Freshwater Marshes of the United States East Coast: A Community Profile, FWS/OBS - 83/17, U.S. Fish and Wildlife Service, Washington, D.C., 1984.

Table 5.5–22   Annual Productivity of Wetland and Aquatic Plants (g DM/m^2/year)

Species	Productivity			Location
	ABG	BLG	Total	
Emergent Species				
Acorus calamus	450–830			Czech Republic
	712–940			New Jersey
Arundo donax	10,000			Thailand
Bidens laevis	21.3			Virginia
Calamagrostis canadensis	48			Canada, Manitoba
Carex acutiformis	778–790			The Netherlands
C. aquatilis	340			Canada, Alberta
	820	210	1,030	Canada, Quebec
C. atherodes	432			Canada, NWT
	2,858	548	3,406	Iowa
C. diandra	288–535			The Netherlands
C. gracilis	580	250	830	Poland
C. gracilis + vesicaria	530–1,390			Czech Republic
C. lacustris	857	161	1,018	New York
	940	130	1,070	Wisconsin
	965	208	1,173	New York
	1,186	134	1,320	Wisconsin
C. lyngbyei	687–1,322			Canada, B.C.

Table 5.5–22   Annual Productivity of Wetland and Aquatic Plants (g DM/m²/year) (continued)

Species	Productivity			
	ABG	BLG	Total	Location
C. paludosa			710	Germany
C. rostrata	10–14			Sweden
	515			Canada, Alberta
	57	7	64	Sweden
	116			Canada, Manitoba
	540	260	900	New York
	738	180	918	Minnesota
	1,917	345	1,262	The Netherlands
	1,080	260	1,340	New York
C. stricta	32			Virginia
C. subspathacea	80			Hudson Bay
Carex sp.	1,340			New Jersey
Cladium jamaicense	802–2,028			Florida
C. mariscoides	228	47		New York
Cypreus articulatus	923–1677			India
C. papyrus			9,000–15,000	Africa (Equator)
C. papyrus	12,500			Uganda
Distichlis spicata	1,291			Louisiana
D. spicata	1,484			Mississippi
	1,967			Louisiana
	3,108–3,366			Louisiana
Glyceria maxima			751	Germany
	900–1,570			Czech Republic
	1,390–2,860			Czech Republic
Heleocharis palustris			138	Germany
Iris pseudacorus			647	Germany
Juncus effusus	1,670	190	1,860	South Carolina
J. militaris	620	589	1,209	Rhode Island
J. roemerianus	796			North Carolina
	1,360			North Carolina
	1,697			Mississippi
	1,806			Louisiana
	2,156			Georgia
	3,295			Louisiana
	3,029–3,794			Louisiana
Lysimachia thyrsiflora			268	Germany
Lythrum salicaria	1,749			Pennsylvania
	2,104			New Jersey
Mentha aquatica			265	Germany
Menyanthes trifiloata			542	Germany
Panicum hemitomon	1,700			Louisiana
Peltandra virginica	144			Virginia
	269			Pennsylvania
	500–800			New Jersey
Phalaris arunidnacea	800			England
Phragmites australis	551–1,080			England
	183–1,600			Finland
	1,825–2,811			Louisiana
	265–5,280			India
			1,273	Germany
	780	620	1,400	Denmark
P. karka	3,182–7,543			India
Polygonum glabrum	314–4,793			India
Sagittaria falcata	1,389–1,613			Louisiana
	2,310			Louisiana
Sagittaria sp.	628			Pennsylvania

Table 5.5–22   Annual Productivity of Wetland and Aquatic Plants (g DM/m²/year) (continued)

| Species | Productivity | | | Location |
	ABG	BLG	Total	
Schoenoplectus lacustris	1,600–3,000			Czech Republic
Scirpus fluviatillis			1,533	Wisconsin
S. lacustris	785			Germany
	1,330	1,220	2,550	Denmark
	4,600			Worldwide
Scirpus mucronatus	1,696–2,023			India
Sparganium erectum			307	Germany
S. eurycarpum	1,066			Iowa
	924–1,448			Iowa
Spartina alterniflora	300			New Jersey
	63–460			Alabama
	540–580			Mississippi
	758–763			Maine
	973			Georgia
	1,158			Georgia
	1,207			Maryland
	1,296			North Carolina
	1,332			Virginia
	1,964			Mississippi
	2,000			Mississippi
	1,323–2,645			Louisiana
	1,381			Louisiana
	2,523–2,794			Louisiana
	2,883–3,990			Georgia
	2,895			Louisiana
		460–503		North Carolina
		2,500–3,500		Delaware
S. cynosuroides	1,052–1,659			Louisiana
	1,134			Louisiana
S. patens	805			Virginia
	1,296			North Carolina
	1,428			Louisiana
	1,922			Mississippi
	3,053–5,509			Louisiana
	4,159			Louisiana
	4,924–6,163			Louisiana
Typha angustata	1,662–4,577			India
	9,339			India
T. angustifolia	810	1,800	2,610	Denmark
	850	1,800	2,650	United States
	1,059			Germany
	1,445			England
	2,560	2,505	5,065	Texas (1983)
	2,895	2,314	5,209	Texas (1984)
T. domingensis	1,483			South Carolina
	1,112–1,580			Malawi
	1,080–2,832			Florida
	1,077–3,035			Florida
T. elephantina	975–2,464			India
T. glauca	1,360			Minnesota
	2,297			Iowa
	2,320			Minnesota
T. latifolia	330–418			Oregon
	574			South Carolina
	1,070			England

Table 5.5–22   Annual Productivity of Wetland and Aquatic Plants (g DM/m²/year) (continued)

| Species | Productivity | | | |
	ABG	BLG	Total	Location
	520–1,132			South Carolina
	1,358			New York
	1,360			Minnesota
	1,570	1,050	2,620	New Jersey
	1,600			Czech Republic
	1,657			Czech Republic
	3,338–3,560			Czech Republic
Typha sp.	930			Virginia
	987			New Jersey
	1,240			North Dakota
	730–1,336			Oklahoma
	1,119–1,528			New Jersey
	874–2,064			Pennsylvania
	1,680	1,480	3,160	Minnesota
	2,210			Texas
			1,450	New Jersey
	659–1,125			New Jersey
	1,390			New Jersey
	605–1,547			Pennsylvania

Floating-Leaved Species

Species	ABG	BLG	Total	Location
Nelumbo lucifera	68–813			India
Nuphar advena	245			Virginia
	516			New Jersey
	775			New Jersey
	1,166–1,188			Pennsylvania
N. lutea	222			North Carolina
	200–400			USSR
	405			Germany
Nymphaea alba	155–210			India
	293			Germany
N. stellata	262			India
Nymphoides peltata	203–305			India
Trapa natans	115–435			India

Submerged Species

Species	ABG	BLG	Total	Location
Ceratophyllum demersum	211			Germany
	23–565			India
	610–960			USSR
Elodea canadensis	326			Germany
Hydrilla verticillata	194–273			India
	104–387			India
	180–523			India
Isoetes savatieri	52–158			Argentina
Myriophyllum spicatum	334			Germany
	710			India
	210–830			USSR
Najas graminea	144–203			India
	139–555			India
N. major	249–799			India
Potamogeton amphibium	167			Germany
P. crispus	130			Germany
	92–276			India
	166–361			India
P. friesii	91			Germany

Table 5.5–22   Annual Productivity of Wetland and Aquatic Plants (g DM/m²/year) (continued)

Species	ABG	BLG	Total	Location
		Productivity		
P. filiformis	44			United Kingdom
P. lucens	239			Germany
	160–380			India
P. natans	99			Germany
P. pectinatus	253			Germany
P. perfoliatus	580–830			USSR
	332			Germany
P. pusillus	198–279			India
Ranunculus circinatus	413			Germany
Utricularia vulgaris	14–39			Austria
Utricularia sp.	370			Malaysia
Utricularia sp.(+ periphyton)	54			Georgia
Trapa bispinosa	570			India
Vallisneria spiralis	592–709			India

Floating Species

Species	ABG	BLG	Total	Location
Azolla pinnata			278–400	India
Eichhornia crassipes			200–291	India
			1,473	Louisiana
			723–2,067	India
			6,520	Alabama
Lemna spp.			750–800	Czech Republic
Lemna sp.			>1,000	Israel
Lemna sp. + Spirodela sp.			10–802	India
Lemnaceae			750	Czech Republic
Pistia stratiotes			166–394	Argentina
Salvinia natans			266–335	India
Spirodela polyrhiza			12–35	India
S. polyrhiza (W)			1,758–4,400	Louisiana

Bryophytes

Species	ABG	BLG	Total	Location
Aulacomnium palustre	5–35			Canada, Manitoba
Fontinalis antipyretica	55			Germany
Hylocomium splendens	80–120			Finland
Hypnum pratense	31			Canada, Manitoba
Pleurotium schreberi	70–110			Finland
	108			Canada, Manitoba
Polytrichum juniperinum	35			Canada, Manitoba
Sphagnum angustifolium	110–140			Estonia
S. balticum	109			Sweden
	165			Sweden
	190–480			Estonia
S. balticum + majus	210–410			Finland
S. capillifolium	80–135			England
S. cuspidatum	70–90			Estonia
S. cuspidum	260			Norway
	790			England
S. fuscum	70–330			Finland
S. magellanicum	50–70			Sweden
	68			England
	70			Norway

Table 5.5–22   Annual Productivity of Wetland and Aquatic Plants (g DM/m²/year) (continued)

| Species | Productivity | | | |
	ABG	BLG	Total	Location
	95			Sweden
	100			Sweden
	134			Canada
	250–350			Germany
	500			USSR
S. papillosum	35			England
S. recurvum	490–670			Germany
S. rubellum	70–240			Estonia
	130			England
	210–260			Germany
	240–430			England
Sphagnum spp.	50			Ireland

Source: Vymazal, J., Algae and Element Cycling in Wetlands, Lewis Publishers, Boca Raton, FL, 1995, 398. With permission.

Table 5.5–23   Daily Productivity Rate of Aquatic and Wetland Plants (g DM/m²/year)

Species	Productivity Rate	Location
Emergent Species		
Alisma plantago–aquatica	4.44	Iowa
Althernanthera philoxeroides	17	Florida
Carex aquatilis	4	Canada, Alberta
	15.3	Canada, Quebec
C. atherodes	23	Iowa
C. bigelowii	6	New Hampshire
C. lacustris	8.1	Wisconsin
	15	New York
	20.9	New Jersey
C. nebraskensis	5.3	California
C. rostrata	6.0	Canada, Alberta
	10.9	Minnesota
C. stricta	23	New Jersey
Carex spp.	0.56–4.16	The Netherlands
Eleocharis palustris	4.97	Iowa
E. quadrangulata	1.7–18.5	South Carolina
Equisetum fluviatile	4.25–6.37	Iowa
Juncus effusus	3.6–14.2	South Carolina
Justicia americana	10.1–37.1	Alabama
Scirpus americanus	4.67	South Carolina
S. validus	4.0	Iowa
Sparganium ramosum	7.33	India
Typha glauca	3.73	Minnesota
T. latifolia	2.70	Iowa
	19.4	South Carolina
	49	Grand Canyon
	1.6–52.6	Oklahoma
	0.89–3.14	Oregon
Typha sp.	8.4–10.2	Texas, North Dakota
	20.9	New Jersey
Zizania aquatica	20.9	New Jersey

Table 5.5–23   Daily Productivity Rate of Aquatic and Wetland Plants (g DM/m²/year) (continued)

Species	Productivity Rate	Location
Floating-Leaved Species		
Nuphar variegatum	1.37–2.67	Iowa
Submerged Species		
Ceratophyllum demersum	5.7	Sweden
Myriophyllum verticillatum	2.8	Sweden
Potamogeton natans	2.27–3.35	Iowa
P. pectinatus	1.77–2.97	Iowa
Floating Species		
Eichhornia crassipes (W)	0.4–12.4	Czech Republic
E. crassipes	12.7–14.6	Louisiana
E. crassipes (W)	14.7	Czech Republic
E. crassipes (W)	3.0–15	Czech Republic
E. crassipes	12.5–27.6	Alabama
	6.0–33	Florida
E. crassipes (F)	6.3–40	Florida
E. crassipes	40–54	Florida
	60	Subtropic climate
Lemna sp.	14.9	Louisiana
Lemna spp.	3.5	Czech Republic
Pistia stratiotes	10.8	Czech Republic
	15.3	Florida

Source: Vymazal, J., *Algae and Element Cycling in Wetlands,* Lewis Publishers, Boca Raton, FL, 1995, 405. With permission.

Table 5.5–24   Standing Crop of Wetland and Aquatic Macrophytes (g DM/m²/year)

Plant Species	Standing Crop			Locality
	ABG	BLG	Total	
Emergent Species				
Acorus calamus	140–996	312		India
	605			New Jersey
	819			New Jersey
	1,250			Czech Republic
Alisma plantago-aquatica	52–444			Canada, Alberta
Althernanthera philoxeroides	118			South Carolina
	800			United States
Bidens laevis	17.4			Virginia
	22			South Carolina
	282			New Jersey
Bidens sp.	900			Pennsylvania
Bolboschoenus maritimus	167			Czech Republic
	334			Czech Republic
	456	870	1,326	Czech Republic
	480			Czech Republic
	535–613			Czech Republic
	1,659			Czech Republic
Butomus umbellatus	87–94	49	136	India
Carex acuta	605			Sweden

Table 5.5–24 Standing Crop of Wetland and Aquatic Macrophytes (g DM/m²/year) (continued)

Plant Species	Standing Crop			Locality
	ABG	**BLG**	**Total**	**Locality**
C. acutiformis	320–353			The Netherlands
	630			England
	692			Poland
	800			The Netherlands
	1,140	1,190	2,330	The Netherlands
	1,180	1,560	2,740	Germany
C. aquatillis	101	402	503	Canada, Alberta
	327			Canada, Alberta
	380			Canada, Alberta
	806			Canada, Quebec
C. atherodes	319			Canada, NWT
	624			Minnesota
	667			Iowa
	795			Minnesota
C. diandra	112–197			The Netherlands
	350			The Netherlands
	350	780	1,130	The Netherlands
C. elata	890			Sweden
	353–1,205			Czech Republic
C. gracilis	807			Germany
	950			Czech Republic
C. gracilis + vesicaria	402–732			Czech Republic
C. lacustris	465			Michigan
	485			Minnesota
	575			Sweden
	1,008			Czech Republic
	1,037	433	1,470	New York
	1,145	575		New York
		430		New York
		907–1,172		Iowa
C. lasiocarpa	143–305			Sweden
	270			Sweden
	416			Minnesota
	510			England
	367	1,310	1,677	New York
	940	910	1,850	The Netherlands
C. nebraskensis	161			California
	214–310			California
	474			California
C. rostrata	12–25	14–17		Sweden
	66	31	97	Sweden
	10–150			Sweden
	89–143			The Netherlands
	114–235	300		Minnesota
	165–222			Sweden
	150–328	114–852	1,002	Minnesota
	300			The Netherlands
	320–610			Sweden
	380			Finland
	389			Poland
	420			England
	434			Minnesota
	740			Canada, Alberta
	660	1,430	2,090	Minnesota
	975	431	1,406	New York
	650	1,350	2,000	The Netherlands

Table 5.5–24   Standing Crop of Wetland and Aquatic Macrophytes (g DM/m²/year) (continued)

| Plant Species | Standing Crop | | | Locality |
	ABG	BLG	Total	
C. stricta	585			New Jersey
	736			New York
C. trichocarpa	944	263	1,294	New York
Carex spp.	1–1,205			Czech Republic
Carex spp.	506			Michigan
Carex spp.	667			Iowa
Carex spp.		4,289		Michigan
Caladium jamaicense	403–803			Florida
	150–1,000	5,930–8,610		North Carolina
	1,130			Florida
C. mariscoides	224	119		New York
Cyperus papyrus	9,000–15,000			Central Africa
C. serotinus	6–279	232	447	India
Cyperus sp.			780	India
Distichlis spicata	670			New Jersey
	620–740			Louisiana
	280–970			Florida
	991			Louisiana
	950–960	13,990		North Carolina
		12,400		Delaware
Eleocharis acicularis	118–644			Czech Republic
E. acicularis (F)			1,093–1,905	Czech Republic
E. atropurpurea			40	India
E. capitata			78	India
E. equisetoides	377			South Carolina
E. palustris	12–153	149	302	India
	183–447			Canada, Alberta
			1,150–1,322	India
Eleocharis plantaginea	408	110	518–719	India
			2,200	India
E. quadrangulata	725			South Carolina
	881			South Carolina
Eleocharis sp.	50–63			Florida
Equisetum debile	88–113	60	148	India
E. fluviatile	10–21	20		Sweden
	320			Minnesota
	1,244			Czech Republic
E. limosum	590			Czech Republic
Glyceria grandis	656–673			England
G. maxima	78–208			England
	826			Czech Republic
	0.5–932			Czech Republic
	659–970			Czech Republic
	1,122			Czech Republic
	795–1,220			Czech Republic
G. maxima (W)	1,500	1,400	2,900	Poland
G. maxima	2,690			Czech Republic
Hydrocotyle umbellata	188			South Carolina
Impatiens capensis	22			South Carolina
	119			New Jersey
Ipomoea aquatica			251–2,141	India
Juncus articulatus	13	9	22	India
J. effusus	800			England
			1,592	South Carolina
J. kraussi	1,790	1,600	2,390	Australia

Table 5.5–24   Standing Crop of Wetland and Aquatic Macrophytes (g DM/m²/year) (continued)

| Plant Species | Standing Crop | | | Locality |
	ABG	BLG	Total	
J. roemerianus	786			North Carolina
	1,173			North Carolina
	1,240			Louisiana
	480–1,350	9,930–11,980		North Carolina
	1,959			Louisiana
	2,100			North Carolina
		4,060–5,140		Florida
		7,700–12,400		Mississippi
J. squarrosus	690			England
Justicia americana	95	94	189	North Carolina
	2,458			Alabama
Ludwigia peploides (N)	500–700			California
L. peploides (C)	1,900			California
L. uruguayensis	129			South Carolina
Lycopus europeans	76–83	41	117	India
L. rubellus	82			South Carolina
Lythrum salicaria	1,373			Pennsylvania
	2,104			New Jersey
Oenanthe aquatica	285			Czech Republic
Orontium aquaticum	244			South Carolina
Panicum hemitomon	1,075			South Carolina
P. repens	264			India
P. virgatum	652			Maryland
Panicum sp.	74.2			Florida
Papyrus sp.	2,140			Uganda
Peltandra virginica	84			South Carolina
	386			New Jersey
Phalaris arundinacea	203–296			Czech Republic
	566			New Jersey
	10–683			India
Phragmites australis	182			Finland
	745			Czech Republic
	775			Denmark
	872			Czech Republic
	934			Iowa
	942			England
	990			Louisiana
	840–1,000			Czech Republic
	1,000			Finland
	1,020			Poland
	1,115			Czech Republic
	1,260			Poland
	16–1,442			Czech Republic
	1,451			Maryland
	1,500–1,600			Sweden
	1,727			New Jersey
	960–1,880			Czech Republic
P. australis (F)	740–2,366			Czech Republic
P. australis	2,400			Sweden
	468–4,424	992–4,164		England
	8,147			Florida
	8,650			India
	780	620	1,400	Denmark
	1,360	574	1,934	Germany
		1,121–1,565		Iowa
	1,846	3,144	4,990	South Africa

Table 5.5–24   Standing Crop of Wetland and Aquatic Macrophytes (g DM/m²/year) (continued)

Plant Species	Standing Crop			Locality
	ABG	BLG	Total	
	2,050	6,260	8,310	Czech Republic
	1,170–2,068			Czech Republic
	2,960			Czech Republic
	3,250	8,560		Czech Republic
	265–5,280	824–3,048	1,089–8,328	India
P. australis (W)	6,334	5,348	11,168	South Africa
P. australis (C)	3,401	8,886	12,287	Czech Republic
Phragmites karka	10,950			India
	3,181–7,543	1,460–2,188	4,641–9,730	India
Polygonum arifolium	12.1			South Carolina
	31			Virginia
	200			New Jersey
P. glabrum			1,040–5,300	India
P. hydropiper	613			Czech Republic
	98–114	67	183	India
P. punctatum	14.9			South Carolina
Pontenderia cordata	257			South Carolina
	716			South Carolina
Sagittaria falcata	648			Louisiana
S. latifolia	460			Iowa
S. latifolia (W)	12,563			Florida
S. graminea	29			South Carolina
S. sagittifolia	14–142	6–73	20–191	India
Schoenoplectus lacustris	650			Czech Republic
	2,050			Czech Republic
	4,200			Czech Republic
Scirpus acutus	851			Iowa
		1,208–1,870		Iowa
S. americanus	150			South Carolina
	115–185	40–211	396	Canada, Québec
	410			South Carolina
S. articulatus	523	159	682	India
S. fluviatilis	466			Iowa
		1,254–1,424		Iowa
S. lacustris	197–774			India
	1,330	1,220	2,550	Denmark
	1,800			Poland
S. maritimus	6–71	42	101	India
S. palustris	365	333	624	India
S. pungens (W)	394			Florida
S. validus	28			South Carolina
	330			Iowa
	1,381			South Carolina
Sparganium angustifolium	0.03–0.14			Maine
S. erectum	12–117			England
	930			Czech Republic
	86–976	308–697		India
	1,033			Czech Republic
	942–1,610			Czech Republic
	1,880			Czech Republic
	710–985	412–634	1,122–1,619	India
S. eurycarpum	638			Iowa
	770	1,252–1,945		Iowa
	1,035	1,280		Michigan
	1,054			Iowa
S. eurycarpum (F)	637–1,185	681–1,123		Iowa

Table 5.5–24   Standing Crop of Wetland and Aquatic Macrophytes (g DM/m²/year) (continued)

| Plant Species | Standing Crop | | | Locality |
	ABG	BLG	Total	
S. eurycarpum	1,950			Wisconsin
Spartina alterniflora	649			Virginia
	754			Louisiana
	788–1,018			Louisiana
	1,320			North Carolina
	1,473	100–250		Louisiana
	1,592			New Jersey
	2,410			Virginia
S. cynosuroides	808			Louisiana
	826			Georgia
	968			Maryland
	700–970	6,930–9,260		North Carolina
	1,250	6,100–8,200		Georgia
	1,401			Virginia
	2,000			Misslssippi
		2,000–17,500		Georgia
S. patens	640			North Carolina
	805			Virginia
	1,376			Louisiana
	2,194			Louisiana
Typha angustata	305–1,565	100–428	405–1,993	India
	138–2,068			India
	708–6,803	586–4,440	1,808–8,131	India
T. angustifolia (W)	48–65			Michigan
T. angustifolia	800			Poland
	1,118			England
	1,780			Czech Republic
	2,560	1,015	3,575	Texas (1983)
	2,895	401	3,296	Texas (1984)
	3,710			Czech Republic
	4,000			Czech Republic
T. angustifolia (C)	3,039	2,070	5,109	Czech Republic
T. angustifolia + *latifolia*	478–814	3,295–4,889	4,261–6,460	Michigan
T. domingensis	225–590			Florida
	250–600	7,010–11,280		North Carolina
	1,483			South Carolina
T. elephantina	975–2,464	1,542–5,269	2,517–7,733	India
T. glauca	1,156			Iowa
	1,281			Iowa
	1,309–1,477			New York
	1,549	1,167–1,450		Iowa
	2,000	1,431	3,431	Iowa
T. glauca (F)	1,351–2,343	1,300–1,799		Iowa
T. latifolia (W)	48–62			Michigan
T. latifolia	181–322			Canada, Alberta
	574			South Carolina
	684			South Carolina
	456–848	393–807 (R)		Canada, Alberta
	1,070			England
	1,150			Czech Republic
	1,400			Wisconsin
	1,496			Wisconsin
	147–1,527			Oklahoma
	1,483			South Carolina
	1,400–1,600			Czech Republic
	1,620			Czech Republic
	1,566			New Jersey

Table 5.5–24   Standing Crop of Wetland and Aquatic Macrophytes (g DM/m²/year) (continued)

Plant Species	Standing Crop			Locality
	ABG	BLG	Total	
	428–2,252			S.E. United States
	3,600			Czech Republic
	416	556	972	Nebraska
	378	892	1,250	South Dakota
	404	912	1,316	North Dakota
	730	804	1,534	Oklahoma
T. latifolia (GR)	1,363	2,005	3,368	Florida
T. latifolia	215–1,054	395–2,223	610–3,405	Oregon
	1,336	2,646	3,982	Texas
	488–731	3,402–4,303	4,648–5,970	Michigan
T. latifolia (F)	720–4,371	1,444–3,860	2,164–8,231	Czech Republic
Typha sp.	966			Maryland
	987			New Jersey
	1,297			New Jersey
	1,310			Pennsylvania
	2,338			Maryland
T. orientalis (GR)	2,304	1,116	3,420	Australia
Zizania aquatica	560			Virginia
	866			New Jersey
	1,200			New Jersey
	1,390			New Jersey
Zizaniopsis miliacea	442			South Carolina
	1,039			Georgia

Floating-Leaved Species

Plant Species	ABG	BLG	Total	Locality
Nelumbo lutea	184			South Carolina
N. nucifera	200			India
	549–813			India
Nuphar advena	84			South Carolina
	245			Virginia
	516			New Jersey
	529	4,799		New Jersey
	605	1,146		New Jersey
	1,175			Pennsylvania
N. lutea	36	119	155	North Carolina
N. variegatum	89–235			Canada, Alberta
Nymphaea alba	21–246			India
N. odorata	110			Minnesota
	256			South Carolina
N. stellata	18–319			India
Nymphoides peltata	4–112		70–114	India
	228	273	501	The Netherlands

Submerged Species

Plant Species	ABG	BLG	Total	Locality
Batrachium fluitans	160			Czech Republic
B. rionii	90			Czech Republic
Ceratophyllum demersum	500–700			Sweden
			15–447	India
			16–480	India
Elodea canadensis	450			Czech Republic
Heteranthera dubia	5			Québec–Vermont
Hydrilla verticillata			29–470	India
			609	India
Myriophyllum spicatum	0.43–3.05			Austria
	20–60			Québec–Vermont
	220			Wisconsin

Table 5.5–24 Standing Crop of Wetland and Aquatic Macrophytes (g DM/m²/year) (continued)

Plant Species	Standing Crop			
	ABG	BLG	Total	Locality
	402			India
			2–304	Canada, Ontario
			644	India
			115–880	India
M. verticillatum	0.01–0.22			Maine
	310–379	360–580	670–959	India
Najas graminea			66–139	India
			144–203	India
			244	India
Najas major			197–469	India
N. maritima	83–478			Australia
Potamogeton crispus	129			Japan
			23–205	India
	6,410			Australia
P. epihydrus	0.02–0.03			Maine
P. lucens		100–220		Poland
			131–380	India
P. natans	146–295			Canada, Alberta
	1–89		70–298	India
	196–256			Czech Republic
P. pectinatus	17.5			North Baltic Sea
	100			Poland
	170			Czech Republic
	91–297			Canada, Alberta
			379–445	India
	78–950			Australia
	1,313			The Netherlands
P. perfoliatus	0.85			Hungary
	88–986			Australia
		82–225		Poland
P. pusillus			198–279	India
Potomogeton spp.	336			Minnesota
Renunculus aquatilis	12			India
Riccia rhenana	95			Czech Republic
Trapa bispinosa	82–545			India
			436–994	India
T. natans			10–490	India
	107			Czech Republic
	22–500			India
Utricularia intermedia	0.39			Maine
U. vulgaris	1.6–4.0			Maine
Utricularia cf. Vulgaris	80			Czech Republic
Utricularia sp.	2.5–6.9			Czech Republic
Utricularia sp.	143			Florida
Vallisneria americana	2–5			Québec–Vermont
V. spiralis	136–3,632			Australia
V. spiralis			169	India
Zannichelia palustris			285–295	India
Floating Species				
Azolla pinnata			98	India
			70–90	India
Eichhornia crassipes			798	India
			963	Japan
E. crassipes (W)			1,060	Czech Republic

Table 5.5–24   Standing Crop of Wetland and Aquatic Macrophytes (g DM/m²/year) (continued)

| Plant Species | Standing Crop | | | |
	ABG	BLG	Total	Locality
E. crassipes			440–1,276	Louisiana
			2,130	Alabama
E. crassipes (W)			2,970	Iowa
E. crassipes (F)			3,000	Florida
Lamna gibba			9–150	Czech Republic
L. paucicostata			19.2–49.1	Nigeria
L. trisulca			93	India
Lemna sp.			75–295	India
Pistia stratiotes			495	India
Spirodela polyrrhiza			134	Czech Republic
S. polyrrhiza			43–190	India
Bryophytes				
Sphagnum balticum	440			Finland
S. fuscum	344			USSR
Sphagnum sp.	0.48–8.85			Maine
Sphagnum sp.	54–450			Finland

Source: Vymazal, J., *Algae and Element Cycling in Wetlands,* Lewis Publishers, Boca Raton, FL, 1995, 384. With permission.

5.6 BENTHIC FAUNA

Table 5.6–1   Cleavage Patterns and Larval Forms of Marine Invertebrate Taxa

Taxon	Cleavage Pattern	Larval Form
Placozoa		?
Porifera	Holoblastic, equal/unequal	Amphiblastula and parenchymula
Cnidaria	Holoblastic, equal radial	
Hydrozoa		Planula
Scyphozoa		Planula
Anthozoa		Planula
Ctenophora		Cydippid
Platyhelminthes	Spiral, determinate	
Turbellaria		Müller's
Monogenea		Onchomiracidium
Trematoda		Miracidium and cercaria
Cestoda		Onchosphere (coracidium)
Mesozoa		Infusoriform
Rhynchocoela	Holoblastic, spiral, determinate	Pilidium
Gastrotricha		None
Nematoda	Holoblastic, bilateral determinate	None
Rotifera	Holoblastic, modified spiral	None
Acanthocephala		Acanthor
Kinorhyncha		None
Loricifera		Higgins
Priapulida		No name
Gnathostomulida		None
Annelida	Holoblastic, spiral determinate	
Polychaeta		Trochophore
Oligochaeta		None
Mollusca	Holoblastic, spiral, equal/unequal, determinate	
Gastropoda		Veliger
Monoplacophora		?
Polyplacophora		Trochophore
Aplacophora		Trochophore
Bivalvia		Trochophore, veliger, pericalymma
Scaphopoda		Trochophore and veliger
Cephalopoda	Incomplete	None
Arthropoda	Superficial/holoblastic spiral	
Merostomata		Trilobite
Arachnida		None
Pycnogonida		Protonymphon
Crustacea		Nauplius, zoea, megalopa, cypris, gaucothoe, phylosoma
Insecta	Superficial	Larva and pupa
Pogonophora		?
Sipuncula		Trochophore and pelagosphera
Echiura		Trochophore
Tardigrada	Holoblastic, subequal	None
Bryozoa (Ectoprocta)	Holoblastic, equal, radial	Coronate and cyphonautes
Entoprocta	Spiral	Modified trochophore
Phoronida	Holoblastic, radial	Actinotroch
Brachiopoda	Holoblastic, radial, equal/subequal	No name
Chaetognatha	Holoblastic, radial	None
Echinodermata	Holoblastic, radial, equal/unequal	
Asteroidea		Bipinnaria and brachiolaria
Ophiuroidea		Ophiopluteus
Echinoidea		Echinopluteus

Table 5.6–1   Cleavage Patterns and Larval Forms of Marine Invertebrate Taxa (continued)

Taxon	Cleavage Pattern	Larval Form
Holothuroidea		Auricularia and doliolaria
Crinoidea		Doliolaria
Hemichordata	Holoblastic, equal, radial	
Enteropneusta		Tornaria
Pterobranchia		No name
Planktosphaerordae		Planktosphera
Urochordata		
Ascidiacea	Holoblastic, bilateral, determinate	Tadpole
Thaliacea		None
Larvacea		Tadpole
Cephalochordata	Holoblastic, bilateral	Amphioxides

Note: Taxa are listed with corresponding larval forms and cleavage patterns where known. Major taxonomic headings are phyla; where appropriate other taxa are listed within a phylum. "None" indicates
that no larval form appears to exist; "?" indicates that the presence of a larval form is uncertain;
"No name" is substituted for those forms which do not possess a specific name other than larva,
e.g., brachiopod larva.

Source: Levin, L. A. and Bridges, T. S., in *Ecology of Marine Invertebrate Larvae,* McEdward, L., Ed., CRC
Press, Boca Raton, FL, 1995, 1. With permission.

Table 5.6–2   Field Surveys of Fertilization Success in Marine Invertebrates

Taxon	Number Spawning	MD	Mean, %	Reference[a]
Cnidaria				
Gorgonacea				
Briareum asbestinum	Spawning not observed	?	5.3	Brazeau and Lasker, 1992
	Spawning not observed	?	6.3	Brazeau and Lasker, 1992
	Spawning not observed	?	0.01	Brazeau and Lasker, 1992
Scleractinia				
Montipora digitata	None observed	?	36	Oliver and Babcock, 1992
	Many	?	75	Oliver and Babcock, 1992
	Many	?	32	Oliver and Babcock, 1992
Echinodermata				
Asteroidea				
Acanthaster planci	Peak major spawning	?	82.6 (SD 5.3)	Babcock and Mundy, 1992
	End major spawning	?	23.2 (SD 40.4)	Babcock and Mundy, 1992
	Minor spawning	10–15	23.1 (SD 15.7)	Babcock and Mundy, 1992
Holothuroidea				
Actinopyga lecanora	Several	?	72 (67–78)	Babcock et al., 1992
Bohadschia argus	4 (2 male)	<l	90 (86–96)	Babcock et al., 1992
	5 (4 male)	<l	81 (73–88)	Babcock et al., 1992
	Several	20–40	1 (0–2)	Babcock et al., 1992
Holothuria coluber	80% of population	?	83, 28, 10, 9[b]	Babcock et al., 1992
Cucumaria miniata	>1000	<l	92 (1–100)	Sewell and Levitan, 1992

Note: "MD" is the minimum distance (meters) between spawning males and females. "Mean" is mean percent
of eggs fertilized (range or standard deviation).

[a] See original source for reference citations.
[b] Samples taken at 10, 30, 35, 40 min after initiation of spawning.

Source: Levitan, D. R., in *Ecology of Marine Invertebrate Larvae,* McEdward, L., Ed., CRC Press, Boca Raton,
FL, 1995, 123. With permission.

Table 5.6–3 Factors Influencing Fertilization in Marine Invertebrates

Gamete	Individual	Population	Environmental
Sperm	Behavior	Density	Topographical complexity
Morphology	Aggregation	Size	Flow
Behavior	Synchrony	Distribution	Advective velocity
Velocity	Spawning posture	Size structure	Turbulence
Longevity	Spawning rate	Age structure	Water depth
Egg	Morphology	Sex ratio	Water quality
Size	Size		Temperature
Jelly coat	Reproductive output		Salinity
Chemotaxis	Age		pH
Sperm receptors	Energy allocation		
General			
Age			
Compatibility			

Source: Levitan, D. R., in *Ecology of Marine Invertebrate Larvae,* McEdward, L., Ed., CRC Press, Boca Raton, FL, 1995, 125. With permission.

Table 5.6–4 Endogenous Rhythms of Marine Animals

Rhythm	Clock	Environmental Period
Tidal	Circatidal	12.4 h
Lunadian	Circalunadian	24.8 h
Daily	Circadian	24.0 h
Biweekly	Circasemilunar	14.8 d
Monthly	Circalunar	29.5 d
Annual	Circannual	365.0 d

Note: Reproductive rhythms of animals are timed by endogenous clocks that approximate periodicities of environmental cycles.

Source: Morgan, S. G., in *Ecology of Marine Invertebrate Larvae,* McEdward, L., Ed., CRC Press, Boca Raton, FL, 1995, 160. With permission.

Table 5.6–5 Review of Hatching Rhythms of Crabs with Respect to Vertical Zonation and Lunar, Tidal Amplitude, Tidal, and Light Cycles

Species	Family	Lunar/Tidal Amplitude	Tidal	Light
Supratidal-High Intertidal				
Gecarcoidea natalis	Gecarcinidae	Annually	HT	N (Early)
Gecarcinus lateralis	Gecarcinidae	Monthly/biweekly	HT	N (Early)
Cardisoma guanhumi	Gecarcinidae	Monthly/biweekly	—	—
Aratus pisoni	Grapsidae	Biweekly	—	—
Sesarma intermedia	Grapsidae	Biweekly	HT	N (Early)
S. haematocheir	Grapsidae	Biweekly	HT	N (Early)
S. dehaani	Grapsidae	Biweekly	HT	N (Early)
S. cinereum	Grapsidae	Biweekly	HT	N (Early)
Uca rapax	Ocypodidae	Biweekly	Ebb	N (Late)

Table 5.6–5   Review of Hatching Rhythms of Crabs with Respect to Vertical Zonation and Lunar, Tidal Amplitude, Tidal, and Light Cycles (continued)

Species	Family	Lunar/Tidal Amplitude	Tidal	Light
High Intertidal				
Sesarma rhizophorae	Grapsidae	Monthly	HT	N (Late)
Chiromanthes onychophorum	Grapsidae	Monthly	HT	N
Uca rosea	Ocypodidae	Monthly	HT	N
U. oerstedi	Ocypodidae	Monthly	HT	N (Late)
U. galapagensis	Ocypodidae	Monthly	HT	N (Late)
Intertidal				
Uca minax	Ocypodidae	Biweekly	HT	N (Early)
U. pugilator	Ocypodidae	Biweekly	HT	N (Early/late)
U. pugnax	Ocypodidae	Biweekly	HT	N (Early)
U. tangeri	Ocypodidae	Biweekly	HT	N (Early)
U. beebei	Ocypididae	Biweekly	HT	D (Dawn)
U. dussumieri	Ocypodidae	Biweekly	HT	N
Metaplax elegans	Grapsidae	Biweekly	HT	N
Sesarma reticulatum	Grapsidae	Biweekly	HT	N (Early)
Pachygrapsus transversus	Grapsidae	Biweekly	HT	D and N (Crepuscular)
P. marmoratus	Grapsidae	Biweekly	—	—
Eurypanopeus transversus	Grapsidae	Biweekly?	Flood	N (Late)
Low Intertidal-Subtidal				
Cataleptodius floridanus	Xanthidae	Biweekly	HT	N (Dusk)
C. taboganus	Xanthidae	Biweekly	HT	N (Dusk)
Xanthodius sternberghii	Xanthidae	Biweekly	HT	N (Dusk)
Eurypanopeus planus	Xanthidae	Asynchronous	HT	D and N (Crepuscular)
E. depressus	Xanthidae	Asynchronous	—	D and N
Rhithropanopeus harrisii	Xanthidae	Asynchronous	HT	N (Early)
Dyspanopeus sayi	Xanthidae	Asynchronous	HT	D and N
D. texana	Xanthidae	Asynchronous	—	D and N
Panopeus herbstii	Xanthidae	Biweekly	HT	N (Early)
P. simpsoni	Xanthidae	Biweekly	—	N
P. obesus	Xanthidae	Biweekly	—	D and N
Xantho pilipes	Xanthidae	Asynchronous	—	—
Pinnixa chaetopterana	Pinnotheridae	Biweekly	HT	N (Early)
Pinnotheres ostreum	Pinnotheridae	Asynchronous	—	—
Petrolisthes armatus	Porcellanidae	Biweekly	HT	N (Crepuscular)
Subtidal				
Pinnotheres maculatus	Pinnotheridae	Asynchronous	—	—
Carcinus maenus	Portunidae	Biweekly	—	—
Callinectes arcuatus	Portunidae	Biweekly	—	—
C. sapidus	Portunidae	Asynchronous	—	N
Microphyrs bicornutus	Majidae	Asynchronous	—	D and N
Leucosilia jurinei	Leucosiidae	Asynchronous	—	D and N

Note: HT, near the time of high slack tide; N, night; D, day.

Source: Morgan, S. G., in Ecology of Marine Invertebrate Larvae, McEdward, L., Ed., CRC Press, Boca Raton, FL, 1995, 178. With permission.

**Table 5.6–6 Classification Schemes for Invertebrate
Larval Development**

Nutritional Modes	Site of Development
Planktotrophy	Planktonic
Facultative Planktotrophy	Demersal
Maternally derived	Benthic
Lecithotrophy	Aparental
Adelphophagy	Solitary
Translocation	Encapsulated
Osmotrophy	Parental
Autotrophy	Internal brooding
Photoautotrophy	External brooding
Chemoautotrophy	
Somatoautotrophy	

Dispersal Potential	Morphogenesis
Teleplanic	Indirect
Actaeplanic	Free-living
Anchiplanic	Contained
Aplanic	Direct

Source: Levin, L. A. and Bridges, T. S., in *Ecology of Marine Invertebrate Larvae,* McEdward, L., Ed., CRC Press, Boca Raton, FL, 1995, 11. With permission.

Table 5.6–7 Some Common Scalar and Vector Cues Used by Swimming Larvae and the General Categories of Behavioral Responses Associated with Each

Type of Cue	Cue	Category of Response
Scalar	Pressure	Barokinesis
	Salinity	Halokinesis
	Temperature	Thermokinesis
	Light intensity	Photokinesis
	Solid objects	Thigmokinesis
Vector	Gravity	Geotaxis
	Light	Phototaxis
	Current	Rheotaxis
	Light polarity	Polarotaxis

Source: Young, C. M., in *Ecology of Marine Invertebrate Larvae,* McEdward, L., Ed., CRC Press, Boca Raton, FL, 1995, 257. With permission.

Table 5.6–8 Mortality Rates of Invertebrate Larvae That Develop (Part A) in the Plankton or (Part B) Attached to the Substrate

Genus	Species	Stage of Development	Survival $S = N_t/N_o$	Time (day)	Mortality $M = \ln(S)/-t$	Reference[a]
Part A						
Crassostrea	gigas	Veligers to settled spat	1.80E-02	18	0.2558	Quayle, 1964
Ostrea		Veligers to settled spat	1.80E-02	30	0.1339	Quayle, 1964
	edulis	Larvae to settled spat	2.38E-02	17	0.2170	Korringa, 1941
		Larvae to settled spat	2.50E-02	13	0.2838	Korringa, 1941
Mytilus	edulis	Veliger larvae	2.43E-02	29	0.1282	Jorgensen, 1981
Mercenaria	mercenaria	Larvae to settled spat	2.60E-02			Carriker, 1961
Membranipora	membranacea	Cyphonautes larvae	2.50E-03	33	0.1816	Yoshioka 1982
		Cyphonautes larvae	1.80E-06	37	0.3575	Yoshioka, 1982
Balanus	balanoides	Nauplii and cyprids	3.20E-01	19	0.0600	Pyefinch, 1950
	improvisus	Nauplii and cyprids			0.1450	Bousefield, 1955
Penaeus	duorarum	Zoea larvae	8.04E-01	1	0.2200	Munro et al., 1968
Panulirus	interruptus	Zoea larvae			0.0180	Johnson, 1960
Homarus	americanus	Zoea larvae	1.13E-02			Scarratt, 1964
Cancer	magister	Zoea larvae	4.46E-01	50	0.0161	Lough, 1976
Pandalus	jordani	Zoea larvae	6.70E-03	114	0.0471	Rothlisberg and Miller, 1982
		Zoea larvae	6.80E-04	100	0.0502	Rothlisberg and Miller, 1982
Strongylocentrotus	droebachiensis	Pluteus larvae	1.31E-01	13	0.1563	Rumrill, 1987
	purpuratus	Pluteus larvae	3.46E-01	17	0.0625	Rumrill, 1987
		Pluteus larvae	5.37E-02	11	0.2658	Rumrill, 1987
Paracalanus	parvus	Nauplii	1.90E-01	19	0.0874	Gaudy, 1976
Centropages	typicus	Nauplii	6.50E-03	21	0.2398	Gaudy, 1976
Euterpina	acutifrons	Nauplii	1.56E-02	8	0.2322	Gaudy, 1976
		Nauplius to copepodite	5.08E-02	6	0.4258	D'Apolito and Stancyk, 1979
Acartia	clausi	Nauplii	4.33E-02	14	0.2243	Gaudy, 1976
	tonsa	Nauplii	1.22E-01	3	0.7012	Allan et al., 1976
		Nauplii			0.8200	Heinle, 1966
Oithona	helgolandica	Nauplii	8.50E-02	20	0.1233	Gaudy, 1976
	nana	Nauplii	5.50E-01	15	0.0399	Gaudy, 1976
Eurytemora	affinis	Nauplii	3.57E-01	1	1.0100	Allan et al., 1976
Calanus	helgolandicus	Nauplii			0.4200	Mullin and Brooks, 1979

Part B

Leptasterias	hexactis	External brooded embryos	0.5550	60	0.0098	Menge, 1975
Ophioplocus	esmarki	Internal brooded embryos	0.7616	59	0.0046	Rumrill, 1983
Thais	lamellosa	Encapsulated embryos	0.0042	900	0.0061	Spight, 1975
Conus	pennaceus	Embryos-juveniles	0.0007	365	0.0199	Perron, 1986
Cerithidea	californica	Encapsulated embryos	0.4230	21	0.0410	Race, 1982
Neptunea	pribiloffensis	Encapsulated embryos	0.8694	28	0.0050	Shimek, 1981
Ilyanassa	obsoleta	Encapsulated embryos	0.4800	10	0.0734	Brenchley, 1982
Gemma	gemma	Brooded embryos	0.2080	30	0.0523	Sellmer, 1967
Axiothella	mucosa	Brooded embryos	0.1111			Wilson, 1986

Note: Mortality rates in the plankton were estimated by tracking larval cohorts. Most mortality rates were calculated using the equation, $M = \ln (N_0/N_t)/-t$, where M is the instanteous mortality rate of larvae; N_t is the abundance of larvae within a specific water mass after time interval, t; and N_0 is the initial number of larvae.

[a] See original source for reference citations.

Source: Rumrill, S. S., *Ophelia*, 32, 163–198, 1990. With permission. Morgan, S. G., in *Ecology of Marine Invertebrate Larvae*, McEdward, L., Ed., CRC Press, Boca Raton, FL, 1995, 282. With permission.

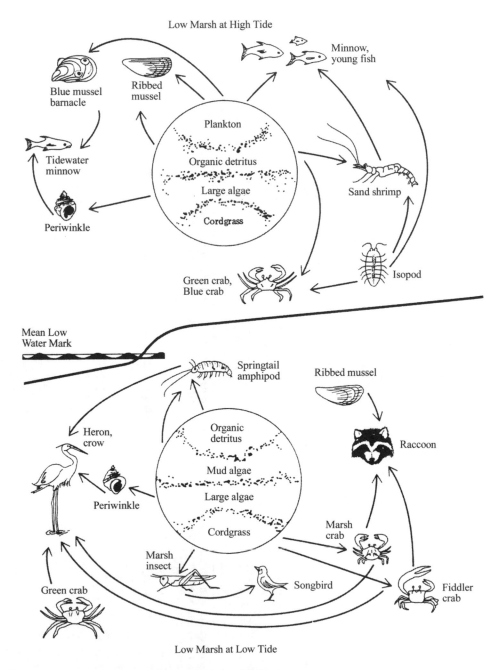

Figure 5.6–1   Characteristic animals present in a salt marsh at low and high tides on the Atlantic coast of North America. (From Nybakken, J. W., *Marine Biology: An Ecological Approach,* Harper & Row, New York, 1988, 363. With permission.)

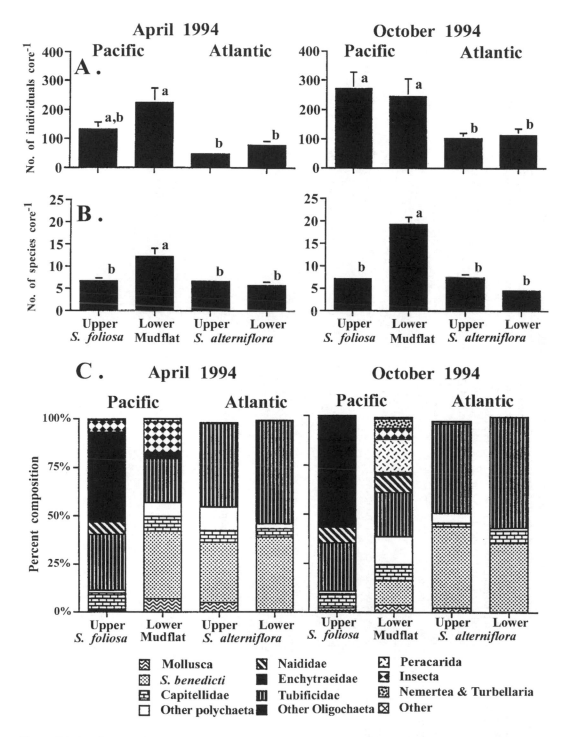

Figure 5.6–2 Comparison of macrobenthos (≥0.3 mm) in the Northern Wildlife Preserve, Mission Bay, California (Pacific), and the Port Marsh, Newport River, North Carolina (Atlantic) marsh, at two elevations on two sampling dates. Upper elevation transects have *Spartina foliosa* in Mission Bay and *S. alterniflora* in the Port Marsh. Lower elevation transects are mudflats in Mission Bay and *S. alterniflora* in the Port Marsh. (A) Mean density (±1 SE) (number of individuals 18.05/cm²). (B) Mean species richness (±1 SD) (number of species 18.05/cm²) (C). Representation of major taxa as a fraction of total macrofauna. (From Levin, L. A. et al., *Estuaries*, 21, 139, 1998. With permission.)

Table 5.6–9 Densities of Meiobenthos (Number of Individuals 10/cm²) in Some Tropical Mangrove and Coral Reef Sediments

Location	Habitat and Season	Total Faunal Densities	Sediment Depth (cm)	Percent of Dominant Taxon	Ref.[a]
Tuamotu, French Polynesia	Mururoa Lagoon, summer	440 (390–1293)	0–1	Nematodes, 52–56%	96
	Maturei Vavao Atoll lagoon, autumn	111 (36–240)	0–5	Nematodes, 16–61%	55
Bermuda	Beach and lagoon, carbonate sands (all seasons)	522 (122–1333)	0–10	Nematodes, $\bar{x}$ = 73%	103
Moorea Island, French Polynesia	Reef flat, winter	79 (63–107)	0–3	Polychaetes, $\bar{x}$ = 42%	98
	Lagoon channel	60			
	Reef flat	57 (39–93)			
	Deep outer reef slope	120			
Grand Recif, Madagascar	Inner reef flat, winter	327 (263–610)	0–1.5	Copepods, $\bar{x}$ = 61%	98
Godavari River, India	Low intertidal, all seasons mangroves	2131	0–20	Nematodes, $\bar{x}$ = 86%	105
Lakshadweep, S. Arabian Sea	Carbonate beach sand, winter				
	High beach	25 (7–48)	0–5	Nematodes, $\bar{x}$ = 31%	100
	Mid-beach	35 (3–60)			
	Low beach	28 (3–48)			
Bay of Bengal, India	Mangroves, low intertidal				
	Winter	156 (35–270)	0–9	Nematodes, 50–67%	110
	Spring	193 (90–280)			
	Summer	173 (80–280)			
	Autumn, monsoon	88 (60–160)			70
Cebu, Philippines	Beach-reef carbonates, autumn	3139 (744–8769)	0–5	Nematodes, $\bar{x}$ = 39%	117
Great Barrier Reef, Australia	Lagoon, winter		0–5		118
	Lizard Island	1182		Nematodes, 85%	
	Orpheus Island	445		Copepods, 47%	
	Davies Reef	223		Copepods, 82%	
Mangroves, South Cuba	Tolete Lagoon		0–5	Nematodes, $\bar{x}$ = 54%	119
	Dry season	196			
	Wet season	115			
	Basto Lagoon				
	Dry season	245			
	Wet season	36			

Location	Habitat	Mean (and range)		Dominant taxa	
Davies Reef, Great Barrier Reef	Reef front, winter	477 (280–511)			
	Reef flat, winter	615 (415–1010)	0–5	Copepods, $\bar{x} = 86\%$ Copepods, $\bar{x} = 51\%$	33
Isla del Caro Reefs, Costa Rica	Shallow lagoon, winter	631 (622–640)		Copepods, $\bar{x} = 48\%$	
	Deep lagoon, winter	231 (220–255)		Nematodes, $\bar{x} = 50\%$	
	Reef flat				
	Winter	248 (99–330)	0–6	Foraminfera, $\bar{x} = 21\%$	232
	Spring	317 (270–363)			
	Summer	221 (117–325)		Copepods, $\bar{x} = 20\%$	
	Autumn	501 (427–575)			
Cape York Peninsula, Australia	Mangrove estuaries				
	Summer wet	776 (217–2454)	0–5	Turbellaria, $\bar{x} = 70\%$	19
	Winter dry	330 (66–1660)		Turbellaria, $\bar{x} = 46\%$	
Hinchinbrook Island, Australia	Mangroves, all seasons				
	Low intertidal	767 (347–1840)	0–5	Turbellaria, $\bar{x} = 68\%$	19
	Mid-intertidal	94 (57–187)		Nematoda, $\bar{x} = 86\%$	
	High intertidal	42 (14–85)		Nematoda, $\bar{x} = 67\%$	

Note: Values depict mean (and range) except where noted.

[a] Full citations of references in original source.

Source: Alongi, D. M., *Rev. Aquat. Sci.,* 1(2), 255, 1989. With permission.

Table 5.6–10   Annual Production of Macrobenthos (Production in Dry Weight Unless Otherwise Stated)

Species	Production (m²/year)	P/$\overline{\text{B}}$	Maximum Age (years)	Locality	Ref.[a]
Nephtys incisa	9.34 g	2.16	3	Long Island Sound, U.S., 4–30 m	190
Cistenoides gouldii	1.70 g	1.94	2	Long Island Sound, U.S., 4–30 m	190
Yoldia limatula	3.21 g	2.28	2	Long Island Sound, U.S., 4–30 m	190
Pandora gouldiana	6.13 g	1.99	2	Long Island Sound, U.S., 4–30 m	190
Moira atropos	2.52 g	0.70	6	Biscayne Bay, Florida, U.S., 3 m	191
Tagelus divisus	21.0 g	1.78	2	Biscayne Bay, Florida, U.S., L. W. S	192
Ampharete acutifrons	0.719 g (wet)	4.58	1	Long Island Sound, U.S., 917 m	193
Neomysis americana	36.2 mg	3.66	1?	Long Island Sound, U.S., 917 m	193
Crangon septemspinosa	0.519 g	3.82	3	Long Island Sound, U.S., 917 m	193
Asterias forbesi	4.52 g	2.64	3	Long Island Sound, U.S., 917 m	193
Tellina martinicensis	0.23 g	2.40	2	Biscayne Bay, Florida, U.S., 3 m	194
Chione cancellata	8.9 g	0.42	7	Biscayne Bay, Florida, U.S., M.L.W.S.	195
Dosinia elegans	0.13 g	1.25	2	Biscayne Bay, Florida, U.S., 3 m	196
Pectinaria hyperborea	10.6 g	4.60	2	St. Margaret's Bay, Nova Scotia, 60 m	197
Scrobicularia plana	60 kcal	0.29	7?	North Wales, lower shore	198
	13.3 kcal	0.67	4	North Wales, upper shore	198
Anodontia alba	14.09 g	1.43	4 + (?)	Biscayne Bay, Florida, U.S., low water	199
Strongylocentrotus droebachiensis	401.0 kcal	0.80	6	St. Margaret's Bay, Nova Scotia, intertidal	200
Neanthes virens	45.2 kcal	1.62	3	Thames estuary, U.K., intertidal	201
Ammotrypane aulogaster	359 mg	2.08	?	Northumberland, U.K., 80 m	202
Heteromastus filiformis	297 mg	1.01	2	Northumberland, U.K., 80 m	202
Spiophanes kroyeri	196 mg	1.40	3	Northumberland, U.K., 80 m	202
Glycera rouxi	192 mg	0.37	5	Northumberland, U.K., 80 m	202
Calocaris macandreae	142 mg	0.12	9.5	Northumberland, U.K., 80 m	202
Abra nitida	118 mg	1.11	3	Northumberland, U.K., 80 m	202
Lumbrineris fragilis	78 mg	1.34	3	Northumberland, U.K., 80 m	202
Chaetozone setosa	50 mg	1.28	3	Northumberland, U.K., 80 m	202
Brissopsis lyrifera	108 mg	0.30	4	Northumberland, U.K., 80 m	202
Mya arenaria	11.6 g	2.54	3	Petpeswick Inlet, E. Canada, intertidal	203
Macoma balthica	1.93 g	1.53	3	Petpeswick Inlet, E. Canada, intertidal	203
Littorina saxatilis	3.25 g	4.11	1	Petpeswick Inlet, E. Canada, intertidal	203
Macoma balthica	10.07 g	2.07	6	Ythan Estuary, Scotland, intertidal	204
Nephtys hombergi	7.34 g	1.90	3	Lynher Estuary, U.K., intertidal	205
Ampharete acutifrons	2.32 g	5.50	1	Lynher Estuary, U.K., intertidal	205
Mya arenaria	2.66 g	0.50	8	Lynher Estuary, U.K., intertidal	205

Species	Weight			Location	Ref.
Scrobicularia plana	0.48 g	0.20	9	Lynher Estuary, U.K., intertidal	205
Macoma balthica	0.31 g	0.90	6	Lynher Estuary, U.K., intertidal	205
Cerastoderma edule	0.21 g	0.20	7	Lynher Estuary, U.K., intertidal	205
Ampelisca brevicornis	4.26 g (wet)	3.95	1.25	Helgoland Bight, 28 m	206
	2.43 g (wet)	3.68	1.25		
Pectinaria californiensis	2.02 g C	5.30	1.2	Puget Sound, Washington, U.S., 34 m	207
	2.798 g C	3.30	2.1	Puget Sound, Washington, U.S., 203 m	207
	3.471 g C	4.10	1.8	Puget Sound, Washington, U.S., 254 m	207
	1.386 g C	5.50	1.9	Puget Sound, Washington, U.S., 207 m	207
	4.816 g C	3.40	2.4	Puget Sound, Washington, U.S., 71 m	207
Cerastoderma edule	29.25 g	1.59	5	Southampton Water, U.K., intertidal	208
	71.36 g	1.10	5		
	46.44 g	2.61	5		
Mercenaria mercenaria	3.99 g	0.52	8	Southampton Water, U.K., intertidal	208
	14.00 g	0.28	8		
	6.19 g	0.17	9		
Venerupis aurea	0.70 g	1.11	5	Southampton Water, U.K., intertidal	208
	1.25 g	1.10	5		
Crassostrea virginica	3828 kcal	1.87	?	South Carolina, U.S., intertidal	209
Littorina littorea	6.13 g	0.61	?	Grevelingen Estuary, Netherlands, intertidal	210
Hydrobia ulvae	7.23 g	1.78	1	Grevelingen Estuary, Netherlands, intertidal	210
	8.80 g	1.24	1		
	12.79 g	1.36	2		
Cardium edule	10.21 g	0.69	3.5	Grevelingen Estuary, Netherlands, intertidal	210
	119.82 g	2.56	3.5		
	51.76 g	1.13	3.5		
Macoma balthica	3.40 g	1.93	8.10	Grevelingen Estuary, Netherlands, intertidal	210
	0.95 g	1.00	8.10		
	0.07 g	0.30	8.10		
	−0.74 g	−0.25	8.10		
Arenicola marina	3.79 g	1.14	3	Grevelingen Estuary, Netherlands, intertidal	210
	6.26 g	0.72	3		
	3.32 g	0.99	3		

(continued)

Table 5.6–10 Annual Production of Macrobenthos (Production in Dry Weight Unless Otherwise Stated) (continued)

Species	Production (m²/year)	P/B̄	Maximum Age (years)	Locality	Ref.[a]
Pontoporeia affinis	3.17 g	1.90	3	North Baltic, 64 m	211
P. femorata	3.03 g	1.43	3	North Baltic, 64 m	211
Harmothoe sarsi	0.23 g	1.99	3	North Baltic, 64 m	211
Pharus legumen	16.12 g	0.56	6	Carmarthen Bay, South Wales, 13.5 m	212
Spiophanes bombyx	3.35 g	4.86	?	Carmarthen Bay, South Wales, 13.5 m	212
Ensis siliqua	1.37 g	0.27	10	Carmarthen Bay, South Wales, 13.5 m	212
Donax vittatus	0.72 g	2.10	2.5	Carmarthen Bay, South Wales, 13.5 m	212
Magelona papillicornis	0.69 g	1.10	3	Carmarthen Bay, South Wales, 13.5 m	212
Venus striatula	0.62 g	0.41	10	Carmarthen Bay, South Wales, 13.5 m	212
Ophiura texturata	0.46 g	0.68	3	Carmarthen Bay, South Wales, 13.5 m	212
Tellina fabula	0.29 g	0.90	6	Carmarthen Bay, South Wales, 13.5 m	212
Glycera alba	0.28 g	0.97	3	Carmarthen Bay, South Wales, 13.5 m	212
Sigalion mathildae	0.17 g	0.44	?	Carmarthen Bay, South Wales, 13.5 m	212
Tharyx marioni	0.015 g	0.79	2	Carmarthen Bay, South Wales, 13.5 m	212
Astropecten irregularis	0.0004 g	0.005	?	Carmarthen Bay, South Wales, 13.5 m	212
Echinocardium cordatum	–0.012 g	–0.02	3	Carmarthen Bay, South Wales, 13.5 m	212

[a] References from Kennish, M. J., *Ecology of Estuaries*, Vol. 1, CRC Press, Boca Raton, FL, 1986.

Source: Modified from Warwick, R. M., in *Marine Benthic Dynamics*, Tenore, K. R. and Coull, B. C., Eds., University of South Carolina Press, Columbia, 1980, 1. With permission.

Table 5.6-11 **Production Estimates for Estuarine Macrobenthic Assemblages in Selected Estuaries of Europe and the United States[a]**

Area	Depth	Production	Ref.[b]
Long Island Sound	Subtidal Subtidal	21.4 (infauna) 5.8 (epifauna)	Sanders (1956), Richards and Riley (1967)
Firemore Bay, Scotland	Whole bay	2.5 (g carbon)	McIntyre and Eleftheriou (1968)
Kiel Bight, Germany	15 m	17.9	Arntz and Brunswig (1975)
Tamar Estuary, England	Intertidal	13.3	Warwick and Price (1975)
Southampton Water, England	Intertidal	152–225 (?)	Hibbert (1976)
Grevelingen Estuary, the Netherlands	Whole estuary	50.3–57.4	Wolff and De Wolf (1977)

[a] Values expressed in grams ash-free dry weight per square meter per year.
[b] Complete citations in source.

Source: Wolff, W. J., *Estuaries and Enclosed Seas,* Ketchum, B. H., Ed., Elsevier, Amsterdam, 1983, 151.

Table 5.6–12 **Biomass Values of Macrobenthic Species Assemblages in Selected European and U.S. Estuaries**

Area	Depth	Biomass (g m²)	Ref.[a]
Wadden Sea, Denmark	Intertidal	174–497 (wet)	Smidt (1951)
Ringkøbing Fjord, Denmark	Intertidal	267–450 (wet)	Smidt (1951)
Long Island Sound	Subtidal	54.6 (dry)	Sanders (1956)
Barnstable Harbor	Intertidal	38–40 (dry)	Sanders et al. (1962)
North Carolina estuary		8.1 (dry tissue)	Williams and Thomas (1967)
Puget Sound	12–195 m	8–9 (ash-free)	Lie (1968)
Firemore Bay, Scotland	Beach	1.3 (dry)	McIntyre and Eleftheriou (1968)
	Subtidal	3.7 (dry)	McIntyre and Eleftheriou (1968)
Biscayne Bay	Intertidal	30 (dry)	Moore et al. (1968)
Narragansett Bay	Intertidal musselbed	1852 (ash-free)	Nixon et al. (1971)
Lower Mystic River	15 m	1.0–50 (wet)	Rowe et al. (1972)
Balgzand, the Netherlands	Intertidal	9.6–25.1 (ash-free)	Beukema (1974)
Kiel Bay, Germany	>15 m	26.3 (ash-free)	Arntz and Brunswig (1975)
Orwell Estuary, England	Intertidal	46.0 (ash-free)	Kay and Knights (1975)
Stour Estuary, England	Intertidal	28.1–38.7 (ash-free)	Kay and Knights (1975)
Colne Estuary, England	Intertidal	7.8 (ash-free)	Kay and Knights (1975)
Dengie Flats, England	Intertidal	18.5 (ash-free)	Kay and Knights (1975)
Crouch Estuary, England	Intertidal	8.7 (ash-free)	Kay and Knights (1975)
Roach Estuary, England	Intertidal	8.7 (ash-free)	Kay and Knights (1975)
Foulness Flats, England	Intertidal	19.6 (ash-free)	Kay and Knights (1975)
Thames Estuary, England	Intertidal	18.8 (ash-free)	Kay and Knights (1975)
Medway Estuary, England	Intertidal	24.1 (ash-free)	Kay and Knights (1975)
Swale Estuary, England	Intertidal	32.8 (ash-free)	Kay and Knights (1975)
Tamar Estuary, England	Intertidal	13 (ash-free)	Warwick and Price (1975)
Southampton Water, England	Intertidal	90–190 (ash-free)	Hibbert (1976)
Wadden Sea, the Netherlands	Intertidal	27 (ash-free)	Beukema (1976)
Grevelingen Estuary, the Netherlands	Whole estuary	20.8 (ash-free)	Wolff and De Wolf (1977)
Byfjord, Sweden	Subtidal	4.5 (dry)	Rosenberg et al. (1977)
Delaware River	Intertidal fresh water	7.0 (dry)	Crumb (1977)
San Francisco Bay	Intertidal	10.1–30.5 (ash-free)	Nichols (1977)

[a] Complete citations in source.

Source: Wolff, W. J., *Estuaries and Enclosed Seas,* Ketchum, B. H., Ed., Elsevier, Amsterdam, 1983, 151. With permission.

Table 5.6–13 Densities of Macrobenthos (Number of Individuals/m²) in Some Tropical Mangrove, Coral Reef, and Adjacent Mudflat and Sand Flat Sediments

Location	Habitat	Season	Total Densities	Species Richness (no./site)	Ref.[a]
Selangor, Malaysia	Mangrove, low intertidal	All	70 (±102)	0–11	122
	Epifauna				
	Infauna		137 (±91)	7–24	
Morrumbere Estuary	Mangroves, low intertidal	All	170 (±152)	31–74	124
	Sand flat		242 (±235)	42–103	
Phuket Island, Thailand	Low intertidal mangroves	Autumm	80 (±28)	26	125
	Mid-intertidal mangroves		218 (±34)	92	
	High intertidal mangroves		129 (±65)	60	
	Mudflat		52	36	
	Sand flat		147(±20)	47	
Surin Island, Thailand	Low intertidal mangroves	Spring	4	5	126
	Mid-intertidal mangroves		10	8	
	High intertidal mangroves		28	34	
	Mudflat		26	11	
	Sand flat		43	11	
Kuala Lumpur, Malaysia	Mudflat near mangroves	All	304 (±247)	22	127
Northwest Cape, W. Australia	Mudflat	Spring	992 (±722)	122	128
	Avicennia forest		257 (±390)	59	
	Rhizophora forest		473 (±319)	31	
	High intertidal flat		1 (±3)		
Cochin Estuary, India	Low intertidal mangroves	Premonsoon	5,872	N.A.	123
		Monsoon	420		
	Mudflat	Premonsoon	16,000		
		Monsoon	1,036		
Ka Yao Yai, Thailand	Low intertidal mangroves	All	49	43	129
	Mid-intertidal mangroves		107	55	
	High intertidal mangroves		142–178	70	
	Mudflat		247	109	
	Sand flat		190	48	

Location	Habitat	Season			Ref.
Maturei Vavao, Polynesia	Coral lagoon	Autumn	1,138 (249–3,264)	N.A.	55
Discovery Bay, Jamaica	Coral lagoon	Spring	1,610 (918–2,660)	38(26–50)	130
Grand Recif, Madagascar	Coral lagoon	Winter	632 (84–1,752)	57(27–160)	97
Florida Keys, U.S.	Carbonate lagoon	Spring	5,528 (275–12,825)		120
		Summer	7,181 (725–27,020)	55(33–76) 58 (24–76)	
Red Sea, Sudan	Fore reef	Autumn	30,599–42,840	35–79	131
	Reef lagoon		5,981–12,125	18–21	
Gulf of Aquaba, Red Sea	Fringing reef	Winter	673	57	132
	Lagoon	Summer	449	41	
		Autumn	307	49	
Moorea, Polynesia	Fringing reef	Summer	2,242–4,866	N.A.	98
	Reef flat		496–1,824		
	Reef slope		632		
	Lagoon		1,335		
Carrie Bow Cay, Belize	Carbonate	Spring	16,750	32–133	121
	Sand flat				

a Full citations in source.

Note: Values depict means and SD or ranges in parentheses; N.A. = information not provided.

Source: Alongi, D. M., *Rev. Aquat. Sci.,* 1(2), 259, 1989. With permission.

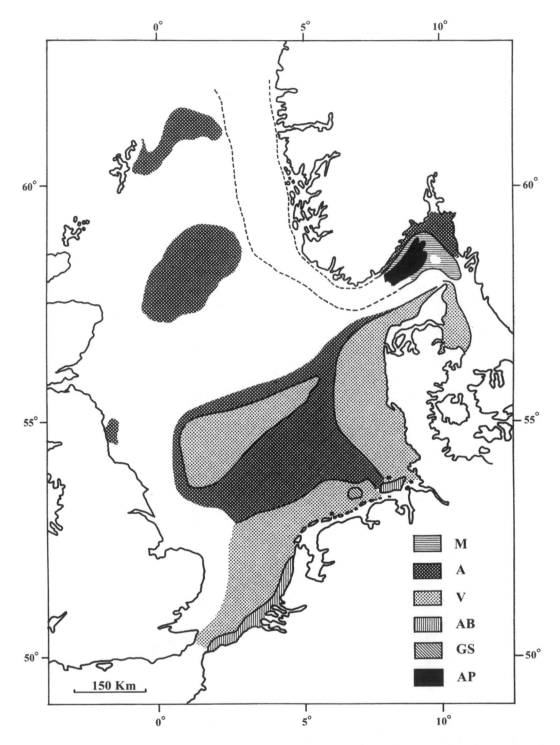

Figure 5.6–3   Distribution of benthic communities in the North Sea. M: *Maldane sarsi–Ophiura sarsi* community; A: *Amphiura filiformis* community; V: *Venus* community; AB: *Abra alba* community; GS: *Goniadella–Spisula* community; and AP: *Amphilepis–Pecten* community. (From Eisma, D., *Rev. Aquat. Sci.,* 3, 211, 1990. With permission.)

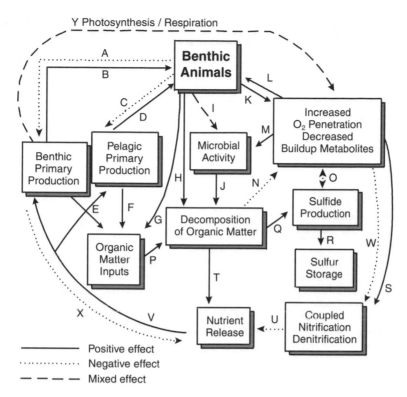

Figure 5.6–4   Model of the interrelationships between benthic organisms and biogeochemical processes in marine sediments. Solid lines represent positive interactions, dashed lines depict negative interactions, and dotted lines represent mixed outcomes. Letters are keyed to processes discussed in the source. (From Alongi, D. M., *Coastal Ecosystem Processes,* CRC Press, Boca Raton, FL, 1998, 7. With permission.)

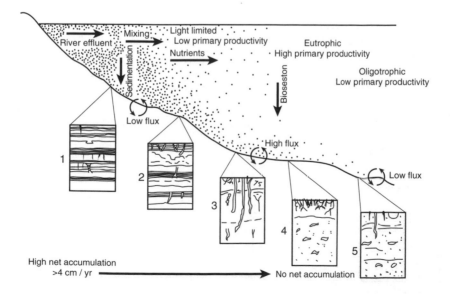

Figure 5.6–5   Idealized model of infaunal responses and sedimentary fabric to discharge from large tidal rivers. (From Alongi, D. M., *Coastal Ecosystem Processes,* CRC Press, Boca Raton, FL, 1998, 248. With permission.)

Table 5.6–14   Biotic Effects on Habitat Characteristics Proposed by Functional-Group Hypotheses

Habitat Alteration	Functional-Group Hypothesis Proposing Effect	Examples[a]
Bioturbation of sediments (sediment reworking and destabilization)	Adult-larval interactions, trophic group amensalism, mobility-mode interactions	*Callianassa californiensis* (50 ml/ind./d)
		Callianassa spp. (up to 1300 ml/ind./d or 3.9 kg/m²/d)
		Saccoglossus kowalevskii (36 ml/m²/h)
		Enteropneust sp. (up to 700 ml/ind./d or 600 ml/m²/d)
		Arenicola marina (2500 ml/m²/d)
Resuspension of sediments	Trophic-group amensalism	*Callianassa californiensis* (150g/m²/d)
		Callianassa spp. (up to 1300 ml/ind./d or 3.9 kg/m²/d)
		Upogebia pugettensis (6 g/m²/d)
		Enteropneust sp. (600 ml/m²/d)
		Arenicola marina (400 1/m²/yr)
Sediment stabilization	Mobility-mode interactions	*Zostera marina* (reduced erosion in experimental aquaria)
		Tube mat (reduced erosion)
		Saltmarsh plants (inhibition of bioturbator species)
		Diopatra cuprea tubes (inhibition of bioturbator species)

[a] Sediment reworking and resuspension rates given in parentheses.
Source: Posey, M. H., *Rev. Aquat. Sci.,* 2(3,4), 345, 1990. With permission.

Table 5.6–15   Rates of Benthic Nitrogen Recycling in Some Representative Coastal Environments

Location	Benthic N Flux NH$_4^+$	Benthic N Flux NO$_x$	Denitrification	% of Plankton N Requirements
Belgian coastal zone	25	32	17 (23%)	78
North Sea Bight	13	17	5 (15%)	38
Great Belt (Denmark)	10	7	49 (16%)	36
Kattegat (Denmark)	13	5	69 (22%)	42
E. Kattegat (Denmark)	20	5	28 (8%)	53
Limfjord (Denmark)	25	13	188 (26%)	55
La Jolla Bight (U.S.)	12	2	—	7
Narragansett Bay (U.S.)	34	4	33 (16%)	—
South River (U.S.)	38	0.5	—	29
Neuse River (U.S.)	76	1	—	26
Georgia Bight (U.S.)	55	3	—	16
Buzzards Bay (U.S.)	18	1	—	70
Cap Blanc (Africa)	78	58	—	39
Missionary Bay (Australia)	6	1	—	6
Great Barrier Reef shelf (Australia)	12	1	—	13

Note: Values are expressed as mg N/m²/day. Values in parentheses represent percentage of total benthic nitrogen mineralization.

Source: Alongi, D. M., *Coastal Ecosystem Processes,* CRC Press, Boca Raton, FL, 1998, 210. With permission.

5.7 NEKTON

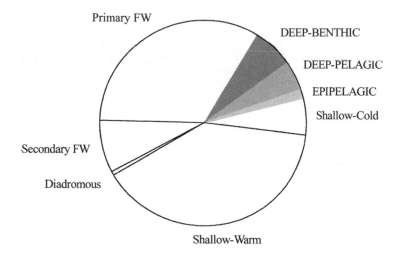

Figure 5.7–1   The relative numbers of fish species in the world, according to habitat. FW = freshwater species. (From Haedrich, R. L., in *Deep-Sea Fishes,* Randall, D. J. and Farrell, A. P., Eds., Academic Press, New York, 1997, 80. With permission.)

Table 5.7–1   Feeding Strategies of Representative Fishes of Barnegat Bay, NJ

Species	Feeding Strategy
Alosa aestivalis (blue herring)	Primarily planktivorous; consumes copepods, mysids, ostracods, small fishes
Ammodytes americanus (sand lance)	Principally feeds on small crustaceans
Anchoa mitchilli (bay anchovy)	Primarily planktivorous; consumes mysids, copepods, organic detritus, small fishes
Anguilla rostrata (American eel)	Omnivorous; consumes annelids, crustaceans, echinoderms, mollusks, eelgrass, small fishes
Bairdiella chrysoura (silver perch)	Omnivorous; consumes calanoid copepods, annelids, fishes, organic detritus
Brevoortia tyrannus (Atlantic menhaden)	Planktivorous; detritivorous; principally filter-feeds on phytoplankton; also feeds on zooplankton and organic detritus
Caranx hippos (crevalle jack)	Fish predator
Centropristis striata (black sea bass)	Omnivorous; benthophagous; feeds on crustaceans, mollusks, fishes, plants
Cynoscion regalis (weakfish)	Primarily a fish predator; mostly consumes anchovies and silversides; also feeds on small crustaceans
Fundulus heteroclitus (mummichog)	Omnivorous; consumes small animals and plants
Gasterosteus aculeatus (threespine stickleback)	Omnivorous; feeds on algae, copepods, fish eggs
Gobiosoma spp. (gobies)	Consume small invertebrates and fishes
Leiostomus xanthurus (spot)	Detritivore; benthophagous; consumes organic detritus and microbenthos
Menidia menidia (Atlantic silverside)	Omnivorous; feeds on small crustaceans, annelids, algae, insects
Morone americana (white perch)	Omnivorous; consumes annelids, crustaceans, fishes, organic detritus
M. saxatilis (striped bass)	Primarily a fish predator; also feeds on annelids, crustaceans, mollusks
Opsanus tau (oyster toadfish)	Omnivorous; principally consumes annelids, crustaceans, fishes
Paralichthys dentatus (summer flounder)	Predator; feeds on fishes, crustaceans, mollusks
Pomatomus saltatrix (bluefish)	Fish predator; consumes large numbers of Atlantic menhaden and bay anchovy
Pseudopleuronectes americanus (winter flounder)	Benthophagous; feeds on annelids, crustaceans, mollusks
Sphoeroides maculatus (northern puffer)	Primarily consumes small crustaceans
Stenotomus chrysops (scup)	Benthophagous; feeds on annelids, crustaceans, small fishes
Syngnathus sp. (pipefish)	Primarily feeds on amphipods and copepods
Tautog onitis (tautog)	Consumes crustaceans, mollusks, and other invertebrates.

Source: Kennish, M. J. and Loveland, R. E., in *Ecology of Barnegat Bay, New Jersey,* Kennish, M. J. and Lutz, R. A., Eds., Springer-Verlag, New York, 1984, 302. With permission.

Table 5.7–2   Primary Foods of Some Estuarine Fishes along the U.S. Atlantic and Gulf Coasts

Scientific Name	Common Name	Food Habits
Megalops atlanticus	Tarpon	Juveniles consume copepods, ostracods, grass shrimp, and fish
Brevoortia tyrannus	Atlantic menhaden	Algae, planktonic crustacea
Harengula jaguana	Scaled sardine	Harpacticoid copepods, amphipods, mysids, isopods, and chironomid larvae
Anchoa hepsetus	Striped anchovy	Copepods, mysids, isopods, mollusks, fish, zooplankton
A. mitchilli	Bay anchovy	Zooplankton, fish, decapods, amphipods, mysids, detritus
Synodus foetens	Inshore lizardfish	Fish
Arius felis	Sea catfish	Amphipods, decapods, insects, mollusks, copepods, schizopods, isopods, hydroids
Opsanus tau	Oyster toadfish	Crustaceans, mollusks, polychaetes

Table 5.7–2 Primary Foods of Some Estuarine Fishes along the U.S. Atlantic and Gulf Coasts (continued)

Scientific Name	Common Name	Food Habits
Fundulus majalis	Striped killifish	Mollusks, crustaceans, insects, fish
F. pulvereus	Bayou killifish	Insects, isopods
F. similis	Longnose killifish	Harpacticoid copepods, ostracods, barnacle larvae, insects, isopods, amphipods
Cyprinodon variegatus	Sheepshead minnow	Plant detritus, small crustaceans, nematodes, diatoms, blue-green algae, filamentous algae, formas, insects
Adinia xenica	Diamond killifish	Plant detritus, filamentous algae, amphipods, insects, small copepods, diatoms
Lucania parva	Rainwater killifish	Insects, crustacean larvae, annelids, mysids, amphipods, cumaceans, copepods, plant detritus, small mollusks
Gambusia affinis	Mosquito fish	Amphipods, chironomids, insects, algae
Poecilia latipinna	Sailfin molly	Algae, diatoms, vascular plant detritus, inorganic matter
Menidia beryllina	Tidewater silverside	Isopods, amphipods, copepods, mysids, detritus, algae, insects, barnacle larvae
Membras martinica	Rough silverside	Copepods, barnacle larvae, amphipods, insects, shrimp fish
Morone saxatilis	Striped bass	Fish, crustaceans
Lutjanus griseus	Gray snapper	Fish, crustaceans
Diapterus plumieri	Striped mojarra	Mysids, amphipods, mollusks, ostracods, detritus, copepods
Eucinostomus gula	Silver jenny	Copepods, amphipods, mollusks, detritus, mysids
E. argenteus	Spotfin mojarra	Amphipods, copepods, mysids, mollusks, detritus
Archosargus probatocephalus	Sheepshead	Shrimp, mollusks, small fish, crabs, other crustaceans, algae, plant detritus
Lagodon rhomboides	Pinfish	Fish, crustaceans, vascular plants, algae, detritus, copepods, mysids, mollusks
Bairdiella chrysoura	Silver perch	Decapods, schizopods, copepods, mysids, amphipods, polychaetes, ectoprocts, fish, detritus
Cynoscion nebulosus	Spotted seatrout	Copepods, decapods, mysids, carideans, fish, mollusks
C. regalis	Weakfish	Polychaetes, copepods amphipods, mysids, stomatopods, decapods, fishes.
Leiostomus xanthurus	Spot	Polychaetes, copepods, isopods, amphipods, mysids, cumacea, fishes
Micropogonias undulatus	Atlantic croaker	Polychaetes, mollusks, amphipods, isopods, copepods, decapods, stomatopods, mysids, cumacea, ascidians, fish
Stellifer lanceolatus	Star drum	Amphipods, isopods, copepods, cumaceans, mysids, stomatopods, decapods, fish
Chaetodipterus faber	Atlantic spadefish	Small crustaceans, annelids, detritus, ctenophores
Mugil cephalus	Striped mullet	Algae, detritus, vascular plants, crustaceans, bacteria diatoms
Gobiosoma robustum	Code goby	Amphipods, mysids, insect larvae, cladocerans, algae, detritus, mollusks
Ancylopsetta quadrocellata	Ocellated flounder	Mysids, copepods, other crustaceans, polychaetes, fish
Citharichthys spilopterus	Bay whiff	Mysids, other crustaceans, fish
Etropus crossotus	Fringed flounder	Polychaetes, mollusks, copepods, isopods
Scophthalmus aquosus	Windowpane	Mysids, other crustaceans, fish
Trinectes maculatus	Hogchoker	Annelids, algae, amphipods, detritus, foraminifera, plant seeds, copepods, insect larvae, mollusks, cumaceans
Symphurus plagiusa	Blackcheek tonguefish	Mollusks and crustaceans

Source: Stickney, R. R., *Estuarine Ecology of the Southeastern United States and Gulf of Mexico,* Texas A&M University, College Station, 1984. With permission.

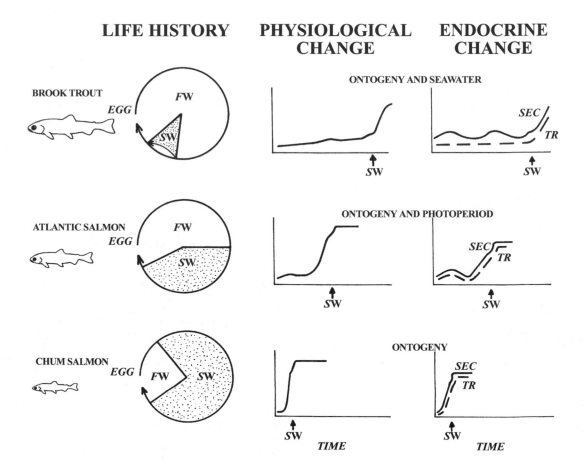

Figure 5.7–2   Generalized life cycle of a salmonid, as exemplified by the Atlantic salmon, *Salmo salar* L. (From
McCormick, S. D., *Estuaries,* 17, 77, 1994. With permission.)

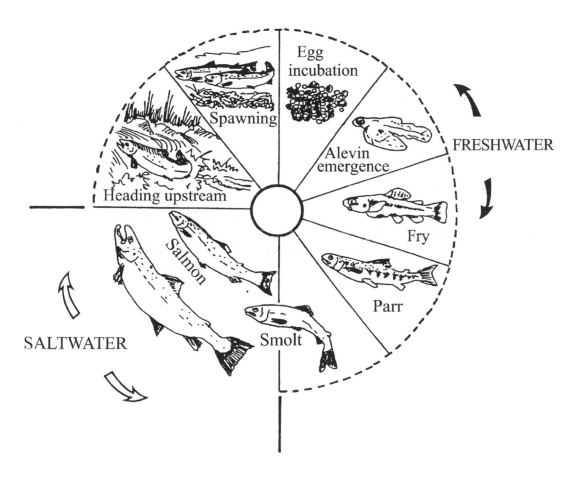

LIFE CYCLE OF
THE ATLANTIC SALMON

Figure 5.7–3 Comparison of life history, anadromy, and ontogeny of physiological changes (largely unknown) and endocrine changes (hypothesized) associated with ontogeny of salinity tolerance in three representative species of anadromous salmonids (brook trout, *Salvelinus fontinalis;* Atlantic salmon, *Salmo salar,* and chum salmon, *Oncorhynchus keta*). The analysis indicates that the ontogeny of salinity tolerance occurs late in development in brook trout, earlier in Atlantic salmon, and earlier still in chum salmon. The factors regulating physiological and endocrine change have shifted from primarily salinity in brook trout, to photoperiod in Atlantic salmon, to ontogeny (genetic developmental program) in chum salmon. Salinity retains at least some function in differentiation of salt secretory mechanisms in all species. (From McCormick, S. D., *Estuaries,* 17, 29, 1994. With permission.)

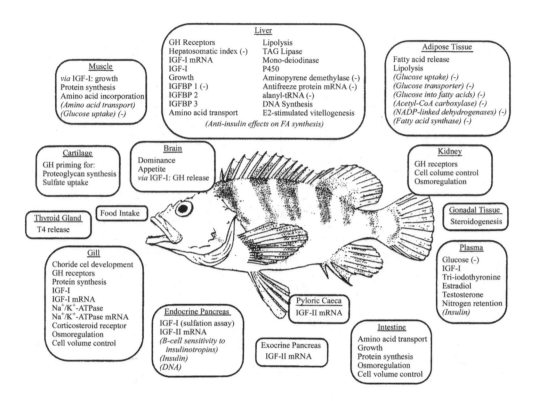

Figure 5.7–4 Model of the targets for growth hormone action in fish. Target tissues for GH and specific actions of GH in these tissues are presented. In some tissues, indirect actions mediated most likely via hepatically produced IGF-I are included. To provide a larger framework, the figure also includes, in parentheses, some specific actions of GH in the corresponding mammalian tissue: (-) signifies a GH-dependent decrease; all other variables are increased by GH. GH-dependent expression of IGF-I in tissues other than liver, gill, and endocrine pancreas has been described, but is controversial. Abbreviations: E2, estradiol; GH, growth hormone; IGF-I, insulin-like growth factor-I; T4, tetraiodothyronine; and T3, triiodothyronine. (From Mommsen, T. P., in *The Physiology of Fishes,* 2nd ed., Evans, D. H., Ed., CRC Press, Boca Raton, FL, 1997, 83. With permission.)

Table 5.7–3   Density of Various Fish Tissues

Tissue	*Pleuronectes platessa* (kg/l)	*Myoxocephalus scropius* (kg/l)	*Scyliorhinus canicula* (kg/l)
Skin	1.054	1.070	1.128
Fins	1.092	1.151	—
Muscle	1.048	1.062	1.071
Liver	1.040	1.062	1.072
Head	1.300	1.530	1.165
Axial skeleton	1.299	1.532	1.128

Source: Pelster, B., in *The Physiology of Fishes,* 2nd ed., Evans, D. H., Ed., CRC Press, Boca Raton, FL, 1997, 26. With permission.

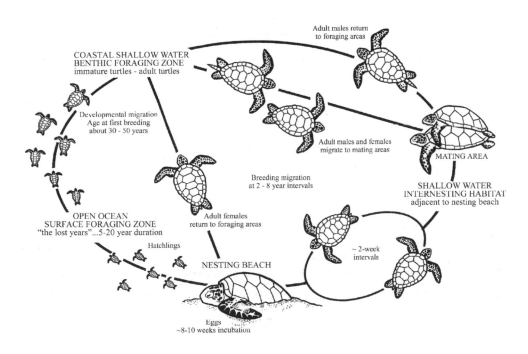

Figure 5.7–5   Generalized life cycle of sea turtles; individual species vary in the duration of phases. *Dermochelys coriacea* and at least some populations of *Lepidochelys olivacea* remain pelagic foragers throughout their lives. (From Miller, J. D., in *The Biology of Sea Turtles,* Lutz, P. L. and Musick, J. A., Eds., CRC Press, Boca Raton, FL, 1997, 53. With permission.)

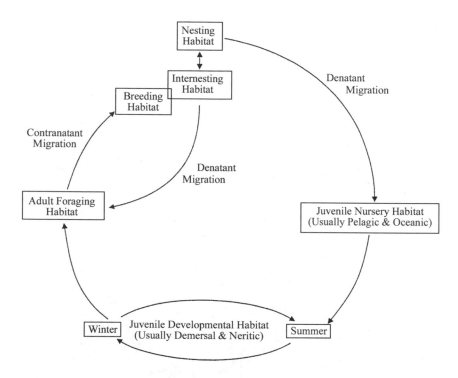

Figure 5.7–6   Conceptual model of ontogenetic habitat stages in sea turtles. (From Musick, J. A. and Limpus, C. J., in *The Biology of Sea Turtles,* Lutz, P. L. and Musick, J. A., Eds., CRC Press, Boca Raton, FL, 1997, 138. With permission.)

Table 5.7–4   Reproductive Characteristics of Marine Turtles: Reproducing Females

Species	Dominant Period of Nesting	Carapace Length (cm)	Clutch Frequency (clutches/season)	Renesting Interval (days) (mean, range)	Remigration Interval (years)
Dermochelys coriacea	Night	148.7 (1.7) 11	6.17 (0.47) 4	9/5, 9–10	2.28 (0.14) 5
Chelonia mydas	Night	99.1 (1.9) 22	2.93 (0.28) 9	12, 10–17	2.86 (0.23) 9
Natator depressus	Night and day	90.7 (0.9) 6	2.84 (—) 1	16, 2–23	2.65 (—) 1
Lepidochelys kempi	Day	64.6 (—) 1	1.80 (—) 1	20–28[a]	1.50 (—) 1
L. olivacea	Night	66.0 (1.1) 8	2.21 (0.79) 2	17, 22, 30, 45[a]	1.70 (0.30) 2
Eretmochelys imbricata	Night and day	78.6 (1.7) 15	2.74 (0.22) 5	14.5, 11–28	2.90 (0.11) 5
Caretta caretta	Night	87.0 (1.6) 18	3.49 (0.20) 4	14, 13–17	2.59 (0.15) 5

[a] Interval between *arribadas*.

Note: Values are means of means of populations (standard deviation), number of populations included. No account has been made of the number of samples from each population.

Source: Miller, J. D., in *The Biology of Sea Turtles,* Lutz, P. L. and Musick, J. A., Eds., CRC Press, Boca Raton, FL, 1997, 56. With permission.

Table 5.7–5 Reproductive Characteristics of Marine Turtles: Nesting Period and Sizes of Eggs and Hatchlings

Species	Clutch Count (No. of eggs)	Egg Weight (g)	Egg Diameter (mm)	Egg Volume	Hatchling Weight (g)
Dermochelys coriacea	81.5 (3.6) 12	75.9 (4.2) 4	53.4 (0.5) 9	79.7 (2.4) 9	44.4 (4.16) 5
Chelonia mydas	112.8 (3.7) 24	46.1 (1.6) 10	44.9 (0.7) 17	45.8 (1.2) 17	24.6 (0.91) 11
Natator depressus	52.8 (0.9) 6	51.4 (0.4) 3	51.5 (0.3) 6	70.8 (1.1) 6	39.3 (2.42) 3
Lepidochelys kempi	110.0 (—) 1	30.0 (—) 1	38.9 (—) 1	30.8 (—) 1	17.3 (—) 1
L. olivacea	109.9 (1.8) 11	35.7 (—) 1	39.3 (0.4) 6	31.8 (1.1) 6	17 (—) 1
Eretmochelys imbricata	130.0 (6.8) 17	26.6 (0.9) 5	37.8 (0.5) 1	28.7 (1.3) 11	14.8 (0.61) 5
Caretta caretta	112.4 (2.2) 19	32.7 (2.8) 7	40.9 (0.4) 14	36.2 (1.1) 14	19.9 (0.68) 7

Note: Values are means of means of populations (standard deviation), number of populations included.

Source: Miller, J. D., in *The Biology of Sea Turtles,* Lutz, P. L. and Musick, J. A., Eds., CRC Press, Boca Raton, FL, 1997, 65. With permission.

Table 5.7–6 Characteristics of the Great Whales

	Distribution	Breeding Grounds	Average Weight (tons)	Greatest Length (m)	Food
Toothed Whales					
Sperm	Worldwide; breeding herds in tropic and temperate regions	Oceanic	35	18	Squid, fish
Baleen Whales					
Blue	Worldwlde; large north–south migrations	Oceanic	84	30	Krill
Finback	Worldwide; large north–south migrations	Oceanic	50	25	Krill and other plankton, fish
Humpback	Worldwide; large north–south migrations along coasts	Coastal	33	15	Krill, fish
Right	Worldwide; cool temperate	Coastal	(50)	17	Copepods and other plankton
Sei	Worldwide; large north–south migrations	Oceanic	17	15	Copepods and other plankton, fish
Gray	North Pacific; large north–south migrations along coasts	Coastal	20	12	Benthic invertebrates
Bowhead	Arctic; close to edge of ice	Unknown	(50)	18	Krill
Bryde's	Worldwide; tropic and warm temperate regions	Oceanic	17	15	Krill
Minke	Worldwide; north–south migrations	Oceanic	10	9	Krill

Source: Allen, K. R., Conservation and Management of Whales, Washington Sea Grant Program, University of Washington, Seattle, 1980. With permission.

Table 5.7–7   Global Population Estimates of Great Whales[a]

Species	Initial	Present	Percent of Initial	Status[b]
Right whale	100,000	>2000	2%	Severely depleted
Blue whale	>196,000	<11,000	<5%	Severely depleted
Fin whale	>464,000	62,000	<13%	Severely depleted
Bowhead whale	>55,000	8000	<15%	Severely depleted
Humpback whale	>120,000	>27,000	<23%	Severely depleted
Sei whale	>105,000	>36,000	34%	Depleted
Sperm whale	2,770,000	1,810,000	65%	At or above optimum sustainable population
Gray whale	>20,000	20,869	100%	Recovered
Minke whale	900,000	900,000	100%	Sustained

[a] Population estimates are from *Endangered Whales: Status Update,* 1994, National Marine Mammal Laboratory, National Marine Fisheries Service, National Oceanic and Atmospheric Administration.
[b] The U.S. Marine Mammal Protection Act defines a stock below its optimal sustainable population as being depleted, and this is thought to occur at or above 60% of maximum population size.

Source: National Marine Fisheries Service, *Endangered Whales: Status Update*, 1994, National Marine Mammal Laboratory, National Marine Fisheries Services, National Oceanic and Atmospheric Administration, Washington, D.C., 1994.

5.8 FISHERIES

Table 5.8–1   Food Production in the Sea

Area	Plant Production (metric tons of carbon/year)	Efficiency of Energy Transfer per Trophic Level	Trophic Level Harvested	Fish Production (metric tons/year)
Open ocean	39.9×10^9	10%	5	4.0×10^6
Coastal regions	8.6×10^9	15%	4	29.0×10^6
Upwelling areas	0.23×10^9	20%	2.5	46.0×10^6

Source: Smith, S. and Hollibaugh, J., *Rev. Geophys.,* 31, 75, 1993. With permission.

Table 5.8–2   U.S. Domestic Commercial Fish Landings in 1987, by Distance from Shore or off Foreign Coasts[a]

Distance from Shore	Fishes	
	Weight	Dollar Value
0 to 4.8 km	3431	897
4.8 to 320 km	5748	897
International waters (including foreign coasts)	604	276
Totals	9773	2070

[a] In millions of pounds and millions of dollars.

Source: Fisheries of the United States, 1987, Current Fishery Statistics 8700, U.S. Department of Commerce, NOAA, NMFS, May 1988.

5.9 FOOD WEBS

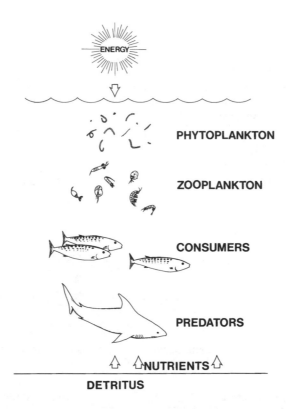

Figure 5.9–1   The trophic web in the ocean. The diagram emphasizes that the light energy for primary production decreases with depth and is thus limited to surface waters, whereas the nutrients are reworked from detritus and are most accessible near the bottom. The two regions overlap in high productivity areas. (From Beer, T., *Environmental Oceanography,* 2nd ed., CRC Press, Boca Raton, FL, 1997, 317. With permission.)

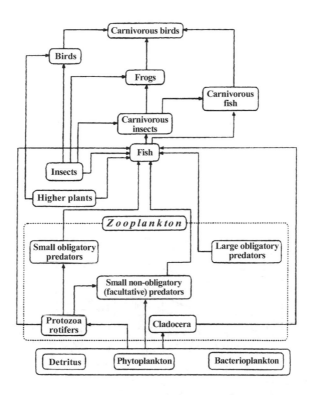

Figure 5.9–2 Generalized flow diagram of the aquatic food chain (trophic levels). (From Bukata, R., et. al., *Optical Properties and Remote Sensing of Inland and Coastal Waters,* CRC Press, Boca Raton, FL, 1995, 116. With permission.)

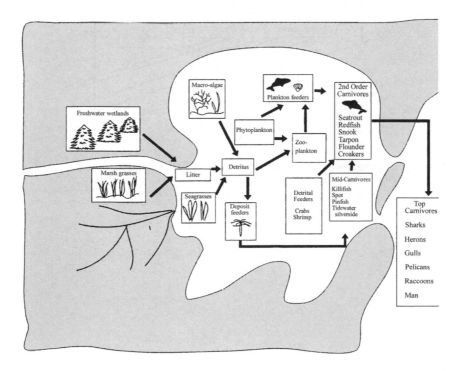

Figure 5.9–3 Conceptual diagram of a marsh-estuarine food chain. (From Kruczynski, W. L. and Ruth, B. F., in *Ecology and Management of Tidal Marshes: A Model from the Gulf of Maxico,* Coultas, C. L. and Hsieh, Y.-P., Eds., St. Lucie Press, Delray Beach, FL, 1997, 137. With permission.)

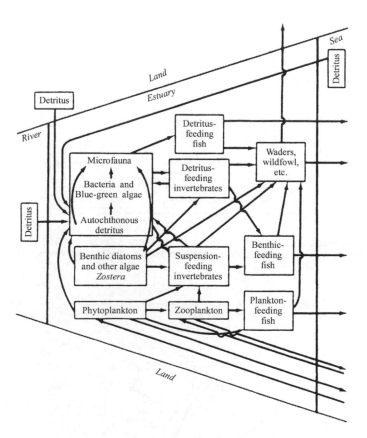

Figure 5.9–4   A simplified estuarine food web. (From Barnes, R. S. K., *Estuarine Biology*, Edward Arnold, London, 1974. With permission.)

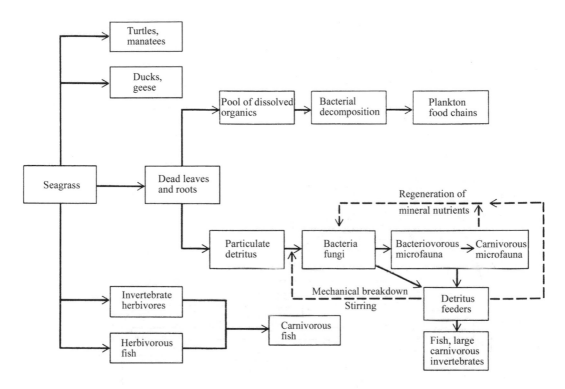

Figure 5.9–5   The pathway for the channeling of eelgrass into the marine food web. (From Nybakken, J. W., *Marine Biology: An Ecological Approach,* 2nd ed., Harper & Row, New York, 1988, 228. With permission.)

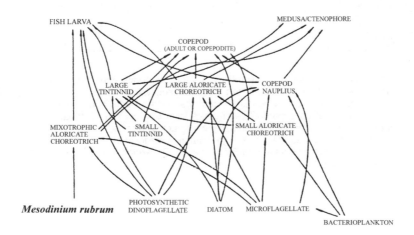

Figure 5.9–6   Trophic pathways involving marine ciliates. (From Pierce. R. W. and Turner, J. T., *Rev. Aquat. Sci.*, 6, 139, 1992. With permission.)

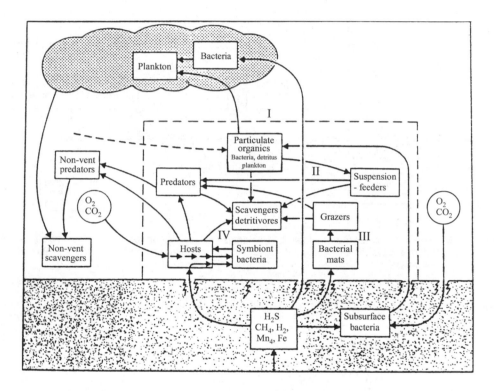

Figure 5.9–7   The trophic interactions of a hydrothermal vent community. Four distinct carbon-energy pathways
are recognizable: I, immediate exit of either reduced compounds or chemosynthetic bacteria
such that primary production is concentrated in the hydrothermal plume; II, the suspension-
feeder loop that uses particulate organics from both *in situ* bacterial production and external
input; III, the localized grazer-scavenger interaction dependent upon chemosynthetic bacterial
mats; and IV, the greatest biomass generation in the form of symbiotic chemosynthetic microbes
inside invertebrate hosts. (From Tunnicliffe, V., *Oceanogr. Mar. Biol. Annu. Rev.,* 29, 319, 1991.
With permission.)

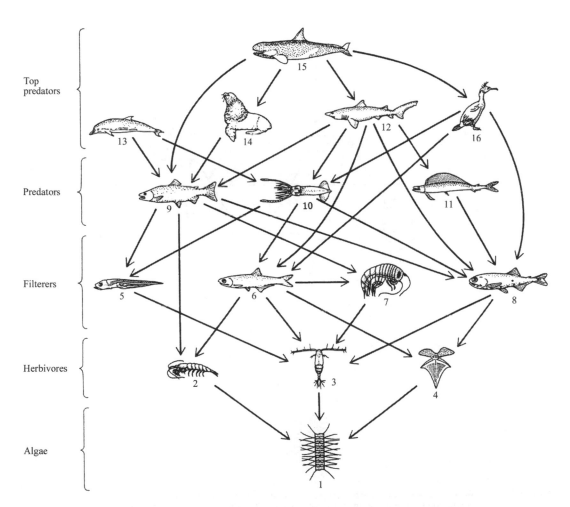

Figure 5.9–8 Food web and trophic structure of the pelagic community in cold temperate waters. Algae:
(1) diatoms. Herbivores: (2) euphausiids, (3) copepods, (4) pteropods. Filterers: (5) juvenile fish,
(6) anchovy, (7) hyperiid amphipods, (8) lantern fish. Predators: (9) salmon, (10) squid,
(11) lancetfish. Top predators: (12) mackerel shark, (13) porpoise, (14) seals and sea lions,
(15) killer whale, (16) marine birds. (From Nybakken, J. W., *Marine Biology: An Ecological
Approach,* 2nd ed., Harper & Row, New York, 1988, 140. With permission.)

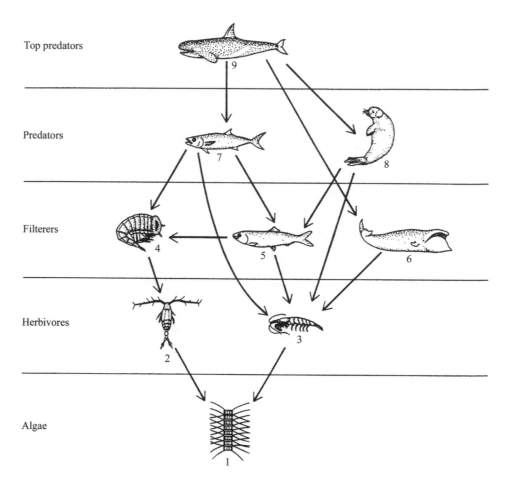

Figure 5.9–9   Food web and trophic structure of the pelagic community in Antarctic Seas. Algae: (1) diatoms. Herbivores: (2) copepods, (3) euphausiids. Filterers: (4) hyperiid amphipods, (5) planktivorous fish, (6) baleen whales. Predators: (7) predatory fish, (8) seals and sea lions. Top predators: (9) killer whales. (From Nybakken, J. W., *Marine Biology: An Ecological Approach,* 2nd ed., Harper & Row, New York, 1988, 141. With permission.)

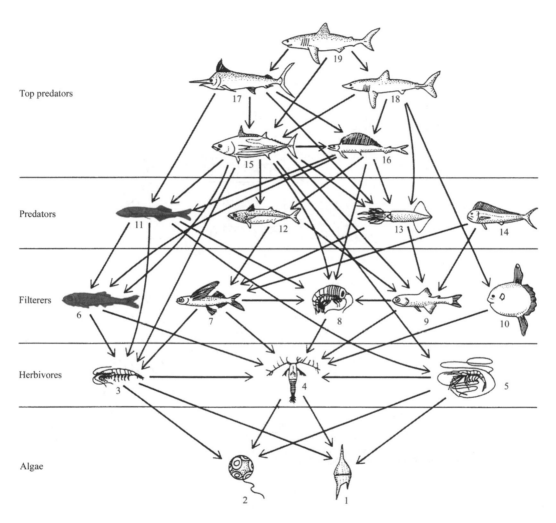

Figure 5.9–10   Food web and trophic structure of the pelagic community in tropical seas. Algae: (1) coccolithophores, (2) dinoflagellates. Herbivores: (3) euphausiids, (4) copepods, (5) shrimp. Filterers:
(6) vertically migrating mesopelagic fishes, (7) flying fishes, (8) hyperiid amphipods,) (9) lanternfish, (10) ocean sunfish. Predators: (11) mesopelagic fishes, (12) snake mackerel,
(13) squid, (14) dolphin (*Coryphaena*). Top predators: (15) tuna, (16) lancetfish, (17) marlin,
(18) medium-sized sharks, (19) large sharks. (From Nybakken, J. W., *Marine Biology:
An Ecological Approach,* 2nd ed., Harper & Row, New York, 1988, 142. With permission.)

5.10 CARBON FLOW

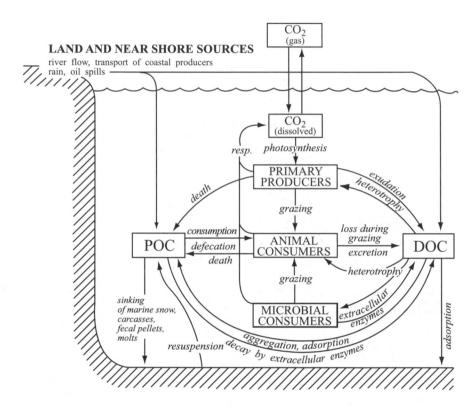

Figure 5.10–1   Carbon transfer in aerobic marine environments. The boxes represent pools and the arrows, processes. The inorganic parts of the cycle are simplified. Organic aggregates and debris comprise marine snow. Some resuspension of dissolved organic carbon from sediments into the overlying water is not shown in the diagram. (From Valiela. I., *Marine Ecological Processes,* 2nd ed., Springer-Verlag, New York, 1995. With permission.)

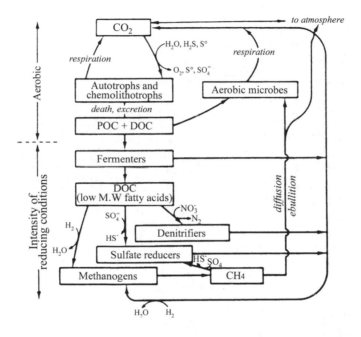

Figure 5.10–2   Carbon transformation in the transition from aerobic to anaerobic situations. The gradient from aerobic to anaerobic can be thought of as representing a sediment profile, with increased reduction and different microbial processes deeper in the sediment. Boxes represent pools or operations that carry out processes; arrows are processes that can be biochemical transformations or physical transport. Elements other than carbon are shown, where relevant, to indicate the couplings to other nutrient cycles. Some arrows indicate oxidizing and some reducing pathways. (From Valiela, I., *Marine Ecological Processes,* 2nd ed., Springer-Verlag, New York, 1995. With permission.)

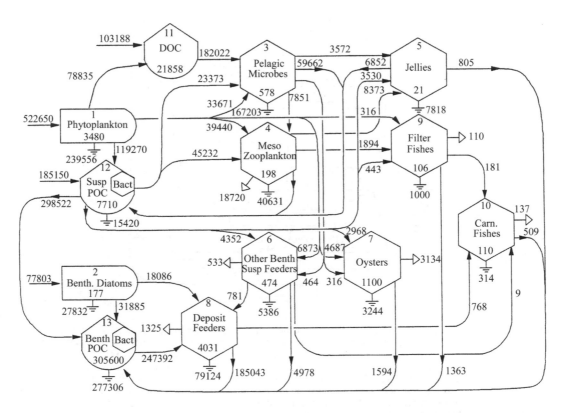

Figure 5.10–3   Carbon flows (mg/m²/year) and stocks (mg/m²) for 13 components of the mesohaline Chesapeake Bay ecosystem. Ground symbols represent respirations; open arrows, exports. (From Ulanowicz, R. E. and Tuttle, J. H., *Estuaries,* 15, 298, 1992. With permission.)

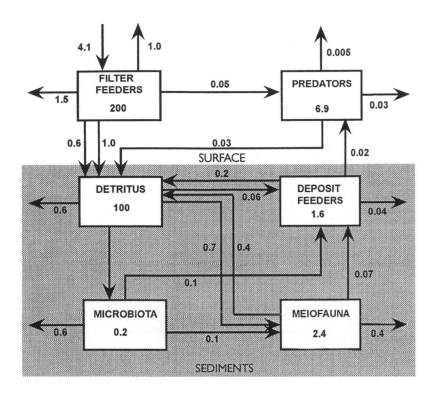

Figure 5.10–4   Carbon flow through an intertidal oyster reef in South Carolina. Flow = g C/m²/day; states = g C/m². (From Dame, R. F., *Ecology of Marine Bivalves: An Ecosystem Approach,* CRC Press, Boca Raton, FL, 1996, 151. With permission.)

Table 5.10–1   Some Physical Characteristics and Rates of Carbon Flow in Several Saltmarsh and Mangrove (Missionary Bay and Rookery Bay) Ecosystems[a]

	Sippewissett	Flax Pond	Sapelo Island	Barataria	Missionary Bay	Rookery Bay
Total area (m²)	4.8×10^5	5.7×10^5	1.3×10^7	5.1×10^9	6.4×10^7	4.0×10^5
Tidal range (m)	1.1	1.8	1.4–3.2	0–0.3	2	0.55
Water volume (m³)	4.5×10^4	2.8×10^4	8.4×10^6	4.2×10^9	1.5×10^7	3.8×10^7
Rainfall (m)	1.2	1.0	1.5	1.5	2.5	1.3
Gpp	1695–2140	1275	3941	4261	—	3318
R_{auto}	906	696–1190	2149	2287	—	2219
R_{auto}/GPP (%)	42–53	40–68	55	54	—	67
NPP	941–1386	535–1029	1791	1974	1500	1099
R_{hetero}	772	235–729	710	1773	130	197
R_{hetero}/NPP (%)	56–82	44–68	40	89	9	18
NPP – R_{hetero}	169–614	280	1081	201	1370	902
$Exp_{measured}$	80	100	379	9–30	332	64
$Exp_{measured}$/NPP (%)	6–9	10–19	21	1	18	6
$Exp_{calculated}$	80–525	100	1052	27	325	64
$Exp_{calculated}$/NPP(%)	6–38	10–19	59	1	22	6

[a] Units are g C/m²/year.

Note: Gpp = gross primary production; R_{auto} = plant respiration; NPP = net primary production; R_{hetero} = heterotrophic respiration; $Exp_{measured}$ = directly measured export; $Exp_{calculated}$ = export calculated from mass balance.

Source: Alongi, D. M., *Coastal Ecosystem Processes,* CRC Press, Boca Raton, FL, 1998, 88. With permission.

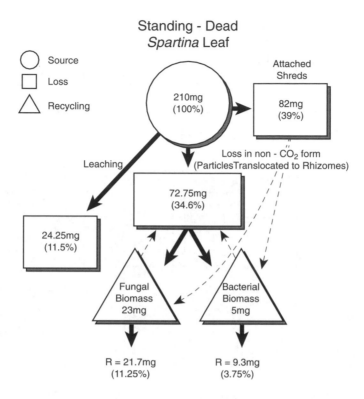

Figure 5.10–5   Carbon flow from decomposition of standing-dead leaves of *Spartina alterniflora* in a Georgia
marsh in autumn. Units are in mg C. Values in parentheses represent original carbon. R =
respiration. (From Alongi, D. M., *Coastal Ecosystem Processes,* CRC Press, Boca Raton, FL,
1998, 69. With permission.)

**Table 5.10–2   Contribution of Bacterial Decomposition Pathways to Total
Carbon Oxidation in Two Saltmarsh and One Mangrove
Ecosystems (Western Australia)**

	Sapelo Island	Sippewissett	Western Australia
Aerobic respiration	390	390	60
Denitrification	10	3	6
Manganese and iron reduction	Negligible	Negligible	Negligible
Sulfate reduction	850	1800	200
Methanogenesis	40	1–8	5
Total	**1290**	**2201**	**271**

[a] Units are g C/m²/year.

Source: Alongi, D. M., *Coastal Ecosystem Processes,* CRC Press, Boca Raton,
FL, 1998, 72. With permission.

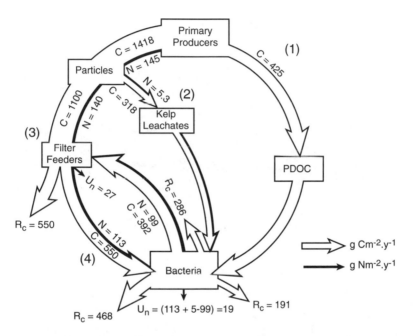

Figure 5.10–6   Relative flux of carbon and nitrogen through a kelp bed community off the Cape Peninsula, South Africa. Numbers in parentheses denote stages. R_c = respired carbon; U_n = nitrogen excreted but unaccounted for by bacteria. (From Alongi, D. M., *Coastal Ecosystem Processes*, CRC Press, Boca Raton, FL, 1998, 134. With permission.)

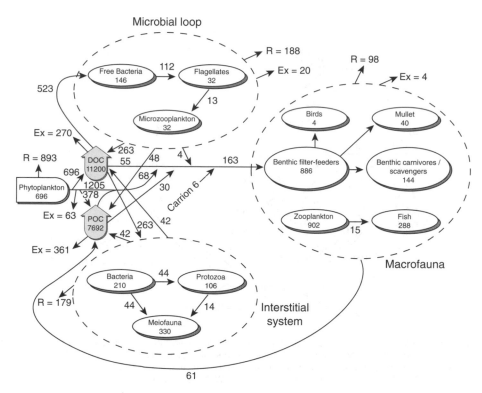

Figure 5.10–7   The main pathways of carbon flow in the Sundays River sandy beach/surf-zone ecosystem, South Africa. Fluxes are mg/C²/m/d¹, and stocking stock units are mg C/m². R = respiration; Ex = export. (From Alongi, D. M., *Coastal Ecosystem Processes,* CRC Press, Boca Raton, FL, 1998, 33. With permission.)

5.11 COASTAL SYSTEMS

Table 5.11–1   A Classification of Coastal Ecological Systems and Subsystems According to Characteristic Energy Sources

Category	Name of Type	Characteristic Energy Source or Stress
1. Naturally stressed systems of wide latitudinal range		High stress energies
	Rocky sea fronts and intertidal rocks	Breaking waves
	High energy beaches	Breaking waves
	High velocity surfaces	Strong tidal currents
	Oscillating temperature channels	Shocks of extreme temperature range
	Sedimentary deltas	High rate of sedimentation
	Hypersaline lagoons	Briny salinities
	Blue-green algal mats	Temperature variation and low nighttime oxygen
2. Natural tropical ecosystems of high diversity		Light and little stress
	Mangroves	Light and tide
	Coral reefs	Light and current
	Tropical meadows	Light and current
	Tropical inshore plankton	Organic supplements
	Blue water coasts	Light and low nutrient
3. Natural temperate ecosystems with seasonal programming		Sharp seasonal programming and migrant stocks
	Tidepools	Spray in rocks, winter cold
	Bird and mammal islands	Bird and mammal colonies
	Landlocked sea waters	Little tide, migrations
	Marshes	Lightly tidal regimes and winter cold
	Oyster reefs	Current and tide
	Worm and clam flats	Waves and current, intermittent flow
	Temperate grass flats	Light and current
	Shallow salt ponds	Small waves: light energy concentrated in shallow zone
	Oligohaline systems	Saltwater shock zone, winter cold
	Medium-salinity plankton estuary	Mixing intermediate salinities with some stratification
	Sheltered and stratified estuary	Geomorphological isolation by sill
	Kelp beds	Swells, light and high salinity
	Neutral embayments	Shelfwaters at the shore
	Coastal plankton	Eddies of larger oceanic systems
4. Natural Arctic ecosystems with ice stress		Winter ice, sharp migrations and seasonal programming
	Glacial fjords	Icebergs
	Turbid outwash fjords	Outflow of turbid icewater
	Ice-stressed coasts	Winter exposure to freezing
	Inshore Arctic ecosystems with ice stress	Ice, low light
	Sea and under-ice plankton	Low light

Table 5.11–1   A Classification of Coastal Ecological Systems and Subsystems According to Characteristic Energy Sources (continued)

Category	Name of Type	Characteristic Energy Source or Stress
5. Emerging new systems associated with humans		New but characteristic anthropogenic energy sources and/or stresses
	Sewage wastes	Organic and inorganic enrichment
	Seafood wastes	Organic and inorganic enrichment
	Pesticides	Organic poison
	Dredging spoils	Heavy sedimentation by humans
	Impoundments	Blocking of current
	Thermal pollution	High and variable temperature discharges
	Pulp mill wastes	Wastes of wood processing
	Sugarcane wastes	Organics, fibers, soils of sugar industry wastes
	Phosphate wastes	Wastes of phosphate mining
	Acid waters	Release or generation of low pH
	Oil shores	Petroleum spills
	Pilings	Treated wood substrates
	Saline	Brine complex of salt manufacture
	Brine pollution	Stress of high salt wastes and odd element ratios
	Petrochemicals	Refinery and petrochemical manufacturing wastes
	Radioactive stress	Radioactivity
	Multiple stress	Alternating stress of many kinds of wastes in drifting patches
	Artificial reef	Strong currents
6. Migrating subsystems that organize areas		Some energies taxed from each system

Source: Odum, H. T. and Copeland, B. J., in *Coastal Ecological Systems* of *The United States*. Vol. 1. Odum, H. T. et al., Eds., The Conservation Foundation, Washington, D.C., 1974, 5. With permission.

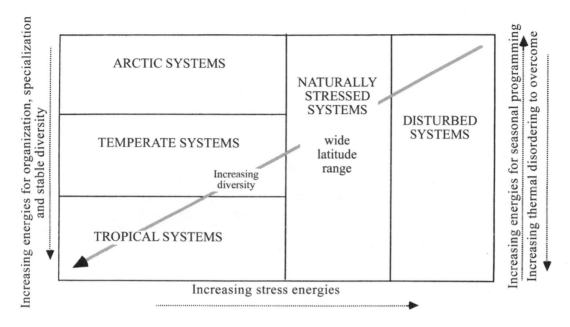

Figure 5.11–1   Systems analysis diagram applied to the classification of estuaries. (From Odum, H. T. and Copeland, B. J., in *Coastal Ecological Systems of the United States.* Vol. 1, Odum, H. T. et al., Eds., The Conservation Foundation, Washington, D.C., 1974, 5. With permission.)

Table 5.11–2   Physical, Chemical (Salinity), and Biological Zones in an Estuary

Estuary Division	Substrate	Salinity Range (10^{-3})	Zone	Organism Type	No. of Species
River	Gravels	<0.05	Limnetic	Freshwater	>50
Head	Becoming finer	0.5–5.0	Oligohaline	Oligohaline, freshwater migrants	20
Upper reaches	Mud, currents minimal	5.0–18.0	Mesohaline	True estuarine, limit of nontransient migrants	12
Middle reaches	Mud, some sand	18.0–25.0	Polyhaline	Estuarine, euryhaline	20
Lower reaches	Sand/mud, depending on tidal currents	25.0–30.0	Polyhaline	Estuarine, euryhaline, marine migrants	40
Mouth	Clean sand/rock	30.0–35.0	Euhaline	Stenohaline, all marine	>100

Source: Wilson, J. G. and Jeffrey, D. W., in *Biomonitoring of Coastal Waters and Estuaries,* Kramer, K. J. M., Ed., CRC Press, Boca Raton, FL, 1994, 313. With permission.

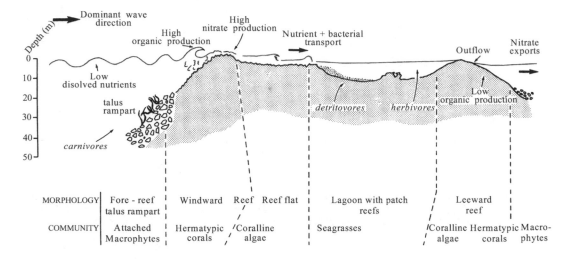

Figure 5.11–2   Cross section of a coral reef, marking the various energy zones. (From Carter, R. W. G., *Coastal Environments: An Introduction to the Physical, Ecological, and Cultural Systems of Coastlines*, Academic Press, London, 1988, 302. With permission.)

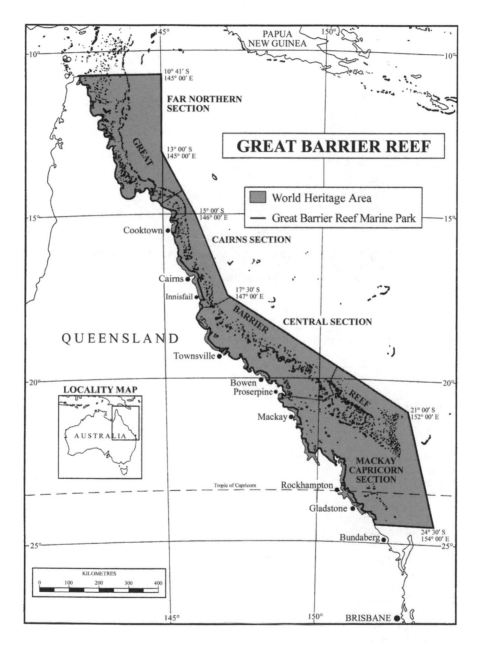

Figure 5.11–3 Map of the Great Barrier Reef Marine Park, showing the location of the Great Barrier Reef along the coast of Australia. (From Clark, J. R., *Coastal Zone Management Handbook,* Lewis Publishers, Boca Raton, FL, 1996, 488. With permission.)

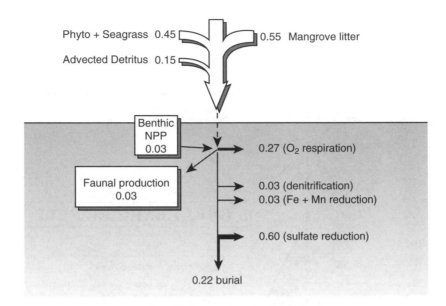

Figure 5.11–4   Preliminary carbon budget (g C/m²/day) for sediments of the coastal zone within the central Great Barrier Reef lagoon. (From Alongi, D. M., *Coastal Ecosystem Processes,* CRC Press, Boca Raton, FL, 1998, 295. With permission.)

Table 5.11–3   **Ranges of Gross (P_G) and Net (P_N) Primary Productivity (g C/m²/day) and Production-to-Respiration (P:R) Ratios in Coral Reefs**

	P_G	P_N	P:R Ratio
Major Reef Zones and Entire Reefs			
Ocean water ($n = 8$)[a]	?	0.01–0.65	?
Outer reef slope			
Fore-reef ($n = 4$)	2.0–7.0	−1.0–5.1	0.5–5.5
Reef flat			
Reef crest ($n = 2$)	2.0–7.0	0.3–1.5	1.0–4.0
Back reef ($n = 25$)	2.6–40.0	−1.7–27.0	0.7–3.2
Lagoon			
Sand, shallow	0.9–12.9	−0.5–3.4	0.7–1.4
Patch reefs ($n = 8$)			
Water ($n = 21$)	0.01–2.0	−1.3–1.4	0.1–1.4
Entire reef ($n = 5$)	2.3–6.0	−0.01–0.17	1.0
Producer Group[b]			
Symbionts			
Corals ($n = 10$)	0.77–10.2	8.0–40.0	0.5–5.0
Endo- and epilithic algae			
Corallines ($n = 7$)	0.8–2.8	0.06–11.7	1.0–5.4
Turfs ($n = 13$)	0.9–12.1	17.0–280.0	1.2–6.7
Macroalgae ($n = 6$)	2.3–39.4	2.5–118.0	1.2–6.3
Epipelic and rhizobenthic assemblages			
Microalgae ($n = 9$)	0.08–3.7	?–363.0	1.1–10.3
Seagrasses ($n = 5$)	3.0–16.0	4.0–8.8	1.5–2.5
Macroalgae ($n = 5$)	?–4.0	0.2–40.0	1.9–2.8

[a] n = number of empirical measurements.
[b] P_{Nsp} is weight-specific primary productivity in mg C dry weight per day.

Source: Alongi, D. M., *Coastal Ecosystem Processes,* CRC Press, Boca Raton, FL, 1998, 146. With permission.

Table 5.11–4   Some Major Environmental Characteristics Unique or Peculiar to the Tropical Coastal Oceans

Habitats occurring mainly in tropics	Mangroves, coral reefs, hypersaline lagoons, and stromatolites
Temperature	Higher; narrow annual range; thermal maximum closer to ambient
Light intensity	Higher; narrow annual range
Climate	Monsoonal and dry; two seasons rather than four; greater incidence of storms (cyclones, typhoons, etc.)
Geological characteristics	Most of world's sediment discharge from continents; mud and coral more abundant inshore; shelves wide and shallow, with several dominated by carbonates; mixed terrigenous–carbonate facies; migrating fluid mudbanks; sabkhas
Hydrological and chemical	Lower dissolved nutrient and gas (O_2, CO_2) concentrations in most habitats; most of world's freshwater discharge to ocean; ENSO phenomenon; lower mean amplitude with slight increase near equator; permanently stratified waters; O_2 minimum layers; estuarization of shelf by river plumes; large buoyancy flux; highly variable salinity; strong tidal fronts; lutoclines and high salinity plugs in dry tropical estuaries

Source: Alongi, D. M., *Coastal Ecosystem Processes,* CRC Press, Boca Raton, FL, 1998, 9.

5.12 DEEP-SEA SYSTEMS

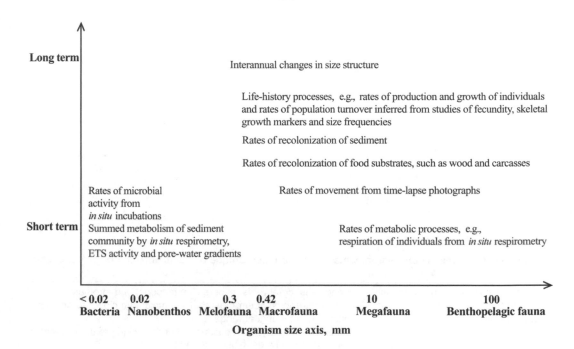

Figure 5.12–1   Schematic presentation of scales of biological rate processes amenable to measurement in the deep sea. (From Gage, J. D., *Rev. Aquat. Sci.,* 5, 49, 1991. With permission).

Table 5.12–1   Rates of Deep-Sea Invertebrates Recorded at Various Depths

Taxon	Depth (m)	Body Size (cm)	Speed cm/min	Speed Body Length/min	Speed cm/h	Speed Body Length/h
Vermes	360		2–9	~1		
"Green worm"						
Crustacea						
Glyphocrangon sculpta	2664	7	(1.67)[a]	(0.24)[a]	Mean 100	Mean 14.3
	2664	7	(4.5)[a]	(0.65)[a]	(270 max)[a]	39
Cancer borealis	360		7.7			
Sclerocrangon sp.	360		1.7–4.5			
"Decapod"	4800		0.4			
Hemichordata sp.	4873		0.2			
Echinodermata						
Asteroidea						
Hymenaster membranaceus	2664	4–9	(5.17)[a]	(0.83)[a]	310	50
Bathybiaster vexillifer	2008	16	(1.67)[a]	(0.1)	100	6
Henricia sp.	360					
Echinoidea						
Echinus affinis	2008	4	(0.78)[a]	(0.2)[a]	47	12
Ophiuroidea						
?*Ophiomyxa tumida*	1180	~4 (max)	6[b]	1.5[b]		
Ophiomusium lymani	1200	20 (max)	0.7–3.4	0.004–0.17		
"Ophiuroid"	4800					
Holothurioidea						
Benthogone rosea	2008	17	(1.48)[a]	(0.09)[a]	89	5
Scotoplanes sp.	1200	ca. 10	0.6	0.06		

Table 5.12–1   Rates of Deep-Sea Invertebrates Recorded at Various Depths (continued)

Taxon	Depth (m)	Body Size (cm)	Speed			
			cm/min	Body Length/min	cm/h	Body Length/h
Gastropoda						
Bathybembix bairdi	1300		1.5			
Neptunia amianta	1300		1.4			
Gastropod sp. "A"	4800		1.0			
Gastropod sp. "A"	4800		0.2			
Gastropod sp. "B"	4800		0.1			
Gastropod sp.	2664	4	(2.57)[a]	(0.64)	154	38

[a] Time-lapse rate, 64 min.
[b] Time-lapse rate, 36 s.

Source: Gage, J. D., *Rev. Aquat. Sci.,* 5, 49, 1991. With permission.

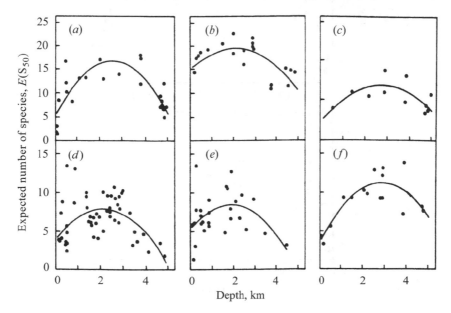

Figure 5.12–2   Deep-sea species diversity patterns for invertebrates and fishes collected in the northwest Atlantic Ocean. (From Rex, M. A., in *Annu. Rev. Ecol. Syst.,* 12, 331, 1981. With permission.)

Figure 5.12–3   Expected species diversity (number of species expected in a random sample of 50 individuals) $E(S_{50})$ vs. depth. (a) gastropods; (b) polychaetes; (c) protobranch bivalves; (b) cumaceans; (e) invertebrate megafauna; and (f) fish. (From Rex, M. A., *Annu. Rev. Ecol. Syst.,* 12, 331, 1981. With permission.)

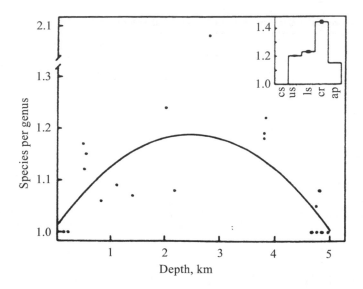

Figure 5.12–4   Ratios of species per genus plotted against depth in the northwest Atlantic Ocean. Inset: Expected ratios for the continental shelf (cs); upper slope (us); lower slope (ls); continental rise (cr); and abyssal plain (ap). (From Rex, M. A., *Annu. Rev. Ecol. Syst.,* 12, 331, 1981. With permission.)

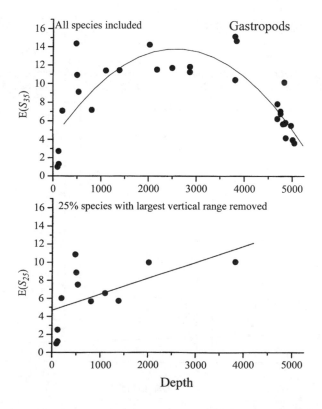

Figure 5.12–5   Species diversity plotted against depth for gastropods in the northwest Atlantic Ocean. Top: all species plotted. Bottom: 25% of the species with the largest vertical range removed. (From Rex, M. A., *Annu. Rev. Ecol. Syst.,* 12, 331, 1981. With permission.)

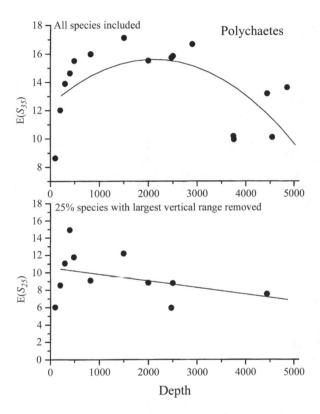

Figure 5.12–6   Species diversity plotted against depth for polychaetes in the northwest Atlantic Ocean. Top: all species plotted. Bottom: 25% of the species with the largest vertical range removed. (From Rex, M. A., *Annu. Rev. Ecol. Syst.,* 12, 331, 1981. With permission.)

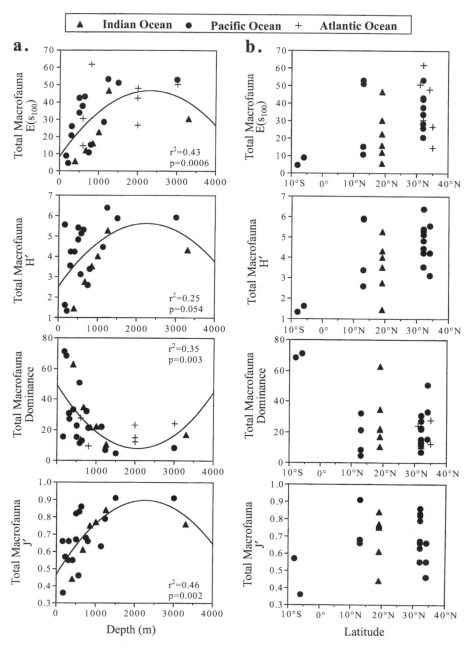

Figure 5.12–7 Relationships of total macrofaunal diversity measures to (a) water depth, (b) latitude, (c) bottom water oxygen, and (d) sediment percent organic-carbon content. (From Levin, L. A. and Gage, J. D., *Deep-Sea Res. II*, 45, 140, 1998. With permission.)

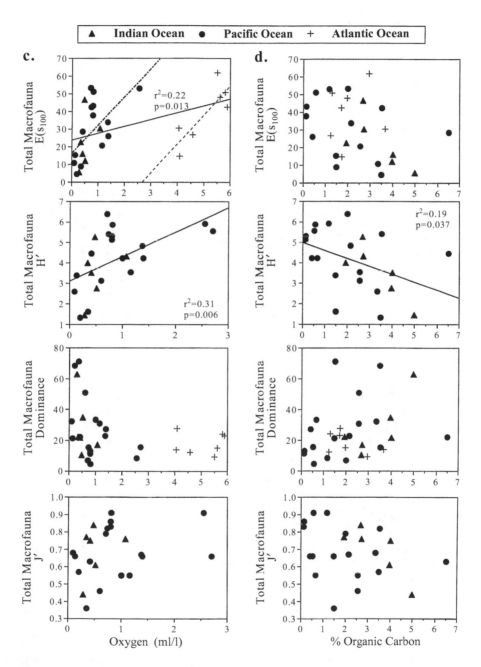

Figure 5.12–7 (continued)

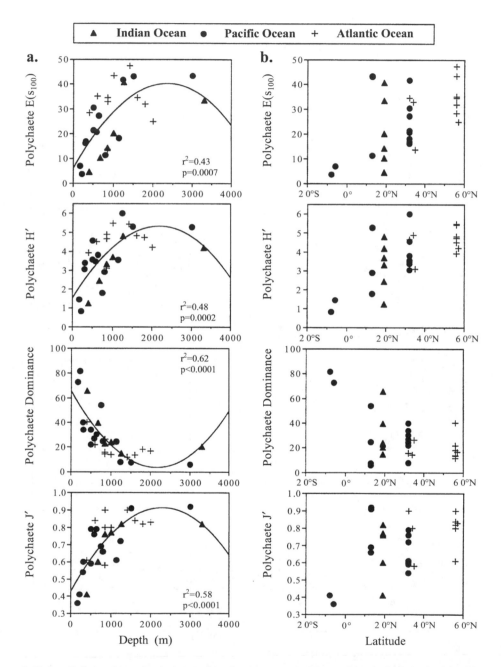

Figure 5.12–8   Relationships of polychaete diversity measures to (a) water depth, (b) latitude, (c) bottom water oxygen, and (d) sediment percent organic-carbon content. (From Levin, L. A. and Gage, J. D., *Deep-Sea Res. II*, 45, 143, 1998. With permission.)

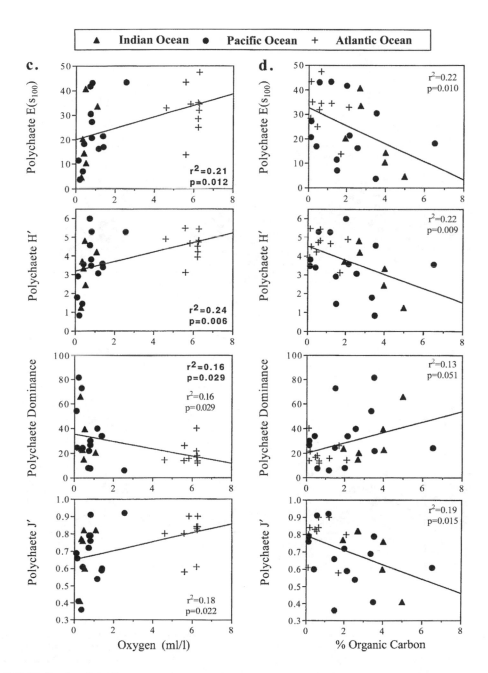

Figure 5.12–8 (continued)

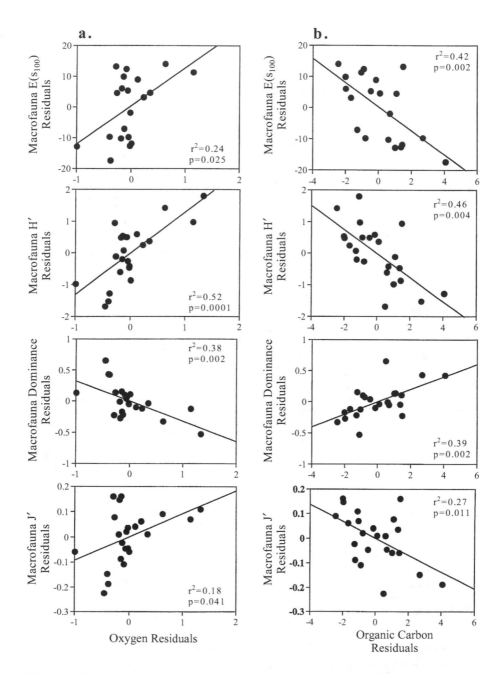

Figure 5.12–9   Relationships for Indo-Pacific stations of macrofaunal species richness, diversity, dominance, or evenness residuals to (a) bottom-water oxygen residuals and (b) organic carbon residuals, after effects of depth and latitude have been removed. (From Levin, L. A. and Gage, J. D., *Deep-Sea Res. II*, 45, 147, 1998. With permission.)

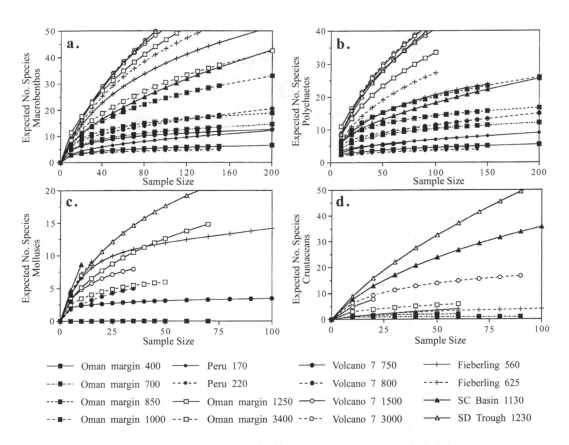

Legend:

- ■— Oman margin 400
- ·■··· Oman margin 700
- —■— Oman margin 850
- — ■ — Oman margin 1000
- — Peru 170
- —◆— Peru 220
- —□— Oman margin 1250
- — □ — Oman margin 3400
- ● Volcano 7 750
- ●— Volcano 7 800
- —○— Volcano 7 1500
- —○— Volcano 7 3000
- —+— Fieberling 560
- —+— Fieberling 625
- —▲— SC Basin 1130
- —△— SD Trough 1230

Figure 5.12–10 Rarefaction curves for (a) total macrobenthos, (b) polychaetes, (c) mollusks, and (d) crustaceans from bathyal stations in the Indian and Pacific Oceans. (From Levin, L. A. and Gage, J. D., *Deep-Sea Res. II*, 45, 151, 1998. With permission.)

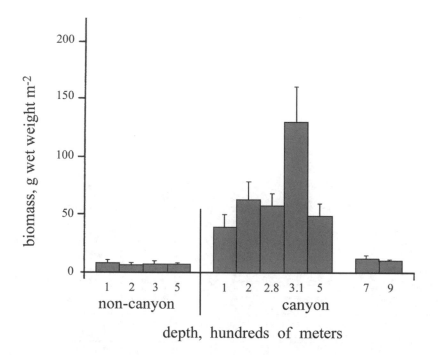

Figure 5.12–11   Mean macrofaunal biomass (wet) in Scripps Canyon from 100 to 280 m, La Jolla Canyon from 300 to 900 m, and outside of the canyons from 100 to 500 m. Bars = 1 standard error. (From Vetter, E. W. and Dayton, P. K., *Deep-Sea Res. II*, 45, 35, 1998. With permission.)

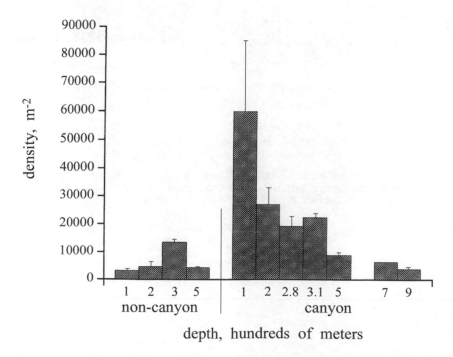

Figure 5.12–12   Mean macrofaunal density in Scripps Canyon from 100 to 280 m, La Jolla Canyon from 300 to 900 m, and outside of the canyons from 100 to 500 m. Bars = 1 standard error. (From Vetter, E. W. and Dayton, P. K., *Deep-Sea Res. II*, 45, 36, 1998. With permission.)

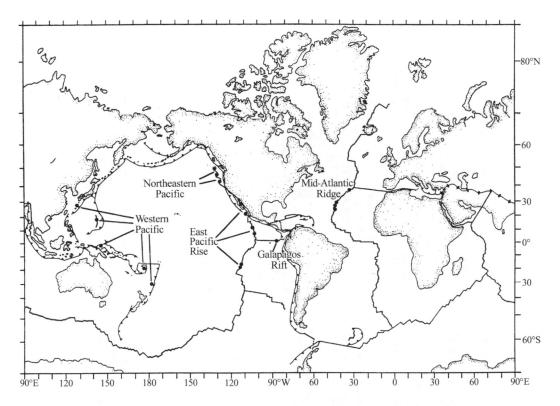

Figure 5.12–13   Map of the mid-ocean ridge system, showing hydrothermal vent locations that have been studied by biologists. (From Hessler, R. R. and Kaharl, V. A., *Seafloor Hydrothermal Systems: Physical, Chemical, Biological, and Geological Interactions,* Humphris, S. E. et. al., Eds., Geophysical Monograph 91, American Geophysical Union, Washington, D.C., 1995, 73. With permission.)

Table 5.12–2   Characteristics of Specific Thermal Environments Associated with Deep-Sea Hydrothermal Vent Systems

Environment[a]	Geographic Location	Temp. Characteristics	Geological, Chemical Properties	Presence of Hyperthermophiles	Other Biological Properties
(A) Subsurface	All vent sites	Expansive vertical and horizontal gradients; two-phase separation at >420°C, possibly associated with brine cells[b]	Lithostatic pressure 3 × hydrostatic; high surface:rock ratio; dynamic changes in solid phase-cracking, chemical alteration[c]	Unknown, requires drilling to sample; inferred from presence in smoker fluids[d] and barophily of isolates[e]	Unknown
(B) Sulfide mounds	JFR	Variable, depending on porosity, SW intrusion; portions bathed by hot (to 357°C) smoker and flange fluids	Formation, hydrothermal circulation unknown; mineralogy presumed similar to smokers and flanges	Unknown; predictable adundance greater than smoker solids due to areally more extensive, stable thermal regimes	Extensive microbial mats; animal species typical of smokers and warm-water vents
(C) Smokers	EPR, JFR, MAR, Guaymas Basin, Okinawa Trough	Sharp local gradients in solids (<20 to >350°C) bathed by hot fluids, evidenced by distinct mineral layers	Cu-Fe sulfides in anhydrite matrix (at 2 to 350°C), chalcopyrite (at <300°C), and Mg silicates at (3 to <200°C)[g];	Known from isolations[f] (moderate thermophiles inferred from EM observations of Fe- and Si-depositing bacteria)[h]	Distinct animal communities, depending on geographic location
(D) Flanges	JFR, MAR, Guaymas Basin, Explorer Ridge	Gradients in solids somewhat similar to smokers; variable temp. (<20 to >350°C) in pooled fluids[j]	Same as smokers, but less distinct mineral layers; sandlike textures of high organic content[i,j]	Known from isolations[f] and archaebacterial lipid analyses[i]	Extensive microbial mats; few and stunted tube worms (Ridgeia) on upper surfaces; some with polychaetes and other animals[i]
(E) Warm water vents	All vent sites	Variable (ambient SW to <50°C); s-min shifts of ≤10°C[k]	Extensive SW dilution	Not yet isolated	Microbial mats, large populations of animals with endosymbionts[l]
(F) Sediments	Guaymas Basin	Highly variable with seafloor location; gradients (40 to 150°C) with sediment depth (measured to 60 cm)[m]	400-m-thick sediment layer; detrital components of clay and diatoms; 2 to 5% organic carbon; mostly hydrocarbons[n]	Known from isolations[f] and sulfate reduction measurements to 110°C[m]	Extensive Beggiatoa mats[m]

(G) Buoyant plumes	All vent sites with smokers or flanges	Steep gradients above outlet (>350 to <5°C); ambient SW temp. above 20 m	Enriched in NH_4, CH_4, H_2, Mn^{o}	Known from isolations, notably MacDonald Seamount volcanic plume[f]	Active microbial oxidation of methane[p]
(H) Horizontal plumes	All vent sites with smokers or flanges	Ambient SW temp.	Enriched in CH_4 and Mn^{o-q}	Unknown; predictable abundance very low; possibly nonculturable	Highest rates of microbial oxidation of methane[p] and Mn^{q}

Note: JFR = Juan de Fuca Ridge. EPR = East Pacific Rise. MAR = Mid-Atlantic Ridge. Full citations to all references in footnotes in original source.

[a] Letter designations match those in Figure 8 (of source).
[b] Bischoff and Rosenbauer (1989); Butterfield (1990).
[c] Norton (1984).
[d] Baross et al. (1982); Baross and Deming (1985); Deming and Baross (1986; 1993); Fiala et al. (1986); Reysenbach and Deming (1991).
[e] Reysenbach and Deming (1991); Pledger and Baross (1991); Prieur et al. (1992); Jannasch et al. (1992); Nelson et al. (1992); Erauso et al. (1993); Deming and Baross (1993).
[f] See Tables 1 and 2 (of source).
[g] Goldfarb et al. (1983); fluid properties described in Table 4 (of source).
[h] Baross and Deming (1985); Juniper and Fouquet (1988); Juniper et al. (1992).
[i] Hedrick et al. (1992).
[j] Delaney et al. (1988, 1992).
[k] Johnson et al. (1988).
[l] Childress (1988).
[m] Belkin and Jannasch (1989); Jorgenson et al. (1990, 1992).
[n] Simoneit (1988); Gieskes et al. (1988); Jones et al. (1989).
[o] Lilley et al. (1983, 1987, 1989).
[p] de Angelis (1989); de Angelis et al. (1993).
[q] Cowen et al. (1986); Winn et al. (1986).

Source: Baross, J. A. and Deming, J. W., in *The Microbiology of Deep-Sea Hydrothermal Vents*, Karl, D. M., Ed., CRC Press, Boca Raton, FL, 1995, 188. With permission.

Table 5.12–3   Summary of Methods Used in the Study of Deep-Sea Hydrothermal Vent Microbial Assemblages

Parameters/processes	Procedures	Reference[a]
Enumeration	Spread plate	Baross et al., 1982; Lilley et al., 1983; Naganuma et al., 1991
	Dilution-MPN	Tuttle et al., 1983; Naganuma et al., 1989; Huber et al., 1990
	Microscopy[b]	Karl et al., 1980; Jannasch and Wirsen, 1981; Baross et al., 1982; Harwood et al., 1982. Lilley et al., 1983; Baross et al., 1984; Chase et al., 1985; Tunnicliffe et al., 1985. Cowen et al., 1986; Deming and Baross, 1986; Winn et al., 1986; Karl et al., 1988a, b, c; Mita et al., 1988; Karl, 1987; Nelson et al., 1989; Naganuma et al., 1989; de Angelis et al., 1991; Naganuma et al., 1991
Biomass	Elemental or macromolecular composition[c]	Comita et al., 1984; Karl et al., 1988c, 1989; Nelson et al., 1989; Straube et al., 1990
	ATP, total adenylates	Karl et al., 1980, 1984; Karl 1985; Winn et al., 1986; Karl et al., 1988a, b, c; Haberstroh and Karl, 1989; Karl et al., 1989
	Lipopolysaccharide	Mita et al., 1989
Viability and physiological potential	autoradiography	Nelson et al., 1989
	Diagnostic enzyme activity	Tuttle et al., 1983; Tuttle, 1985; Wirsen et al., 1986; Jannasch et al., 1989; Nelson et al., 1989
	Enrichment culture	See Table 7 [of source] for detailed listing
Metabolic activity, growth rate, production	^{35}S-oxidation	Tuttle, 1985
	^{35}S-reduction	Jergensen et al., 1990, 1992
	^{54}Mn/^{59}Fe uptake	Cowen et al., 1986
	Labeled organic substrate uptake and turnover[d]	Baross et al., 1982; Tuttle et al., 1983; Tuttle, 1985; Karl et al., 1988a, b, c; Bazylinski et al., 1989; Karl et al., 1989; de Angelis et al., 1991
	^{14}CO$_2$ dark uptake	Karl et al., 1980; Tuttle et al., 1983; Chase et al., 1985; Tunnicliffe et al., 1985; Tuttle, 1985; Wirsen et al., 1986; Jannasch et al., 1989; Nelson et al., 1989
	GTP:ATP ratios	Karl et al., 1980
	Frequency of dividing cells	Naganuma et al., 1989, 1991
	Incorporation of nucleic acid precursors[e]	Karl et al., 1984; Karl, 1985; Winn et al., 1986; Karl et al., 1988a, b, c, 1989
	Colonization of introduced surfaces	Jannasch and Wirsen, 1981
	Production of biogenic gases	Baross et al., 1982; Lilley et al., 1983; Baross et al., 1984
	Growth rate by chemostat washout	Naganuma et al., 1989

[a] Full citations in original source.

[b] Includes brightfield, phase-contrast, epifluorescence, and electron beam.

[c] Includes measurments of particulate C, N, protein, and nucleic acids.

[d] Includes ^{3}H and ^{14}C-labeled CH$_4$, acetate, glutamic acid, amino acid mixtures, glucose, and selected hydrocarbons.

[e] Includes ^{3}H-adenine and ^{3}H-thymidine.

Source: Karl, D. M., in *The Microbiology of Deep-Sea Hydrothermal Vents,* Karl, D. M., Ed., CRC Press, Boca Raton, FL, 1995, 67. With permission.

Table 5.12–4   Classification of Major Physiological Groups of Bacteria on Basis of Electron (e^-) Donors and Major C Sources Used in Metabolism

Type of Metabolism	Electron Donor	C Source
Phototrophy[a]		
Photolithoautotroph	H_2O, H_2S, S^0, H_2	CO_2
Photolithoheterotroph	H_2O, H_2S, S^0, H_2	Organic substrate
Photoorganoautotroph	Organic substrate	CO_2
Photoorganoheterotroph	Organic substrate	Organic substrate
Photomixotroph[b]	Mixed inorganic and organic	Organic substrate
Chemotrophy[c]		
Chemolithoautotroph	Reduced inorganic substrate (e.g., H_2, H_2S, $S_2O_3^{2-}$, S^0, NH_4^+, Fe^{2+})	CO_2
Chemolithoheterotroph	Reduced inorganic substrate (e.g., H_2, H_2S, $S_2O_3^{2-}$, S^0, NH_4^+, Fe^{2+})	Organic substrate
Chemoorganoautotroph	Organic substrate	CO_2
Chemoorganoheterotroph	Organic substrate	Organic substrate
Chemomixotroph[d]	Mixed inorganic and organic	Mixed inorganic and organic

[a] Phototrophy refers to processes dependent upon light energy.

[b] Photomixotrophs (also referred to as facultative photolithoautotrophs) can simultaneously use mixtures of inorganic and organic e^- donors, or mixtures of inorganic and organic C sources, or mixtures of both. The potential for simultaneous utilization of photo- and chemotrophic modes of metabolism is also possible.

[c] Chemotrophy refers to processes dependent upon chemical energy.

[d] Chemomixotrophs (also referred to as facultative chemolithoautotrophs) can simultaneously use mixtures of inorganic and organic e^- donors, or inorganic and organic C sources, or mixtures of both. The potential for simultaneous utilization of photo- and chemotrophic modes of metabolism is also possible.

Source: Karl, D. M., in *The Microbiology of Deep-Sea Hydrothermal Vents,* Karl, D. M., Ed., CRC Press, Boca Raton, FL, 1995, 39. With permission.

Table 5.12–5   Potential Bacterial Metabolic Processes at Deep-Sea Hydrothermal Vents

Conditions	Electron (Energy) Donor	Electron Acceptor	C Source	Metabolic Process
Aerobic	H_2	O_2	CO_2	H oxidation
	HS^-, S^0, $S_2O_3^{2-}$, $S_4O_6^{2-}$	O_2	CO_2	S oxidation
	Fe^{2+}	O_2	CO_2	Fe oxidation
	Mn^{2+}	O_2	?	Mn oxidation
	NH_4^+, NO_2^-	O_2	CO_2	Nitrification
	CH_4 (and other C-1 compounds)	O_2	CH_4, CO_2, CO	Methane (C-1) oxidation
	Organic compounds	O_2	Organic compounds	Heterotrophic metabolism
Anaerobic	H_2	NO_3^-	CO_2	H oxidation
	H_2	S^0, SO_4^{2-}	CO_2	S and sulfate reduction
	H_2	CO_2	CO_2	Methanogenesis
	CH_4	SO_4	?	Methane oxidation
	Organic compounds	NO_3^-	Organic compounds	Denitrification
	Organic compounds	S^0, SO_4^{2-}	Organic compounds	S and sulfate reduction
	Organic compounds	Organic compounds	Organic compounds	Fermentation

Source: Karl, D. M., in *The Microbiology of Deep-Sea Hydrothermal Vents,* Karl, D. M., Ed., CRC Press, Boca Raton, FL, 1995, 41. With permission.

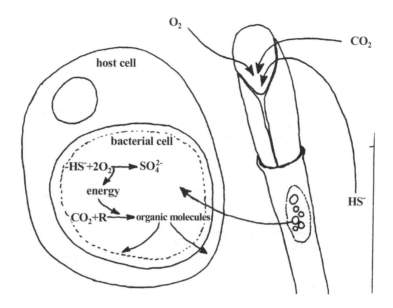

Figure 5.12–14   Sketch of the host bacterial cell that oxidizes H_2S for energy to make organic compounds. (From Millero, F. J., *Chemical Oceanography,* 2nd ed., CRC Press, Boca Raton, FL, 1996, 376. With permission.)

Table 5.12–6   Electron Sources and Types of Chemolithotrophic Bacteria Potentially Occurring at Hydrothermal Vents

Electron Donor	Electron Acceptor	Organisms
S^{2-}, S^0, $S_2O_3^{2-}$	O_2	Sulfur-oxidizing bacteria
S^{2-}, S^0, $S_2O_3^{2-}$	NO_3^-	Denitrifying and sulfur-oxidizing bacteria
H_2	O_2	Hydrogen-oxidizing bacteria
H_2	NO_3^-	Denitrifying hydrogen bacteria
H_2	S^0, SO_4^-	Sulfur- and sulfate-reducing bacteria
H_2	CO_2	Methanogenic and actogenic bacteria
$NH4^+$, NO_2^-	O_2	Nitrifying bacteria
Fe^{2+}, Mn^{2+}	O_2	Iron- and manganese-oxidizing bacteria
CH_4, CO	O_2	Methylotrophic and carbon monoxide-oxidizing bacteria

Source: Millero, F. J., *Chemical Oceanography,* 2nd ed., CRC Press, Boca Raton, FL, 1996, 376. With permission.

Table 5.12–7 Sources and Characteristics of Hyperthermophilic Heterotrophs from Submarine Hydrothermal Vent Environments[a]

Organism	Source	Temp. Growth Range (°C)	Morphology	C Substrates	Electron Adapters	G + C Ratio	Reference
Archaea Domain							
Pyrodictium occultum[b]	Shallow vent field (Italy), MacDonald Seamount	82–110	Disks with fibers	$CO_2 + H_2$ (cell extracts[b] $+ H_2$)	$S^0 (S_2 O_2^{2-})$[b]	62.0	Stetter et al. (1983); (Huber et al.) (1990)
Pyrodictium abyssi	Deep-sea vent field	80–110	Disks with fibers	YE, gelatin starch, formate	Fermentative?, (S^0)[c]	60.0	Stetter et al. (1990); Pley et al. (1991)
Thermodiscus maritimus	Shallow vent field (Italy)	75–98	Disks	YE (H_2 + cell extracts)	Fermentative?, S^0	49.0	Stetter and Zillig (1985)
Thermococcus celer	Shallow vent field, MacDonald Seamount	75–97	Irregular cocci, monopolar, polytrichous flagella	YE, tryptone, casein	Fermentative, S^0	56.6	Zillig et al. (1983a, b)
T. litoralis	Shallow vent field (Italy)	55–98	Irregular cocci, nonmotile	YE, tryptone, casein	Fermentative, S^0	39.1	Belkin and Jannasch (1985); Neuner et al. (1990)
T. stetteri (4 strains)	Shallow vent field (Northern Kurils)	55–98	Irregular cocci, nonmotile or motile with polytrichous flagella	Casein, peptone, trypticase, starch, pectin	S^0	50.2	Miroshnichenko et al. (1989)
Pyrococcus furiosus (and other spp.)	Shallow vent field, MacDonald Seamount	70–103	Cocci, monopolar polytrichous flagella	YE, casein, starch, maltose	Fermentative, S^0	38.0	Fiala and Stetter (1986); Zillig et al. (1987)
Staphylothermus marinus	Deep-sea smoker fluid (11°N, EPR)	65–98	Cocci, nonmotile	YE, peptone, beef extract	S^0	34.5	Zillig et al. (1982); Fiala et al. (1986)
Desulfurococcus (strain S)	Deep-sea smoker wall (11°N, EPR)	50–94	Irregular cocci, monopolar flagella bundle	YE, trypton, casein	S^0	52.0	Jannasch et al. (1988b)
Desulfurococcus (strain SY)	Deep-sea smoker wall (11°N, EPR)	50–96	Irregular cocci, no flagella	YE, tryptone, casein	S^0	52.4	Jannasch et al. (1988b)
Archaeoglobus fulgidus[d]	Shallow vent field, MacDonald Seamount	60–95	Irregular cocci, nonmotile, no flagella	Proteins, sugars, lactate, formate $(CO_2 + H_2)$[d]	$SO_4^{2-}, S_2 O_3^{2-}, O_3^{2-}$	46.0	Stetter (1988); Zellner et al. (1989)
A. profundus[e]	Deep-sea vent sediment (Guaymas Basin)	65–90	Irregular cocci, nonmotile, no flagella	YE, peptone, meat extract, lactate, pyruvate, acetate (H_2)[e]	$SO_4^{2-}, S_2 O_3^{2-}, SO_3^{2-}$	41.0	Burggraf et al. (1990)
Hyperthermus butylicus	Shallow vent field (Azores)	85–108	Irregular cocci, pili, no flagella	Complex organics, peptides	Fermentative?, S^0	55.6	Zillig et al. (1990)

(continued)

Table 5.12–7 Sources and Characteristics of Hyperthermophilic Heterotrophs from Submarine Hydrothermal Vent Environments (continued)

Organism	Source	Temp. Growth Range (°C)	Morphology	C Substrates	Electron Adapters	G + C Ratio	Reference
Thermococcus sp.[f] (ES-1)	Deep-sea vent polychaete (Endeavour, JFR)	50–91	Irregular cocci, nonmotile, no flagella	YE, peptone	Fermentative?, (S⁰)[c]	58.6	Pledger and Baross (1989)
Pyrococcus sp.[f] (ES-4)	Deep-sea flange solids (Endeavour, JFR)	66–110	Irregular cocci, nonmotile, no flagella	YE, peptone, starch, amino acids	Fermentative?, S⁰	55.0	Pledger and Baross (1991)
Thermococcus sp.[f] (AL-1)	Deep-sea smoker fluid (Endeavour, JFR)	55–94	Irregular cocci, nonmotile	YE, tryptone, peptone, casein	Fermentative? (S⁰)[c]	51	Reysenbach and Deming (1991); Reysenbach et al. (1992)
Pyrococcus sp.[f] (AL-2)	Deep-sea flange solids (Endeavour, JFR)	60–108	Irregular cocci, nonmotile	YE, tryptone, peptone, casein	Fermentative?, (S⁰)[c]	41	Reysenbach and Deming (1991); Reysenbach et al. (1992)
Pyrococcus fidjiensis (GE5)	Deep-sea vent sulfides (North Fiji Basin)	67–102	Cocci, polar tuft of flagella	YE, peptides, casamino acids	Fermentative, (S⁰)[c]	44.7	Prieur et al. (1992); Erauso et al. (1993)
Pyrococcus sp. (GB-D)	Deep-sea vent sediment (Guaymas Basin)	65–103	Irregular cocci, polar bundle of flagella	YE complex proteins, trypticase	Fermentative, S⁰, cystine	39.5	Jannasch et al. (1992); Hoaki et al. (1993)

Bacteria Domain

Thermotoga maritima	Shallow vent field (Italy)	55–90	Sheathed rods, no flagella	YE (cell extracts + H₂)	Fermentative?, S⁰	46	Huber et al. (1986, 1989c)
T. neapolitana	Shallow vent field (Italy)	55–90	Sheathed rods, no flagella	YE, starch, sugars, glycogen	Fermentative	40	Belkin et al. (1986); Windberger et al. (1989); Jannasch et al. (1988a)
Aquifex pyrophilus[g]	Shallow vent sediments (Iceland; 106 m)	67–95	Rods with rounded ends, polytrichous flagella	$CO_2 + H_2$ ($S_2O_3^{2-}$, S^0 as electron donors)	O_2,[g]NO_3^-	41.5	Huber et al. (1992); Burggraf et al. (1992)

Note: YE = Yeast extract. EPR = East Pacific Rise; JFR = Juan de Fuca Ridge. Full citations for references are in original source.
[a] Except for *Aquifex pyrophilus*, an obligate chemolithotroph.
[b] Grows chemolithotrophically with S⁰ as electron acceptor; mixotrophically with $S_2O_3^{2-}$.
[c] Stimulates growth, but not required; strains AL-1 and AL-2 cultured under H₂ pressure require S⁰ as detoxifying agent.
[d] Grows chemolithotrophically with $S_2O_3^{2-}$ as electron acceptor.
[e] Grows mixotrophically with obligate requirements for H₂.
[f] Member of order Thermococcales, according to 16S rRNA sequence (Reysenbach et al., 1992).
[g] The only known microaerophilic hyperthermophile.
Source: Baross, J. A. and Deming, J. W., in *The Microbiology of Deep-Sea Hydrothermal Vents*, Karl, D. M., Ed., CRC Press Boca Raton, FL, 1995, 174. With permission.

Table 5.12–8 Total Microbial Production Data from a Variety of Mid-Ocean Ridge Hydrothermal Vents

Location and Incubation Conditions	ALVIN Dive #	Rate of Synthesis[a] (pmol/l/h)		Specific Rate[b] (pmol/ng ATP/h)		C Production[c]	Reference
		RNA	DNA	RNA	DNA		
Galapagos Rift					**Water Column**		
Rose Garden (*in situ*)	988	12.9 (±1.3)	—[d]	2.2	—	—	Karl, 1985; Karl, unpublished
Rose Garden	990	20.9	—	1.7	—	—	
(1 atm, 4°C)							
(1 atm, 27°C)							
Rose Garden (1 atm, 4°C)	990	513.1	—	32.9	—	—	
Total		130	—	7.2	—	—	
<12 μm		65.6	—	7.3	—	—	
21°N							
Niskin baggie (*in situ*)	1212	12.8 (±0.1)	0.40 (±0.16)	3.7	0.13	0.64	Karl, 1985; Karl, unpublished
Pump sample (1 atm)	1214						
3.5°C		28.6	5.57	4.9	0.96	8.85	
17°C		148.2	50	7.1	2.39	79.4	
25°C		187.5	50.8	4.5	1.23	80.7	
35°C		86.2	30.9	4.7	1.67	49.1	
45°C		86.5	27.0	6.0	1.87	42.9	
55°C		4.1	2.71	1.2	0.81	4.31	
Niskin baggie (*in situ*)	1225	176.6 (±65.1)	55.3 (±36.4)	2.5	0.76	87.9	
Syringe sample (*in situ*)	1225	68.6 (±16.5)	3.20 (±0.78)	8.3	0.39	5.08	
11°N							
Niskin bottle (1 atm, 30°C)	1381	161.5	4.6	13.1	0.37	7.31	Karl, unpublished
(1 atm, 30°C)	1385	27.9	1.14	2.36	0.10	1.81	
Endeavour Ridge							
Niskin bottle (1 atm, 12°C)	1419	94.0	1.8	3.8	0.07	2.86	Winn and Karl, unpublished
Guaymas Basin							
Niskin bottle-2 (1 atm, 26°C)	1603	15.0	1.1	1.9	0.14	1.75	Karl, unpublished
Niskin bottle-3 (1 atm, 26°C)	1603	4.3	0.17	0.73	0.03	0.27	
Niskin bottle-4	1603	10.0	0.58	0.52	0.03	0.92	
Pump sample (1 atm)	1605						
4°C		5.4	0.24	0.60	0.27	0.38	
25°C		40.8	2.99	9.49	0.70	4.75	

(continued)

Table 5.12–8 Total Microbial Production Data from a Variety of Mid-Ocean Ridge Hydrothermal Vents (continued)

Location and Incubation Conditions	ALVIN Dive #	Rate of Synthesis[a] (pmol/l/h)		Specific Rate[b] (pmol/ng ATP/h)		C Production[c]	Reference
		RNA	DNA	RNA	DNA		
Control (nonvent) samples[e]							
5 m		148 (±25.6)	20.2 (±3.1)	2.8	0.39	32.1	
100 m		5.1 (±0.3)	0.70 (±0.02)	0.3	0.04	1.11	
2000 M		0.50	0.10	0.24	0.05	0.16	
Sediments/Particles							
21°N							
Black smoker particles							
21°C		909	137	6.8	1.0	228	Karl et al., 1984
50°C		2256	424	22.8	4.3	706	
90°C		652	273	8.7	3.6	456	
Guaymas Basin Heated sediments (no mats)							
0–1 cm		20–100	5–15	0.5–2.5	0.1–0.4	~300	Karl and Novitsky, unpublished

[a] Rates of RNA and DNA synthesis measured using ^{3}H-adenine (Karl and Winn, 1984); values in parentheses are ±1 standard deviation where $n \geq 3$. Units: water column samples, pmol/1/h; sediments and particles, pmol/g (dry wt)/h.

[b] Specific rates are calculated from ATP measured at the end of the incubation period, which ranged from 2 to 18 h.

[c] Carbon production extrapolated from DNA synthesis rates based on the assumptions of Karl and Winn (1984). Units: water column, µg/1/d C; sediments and particles, µg/g (dry wt)/d C.

[d] Not determined.

[e] Control samples collected at the 21°N East Pacific Rise site.

Note: Full citations of references in original source.

Source: Karl, D. M., in *The Microbiology of Deep-Sea Hydrothermal Vents*, Karl, D. M., Ed., CRC Press, Boca Raton, FL, 1995, 88. With permission.

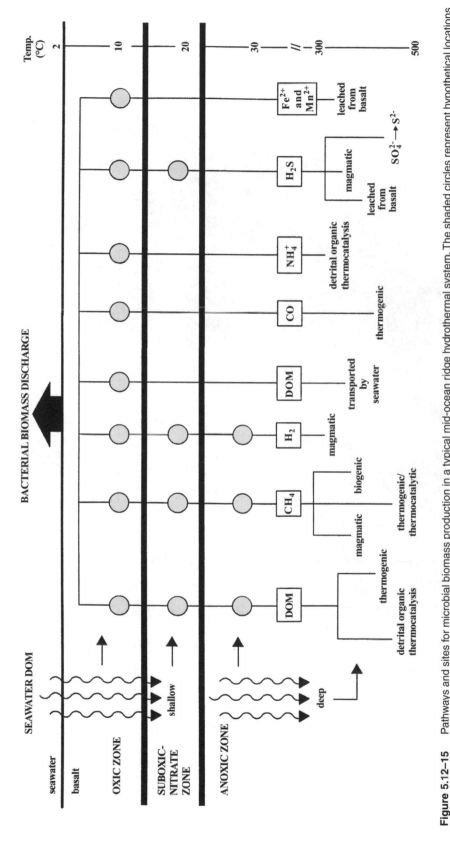

Figure 5.12–15   Pathways and sites for microbial biomass production in a typical mid-ocean ridge hydrothermal system. The shaded circles represent hypothetical locations in the vent system in which bacterial biomass is produced by one of a variety of metabolic processes. Not shown in the diagram are potential reactions with pelagic sediment both at the ridge crest for sediment-covered systems (e.g., Guaymas Basin) or on the ridge flanks of all hydrothermal habitats, or the potential photochemical processes. (From Karl, D. M., in *The Microbiology of Deep-Sea Hydrothermal Vents,* Karl, D. M., Ed., CRC Press, Boca Raton, FL, 1995, 57. With permission).

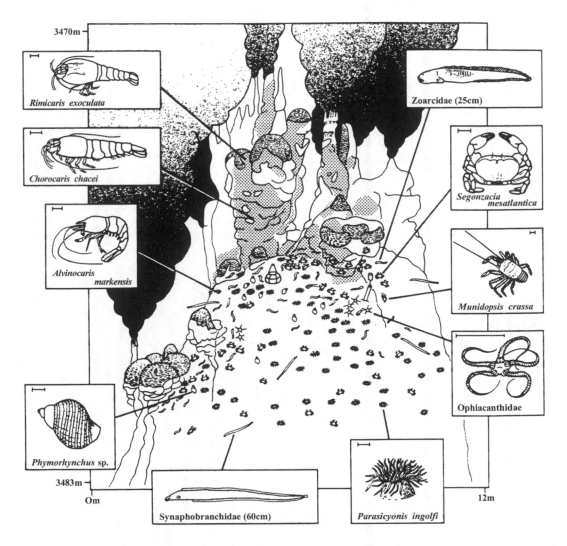

Figure 5.12–16   Snake Pit (Mid-Atlantic Ridge) faunal distributions. (From Van Dover, C. L., in *Hydrothermal Vents and Processes,* Parson, L. M. et al., Eds., Geological Society Special Publication No. 87, Geological Society of London, London, 1995, 267. With permission).

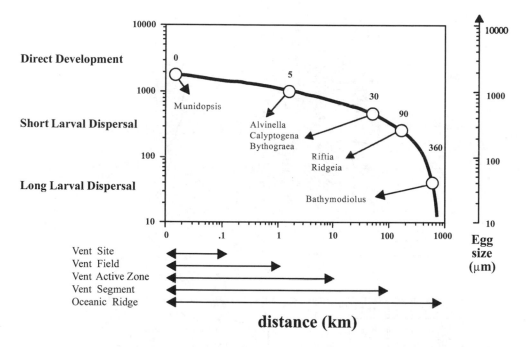

Figure 5.12–17   Relationship between length of the planktonic larval stage of hydrothermal vent organisms and the potential distance traveled in one generation, assuming a residual current speed of 2 to 3 m/s. (From Dixon, D. R. et al., in *Hydrothermal Vents and Processes,* Parson, L. M. et al., Eds., Geological Society Special Publication No. 87, Geological Society of London, London, 1995, 348. With permission.)

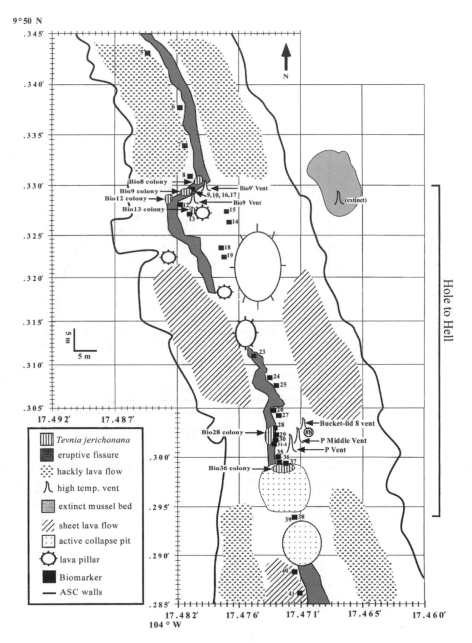

Figure 5.12–18   Map of the northern transect region between 9°50.285'N and 9°50.345'N (a 120-m distance) featuring the location of the six vestimentiferan colonies (as they appeared in March 1992) in close proximity to five high-temperature vents along the central primary eruptive fissure within the Hole-to-Hell area. Unlabeled lava flow types on the axial summit caldera floor are areas where flows are a jumbled mixture of adjacent flows. Axis tick = 1 m. (From Shank, T. M. et al., *Deep-Sea Res. II*, 45, 472, 1998. With permission).

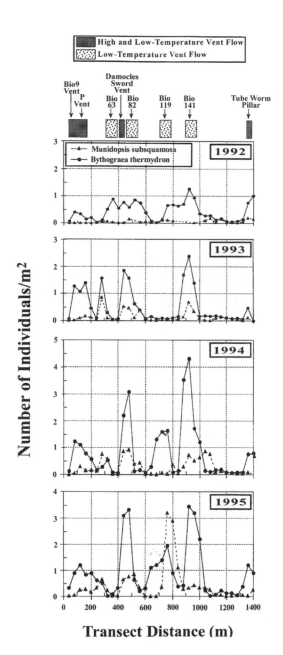

Figure 5.12–19   The temporal distribution and density of brachyuran and galatheid crabs along the transect near 9°50′N on the East Pacific Rise between March 1992 and November 1995. Nematocarcinid shrimp densities (not shown here) closely mirrored those of galatheid crabs and often exceeded galatheid density and abundance in 1992. The major high- and low-temperature venting areas are indicated in the uppermost graph. (From Shank, T. M. et al., *Deep-Sea Res. II*, 45, 482, 1998. With permission).

CHAPTER 6

Marine Pollution and Other Anthropogenic Impacts

I. INTRODUCTION

Estuarine and oceanic environments are subject to a wide array of anthropogenic impacts associated with development of the coastal zone, the input of pollutants, overexploitation of marine resources, electric power generation, mariculture, transportation, and many other activities.[1-3] These impacts account for considerable ecologic and economic losses worldwide.[4] Estuarine and coastal marine waters continue to be the most threatened systems due in large part to a burgeoning human population residing in nearby watersheds and increasing coastal resource use. Approximately 4 billion people worldwide live within 60 km of the coast, and this dense population has been responsible for many ongoing environmental problems observed in estuaries and continental shelf regions. Whereas the health of coastal waters remains a cause of concern, the open ocean appears to be much less impacted, except along frequently used shipping lanes. Nevertheless, even the deep sea is not totally immune to human influence. For example, deep-sea sediments have served as repositories for radioactive material derived from nuclear weapons testing, military projects (e.g., disposal of military hardware), and direct dumping of radioactive waste.[5] Thus, it is necessary to examine human effects on the totality of the world's oceans.

Anthropogenic impacts on estuarine and marine ecosystems can be subdivided into three major categories: (1) pollution; (2) habitat loss and alteration; and (3) resource utilization and over-exploitation. The most conspicuous impacts occur near metropolitan centers and other heavily developed regions in the coastal zone. Examples include Boston Harbor, Newark Bay, New York Bight Apex, and Baltimore Harbor on the East Coast of the United States, Commencement Bay, San Francisco Bay, San Diego Harbor, and Santa Monica Bay on the West Coast, and Galveston Bay, Corpus Christi Bay, and Tampa Bay on the Gulf Coast. Despite significant improvements in water quality, habitat degradation, and resource uses in many U.S. coastal systems during the past three decades related to the enactment of a series of federal environmental legislation (e.g., the National Environmental Policy Act of 1969, Federal Water Pollution Control Act Amendments of 1972, Ocean Dumping Act of 1972, Toxic Substances Act of 1976, Clean Water Act of 1977, Nuclear Waste Policy Act of 1982, Clean Water Act Amendments of 1987, and Ocean Dumping Ban Act of 1988), many problems persist.

Ecological conditions appear to be even more problematic in foreign waters, where marine environmental management programs often are weak or nonexistent. A comprehensive global approach to deal effectively with marine environmental issues and to maintain marine environmental quality has yet to be formulated. Although progress is being made in this regard as demonstrated by the United Nations Convention on the Law of the Sea, the present framework consists of a patchwork of agreements, protocols, and regional and international conventions, as well as national laws, regulations, and guidelines.[6]

II. TYPES OF ANTHROPOGENIC IMPACTS

A. Marine Pollution

The United Nations Joint Group of Experts on the Scientific Aspects of Marine Pollution (GESAMP)[7] has defined marine pollution as "the introduction by man, directly or indirectly, of substances or energy into the marine environment, resulting in such deleterious effects as harm to living resources, hazards to human health, hindrance to marine activities including fisheries, impairment of quality for use of seawater, and reduction of amenities." A less restrictive definition of marine pollution has been proffered by Pinet:[8] "the introduction or extraction by humans of material and energy from the environment, such that concentrations of these substances are raised above or lowered below natural levels to such a degree that environmental conditions change." Pollution differs from contamination. As noted by Clark,[9] contamination occurs "when a man-made input increases the concentration of a substance in seawater, sediments, or organisms above the natural background level for that area and for the organisms."

Pollutants enter estuarine and oceanic waters via five principal pathways: (1) nonpoint-source runoff from land; (2) direct pipeline discharges; (3) riverine influx; (4) dumping and accidental releases from ships at sea; and (5) atmospheric deposition.[10,11] Approximately 80 to 90% of the pollutants in the sea derive from land-based sources.[12] In the United States, both point- and non-point-source pollution contributes to water quality impairment in estuarine and coastal marine waters, although during the past two decades contaminant loadings due to point-source inputs have decreased significantly as a result of tighter state and federal government regulations and improved engineering controls of wastewater discharges. Point-source pollution refers to the discharge of pollutants from a discernible, confined, and discrete conveyance such as a pipe, conduit, channel, or tunnel. In contrast, nonpoint-source pollution pertains to the input of pollutants from dispersed, diffuse, and uncontrolled sources such as general surface runoff, groundwater seepage, and atmospheric fallout.[13] Because of the thousands of land-based nonpoint sources of pollution, management strategies to restore and protect the nation's coastal waters are now focusing on corrective actions in neighboring watersheds. The federal Clean Water Action Plan, a multiyear effort released in February 1998 to address environmental problems in U.S. watersheds, provides an excellent example.

The list of contaminants found in estuarine and marine environments is extensive and includes numerous petroleum hydrocarbons, halogenated hydrocarbons, metals, radionuclides, and many other substances. Contaminant loadings in these environments are commonly associated with the introduction of municipal and industrial wastewaters, sewage sludge, dredged material, spills from coastal installations and vessels, as well as the inputs from stormwater runoff, groundwater seepage, and atmospheric deposition. Many estuaries and nearshore oceanic waters throughout the world are plagued by pollution problems because of the accumulation of contaminants in various media. The tendency is for these substances to accumulate in three marine zones, specifically, the neuston layer, along pycnoclines, and in bottom sediments.[9] In heavily impacted systems, pollutants have caused both acute and insidious alteration of biotic communities and habitats.[14]

1. Nutrient Loading

The enrichment of nutrients in coastal waters contributes greatly to eutrophication problems, which severely degrade water quality and threaten organisms.[15-22] Excess nutrient input to these systems, particularly nitrogen and phosphorus, stimulates production of phytoplankton and benthic flora, and the increased abundance of plant life often leads to widespread and recurring anoxia (0 mg/l dissolved oxygen) and hypoxia (<2 mg/l dissolved oxygen) in response to accelerated benthic respiration. The effects are frequently most pronounced in shallow estuaries and enclosed seas with poor water circulation, where oxygen-depleted waters cannot be effectively reoxygenated.

The affected waterbodies are often poorly flushed shallow coastal bays characterized by low freshwater inflow and restricted tidal ranges.[23] However, factors other than anthropogenic nutrient loading also foster anoxic or hypoxic events. In Chesapeake Bay, for example, oxygen depletion zones appear to be exacerbated by seasonal increases in density stratification of the estuary attributable to the spring freshet, which promotes water column stability and minimizes advective transport of oxygen from the surface to deeper waters. As a consequence, broad expanses of the bay experience recurring anoxia in late spring and summer.[24–26] Major tributaries on the western shore of Chesapeake Bay (e.g., Patuxent, Rappahannock, and York Rivers) also frequently experience hypoxic episodes.

Primary production is usually nitrogen limited in the sea. The three major dissolved inorganic forms of nitrogen in seawater are nitrate (NO_3^-), nitrite (NO_2^-), and ammonia (NH_4-N). Of these three forms, nitrate usually occurs in highest amounts ($\sim$1 μg-atom/l or less and rarely above 25 μg-atom/l). The concentrations of ammonia, urea, and amino acids in marine systems are generally as low or lower than those of nitrate. Phosphate (PO_4^{3-}) levels in seawater range from 0 to $\sim$3 μg-atom/l. However, because phosphate strongly sorbs to particulate matter and forms insoluble salts, it rarely exists in quantities as high as those of nitrate.[27] The concentrations of nitrogen and phosphorus increase substantially in waters that receive inputs of raw sewage or partially treated municipal wastewater discharges. They also are appreciably elevated in many estuaries and coastal embayments where nutrient loading from nonpoint sources in nearby watersheds is high.

Enhanced eutrophication, or hypereutrophication, can substantially alter the structure and function of aquatic ecosystems and produce major trophic imbalances. The eutrophication process affects the trophic interactions of many organisms, including both bottom-up and top-down feeding groups of consumers, thereby altering entire food webs.[28] In these systems, species composition, abundance, distribution and diversity change substantially, and massive kills of fish and benthic invertebrates commonly occur. Shading effects due to extensive phytoplankton blooms and accelerated epiphytic and macroalgal growth create unfavorable conditions for other submerged aquatic vegetation (e.g., seagrass beds) and, in extreme cases, result in the elimination of broad expanses of critical benthic habitat. Opportunists and nuisance organisms often dominate the impacted area, replacing more desirable forms and depressing the species diversity even further. Exotic and toxic algal blooms (i.e., red tides) periodically develop in the more fertile waters, and fishery and shellfishery yields typically decline. Red tides, dominated by dinoflagellates, are especially serious because they occasionally cause acute diseases or even death in marine organisms and humans. Four types of human illnesses have been linked to the consumption of seafood contaminated with algal toxins: (1) paralytic shellfish poisoning (PSP); (2) diarrhetic shellfish poisoning (DSP); (3) neurotoxic shellfish poisoning (NSP); and (4) amnesic shellfish poisoning (ASP).[29] Some toxic dinoflagellates are responsible for epidemic diseases in fish and have been implicated in massive fish kills.[30,31]

The source of nutrient overenrichment problems in coastal waters is usually multifaceted.[1–3,11,28] Large amounts of nutrients entering coastal waters in many parts of the world originate from wastewater discharges and raw or partially treated sewage inputs. Contributions from fertilized agricultural lands remain a universal problem.[32] Stormwater runoff, groundwater seepage, and other nonpoint sources in industrialized and urbanized areas generate some of the most nutrient-enriched systems on earth.[13,27] It is clear that human activities associated with development and land-use practices throughout coastal watersheds facilitate the release of nutrients that ultimately accumulate in coastal waters. Other potentially important, albeit localized, sources of nutrients are boats, marinas, and dredging and (subaqueous) dredged material disposal sites.[8,9,14]

The Committee of the Marine Board of the U.S. National Research Council has targeted three high-priority coastal sites in the United States for the development of systematic programs to assess the general problem of eutrophication. These sites include the New York Bight/Long Island Sound, Chesapeake Bay, and northern Gulf of Mexico (which receives the Mississippi–Atchafalaya outflow). The committee has proposed the collection of the following array of data from these locations:

- Physical measurements (current speeds, tidal direction, density fields, and light fields).
- Chemical measurements (nutrients, dissolved oxygen, chlorophyll, and vertical particle flux).
- Biological measurements (primary productivity, phytoplankton species composition, bacterial biomass and production, benthic respiration, and water column respiration).[1]

2. Organic Carbon Loading

A cause-and-effect relationship has also been established between anthropogenic carbon loading and eutrophication problems in estuarine and coastal marine waters. Anthropogenic carbon inputs (e.g., raw sewage and sewage sludge) can cause anoxia or hypoxia in coastal waters or exacerbate anoxic or hypoxic conditions by raising the biochemical oxygen demand (BOD, the oxygen consumed during the microbial decomposition of organic matter) and the chemical oxygen demand (COD, the oxygen consumed through the oxidation of ammonium and other inorganic reduced compounds).[33] Estuarine and nearshore oceanic waters that have received large volumes of sewage wastes exhibit the highest dissolved and particulate organic carbon levels, occasionally exceeding 100 mg/l. In comparison, the concentrations of dissolved organic carbon in estuarine and nearshore oceanic waters not impacted by these wastes range from ~1 to 5 mg/l, and those in the open ocean range from ~0.4 to 2 mg/l. The levels of particulate organic carbon in these waters, in turn, range from ~0.5 to 5 mg/l in estuaries, ~0.1 to 1 mg/l in nearshore oceanic waters, and 0.1 to 0.5 mg/l in open ocean surface waters. The ratio of dissolved organic carbon to dead particulate organic carbon to living organic carbon in seawater is ~100:10:2.[27]

Decaying vascular plant matter (e.g., saltmarsh grasses, seagrasses, and mangroves), benthic macroalgae, animal remains, and biodeposits (feces and pseudofeces) account for most oxidizable carbon in relatively pristine estuaries. The detritus concentrations in these systems range from ~0.1 to more than 125 mg/l. In oceanic waters, most organic carbon is introduced via phytoplankton production (200 to 300 × 10^14 g C/year), with much smaller amounts delivered by river inflow, groundwater input, and atmospheric deposition (<10 × 10^14 g C/year). Exudation products from producers and consumers may contribute significant amounts of dissolved organic carbon. As much as 10% of net phytoplankton production is exuded in the open ocean.[27,34]

Systems receiving large inputs of anthropogenic organic carbon wastes via pipeline discharges, sewage sludge dumping, riverine inflow, dredged materials, marinas, boats, and other sources often have significant eutrophication problems. Biotic communities are commonly significantly impacted in these systems.[35-38] In the United States, such impacted waters have historically been found near metropolitan centers (e.g., New York, Boston, Baltimore, Los Angeles, and Seattle). This is also true in other countries worldwide.

Chronic organic carbon enrichment degrades many benthic communities as reflected by gross measures of community structure (i.e., species richness, dominance, diversity, and total animal abundance). Weston[39] reported the following changes in macrobenthic communities subjected to persistent organic carbon loading: (1) a decrease in species richness and an increase in total number of individuals attributable to high densities of a few opportunistic species; (2) a general reduction in biomass, although there may be an increase in biomass corresponding to a dense assemblage of opportunists; (3) a decrease in body size of the average species or individual; (4) shifts in the relative dominance of trophic guilds; and (5) a shallowing of that portion of the sediment column occupied by infauna.

Pathogenic microorganisms (i.e., bacteria, viruses, protozoans, and helminths) in estuarine and coastal waters derived from sewage wastes, wildlife populations, and stormwater runoff are potentially deleterious to humans swimming in contaminated waters and consuming contaminated seafood products. More than 100 human enteric pathogens, bacteria, viruses, and parasites have been identified in urban stormwater runoff and treated municipal wastewater.[29] Pathogenic bacteria of the genera *Salmonella* and *Shigella* are responsible for typhoid and dysentery, respectively. Hepatitis A and viral gastroenteritis result from human consumption of raw, viral-tainted shellfish.[33] Because of the potential threat of sewage-contaminated estuarine and marine environments to human health, intense water quality monitoring programs are conducted by coastal states nationwide. These programs entail comprehensive measurements of fecal and total coliform bacteria levels at bathing beaches and in shellfish growing waters.

3. Oil

Major oil spills pose a serious threat to biotic communities and habitats in the coastal zone, although they take place only sporadically. For example, between 1969 and 1995, more than 40 major marine oil spills occurred, and many were extremely detrimental to coastal environments (e.g., *Argo Merchant, Amoco Cadiz,* and *Exxon Valdez* spills).[40] Oil spills account for only a fraction (~10 to 15%) of the total amount of oil entering the sea on an annual basis. The remainder originates from routine human activities and natural sources. Estimates of oil influx to the sea from natural sources alone range from 5 to 10 million metric tons per year.[8] Included here is oil synthesized by phytoplankton and that which seeps out of the seafloor.

Most chronic oil pollution in marine environments derives from leakages of fixed installations (e.g., coastal refineries, offshore production facilities, and marine terminals), operations of petrochemical industries, as well as municipal and industrial wastewaters, urban and suburban runoff, and atmospheric deposition.[3,41] Inputs associated with marine transportation activities (e.g., routine tanker operations, tanker accidents, nontanker accidents, deballasting, and dry docking), in turn, are responsible for about one third of the oil entering oceanic waters.[9,42] A substantial quantity of oil transported to coastal waters originates from nonpoint sources in nearby watersheds. Estuaries receive more than one third of the entire input of oil to the marine hydrosphere.[43]

Polluting oil is extremely dangerous in estuarine and marine environments because it kills organisms directly by smothering them. In addition, toxic chemicals in the oil render habitats inhospitable to colonization by plant and animal populations, altering natural conditions (e.g., pH, dissolved oxygen levels, and food availability). In severe cases, biotic impacts of polluting oil in an area may persist for more than a decade. Such impacts are most often observed in protected intertidal and shallow subtidal habitats exposed to large oil spills or repeated oiling events. Along open coasts where high-energy conditions exist, strong wave action facilitates the physical breakup and dispersal of waste oil. In contrast, saltmarshes, mangroves, and tidal flats in estuaries and other low-energy environments promote the trapping of oil.

Crude oil is composed of thousands of organic compounds, many of which are toxic to marine life. Some of the more toxic components include benzene, toluene, xylene, carboxylic acids, phenols, and sulfur compounds. The toxicity of petroleum hydrocarbons increases along the series from alkanes, cycloalkanes, and alkenes to the aromatics. Although the severity of environmental impact depends greatly on the composition of the oil (e.g., North Sea crude oil, Kuwait crude oil, No. 2 fuel oil, etc.), many other factors also play a significant role:

1. The volume of the oil;
2. Composition of the oil;
3. Form of the oil (i.e., fresh, weathered, or emulsified);
4. Occurrence of the oil (i.e., in solution, suspension, dispersion, or adsorbed onto particulate matter);
5. Duration of exposure;
6. Involvement of neuston, plankton, nekton, or benthos in the spill or release;
7. Juvenile or adult forms involved;
8. Previous history of pollutant exposure of the biota;
9. Season of the year;
10. Natural environmental stresses associated with fluctuations in temperature, salinity, and other variables;
11. Type of habitat affected; and

12. Cleanup operations (e.g., physical methods of oil recovery and the use of chemical dispersants).[3,42,44,45]

The composition and toxicity of polluting oil in the sea change in response to several physical–chemical processes, notably evaporation, photochemical oxidation, emulsification, and dissolution. The greatest change in composition of an oil spill, for example, takes place within 24 to 48 hours of its occurrence in estuarine and marine waters. During this period, the lighter, more toxic and volatile components are lost as a result of evaporation and dissolution. The low-molecular-weight, volatile fractions evaporate, and the water-soluble constituents dissolve in seawater. While photo-oxidation converts some of the hydrocarbons to polar oxidized components, the immiscible components become emulsified. Wave and current action physically disperses the oil, with intense mixing promoting the formation of oil-in-water emulsions or water-in-oil emulsions. "Chocolate mousse," viscous pancakelike masses, develop from 50 to 80% water-in-oil emulsions. Tar balls measuring 1 mm to 25 cm in diameter are the heaviest residues.

Spilled oil gradually increases in density as a result of the aforementioned physical–chemical processes. Eventually the oil settles to the seafloor. Settlement is facilitated by the sorption of hydrocarbon compounds—which are hydrophobic—on particulate matter (e.g., clay, silt, sand, shell fragments, and organic material) suspended in the water column. Once the oil accumulates in bottom sediments, microbial degradation of the hydrocarbons can proceed rapidly. Diverse bacterial populations, including various *Pseudomonas* species, possess the enzymatic capability to degrade petroleum hydrocarbons effectively. Bioremediation techniques have been developed by adding exogenous microbial populations or stimulating indigenous ones to increase biodegradation rates significantly above those occurring naturally. In the case of the *Exxon Valdez* oil spill, for example, the application of nitrogen- and phosphorus-containing fertilizers increased biodegradation rates 3 to 5 times.[46,47] Microbes degrade as much as 40 to 80% of oil spills in estuarine and marine environments.

Because most polluting oil eventually accumulates in bottom sediments or on rocky intertidal substrates and beaches, the most acute biotic impacts typically arise in benthic communities.[48-54] In some areas, subsurface oil persists for years, creating long-term hazardous benthic conditions.[55] The susceptibility of the benthos to polluting oil stems from the immobility of rooted vegetation and the limited mobility of most epifauna and infauna. Large oil spills can completely smother these communities, culminating in the mass mortality of constituent populations. In less catastrophic cases, the structure of the community changes dramatically, as evidenced by large shifts in species composition, abundance, and diversity. Some of the most serious impacts of polluting oil have been recorded in fringing wetland habitats (i.e., salt marshes and mangroves).

Other biotic groups are also not immune to the impacts of polluting oil. Although juvenile and adult fish can avoid oil spills because of their mobility, the early life stages (eggs and larvae) are much more susceptible. Both sublethal and lethal effects of oil on these early life stages occasionally result in long-term reductions in population abundances. For example, some fishery experts ascribe the reductions in abundance of some commercially important fish populations in Prince William Sound during the 1990s to such impacts of the *Exxon Valdez* oil spill in March 1989.

Birds often incur heavy losses during large oil spills. For instance, the *Exxon Valdez* oil spill killed an estimated 100,000 to 300,000 birds. Species that spend much of their time on the sea surface (e.g., auks, diving ducks, and penguins) are particularly vulnerable. Toxins are assimilated by the birds when preening oiled feathers, inhaling fumes from evaporating oil, consuming contaminated food, and drinking contaminated water. Many seabirds drown or experience hypothermia after their feathers are oiled. Others succumb to sublethal effects such as gastrointestinal disorders, immune system depletion, red blood cell damage, and pneumonia.

Marine mammals appear to be less seriously impacted than benthic invertebrates and seabirds to polluting oil. Cetaceans and pinnipeds have a greater capacity for hydrocarbon metabolism. In addition, they are quite mobile and can avoid oil spills. Nevertheless, some serious effects on marine mammals have been demonstrated. For example, the *Exxon Valdez* oil spill adversely affected sea otters and harbor seals over a broad area. Marine mammal casualties may occur soon after exposure to polluting oil as a consequence of acute effects, or over a protracted period of time as a result of multiple sublethal effects (e.g., respiratory deficiencies, blood damage, gastrointestinal problems, renal malfunctions, and other disorders).[56–58]

Oil spill cleanup can produce additional impacts on marine biotic communities. Some dispersants, for instance, are toxic to marine organisms, and are of little value once the oil is stranded on sandy beaches. However, they have proved to be valuable in managing oil spills at sea. Mechanical cleanup along a shoreline can be destructive to benthic habitats. In some cases, the best cleanup strategy may be no spill control, especially when an oil slick remains at sea or is moving away from the shoreline.

4. Toxic Chemicals

a. Polycyclic Aromatic Hydrocarbons

The marine environment receives more than 2.3×10^5 metric tons (mt) of polycyclic aromatic hydrocarbons (PAHs) each year, mainly from anthropogenic sources.[59] PAHs are a widespread group of organic contaminants, found in all the world's oceans. However, the highest concentrations of PAHs occur in bottom sediments, waters, and biota of coastal systems in close proximity to urban and industrialized areas. These hydrocarbon compounds are of major concern because of their potential carcinogenicity, mutagenicity, and teratogenicity to a wide range of organisms, including humans.

PAH compounds consist of hydrogen and carbon arranged in the form of two or more fused benzene rings in linear, angular, or cluster arrangements.[60,61] Some of the more common PAHs are (1) two-ring compounds (biphenyl, naphthalene, 1-methylnaphthalene, 2-methylnaphthalene, 2, 6-dimethylnaphthalene, and acenaphthene); (2) three-ring compounds (fluorene, phenanthrene, 1-methylphenanthrene, and anthracene); (3) four-ring compounds (fluoranthene, pyrene, and benz[*a*]anthracene); and (4) five-ring compounds (chrysene, benzo[*a*]pyrene, benzo[*e*]pyrene, perylene, and dibenz[*a,h*]anthracene). The two- and three-ring (low-molecular-weight) PAHs, although highly toxic, appear to be noncarcinogenic to many marine organisms. The four- to seven-ring PAHs, in contrast, have greater carcinogenic potential but are less toxic. The low-molecular-weight PAHs also are less mutagenic than the high-molecular-weight varieties.[62]

Some PAHs are synthesized in estuarine and marine environments by bacteria, fungi, and algae. Others are released to these environments from natural processes, such as some grass and forest fires and marine seep and volcanic emissions. However, the concentrations of naturally derived PAHs in these environments are low relative to anthropogenic sources.[59] The most important anthropogenic sources include oil spills, fossil fuel combustion, urban and agricultural runoff, and municipal and industrial discharges. Most PAHs enter aquatic environments via oil spills (1.7×10^5 mt/year) and atmospheric deposition (0.5×10^5 mt). Crude oils contain 0.2 to 7% PAHs, and consequently major spills and other transportation losses deliver considerable quantities of these contaminants to the sea.[61] Most of the PAHs in the atmosphere originate from the combustion of fossil fuels, forest and brush fires, and volcanic eruptions. The pyrolysis of organic matter is part of a primary delivery system of PAHs to aquatic environments.[63]

PAHs are hydrophobic. They sorb readily to sediments and other particulate matter and ultimately concentrate on the seafloor.[3,11,14,59,64,65] As a result, benthic organisms often are exposed to the highest concentrations of these contaminants, especially in systems near industrialized metropolitan centers. Once deposited in bottom sediments, the PAHs are more persistent than in the water column, where photochemical or biological oxidation effects rapidly degrade them. They remain essentially unaltered for long periods of time in anoxic sediments. However, bioturbation and the roiling of sediments by wave and current action can remobilize the compounds. Elevated bed shear stresses facilitate particle movement along the benthic boundary layer and the release of the PAHs.

The concentrations of PAHs vary substantially in estuarine and nearshore oceanic sediments. For example, Huggett et al.[66] recorded PAH levels in Chesapeake Bay sediments ranging from 1 to 400 μg/kg; highest values were registered in the northern bay. McLeese et al.[67] documented total PAHs in Boston Harbor sediments amounting to 120 mg/kg, but lower concentrations (160 μg/kg) were observed 64 km offshore. Windsor and Hites[68] also ascertained higher levels of PAHs in bottom sediments of the Gulf of Maine (200 to 870 μg/kg) than in deep-sea sediments directly eastward (18 to 160 μg/kg).

The amounts of PAHs in the water column typically are much lower (by a factor of 1000 or more) than those in bottom sediments. PAH concentrations in marine waters removed from areas of intense anthropogenic activity generally are <1 μg/l. Coastal waters near industrialized centers exhibit higher values (1 to 5 μg/l).

Estuarine and marine organisms also have variable concentrations of PAHs. Three factors strongly influence the body burdens of PAHs in these organisms: (1) the concentrations of PAH compounds in different environmental media and environments; (2) bioavailability of the compounds; and (3) capacity of the organisms to metabolize the contaminants. Although highest PAH levels occur in bottom sediments as noted above, sediment-sorbed PAHs have only limited bioavailability, which ameliorates their toxicity potential, particularly for benthic organisms, which are often continuously exposed to the contaminants. However, some benthic organisms, such as bivalve mollusks and echinoderms, have low mixed function oxygenase enzyme activity for metabolizing PAHs, and thus tend to bioaccumulate them.[69] As a result, mussels and oysters have been used as sentinel organisms to monitor PAH contamination in coastal environments.[70–74]

Both laboratory experiments and field surveys unequivocally demonstrate that PAHs adversely affect estuarine and marine organisms. In a literature review, Eisler[60] chronicled PAH concentrations in estuarine and marine finfish and shellfish ranging from near negligible to 1600 μg/kg fresh weight. PAHs have been implicated in the development of lesions and tumors in fish and neoplasia in bivalves.[29,75] They produce biochemical disruptions and cell damage that lead to mutations, developmental malformations, and cancer.[61] Many marine organisms do not exhibit acute responses to PAH exposure, but suffer sublethal effects (biochemical, behavioral, physiological, and pathological) that frequently require months to develop in individuals, and thus are difficult to assess. Such effects are manifested by impaired metabolic pathways, reduced growth, decreased fecundity, and reproductive failure of individuals.[76] Population and community impacts may not be detected until years after initial PAH exposure.

b. Halogenated Hydrocarbons

Among the most insidious and damaging contaminants in estuarine and marine environments are the halogenated hydrocarbons, a group of ubiquitous, persistent, and toxic substances that pose a threat to biotic communities. Their physicochemical properties account for the deleterious biological impacts observed in aquatic systems. The halogenated hydrocarbons, which contain chlorine, bromine, fluorine, or iodine (the halogens), comprise a wide range of low- to high-molecular-weight compounds. The higher-molecular-weight halogens are particularly serious because of their chemically stable nature, hydrophobicity, lipophilicity, affinity for living systems, bioaccumulative capacity, general toxicity, resistance to degradation, and persistence in the environment. The chlorinated hydrocarbons, contaminants mainly derived from pesticides and industrial chemicals (e.g., polychlorinated biphenyls, chlorinated aromatics, and chlorinated paraffins), provide examples. Because these compounds bioaccumulate in food chains, they attain very high concentrations in marine mammals (cetaceans and pinnipeds). Organochlorines, in general, are suspected etiologic agents, having been implicated in various diseases in animals (e.g., blood disorders, immunosuppression, altered endocrine physiology, aberrant developmental patterns, reproductive abnormalities, skin and liver lesions, and cancer).[3,11,77–81]

Some of the most commonly occurring chlorinated hydrocarbon compounds in estuarine and marine environments are the following:

Insecticides—dichlorodiphenyltrichloroethane (DDT), aldrin, chlordane, dieldrin, endosulfan, endrin, lindane, heptochlor, chlordecone, methoxychlor, mirex, perthane, and toxaphene;

Herbicides—chlorophenoxy compounds (2,4-D and 2,4-T), hexachlorobenzene, and pentachlorophenol; and

Industrial chemicals—polychlorinated biphenyls (PCBs), chlorinated dibenzo-p-dioxins (CDDs), and chlorinated dibenzofurans (CDFs).

Of all halogenated compounds, DDT and PCBs have received the greatest attention in marine studies. These synthetic compounds have been found in biotic, sediment, and water samples from estuarine, nearshore oceanic, and open-ocean environments worldwide.[81–87] They tend to sorb to sediments and other particulates, and ultimately concentrate in benthic habitats, reaching highest levels in shallow-water environments near metropolitan centers.

DDT and PCBs seriously impact marine life. DDT, for example, has been shown to damage the central nervous system of animals exposed to the contaminant, leading to hyperactivity, convulsions, paralysis, and death. It inhibits carbonic anhydrase activity, an enzyme necessary for eggshell production in birds.[88] Some avifauna exposed to DDT (e.g., brown pelicans, *Pelecanus occidentalis*, and double-crested cormorants, *Phalacrocorax auritus*) have exhibited reduced population abundance as a result of this effect.

PCB biotic impacts are most evident in upper-trophic-level organisms, which tend to accumulate the contaminants in their tissues. For example, estuarine and marine fish exposed to high concentrations of PCBs have greater incidence of blood anemia, epidermal lesions, and fin erosion. Similarly, marine birds and mammals with large amounts of PCBs in their tissues have more reproductive abnormalities and depressed reproductive potential.

Total DDT concentrations decrease from <5 ng/l in coastal waters to 0.005 to 0.06 ng/l in the ocean. Coastal and nearshore bottom sediments have total DDT concentrations ranging from <0.01 to >1000 ng/g dry weight. Sentinel organisms (i.e., mussels) from coastal waters of the northwest Atlantic and northeast Pacific (United States) contain total DDT residues varying between 2.8 and 1109 ng/g dry weight.[89] Higher-trophic-level organisms (fish, whales, porpoises, dolphins, and seals) have much higher total DDT burdens. Monitoring programs in the United States have documented peak levels of DDT and other chlorinated hydrocarbons in estuarine and coastal marine systems during the period from 1950 to 1970.[90,91]

The concentrations of PCBs usually range from ~0.1 to 1000 ng/l in estuarine and coastal marine waters to ~0.1 to 150 ng/l in the open ocean. They decline appreciably with increasing depth in the deep sea to ~1.5 to 2.0 pg/l at 3500 to 4000 m.[3,11] However, the contaminants have even been recovered at depths >5000 m in the world's oceans.

In bottom sediments, PCB levels peak in "hot spot" coastal locations. For instance, in U.S. coastal and nearshore bottom sediments, the highest PCB concentrations have been registered in Escambia Bay, Florida (<30 to 480,000 ng/g dry weight), New Bedford Bay, Massachusetts (8400 ng/g dry weight), Palos Verdes, California (80 to 7420 ng/g dry weight), New York Bight, New Jersey (0.5 to 2200 ng/g dry weight), and Hudson–Raritan Bay, New York (286 to 1950 ng/g dry weight).[11,14] These are locations that have historically received significant contaminant concentrations from industrialized centers.

PCB levels in estuarine and marine organisms increase by a factor of 10 to 100 times when proceeding from lower to upper trophic levels. For example, PCB concentrations in marine zooplankton amount to <0.003 to 1 μg/g. In fish and mammals, however, the levels generally rise to 0.03 to 212 μg/g. The most dramatic declines in PCB residues in fish and shellfish from U.S. coastal waters since 1980 have occurred in areas near known industrial contaminant sources and other hot spot locations.[92-94]

c. Heavy Metals

Heavy metals consist of two categories of elements: (1) transitional metals (e.g., cobalt, copper, iron, and manganese) and (2) metalloids (e.g., arsenic, cadmium, lead, mercury, selenium, and tin). The transition metals are essential for metabolism of organisms at low concentrations but may be toxic at high concentrations. The metalloids, although generally not required for metabolic function of organisms, are toxic at low concentrations.[9,14,95] Organometals (e.g., alkylated lead, tributyl tin, and methylmercury) comprise a particularly toxic group of chemicals that not only represent a danger to marine organisms but also a danger to humans consuming contaminated seafood. According to Abel,[41] the approximate order of increasing toxicity of common heavy metals is as follows: cobalt < aluminum < chromium < lead < nickel < zinc < copper < cadmium < and mercury.

A large quantity of heavy metals in the sea derives from natural processes (i.e., weathering of rocks, leaching of soils, eruptions of volcanoes, and emissions of hydrothermal vents). Anthropogenic inputs substantially augment natural loads and, in some industrialized and urbanized coastal systems, can exceed natural concentrations by orders of magnitude. Included here are such diverse sources as municipal and industrial wastewater discharges, leaching of antifouling paints, dredged material disposal, combustion of fossil fuels, mining of metal ores, smelting operations, refining, electroplating, and the manufacture of dyes, paints, and textiles. River inflow, atmospheric deposition, urban runoff, and other nonpoint sources deliver most of the anthropogenic metal burden to the sea.

Waters most greatly impacted by heavy metal pollution are industrialized or urbanized systems that have historically received municipal and industrial wastewater discharges or sewage sludge. In U.S. waters, Boston Harbor, Newark Bay, New York Bight Apex, Southern California Bight, and Commencement Bay are examples.[3,14] In Europe, the Rhine–Waal/Meuse/Scheldt estuaries, Thames estuary, Liverpool Bay, and German Bight have substantial enrichment of heavy metals.[96] However, even deep-sea areas may be impacted by heavy metals derived from human activities.[97]

Heavy metals pose a significant threat to aquatic organisms because they are potentially toxic above a threshold availability, acting as enzyme inhibitors. Estuarine and marine biota exposed to elevated heavy metal concentrations often experience serious physiological, reproductive, and developmental changes. Aberrant respiratory metabolism, feeding behavior, and digestion frequently develop in impacted animal populations. Growth inhibition is widespread across many different taxonomic groups. Pathological responses may include tissue inflammation or degeneration, lack of tissue repair, neoplasm formation, and genetic derangement.[11,98] High burdens of certain heavy metals (e.g., arsenic and mercury) in some marine resource species remain a concern because they can give rise to human health disorders, such as skin lesions (hyperkeratosis, hyperpigmentation, and skin cancer), peripheral blood pathologies, abnormal neurological function, and chromosome damage.[29]

Among the most toxic agents in estuarine and marine waters are organotin compounds, which have been used as insecticides, bactericides, and fungicides, as heat and light stabilizers in polyvinyl chloride materials, as well as in other industrial applications.[99] However, the greatest biotic impact of organometals in coastal zones has been linked to antifouling paints containing tributyltin (TBT), which are used on marine vessels and structures.[100] The leaching of TBT from the hulls of boats and ships has caused severe impacts on nontarget species, leading to bans on its use in many states.[101] Despite legislative action and diminishing contamination levels, numerous areas worldwide have TBT concentrations above thresholds known to produce bioeffects. In addition, the contaminants continue to be detected in sediments and organisms. Some notable persistent biotic effects include spat failure, reduced growth, and increased mortality in shellfish, as well as the development of imposex in gastropods.[101,102] TBT compounds also accumulate to high concentrations in marine mammals, and their long-term effects on these higher organisms remain a concern.[103]

e debris originates from land runoff, beach litter, recreational and commercial
s and ships, oil platforms, and various coastal installations. Litter is widely distrib-
Lecke-Mitchell and Mullin,[139] for example, observed debris ubiquitously distributed
Mexico. Litter is likewise a widespread problem on beaches where plastic lids, cups,
ensils, as well as drug-related paraphernalia, tobacco-related products, and poll rings
nd.[140] Ribic et al.[141] provided an overview of the marine debris problem in the United

te litter accumulation on beaches and in the sea, greater enforcement of national
l regulations and international treaties on pollution is required. The prohibition of any
stic material to the sea, as specified by the International Convention for the Prevention
from Ships (MARPOL) through its Annex V, was a major initial step in the remedial
these efforts to be successful, monitoring programs must also continue in order to
lem areas and trends in debris accumulation.[1]

ing and Dredged Material Disposal

nical and hydraulic dredging operations are regularly conducted in rivers, estuaries,
d ports worldwide to maintain navigable waterways. In addition to maintenance dredging,
ally takes place repeatedly at periodic intervals along specific sites, some dredging is
taken to remove mineral resources (e.g., sand and gravel) from the seafloor. This dredging
ccurs with greater frequency than maintenance projects.[142] Other dredging programs are
to the construction of installations and lagoons, and hence by definition are also not
nce operations. Annual dredging by the U.S. Army Corps of Engineers removes ~2.9 ×
sediment in maintenance projects and ~7.8 × 10⁷ m of sediment in new projects from
stal waters and the Great Lakes.[143] The cost of this dredging is ~$725 million.[8]
t dredged sediment is dumped in estuaries or at sea because it is less costly and a more
mentally attractive option than land-based disposal.[143] There is three times more dredged
t (by volume) dumped at sea than on land in the United States. This sediment constitutes
0% (by volume) of all waste material dumped at sea.[8]
dging and dredged material disposal cause three principal environmental impacts: (1) destruction
benthic habitat; (2) increased mortality of the benthos; and (3) impairment of water quality.
st acute impact of dredging is the physical removal of bottom sediments and the entrainment
anisms by the dredge. In some cases, mortality of the benthos approaches 100% because of
by the dredge or smothering by the sediment when the organisms are picked up or dumped.
baqueous disposal of the organisms at sites remote from the dredged habitats also increases
lity. Recovery of benthic communities at dredged sites in estuarine and coastal marine waters
erally protracted, often requiring a year or more to complete and involving a succession of
isms from opportunistic, pioneering forms to equilibrium assemblages.[144–146]
rganisms may also be adversely affected by changes in water quality during dredging and
ged material disposal operations. More specifically, these operations commonly increase nutri-
concentrations and turbidity levels in the water column, while decreasing dissolved oxygen
ls. All of these effects, however, are ephemeral. In urbanized and heavily industrialized estuaries
harbors, dredging often remobilizes toxic chemicals (e.g., petroleum hydrocarbons, halogenated
rocarbons, and heavy metals) from bottom sediments. Prior to dredging sediments that contain
ge quantities of pollutants, it is necessary to characterize the material chemically. Four frequently
ed procedures are (1) bulk or total sediment analysis; (2) elutriate testing; (3) bioassay testing
quid-phase, suspended-particulate-phase, and solid-phase bioassays); and (4) selective chemical
aching.[147] These procedures thoroughly test for potential environmental impacts, not only for the
esence of possible chemical contaminants, but also for possible biological effects.
Bulk or total sediment analysis provides a basis for estimating mass loads of wastes to estuarine
nd marine environments and, therefore, a basis for assessing the amount of substances that possibly

Heavy metals are also detrimental to aquatic habitats. They rapidly sorb to particulate matter and accumulate in bottom sediments primarily in estuarine and inner-continental shelf zones. The concentrations of heavy metals are three to five orders of magnitude greater in the bottom sediments of estuaries than in overlying waters.[104,105] Being nonbiodegradable, heavy metals persist for long periods of time in benthic habitats and can significantly compromise the quality of coastal sediments.[11,106–111] In addition, the bottom sediments serve as a repository of the contaminants and a source for later remobilization into the water column.[112,113]

Several factors influence the concentration and bioavailability of heavy metals in estuarine sediments. Most significant in this regard are (1) the control exerted by major sediment components (e.g., iron oxides and organics) to which the metals are preferentially bound; (2) the mobilization of the metals to interstitial waters and their chemical speciation; (3) the transformation of the metals; (4) the competition between sediment metals for uptake sites in organisms; and (5) the influence of bioturbation, salinity, redox, or pH on these processes.[98] Bottom sediments are long-term integrators of metal inputs, and historical pollution records may be obtained at specific sites by examining the distribution of the metals in sediment cores.[108,114]

Estuarine and marine organisms accumulate heavy metals in their tissues by the direct uptake of the contaminants from bottom sediments or interstitial waters, by the removal of metals from solution, and by the ingestion of food and suspended particulates containing sorbed metals. The capacity of different organisms to store, remove, or detoxify the metals varies substantially. Some taxa contain significant amounts of metallothioneins (low-molecular-weight, sulfhydryl-rich, metal-binding proteins) and lysosomes (cellular structures involved in intracellular digestion and transport), which play an important role in the sequestration and detoxification of metals.[115-118] Toxic effects may nevertheless develop in some organisms when they inhabit areas with excessive metal enrichment because of the limited capacity of the metallothioneins and lysosomes to sequester the contaminants.

Bioaccumulation of heavy metals is apparent in many estuarine and marine organisms, although evidence of a general biomagnification effect is lacking. Mollusks tend to concentrate the contaminants, and thus are useful as bioindicators. Mussels (e.g., *Mytilus* spp.) and oysters (e.g., *Crassostrea virginica* and *Ostrea sandivicensis*) have been particularly valuable as sentinel organisms in monitoring and assessing heavy metal contamination in estuarine and coastal marine environments.[119,120] They comprise a major part of national and international coastal environmental monitoring programs, such as the National Oceanic and Atmospheric Administration (NOAA) National Status and Trends Program and the Worldwide Mussel Watch Program.[121]

d. Radioactive Substances

Since 1945, the anthropogenic input of radioactive materials in the sea has varied considerably. Nuclear weapons testing delivered most of the artificial radionuclidies to marine waters between 1945 and 1980 via atmospheric fallout. During these 35 years, more than 1200 nuclear weapons tests were conducted worldwide. The increase in the number of nuclear power plants and nuclear fuel reprocessing facilities in operation after 1970 and the accelerated use of radioactive substances in agriculture, industry, medicine, and science after 1975 produced a large amount of radioactive waste, much of which was released or dumped into estuarine and oceanic environments. Nuclear accidents on land (e.g., Three Mile Island and Chernobyl nuclear power plants), as well as the disposal of military hardware, have added to the nuclear waste flux. Since 1980, a significant fraction of radioactive waste in the sea has derived from the nuclear fuel cycle (i.e., mining, milling, conversion, isotopic enrichment, fuel element fabrication, reactor operation, and fuel reprocessing). Aside from artificial radionuclides coupled to human activities, cosmogenic and primordial radionuclides (e.g., ^{40}K, ^{238}U, and ^{232}Th) account for natural background radiation in the sea.

Radioactive waste is subdivided into six categories: (1) high-level wastes; (2) transuranic wastes; (3) low-level wastes; (4) uranium and mill tailings; (5) decontamination and decommissioning wastes from nuclear reactors; and (6) gaseous effluents. Provisions by the Convention on the Prevention of Marine Pollution by Dumping of Wastes and Other Matter of 1972—the London Convention—and the U.S. Marine Protection, Research, and Sanctuaries Act of 1972 have prohibited the dumping of high-level wastes in the sea for years. However, noncompliance has occurred. For example, the former Soviet Union has recently revealed the dumping of high-level radioactive wastes in shallow waters of the Kara and Barents Seas.[122] Radionuclide leakage from these wastes could be disastrous for coastal ecosystems.

Previous dumping of radioactive wastes was largely conducted at specific deep-sea sites to minimize contamination of the food web. For instance, the United States dumped most of its low-level radioactive wastes at a 2.8-km-deep site in the northwest Atlantic Ocean. European countries, in turn, dumped most of their low-level radioactive wastes at the 4.4-km-deep Northeast Atlantic Dump Site. In February 1995, the 16th Consultative Meeting of Contracting Parties to the London Convention imposed a ban on the dumping of low-level radioactive wastes in the sea.[123]

Estuarine and marine organisms accumulate radionuclides in their tissues via uptake from water, sediments, or the consumption of other organisms. Highest uptake occurs near nuclear power plants, nuclear fuel processing facilities, and installations that produce nuclear explosives. Organismal exposure to radiation can result in aberrant growth and development, physiological and genetic changes, diseases (e.g., cancer), and death. The severity of impacts depends on the dose rate, total dose, type of radiation, and exposure period. The uptake of radionuclides by marine organisms and their potential biomagnification through food chains is a major human health concern because radiation damages reproductive and somatic cells leading to chromosome aberrations that may cause cancer or other malfunctions.[124,125] Hence, estuarine and marine organisms of direct dietary importance to humans (e.g., crabs, clams, mussels, oysters, and fish) have been monitored continuously for radionuclides in many regions.[126] Bivalves are particularly valuable because they concentrate the contaminants, and thus have proved to be useful in monitoring artificial radionuclides in U.S. estuarine and coastal marine systems.[127]

Primarily based on laboratory experiments, higher organisms are less tolerant to ionizing radiation than lower organisms. The following order of radiation sensitivity has been demonstrated among major taxonomic groups: mammals > birds > fish > crustaceans > algae > bacteria. Gametes and larval stages are more sensitive than adults.[9] Because of the greater sensitivity of marine mammals to radiation, they have been the subject of detailed investigations during the past decade.[128-130]

B. Other Anthropogenic Impacts

1. Coastal Development

Many environmental problems occurring in the coastal zone are coupled to development in nearby watersheds. U.S. population growth within 80 km of the shoreline of the ocean more than doubled between 1940 and 1980 and continues to accelerate at an alarming rate. Estimates indicate that ~75% of the U.S. population lives within 80 km of a coastline.[131] Increased population growth and land development along the immediate shoreline and in coastal watersheds place multiple stresses on natural ecosystems (e.g., wetlands, estuaries, beaches, and the nearshore ocean). Coastal development has been rapid and poorly planned in many areas with landscapes and habitats often neglected in the process. "Urban sprawl" now characterizes many regions; sprawl development contributes to habitat fragmentation, ecosystem isolation, and functional degradation of upland and wetland complexes. The net effect is considerable habitat loss and alteration and an array of impacts on biotic communities, some of which may be irreversible.

As coastal development escalates, the amount of impervious cover (i.e., watershed areas covered by buildings, concrete, asphalt, and other impenetrable surfaces) also increases, culminating in greater urban and suburban runoff and associated nonpoint-source pollutant inputs into streams,

rivers, estuaries, and the coastal ocean, which degra
degraded water quality, land drainage patterns and the
are frequently altered. The modification of river catchm
fresh water reaching the coastal zone. The removal of upla
of impervious surfaces, the diversion of fresh water fr
domestic uses, and the dredging of waterways can signi
estuaries and coastal embayments. All of these changes p
of the coastal zone.

Although destruction of wetlands has decreased substa
past 25 years, human activities continue to affect these critic
of saltmarshes, for example, is often associated with grid di
marsh diking for construction of causeways for roads and railr
conversion, for the establishment of pleasure boating and swi
impoundments for wildlife. Ditching facilitates tidal water fl
modify the physical characteristics of the marsh habitat. Dikin
communities. Current management strategies entail the applic
control methods, the restoration of impounded wetlands, and the
space.

To protect shoreline property, marine engineering structure
example, jetties, groins, and seawalls are built to control beach ero
from storm damage.[8,132,133] Bulkheads, revetments, and riprap are
of these features, however, also modify or destroy natural habitat.
recreational activities, such as marinas, boat ramps, docks, and pie
in localized areas.

Another consequence of coastal development is resource exploi
in coastal regions have led to intense fishing pressure on various livin
commercial fishermen have targeted the same species in most areas.
significantly to declining catches of many species in U.S. coastal waters.
of exploitation and deleterious habitat changes has seriously impacted A
regalis), summer flounder (*Paralichthys dentatus*), and red drum (*S*
southeast Atlantic coastal zone.[134] Improved management of fish stocks
enforcement of regulations by state and federal agencies to conserve the

Apart from exploiting living resources, humans mine minerals from
utilize water resources in electric power generation. Some of the largest
the coastal zone because of the substantial volumes of cooling water they req
(see Section 6.II.B.7).[135,136] The exploitation of resources in estuarine and oce
results in some environmental impacts. The goals are to minimize adverse
sustainable development of these systems.[131]

2. Marine Debris

An escalating global problem in estuarine and oceanic environments is t
marine debris or litter, which is an indicator of the burgeoning population
development of the coastal zone. Debris on beaches and in the sea is not
displeasing but also potentially detrimental to numerous organisms. Fish, turtles,
and birds are susceptible to debris in the sea, notably large plastics, packing bands,
line, and fishing nets. These discarded items commonly entangle, trap, and suffo
Smaller objects (e.g., plastic pellets, plastic fragments, and bits of packaging
ingested, often obstruct the gut. According to Shaw and Day,[137] marine organi
selectively remove plastic particles, which they mistake for prey items. Some organi
material by consuming prey with plastics in their gut.[138]

Heavy metals are also detrimental to aquatic habitats. They rapidly sorb to particulate matter and accumulate in bottom sediments primarily in estuarine and inner-continental shelf zones. The concentrations of heavy metals are three to five orders of magnitude greater in the bottom sediments of estuaries than in overlying waters.[104,105] Being nonbiodegradable, heavy metals persist for long periods of time in benthic habitats and can significantly compromise the quality of coastal sediments.[11,106–111] In addition, the bottom sediments serve as a repository of the contaminants and a source for later remobilization into the water column.[112,113]

Several factors influence the concentration and bioavailability of heavy metals in estuarine sediments. Most significant in this regard are (1) the control exerted by major sediment components (e.g., iron oxides and organics) to which the metals are preferentially bound; (2) the mobilization of the metals to interstitial waters and their chemical speciation; (3) the transformation of the metals; (4) the competition between sediment metals for uptake sites in organisms; and (5) the influence of bioturbation, salinity, redox, or pH on these processes.[98] Bottom sediments are long-term integrators of metal inputs, and historical pollution records may be obtained at specific sites by examining the distribution of the metals in sediment cores.[108,114]

Estuarine and marine organisms accumulate heavy metals in their tissues by the direct uptake of the contaminants from bottom sediments or interstitial waters, by the removal of metals from solution, and by the ingestion of food and suspended particulates containing sorbed metals. The capacity of different organisms to store, remove, or detoxify the metals varies substantially. Some taxa contain significant amounts of metallothioneins (low-molecular-weight, sulfhydryl-rich, metal-binding proteins) and lysosomes (cellular structures involved in intracellular digestion and transport), which play an important role in the sequestration and detoxification of metals.[115-118] Toxic effects may nevertheless develop in some organisms when they inhabit areas with excessive metal enrichment because of the limited capacity of the metallothioneins and lysosomes to sequester the contaminants.

Bioaccumulation of heavy metals is apparent in many estuarine and marine organisms, although evidence of a general biomagnification effect is lacking. Mollusks tend to concentrate the contaminants, and thus are useful as bioindicators. Mussels (e.g., *Mytilus* spp.) and oysters (e.g., *Crassostrea virginica* and *Ostrea sandivicensis*) have been particularly valuable as sentinel organisms in monitoring and assessing heavy metal contamination in estuarine and coastal marine environments.[119,120] They comprise a major part of national and international coastal environmental monitoring programs, such as the National Oceanic and Atmospheric Administration (NOAA) National Status and Trends Program and the Worldwide Mussel Watch Program.[121]

d. Radioactive Substances

Since 1945, the anthropogenic input of radioactive materials in the sea has varied considerably. Nuclear weapons testing delivered most of the artificial radionuclidies to marine waters between 1945 and 1980 via atmospheric fallout. During these 35 years, more than 1200 nuclear weapons tests were conducted worldwide. The increase in the number of nuclear power plants and nuclear fuel reprocessing facilities in operation after 1970 and the accelerated use of radioactive substances in agriculture, industry, medicine, and science after 1975 produced a large amount of radioactive waste, much of which was released or dumped into estuarine and oceanic environments. Nuclear accidents on land (e.g., Three Mile Island and Chernobyl nuclear power plants), as well as the disposal of military hardware, have added to the nuclear waste flux. Since 1980, a significant fraction of radioactive waste in the sea has derived from the nuclear fuel cycle (i.e., mining, milling, conversion, isotopic enrichment, fuel element fabrication, reactor operation, and fuel reprocessing). Aside from artificial radionuclides coupled to human activities, cosmogenic and primordial radionuclides (e.g., ^{40}K, ^{238}U, and ^{232}Th) account for natural background radiation in the sea.

Radioactive waste is subdivided into six categories: (1) high-level wastes; (2) transuranic wastes; (3) low-level wastes; (4) uranium and mill tailings; (5) decontamination and decommissioning wastes from nuclear reactors; and (6) gaseous effluents. Provisions by the Convention on the Prevention of Marine Pollution by Dumping of Wastes and Other Matter of 1972—the London Convention—and the U.S. Marine Protection, Research, and Sanctuaries Act of 1972 have prohibited the dumping of high-level wastes in the sea for years. However, noncompliance has occurred. For example, the former Soviet Union has recently revealed the dumping of high-level radioactive wastes in shallow waters of the Kara and Barents Seas.[122] Radionuclide leakage from these wastes could be disastrous for coastal ecosystems.

Previous dumping of radioactive wastes was largely conducted at specific deep-sea sites to minimize contamination of the food web. For instance, the United States dumped most of its low-level radioactive wastes at a 2.8-km-deep site in the northwest Atlantic Ocean. European countries, in turn, dumped most of their low-level radioactive wastes at the 4.4-km-deep Northeast Atlantic Dump Site. In February 1995, the 16th Consultative Meeting of Contracting Parties to the London Convention imposed a ban on the dumping of low-level radioactive wastes in the sea.[123]

Estuarine and marine organisms accumulate radionuclides in their tissues via uptake from water, sediments, or the consumption of other organisms. Highest uptake occurs near nuclear power plants, nuclear fuel processing facilities, and installations that produce nuclear explosives. Organismal exposure to radiation can result in aberrant growth and development, physiological and genetic changes, diseases (e.g., cancer), and death. The severity of impacts depends on the dose rate, total dose, type of radiation, and exposure period. The uptake of radionuclides by marine organisms and their potential biomagnification through food chains is a major human health concern because radiation damages reproductive and somatic cells leading to chromosome aberrations that may cause cancer or other malfunctions.[124,125] Hence, estuarine and marine organisms of direct dietary importance to humans (e.g., crabs, clams, mussels, oysters, and fish) have been monitored continuously for radionuclides in many regions.[126] Bivalves are particularly valuable because they concentrate the contaminants, and thus have proved to be useful in monitoring artificial radionuclides in U.S. estuarine and coastal marine systems.[127]

Primarily based on laboratory experiments, higher organisms are less tolerant to ionizing radiation than lower organisms. The following order of radiation sensitivity has been demonstrated among major taxonomic groups: mammals > birds > fish > crustaceans > algae > bacteria. Gametes and larval stages are more sensitive than adults.[9] Because of the greater sensitivity of marine mammals to radiation, they have been the subject of detailed investigations during the past decade.[128-130]

B. Other Anthropogenic Impacts

1. Coastal Development

Many environmental problems occurring in the coastal zone are coupled to development in nearby watersheds. U.S. population growth within 80 km of the shoreline of the ocean more than doubled between 1940 and 1980 and continues to accelerate at an alarming rate. Estimates indicate that ~75% of the U.S. population lives within 80 km of a coastline.[131] Increased population growth and land development along the immediate shoreline and in coastal watersheds place multiple stresses on natural ecosystems (e.g., wetlands, estuaries, beaches, and the nearshore ocean). Coastal development has been rapid and poorly planned in many areas with landscapes and habitats often neglected in the process. "Urban sprawl" now characterizes many regions; sprawl development contributes to habitat fragmentation, ecosystem isolation, and functional degradation of upland and wetland complexes. The net effect is considerable habitat loss and alteration and an array of impacts on biotic communities, some of which may be irreversible.

As coastal development escalates, the amount of impervious cover (i.e., watershed areas covered by buildings, concrete, asphalt, and other impenetrable surfaces) also increases, culminating in greater urban and suburban runoff and associated nonpoint-source pollutant inputs into streams, rivers, estuaries, and the coastal ocean, which degrades their water quality. In addition to the degraded water quality, land drainage patterns and the hydrologic regimes of rivers and estuaries are frequently altered. The modification of river catchments influences the quantity and quality of fresh water reaching the coastal zone. The removal of upland forests and understory, the construction of impervious surfaces, the diversion of fresh water from rivers for agricultural irrigation and domestic uses, and the dredging of waterways can significantly shift the salinity distribution in estuaries and coastal embayments. All of these changes potentially diminish the ecological value of the coastal zone.

Although destruction of wetlands has decreased substantially in the United States during the past 25 years, human activities continue to affect these critically important habitats. The alteration of saltmarshes, for example, is often associated with grid ditching for mosquito control, and with marsh diking for construction of causeways for roads and railroads, for flood control, for agriculture conversion, for the establishment of pleasure boating and swimming areas, and for the creation of impoundments for wildlife. Ditching facilitates tidal water flow into high marsh areas and can modify the physical characteristics of the marsh habitat. Diking often alters the structure of biotic communities. Current management strategies entail the application of less damaging mosquito control methods, the restoration of impounded wetlands, and the preservation of wetlands as open space.

To protect shoreline property, marine engineering structures are routinely constructed. For example, jetties, groins, and seawalls are built to control beach erosion and protect human property from storm damage.[8,132,133] Bulkheads, revetments, and riprap are used for bank stabilization. All of these features, however, also modify or destroy natural habitat. Other structures associated with recreational activities, such as marinas, boat ramps, docks, and piers, also impact habitats, albeit in localized areas.

Another consequence of coastal development is resource exploitation. Demographic changes in coastal regions have led to intense fishing pressure on various living resources. Recreational and commercial fishermen have targeted the same species in most areas. Overfishing has contributed significantly to declining catches of many species in U.S. coastal waters. For instance, a combination of exploitation and deleterious habitat changes has seriously impacted Atlantic weakfish (*Cynoscion regalis*), summer flounder (*Paralichthys dentatus*), and red drum (*Sciaenops ocellatus*) in the southeast Atlantic coastal zone.[134] Improved management of fish stocks is necessary, including the enforcement of regulations by state and federal agencies to conserve the resources.

Apart from exploiting living resources, humans mine minerals from the seafloor. They also utilize water resources in electric power generation. Some of the largest power plants are sited in the coastal zone because of the substantial volumes of cooling water they require for their condensers (see Section 6.II.B.7).[135,136] The exploitation of resources in estuarine and oceanic systems invariably results in some environmental impacts. The goals are to minimize adverse effects and to achieve sustainable development of these systems.[131]

2. Marine Debris

An escalating global problem in estuarine and oceanic environments is the accumulation of marine debris or litter, which is an indicator of the burgeoning population growth and rapid development of the coastal zone. Debris on beaches and in the sea is not only aesthetically displeasing but also potentially detrimental to numerous organisms. Fish, turtles, marine mammals, and birds are susceptible to debris in the sea, notably large plastics, packing bands, synthetic fishing line, and fishing nets. These discarded items commonly entangle, trap, and suffocate the animals. Smaller objects (e.g., plastic pellets, plastic fragments, and bits of packaging material), when ingested, often obstruct the gut. According to Shaw and Day,[137] marine organisms appear to selectively remove plastic particles, which they mistake for prey items. Some organisms ingest this material by consuming prey with plastics in their gut.[138]

Most marine debris originates from land runoff, beach litter, recreational and commercial fishermen, boats and ships, oil platforms, and various coastal installations. Litter is widely distributed in the sea. Lecke-Mitchell and Mullin,[139] for example, observed debris ubiquitously distributed in the Gulf of Mexico. Litter is likewise a widespread problem on beaches where plastic lids, cups, stirrers, and utensils, as well as drug-related paraphernalia, tobacco-related products, and poll rings are usually found.[140] Ribic et al.[141] provided an overview of the marine debris problem in the United States.

To mitigate litter accumulation on beaches and in the sea, greater enforcement of national environmental regulations and international treaties on pollution is required. The prohibition of any release of plastic material to the sea, as specified by the International Convention for the Prevention of Pollution from Ships (MARPOL) through its Annex V, was a major initial step in the remedial process. For these efforts to be successful, monitoring programs must also continue in order to identify problem areas and trends in debris accumulation.[1]

3. Dredging and Dredged Material Disposal

Mechanical and hydraulic dredging operations are regularly conducted in rivers, estuaries, harbors, and ports worldwide to maintain navigable waterways. In addition to maintenance dredging, which usually takes place repeatedly at periodic intervals along specific sites, some dredging is also undertaken to remove mineral resources (e.g., sand and gravel) from the seafloor. This dredging typically occurs with greater frequency than maintenance projects.[142] Other dredging programs are dedicated to the construction of installations and lagoons, and hence by definition are also not maintenance operations. Annual dredging by the U.S. Army Corps of Engineers removes $\sim 2.9 \times 10^8$ m of sediment in maintenance projects and $\sim 7.8 \times 10^7$ m of sediment in new projects from U.S. coastal waters and the Great Lakes.[143] The cost of this dredging is $\sim\$725$ million.[8]

Most dredged sediment is dumped in estuaries or at sea because it is less costly and a more environmentally attractive option than land-based disposal.[143] There is three times more dredged sediment (by volume) dumped at sea than on land in the United States. This sediment constitutes 80 to 90% (by volume) of all waste material dumped at sea.[8]

Dredging and dredged material disposal cause three principal environmental impacts: (1) destruction of the benthic habitat; (2) increased mortality of the benthos; and (3) impairment of water quality. The most acute impact of dredging is the physical removal of bottom sediments and the entrainment of organisms by the dredge. In some cases, mortality of the benthos approaches 100% because of injury by the dredge or smothering by the sediment when the organisms are picked up or dumped. The subaqueous disposal of the organisms at sites remote from the dredged habitats also increases mortality. Recovery of benthic communities at dredged sites in estuarine and coastal marine waters is generally protracted, often requiring a year or more to complete and involving a succession of organisms from opportunistic, pioneering forms to equilibrium assemblages.[144-146]

Organisms may also be adversely affected by changes in water quality during dredging and dredged material disposal operations. More specifically, these operations commonly increase nutrient concentrations and turbidity levels in the water column, while decreasing dissolved oxygen levels. All of these effects, however, are ephemeral. In urbanized and heavily industrialized estuaries and harbors, dredging often remobilizes toxic chemicals (e.g., petroleum hydrocarbons, halogenated hydrocarbons, and heavy metals) from bottom sediments. Prior to dredging sediments that contain large quantities of pollutants, it is necessary to characterize the material chemically. Four frequently used procedures are (1) bulk or total sediment analysis; (2) elutriate testing; (3) bioassay testing (liquid-phase, suspended-particulate-phase, and solid-phase bioassays); and (4) selective chemical leaching.[147] These procedures thoroughly test for potential environmental impacts, not only for the presence of possible chemical contaminants, but also for possible biological effects.

Bulk or total sediment analysis provides a basis for estimating mass loads of wastes to estuarine and marine environments and, therefore, a basis for assessing the amount of substances that possibly